OXFORD MATHEMATICAL MONOGRAPHS

OXFORD MATHEMATICAL MONOGRAPHS

The Theory of Infinite Soluble Groups

JOHN C. LENNOX

Green College, University of Oxford, England

and

DEREK J. S. ROBINSON

University of Illinois, Urbana, Illinois, USA

CLARENDON PRESS · OXFORD

2004

OXFORD

UNIVERSITY PRESS

Great Clarendon Street, Oxford OX2 6DP

Oxford University Press is a department of the University of Oxford.
It furthers the University's objective of excellence in research, scholarship,
and education by publishing worldwide in

Oxford New York

Auckland Bangkok Buenos Aires Cape Town Chennai
Dar es Salaam Delhi Hong Kong Istanbul Karachi Kolkata
Kuala Lumpur Madrid Melbourne Mexico City Mumbai Nairobi
São Paulo Shanghai Taipei Tokyo Toronto

Oxford is a registered trade mark of Oxford University Press
in the UK and in certain other countries

Published in the United States
by Oxford University Press Inc., New York

A catalogue record for this title is available from the British Library
Library of Congress Cataloging in Publication Data
(Data available)
ISBN 0 19 850728 3 (Hbk)

10 9 8 7 6 5 4 3 2 1

Typeset by Newgen Imaging Systems (P) Ltd., Chennai, India
Printed in Great Britain
on acid-free paper by
Biddles Ltd., King's Lynn, Norfolk

In memory of Philip Hall
1904–1982

"To leave no rubs nor botches in the work ..."
Shakespeare, Macbeth,
Act 3, Scene 1.

CONTENTS

LIST OF SYMBOLS

Special sets

$\mathbb{Z}, \mathbb{Q}, \mathbb{R}$ sets of integers, rational numbers, real numbers.
$\mathbb{Q}_\pi$ set of π-adic rational numbers.

Groups

$x, y, \ldots$ elements of a set, group
$x^y = y^{-1}xy$.
$[x, y] = x^{-1}y^{-1}xy$.
$[x_1, x_2, \ldots, x_n]$ left normed commutator of length n.
$[x,_n y] = [x, \underbrace{y, y, \ldots, y}_{n}]$.
$G, H, \ldots$ sets, groups.
$< X_\lambda \mid \lambda \in \Lambda >$ subgroup generated by subsets X_λ.
$H^x = x^{-1}Hx$.
$< X \mid R >$ group presentation with generators X and relations R.
$[X_1, X_2, \ldots, X_n]$ commutator subgroup.
$G^n = < g^n \mid g \in G >$.
$H \triangleleft G$, $H \triangleleft^d G$ H is normal, respectively subnormal in $\leq d$ steps in G.
X^H the normal closure of X in $< H, X >$.
$H^{G,i}$ the ith successive normal closure of H in G.
$H \ltimes K$ semidirect product.
$C_G(H)$, $N_G(H)$ centraliser and normaliser.
G' derived subgroup.
$G_{ab} = G/G'$.
$G^{(n)}$ terms of the derived series.
$\gamma_i(G)$ terms of the lower central series.
$Z(G)$ centre of G.
$Z_i(G)$ terms of the upper central series.
$A_n(H) = A \cap Z_n(HA)$, where $A \triangleleft HA$.
$\mathrm{Aut}(G), \mathrm{Inn}(G)$ automorphism group, inner automorphism group.
$\mathrm{Fit}(G)$ Fitting subgroup
$\mathrm{Sol}(G)$ soluble radical.
$HP(G)$ Hirsch-Plotkin radical.
$\mathrm{Frat}(G)$ Frattini subgroup
$\mathrm{FFrat}(G)$ Fitting mod Frattini subgroup.
$\mathrm{Cch}(G)$ intersection of the centralisers of the chief factors.
$\tau(G)$ maximum normal torsion subgroup.
$\pi(G)$ set of primes diving the orders of elements of G.

$I_{\pi,G}(H)$, $I_G(H)$, $I(H)$ isolators.
$h(G)$ Hirsch length.
$m(G)$ minimality of a soluble minimax group.
$r_0(A)$, $r_p(A)$, $r(A)$, $\hat{r}(A)$ ranks of an abelian group.
$hd(G)$, $cd(G)$ homological and cohomological dimensions.
$H\ wrK$, $H\ WrK$ wreath products.
V   a variety of groups.
A   the variety of abelian groups.
$\mathrm{GL}_n(R)$ general linear group.
$T_n(R)$, $U_n(R)$ triangular and unitriangular groups.

Rings

$M_n(R)$ ring of $n \times n$-matrices over a ring R.
$R^* = U(R)$ the group of units of a ring R.
$\mathbb{R}G$ group ring of a group G over a ring R.
I_G, $\bar{I}_G$ augmentation ideals.
$^\circ L = \bigcap_{g \in G} L^g$.

Modules

$Ann_R(M)$ annihilator of a module M.
$\pi_S(M)$ associated set of prime ideals of an S-module M.
M^G, M_G largest trivial submodule, respectively quotient module, of a G-module M
$\mathrm{Der}(G,A)$, $\mathrm{Inn}(G,A)$ group of derivations, respectively inner derivations, of a G-m
M.
$M(J,\pi)$ a special class of modules (p.66).
C_A centraliser ring of module A.
$S(Q)$ valuation sphere of a finitely generated abelian group Q.
Q_ν $\{q \mid q \in Q, q^\nu \geq 0\}$.
$\mathbb{Z}Q_\nu$ monoid ring.
Σ_A Bieri-Strebel invariant of module A.

INTRODUCTION

The origins of the theory of infinite soluble groups may be traced back to the year 1938, when K. A. Hirsch published the first in a sequence of five papers entitled 'S-groups'. The groups rejoicing under this somewhat colourless name are now known as 'polycyclic groups'. The basic theory of these groups had formed the subject of Hirsch's Ph.D. dissertation at Cambridge in 1937. Probably the topic was suggested by Hirsch's supervisor Philip Hall: it should be remembered that Hall had just completed his great sequence of papers on finite soluble groups and this may well have provided the motivation for the new field of research.

As events turned out, the finite and infinite theories were to follow very different paths. In the theory of infinite soluble groups it became clear that the decisive influences come from ring theory, linear group theory, and, as it turned out later, from homological algebra. Thus in many ways it is the infinite theory that has proved to be the one with the wider connections in algebra.

Hirsch's early work was followed in 1951 by a famous paper of A. I. Mal'cev which established the basic theory of soluble linear groups and soluble groups of finite rank: this work was to determine the direction of research for many years. Then in the 1950s there appeared P. Hall's celebrated series of papers on finitely generated soluble groups. Since that time there has emerged a very large body of work on infinite soluble groups, with contributions by some of the best known names in twentieth century group theory.

It was after contemplating this vast body of work that we decided there was a pressing need to write a book which would give a definitive account of what had been accomplished by so many mathematicians. The decision to begin the book was made during the Ravello Conference in 1994, while the authors were seated in a café in the Piazza del Duomo in Amalfi, although it was some time later that we actually got our act together and started work. It proved to be an even more formidable enterprise than we had expected—a glance at the bibliography will reveal around 900 items and we make no claim of completeness.

Our object has been to present in a unified form a comprehensible account of almost 70 years of work. The hope is that in the future, when current experts have left the scene, should a mathematician be faced with a problem about infinite soluble groups, the book will at least show him or her where to look. We also hope very much that the book will provide a medium for younger algebraists to learn the subject and generally give a stimulus to research. If any of these goals are met, we shall consider ourselves well rewarded for our trouble.

We have aimed to cover all the major areas of the subject, which include finitely generated soluble groups, soluble groups of finite rank, modules over group rings, algorithmic problems, applications of cohomology, and finitely presented groups. To keep the project within reasonable limits, we have stayed fairly strictly within the boundaries of soluble group theory. Even so we could not cover everything and, inevitably, some readers will find their favourite topics ignored or neglected. To all such we apologize—but in the final event the discretion of the authors has been exercised. Some theorems are only mentioned in passing, either because their proofs would occupy too much space or because they seem peripheral, again a judgement call by us. But that said, we believe that our choice of what to present is a reasonable one, with which many will agree.

Certain topics have not been addressed at all, the reason being that they are subjects in their own right and in many cases monographs are already available, on which we could hardly hope to improve. For example, ordered groups, profinite groups, subgroup lattices, varieties of groups, and products of subgroups fall within this category.

Generally speaking the reader is assumed to have acquired the knowledge one would expect to obtain from a first graduate course in group theory at a North American University. The subject is developed from scratch, although admittedly we go rather quickly over foundational material.

There are many acknowledgements to be made on completing a project such as this. We are grateful to our many colleagues and co-workers, from whom we have learned much; especially we thank Frank Cannonito and Ralph Strebel, who read certain chapters of the book at a preliminary stage. We have been greatly influenced, like many others, by Philip Hall's lectures at Cambridge during the 1960s. We are grateful to the Department of Mathematics of the University of Illinois in Urbana-Champaign, and to Professor Joe Rosenblatt in particular, for valuable assistance with the production of the typescript. We were most fortunate in our typist Ms. Sara Nelson, whom we thank for her excellent work. We thank Oxford University Press for their willingness to publish the book and for displaying patience in the face of missed deadlines. Finally, we thank our families for their understanding and support.

This work is dedicated to the memory of Philip Hall, the master from whom we learned so much, and who was also a man of great humanity and charm.

Oxford and Urbana
October 2003

1

BASIC RESULTS ON SOLUBLE AND
NILPOTENT GROUPS

The central concept in this book is that of a *soluble group,* that is, a group which is built up from abelian groups by repeatedly forming group extensions. The word 'soluble' stems from the intimate connections between groups and the solution of polynomial equations by radicals, the principal result being that an equation is soluble by radicals if and only if the Galois group of the equation is soluble.

1.1 Definition and elementary properties of soluble groups

A group G is said to be *soluble* (or solvable) if it has an abelian series, that is, a series $1 = G_0 \triangleleft G_1 \triangleleft \cdots \triangleleft G_n = G$ in which each factor G_{i+1}/G_i is abelian. Thus solubility is a generalization of commutativity, abelian groups being the simplest kind of soluble groups. The smallest example of a non-abelian soluble group is S_3, the symmetric group on 3 letters, which is an extension of a cyclic group of order 3 by a cyclic group of order 2. It is clear that an abelian series in a finite soluble group can be refined to a composition series whose factors are abelian simple groups, that is, cyclic groups of prime order. This means that a finite group is soluble if and only if it has a series whose factors are cyclic groups of prime order.

The class of soluble groups turns out to be very large. For example, all finite groups of odd order are soluble, a famous and difficult result of Feit and Thompson (1961). The literature on finite soluble groups is vast and it is comprehensively covered in the monograph by Doerk and Hawkes (1992), which is, in many ways, a companion volume to the present work.

The length of a shortest abelian series of a soluble group G is called the *derived length* of G. Thus G has derived length 0 if it is of order 1 and the groups of derived length at most 1 are precisely the abelian groups. The term *metabelian* is frequently used for soluble groups of derived length at most 2: for instance S_3 is metabelian.

We note without proof some of the most elementary properties of the class of soluble groups: for details of the proofs see Robinson (1996).

1.1.1 *The class of soluble groups is closed with respect to the formation of extensions, subgroups, and homomorphic images of its members.*

From this there follows quickly.

1.1.2 *The product of two soluble normal subgroups is soluble.*

Thus the product of finitely many soluble normal subgroups is soluble. It follows that every group G which satisfies the maximal condition on normal subgroups, (and hence, in particular, every finite group), has a unique maximum soluble normal subgroup S, namely, the product of all the soluble normal subgroups of G. This subgroup S is called the *soluble radical* of G and is denoted by Sol (G). The quotient group G/S is *semisimple*, that is, it has no non-trivial soluble normal subgroups. Thus every finite group is an extension of a soluble group by a semisimple group.

Commutators

In working with soluble groups it is important to have the calculus of commutators at our disposal. Suppose that G is any group and $x_1, x_2, \ldots, x_n$ are arbitrary elements of G. Then the *commutator* of x_1 and x_2 is $[x_1, x_2] = x_1^{-1} x_2^{-1} x_1 x_2 = x_1^{-1} x_1^{x_2}$ where $x_1^{x_2} = x_2^{-1} x_1 x_2$. Thus G is abelian if and only if all commutators of elements of G are equal to the identity.

We shall need to work with commutators involving more than 2 elements. Accordingly we define a *simple commutator of weight* $n \geq 2$ recursively as follows:

$$[x_1, x_2, \ldots, x_n] = [[x_1, \ldots, x_{n-1}], x_n],$$

where $[x_1] = x$ by convention. We shall also use $[x, {}_n y]$ to denote

$$[x, \underbrace{y, y, \ldots, y}_{n}].$$

Manipulation of commutators is greatly facilitated by use of the following basic properties.

1.1.3 *Suppose that* x, y, z *are elements of a group. Then*

(i) $[x, y] = [y, x]^{-1}$;

(ii) $[xy, z] = [x, z]^y [y, z]$ *and* $[x, yz] = [x, z][x, y]^z$;

(iii) $[x, y^{-1}] = \left([x, y]^{y^{-1}}\right)^{-1}$ *and* $[x^{-1}, y] = \left([x, y]^{x^{-1}}\right)^{-1}$;

(iv) $[x, y^{-1}, z]^y [y, z^{-1}, x]^z [z, x^{-1}, y]^x = 1$, *(the Hall–Witt identity)*.

Proof The first three parts are easy to prove and the proofs are omitted. The final part is most conveniently established by setting $u = xzx^{-1}yx$, $v = yxy^{-1}zy$, and $w = zyz^{-1}xz$, and noting that $[x, y^{-1}, z]^y = u^{-1}v$, $[y, z^{-1}, x]^z = v^{-1}w$, and $[z, x^{-1}, y]^x = w^{-1}u$. The identity is now obvious.   ∎

Commutator subgroups

We now widen the concept of a commutator by defining commutators of subsets. Suppose that $X_1, X_2, \ldots$ are non-empty subsets of a group G. Define the *commutator subgroup* of X_1 and X_2 to be

$$[X_1, X_2] = \langle [x_1, x_2] \mid x_1 \in X_1, x_2 \in X_2 \rangle.$$

It should be noted that, unlike the commutator of two elements, the commutator of two subsets is symmetric: $[X_1, X_2] = [X_2, X_1]$.

For $u \geq 2$ we define recursively

$$[X_1, X_2, \ldots, X_n] = [[X_1, X_2, \ldots, X_{n-1}], X_n],$$

where $[X_1] = \langle X_1 \rangle$. Also we will abbreviate $[X, \underbrace{Y, \ldots, Y}_{n}]$ to $[X, {}_n Y]$.

From 1.1.3 there follows:

1.1.4 *If X_1, X_2, X_3 are normal in G, then $[X_1, X_2 X_3] = [X_1, X_2][X_1, X_3]$ and $[X_1 X_2, X_3] = [X_1, X_3][X_2, X_3]$.*

If H is a subgroup of a group G, the *normal closure of H in G* is defined to be the smallest normal subgroup of G which contains H. This is generated by all conjugates of H by elements of G and is denoted by H^G. More generally, if X and Y are subsets of G, define

$$X^Y = \langle x^y = y^{-1} x y \mid x \in X, y \in Y \rangle.$$

Thus if X is a subset and H a subgroup of a group, then $X \subseteq X^H \triangleleft \langle X, H \rangle$. Therefore $X^H = X^{\langle X, H \rangle}$ is the normal closure of X in $\langle X, H \rangle$, so the notation is consistent with that for normal closures.

The following elementary facts will be used repeatedly—for proofs see Robinson (1996).

1.1.5 *Suppose that X and Y are subsets and H is a subgroup of a group. Then*

(i) $X^H = \langle X, [X, H] \rangle$;
(ii) $[X, H]^H = [X, H]$;
(iii) *if $H = \langle Y \rangle$, then $[X, H] = [X, Y]^H$.*

From 1.1.5 (iii) there follows at once:

1.1.6 *Suppose that H and K are subgroups of a group where $H = \langle X \rangle$ and $K = \langle Y \rangle$. Then $[H, K] = [X, Y]^{HK}$.*

The commutator concept enables us to get a tighter grip on the nature of solubility by constructing canonical abelian series for such groups. If G is any group, we define the *derived* or *commutator subgroup* to be $G' = [G, G]$. By repeatedly forming derived subgroups we obtain a descending sequence of subgroups

$$G = G^{(0)} \geq G^{(1)} \geq G^{(2)} \geq \cdots \geq G^{(n)} \geq \cdots,$$

where $G^{(n+1)} = (G^{(n)})'$, called the *derived series* of G. (Strictly speaking this is not a series since it need not in general reach 1 or even terminate finitely.) It is clear that $G^{(n)}$ is a normal (even fully invariant) subgroup of G for each n. Each of the factors $G^{(n)}/G^{(n+1)}$ in the series is abelian and the first of these, the *abelianization* G/G' of G, is of particular importance. It is often written G_{ab}, since it is the largest abelian quotient of G.

The relationship between the derived series and an arbitrary abelian series for a soluble group is given by:

1.1.7 *Suppose that* $1 = G_0 \triangleleft G_1 \triangleleft \cdots \triangleleft G_n = G$ *is an abelian series of a soluble group* G; *then* $G^{(i)} \leq G_{n-i}$, *where* $0 \leq i \leq n$. *In particular* $G^{(n)} = 1$, *so the derived length of* G *is precisely the length of the derived series of* G.

Proof We proceed by induction on i, the inclusion clearly holding for $i = 0$. Assume it is true for i. Then $G^{(i+1)} = \left(G^{(i)} \right)' \leq \left(G_{n-i} \right)'$ by the induction hypothesis. But $\left(G_{n-i} \right)' \leq G_{n-(i+1)}$, since the given series has abelian factors.   ∎

It follows at once from this that no abelian series can be shorter than the derived series. Also a group is soluble if and only if its derived series reaches the identity subgroup after a finite number of steps.

1.2   Definition and elementary properties of nilpotent groups

It is impossible to discuss soluble groups at any depth without introducing early on the special case of nilpotent groups. A group G is called *nilpotent* if it has a *central series*, that is, a normal series $1 = G_0 \leq G_1 \leq \cdots \leq G_n = G$, such that G_{i+1}/G_i is contained in the centre of G/G_i for $i = 0, \ldots, n - 1$. The length of a shortest central series is called the *nilpotent class* of G.

Evidently a central series is an abelian series and so nilpotent groups are soluble. A nilpotent group of class 0 has order 1 and the nilpotent groups of class at most 1 are precisely the abelian groups. The smallest example of a non-nilpotent soluble group is S_3: its centre is the identity.

1.2.1 *The class of nilpotent groups is closed under the formation of subgroups, homomorphic images, and finite direct products.*

The routine proof is omitted. It should be noted that the class of nilpotent groups is not closed under the formation of group extensions. Again S_3 provides a suitable counterexample.

Just as solubility leads to the definition of a canonical series in a group, so does nilpotency, but in this case there are two canonical series. The first of these has in common with the derived series that it is a descending sequence of commutator subgroups. It is formed as follows. Suppose G is any group and set

$$\gamma_1(G) = G \quad \text{and} \quad \gamma_{n+1}(G) = [\gamma_n(G), G],$$

for $n \geq 1$. The series $G = \gamma_1(G) \geq \gamma_2(G) \geq \cdots \geq \gamma_n(G) \geq \cdots$ is called the *lower central series* of G. We note that $\gamma_n(G)$ is normal and indeed fully invariant in G, and that $\gamma_n(G)/\gamma_{n+1}(G)$ is contained in the centre of $G/\gamma_{n+1}(G)$. Of course, like the derived series, the lower central series does not always reach 1 or even terminate.

The second canonical series associated with nilpotency is the *upper central series*

$$1 = Z_0(G) \leq Z_1(G) \leq \cdots \leq Z_n(G) \leq \cdots$$

defined by $Z_{n+1}(G)/Z_n(G) = Z(G/Z_n(G))$. Thus $Z_1(G)$ is the centre $Z(G)$ of G.

We note that each $Z_n(G)$ is characteristic, but not necessarily fully invariant in G. This series is an ascending series, which, in general, may not reach G or even terminate finitely. Indeed, if $Z_1(G) = 1$, as is the case with S_3, the series does not grow at all. If the series terminates, as it must, for example, in the case of finite groups or groups satisfying the maximal condition for subgroups, then the subgroup in which it terminates is called the *hypercentre* of G.

The relationship between these canonical series and nilpotency is enshrined in:

1.2.2 *Suppose that $1 = G_0 \triangleleft G_1 \leq \cdots \triangleleft G_n = G$ is a central series of a nilpotent group G. Then*

  (i) *$\gamma_i(G) \leq G_{n-i+1}$, for $0 < i \leq n+1$, so that $\gamma_{n+1}(G) = 1$;*
  (ii) *$G_i \leq Z_i(G)$, for $0 \leq i \leq n$, so that $Z_n(G) = G$;*
  (iii) *the lengths of the lower and upper central series are equal and their common value is the nilpotent class of G.*

Proof Here (i) and (ii) are easily proved by induction on i, while (iii) follows at once from them. ∎

Thus a group is nilpotent if and only if its lower central series reaches 1 in a finite number of steps, or, equivalently, if its upper central series reaches the whole group in a finite number of steps.

A rich source of nilpotent groups is provided by the class of groups of prime power orders. As is well known, the centre of a non-trivial group in this class is non-trivial, and by an elementary induction argument, it follows that all such groups are nilpotent. For further details of the central role that finite p-groups play in the theory of finite groups in general, and finite soluble groups in particular, see the book by Doerk and Hawkes (1992).

The following result, discovered independently by Kalužnin (1953) and P. Hall (1958), finds many applications in commutator calculations, in particular to the upper and lower central series.

1.2.3 (The three subgroup lemma) *Suppose that H, K, L are subgroups of a group G. If any two of $[H, K, L]$, $[K, L, H]$, $[L, H, K]$ are contained in a normal subgroup of G, then so is the third.*

This result, which follows easily from the Hall–Witt identity (1.1.3(iv)), finds immediate application in the simple induction arguments which establish a number of useful properties of the canonical central series of a group.

1.2.4 *Suppose that G is any group and i, j are positive integers. Then*

(i) $[\gamma_i(G), \gamma_j(G)] \leq \gamma_{i+j}(G)$;
(ii) $\gamma_i(\gamma_j(G)) \leq \gamma_{ij}(G)$;
(iii) $[\gamma_i(G), Z_j(G)] \leq Z_{j-i}(G)$, *if $j \geq i$*;
(iv) $Z_i(G/Z_j(G)) = Z_{i+j}(G)/Z_j(G)$.

Relation (ii) leads directly to an estimate for the derived length of a nilpotent group via:

1.2.5 *If G is any group, then $G^{(i)} \leq \gamma_{2^i}(G)$. Thus if G is nilpotent of class $c > 0$, its derived length is at most $[\log_2 c] + 1$.*

A useful source of nilpotent groups is groups of unitriangular matrices. Suppose that R is a commutative ring with identity. Then the group

$$U = U_n(R)$$

of all $n \times n$ (upper) unitriangular matrices over R, that is, matrices with 1's on the diagonal and 0's below it, is nilpotent of class $n - 1$.

This example is so important that it will be set in a wider context. Suppose that E is a ring with identity and S is a subring of E. Denote by S^i the set of all finite sums, $\sum \ell_{x_1 \ldots x_i} x_1 \ldots x_i$, where $x_j \in S$ and $\ell_{x_1 \ldots x_i} \in \mathbb{Z}$. Then S^i is clearly a subring of S. Now assume that S is a nilpotent subring, that is, $S^n = 0$ for some $n > 0$, and define $U = \{1 + x \mid x \in S\}$. Then there is the following result.

1.2.6

(i) U is a group with respect to the ring multiplication in S.
(ii) If $U_i = \{1 + x \mid x \in S^i\}$, then $1 = U_n \triangleleft U_{n-1} \triangleleft \cdots \triangleleft U_1 = U$ is a central series of U. Thus U is nilpotent of class at most $n - 1$.

Proof

(i) U is clearly closed under ring multiplication. The existence of an inverse for $1 + x$ follows at once from the fact that

$$(1 + x)\left(1 - x + x^2 - \cdots + (-1)^{n-1} x^{n-1}\right) = 1 - x^n = 1,$$

since $S^n = 0$.
(ii) Suppose that $x \in S^i$ and $y \in S^j$. Then

$$[1 + x, 1 + y] = ((1 + y)(1 + x))^{-1}(1 + x)(1 + y)$$

$$= (1 + y + x + yx)^{-1}(1 + x + y + xy).$$

Now put $u = x + y + xy$ and $v = y + x + yx$. Then

$$[1 + x, 1 + y] = (1 - v + \cdots + (-1)^{n-1} v^{n-1})(1 + u)$$

$$= 1 + (1 - v + \cdots + (-1)^{n-2} v^{n-2})(u - v) + (-1)^{n-1} v^{n-1} u.$$

Since $u - v = xy - yx \in S^{i+j}$ and $v^{n-1}u \in S^n = 0$, we have $[1+x, 1+y] \in 1 + S^{i+j}$. It follows that $[U_i, U_j] \leq U_{i+j}$, so that in particular $[U_i, U] \leq U_{i+1}$ and the U_i form a central series, as required. ∎

Now take for E the ring of all $n \times n$ matrices over a commutative ring R with identity, and for S, the subring of all matrices with 0's on and below the main diagonal. Then an easy calculation shows that S^i is the set of all matrices in S whose first $i - 1$ superdiagonals are 0, so that $S^n = 0$. The group $U = U_n(R)$ is therefore nilpotent of class at most $n - 1$. The subgroup U_i consists of all unitriangular matrices where the first $i - 1$ superdiagonals are zero, so that

$$U_i / U_{i+1} \cong \underbrace{R \oplus \cdots \oplus R}_{n-i}.$$

Furthermore, it can be shown by induction on $n - i$ that $U_i = Z_{n-i}(U)$, so that U is nilpotent of class exactly $n - 1$. If we specialize to the case $R = \mathbb{Z}$, then we see that U is a torsion-free nilpotent group. It is in fact also finitely generated: indeed, on taking E_{ij} to be the matrix with 1 in the (i, j)th position and zeros elsewhere, we see that the set $\{I + E_{ii+1} \mid i = 1, 2, \ldots, n - 1\}$ generates U.

This is a highly significant example since it turns out in fact that any finitely generated torsion-free nilpotent group can be embedded as a subgroup of $U_n(\mathbb{Z})$ for some n—see 3.3.4 below.

We may, of course, think of our original ring E of all $n \times n$ matrices over R as the endomorphism ring of a free R-module M of rank n. Furthermore, if we take M_i to be the submodule of M consisting of all elements whose components in the first $n - i$ summands are 0, then

$$0 = M_0 < M_1 < \cdots M_r = M$$

and

$$S = \{\alpha \mid \alpha \in E, M_i\alpha \leq M_{i-1}\}.$$

Each M_i is a U-module, where $U = U_n(R)$, and it follows that U acts trivially, that is, like the identity automorphism, on the quotient modules M_i/M_{i+1}. Under these circumstances U is said to *stabilize* the series of $M_i's$.

Stability groups

Generalizing this concept to groups acting on groups rather than modules provides us with an important insight into the reason for the nilpotency of U.

Let H be any group and let $1 = H_0 \triangleleft H_1 \triangleleft \cdots \triangleleft H_m = H$ be a series of subgroups of H. Following Kalužnin (1953), we say that an automorphism α of H *stabilizes* the series if $(xH_i)^\alpha = xH_i$ for $i = 1, \ldots, m$ and all $x \in H_{i-1}$. The set of all such α is the *stability group* of the series: it is a subgroup of the automorphism group $\text{Aut}(H)$.

1.2.7 (P. Hall 1958) *The stability group of any series of length $m \geq 1$ is nilpotent of class at most $\binom{m}{2} = \frac{1}{2}m(m - 1)$.*

Kalužnin (1953) had previously established nilpotency in the case of a normal series of length m, when the class is at most $m - 1$.

Proof of 1.2.7 Let G be the stability group of the series

$$1 = H_0 \triangleleft H_1 \triangleleft \cdots \triangleleft H_m = H.$$

We argue by induction on m: the result is trivial for $m = 1$, so suppose that $m > 1$. Set $K = C_G(H_{m-1})$. Now G stabilizes the series $1 = H_0 \triangleleft H_2 \triangleleft \cdots \triangleleft H_{m-1}$, so by the induction hypothesis G/K is nilpotent of class at most $\binom{m-1}{2}$. Hence $\gamma_{1+\binom{m-1}{2}}(G) \leq K$. We now put $K_1 = K$ and $K_{i+1} = [K_i, G]$ for $i \geq 1$ and note that it suffices to establish $[H, K_m] = 1$ since then

$$[H, {}_{1+\binom{m}{2}}(G)] = 1.$$

To achieve this we prove that $[H, K_i] \leq H_{m-i}$ by induction on i. For $i = 1$ this simply says that $[H, K] \leq H_{m-1}$, which is certainly true.

Suppose that $[H, K_i] \leq H_{m-i}$ for some $i \geq 1$. We wish to establish $[H, K_{i+1}] \leq H_{m-i-1}$ and to this end, let $h \in H$ and $k_{i+1} \in K_{i+1}$. Then k_{i+1} is a word in commutators of the form $[k_i, g^{-1}]$, $k_i \in K_i$, $g \in G$, and clearly k_{i+1}^h is the same word in elements

$$[k_i, g^{-1}]^h = [k_i, g^{-1}][k_i, g^{-1}, h].$$

But $[G, H, K_i] \leq [H_{m-1}, K] = 1$ and

$$[h, k_i^{-1}, g] \in [H, K_i, G] \leq [H_{m-i}, G] \leq H_{m-i-1}.$$

Using the Hall–Witt identity we obtain $[k_i, g^{-1}, h] \in H_{m-i-1}$. It now follows that $[h, k_{i+1}] \in [K_i, G, H]$ and $[H, K_{i+1}] \leq H_{m-i-1}$.   ∎

We now record without proof some further elementary but useful properties of nilpotent groups: for detailed proofs see Robinson (1996).

1.2.8 *Suppose that G is a nilpotent group.*

(i) *A non-trivial normal subgroup of G intersects the centre of G non-trivially. Hence a minimal normal subgroup of G is contained in the centre of G.*
(ii) *Maximal subgroups of G are normal and have prime index.*
(iii) *A maximal normal abelian subgroup of G is equal to its centralizer in G.*

Fitting's Theorem and the Fitting subgroup

We have already seen that the product of two soluble normal subgroups of a group is soluble (1.1.2). An analogous result was proved by Fitting (1938) for nilpotent normal subgroups of a group.

1.2.9 *Suppose that M and N are normal nilpotent subgroups of a group G with respective nilpotent classes c and d. Then MN is nilpotent of class at most $c + d$.*

Proof Set $L = MN$. We prove by induction on i that $\gamma_i(L)$ is the product of all $[X_1, \ldots, X_i]$, where each X_j is either M or N. This is clear for $i = 1$, so let $i > 1$. By 1.1.4, we have

$$\gamma_{i+1}(L) = [\gamma_i(L), L] = [\gamma_i(L), MN] = [\gamma_i(L), M][\gamma_i(L), N]$$

and by the induction hypothesis we see that $\gamma_{i+1}(L)$ is expressible in the requisite form.

Finally, set $i = c + d + 1$. Then in any $[X_1, \ldots, X_i]$ either M occurs at least $c + 1$ times or N occurs at least $d + 1$ times. In addition $U \triangleleft G$ implies that $[U, G] \leq U$. Hence $[X_1, \ldots, X_i]$ is contained in either $\gamma_{c+1}(M)$ or $\gamma_{d+1}(N)$ and both of these are trivial. The result now follows. ∎

Fitting's Theorem is of great importance in the theory of both finite and infinite groups. The subgroup generated by all normal nilpotent subgroups of a group G is called the *Fitting subgroup* and we use, $\mathrm{Fit}(G)$ to denote it. If G is a non-trivial soluble group, then $\mathrm{Fit}(G)$ is non-trivial since it contains the smallest non-trivial term of the derived series of G, which is, of course, an abelian normal subgroup of G. It is immediate from Fitting's Theorem that $\mathrm{Fit}(G)$ is nilpotent if, for example, G satisfies max $- n$, the maximal condition for normal subgroups.

Suppose N is a normal nilpotent subgroup of a group G and M is a minimal normal subgroup of G. Then $[M, N] = 1$. For if not, $[M, N] = M$ and hence $M \leq \gamma_r(N)$, which is a contradiction since $\gamma_r(N) = 1$ for some r. It follows from this that N centralizes all chief factors of G. Hence $\mathrm{Fit}(G)$ is contained in the subgroup $Cch(G)$, which is defined to be the intersection of the centralizers of all the chief factors of G. Indeed, *for a finite group*, $\mathrm{Fit}(G) = Cch(G)$, since $Cch(G)$ is nilpotent by 1.2.7 or an easy direct argument.

There is a sense in which the Fitting subgroup of a soluble group plays an analogous role to the centre of a nilpotent group (cf. 1.2.8).

1.2.10 *Suppose that G is a soluble group with Fitting subgroup F.*

(i) *If $1 \neq N \triangleleft G$, then N contains a non-trivial abelian normal subgroup of G, so that $N \cap F \neq 1$.*
(ii) *$C_G(F) = Z_1(F)$.*

Proof

(i) Let k be the maximum such that $M = N \cap G^{(k)} \neq 1$. Then $M' \leq N \cap G^{(k+1)} = 1$, so that M is abelian and normal in G.
(ii) Suppose $C = C_G(F) \not\leq F$. By (i) there exists $A/F \triangleleft G/F$ with $F < A \leq CF$ and A/F abelian. But $A = A \cap (CF) = (A \cap C)F$ and $\gamma_3(A \cap C) \leq [A', C] \leq [F, C] = 1$. Hence $A \cap C$ is nilpotent and, since it is normal in G, it is contained in F, a contradiction. Hence $C \leq F$ and so $C = Z_1(F)$. ∎

Inheritance properties of the lower central series

It turns out that the abelianization, or first lower central factor $G_{ab} = G/G'$ of a group G, has a strong influence on the subsequent lower central factors, and thus on G itself if it is nilpotent.

In order to describe what happens we assume that G is a group with an operator domain Ω and write $G_i = \gamma_i(G)$. Since G_i is a fully invariant subgroup of G, it is Ω-admissible. Also G_i/G_{i+1}, being abelian, is a right Ω-module:[1] we ask what influence the structure of the Ω-module $G_{ab} = G_1/G_2$ has on the Ω-modules G_i/G_{i+1} for $i > 1$. Keep in mind that these modules are being written multiplicatively.

Suppose that $g \in G$ and $a \in G_i$, and consider the mapping $(aG_{i+1}, gG') \mapsto [a,g]G_{i+2}$. This is well-defined: for if $x \in G'$, then $[a, gx] = [a,x][a,g]^x \equiv [a,g]$ mod G_{i+2} since $[G_i, G'] \le G_{i+2}$ by 1.2.4(i). Similarly, if $y \in G_{i+1}$, we have $[ay, g] = [a,g]^y[y,g] \equiv [a,g]$ mod G_{i+2}. The mapping is also bilinear: for $[a_1 a_2, g] = [a_1, g][a_1, g, a_2][a_2, g]$ and $[a_1, g, a_2] \in G_{i+2}$. Hence $[a_1 a_2, g] \equiv [a_1, g][a_2, g]$ mod G_{i+2}. Similarly, $[a, g_1 g_2] \equiv [a, g_1][a, g_2]$ mod G_{i+2}. By the fundamental mapping property of the tensor product (over $\mathbb{Z}$), there is an induced homomorphism

$$\varepsilon_i : (G_i/G_{i+1}) \otimes_{\mathbb{Z}} G_{ab} \to G_{i+1}/G_{i+2}$$

in which

$$aG_{i+1} \otimes gG' \mapsto [a,g]G_{i+2}, \quad a \in G_i, \ g \in G.$$

Since $G_{i+1} = [G_i, G]$, it follows that ε_i is surjective. In fact, ε_i is an Ω-homomorphism if we make the tensor product $A \otimes_{\mathbb{Z}} B$ of two Ω-modules into an Ω-module by the diagonal action: thus $(a \otimes b)\omega = (a\omega) \otimes (b\omega)$, for $a \in A, b \in B, \omega \in \Omega$.

We therefore have the crucial result:

1.2.11 (Robinson 1968*b*) *Suppose that G is an Ω-operator group and set $F_i = \gamma_i(G)/\gamma_{i+1}(G)$. Then the mapping*

$$a(\gamma_{i+1}(G)) \otimes gG' \mapsto [a,g]\gamma_{i+2}(G),$$

$a \in \gamma_i(G), g \in G$, determines a well-defined surjective Ω-homomorphism from $F_i \otimes_{\mathbb{Z}} G_{ab}$ to F_{i+1}.

Iteration of this result yields a surjective Ω-homomorphism

$$\underbrace{G_{ab} \otimes_{\mathbb{Z}} \cdots \otimes_{\mathbb{Z}} G_{ab}}_{i} \to F_i.$$

As an immediate corollary we have the very useful theorem:

1.2.12 *Let $\mathcal{P}$ be any group theoretical property which is inherited by images of tensor products, and suppose G is a group such that G_{ab} has $\mathcal{P}$. Then:*

(i) *$\gamma_i(G)/\gamma_{i+1}(G)$ has $\mathcal{P}$ for $i = 2, 3, \ldots$;*
(ii) *if G is nilpotent and $\mathcal{P}$ is extension closed, then G itself has $\mathcal{P}$.*

Proof

(i) Suppose $F_i = \gamma_i(G)/\gamma_{i+1}(G)$ has $\mathcal{P}$. Then F_{i+1}, being an image of $F_i \otimes_{\mathbb{Z}} G_{ab}$ has $\mathcal{P}$, and so each lower central factor of G has $\mathcal{P}$.

---

[1] Here we extend the normal usage of the term 'module' since Ω need not be a ring.

(ii) If in addition G is nilpotent, we have $\gamma_{c+1}(G) = 1$ for some c. Since $\mathcal{P}$ is closed under forming extensions, we deduce that G has $\mathcal{P}$. ■

Finiteness is one obvious candidate for $\mathcal{P}$. Thus *a nilpotent group is finite if its abelianization is finite.* Another candidate for $\mathcal{P}$ is the property of being a π-torsion group, where π is a set of primes. Therefore *a nilpotent group is a π-torsion group if its abelianization is a π-torsion group.*

In fact 1.2.12 enables us to prove that the torsion elements of a nilpotent group, that is, the elements of finite order, form a subgroup. More precisely we have:

1.2.13 *If G is a nilpotent group, then the torsion elements of G form a fully invariant subgroup T such that G/T is torsion-free. Furthermore, the elements of p-power orders form a subgroup T_p and T is the direct product of the T_p for all primes p.*

Proof Suppose that π is a non-empty set of primes and T_π is the subgroup *generated* by all π-elements of G. Then $(T_\pi)_{ab}$ is generated by π-elements and it is abelian, so it is a π-group. From 1.2.12 we deduce that T_π is a π-group. If we take π to be the set of all primes, it follows that $T = T_\pi$ is the set of all elements of finite order and this is therefore a torsion subgroup. Taking π to be $\{p\}$, we see that T_p is a p-group. Clearly $T_p \lhd G$ and hence T is the direct product of the T_p. Finally, G/T is obviously torsion-free. ■

The subgroup T of 1.2.13 is called the *torsion subgroup* of the nilpotent group G, while T_p is the *p-component*.

Observe that the elements of finite order in an infinite soluble group do not in general form a subgroup. For example, the infinite dihedral group is generated by two elements of order 2, but it has elements of infinite order.

We record two further important results which are derivable from 1.2.12.

1.2.14 *Suppose that G is a group for which G_{ab} has finite exponent dividing m. Then,*

(i) *$\gamma_i(G)/\gamma_{i+1}(G)$ has finite exponent dividing m for $i \geq 1$;*
(ii) *if G is nilpotent of class c, then G has finite exponent dividing m^c.*

1.2.15 *Suppose that G is a group for which G_{ab} is finitely generated. Then,*

(i) *$\gamma_i(G)/\gamma_{i+1}(G)$ is finitely generated for $i \geq 1$;*
(ii) *if G is nilpotent, then G is finitely generated.*

As an immediate corollary of 1.2.15 we have:

1.2.16 *Let G be a nilpotent group for which G_{ab} is finitely generated. Then G satisfies max, the maximal condition on subgroups.*

Proof Finitely generated abelian groups satisfy the maximal condition and so each of the lower central factors of G inherits this condition from G_{ab}. The result now follows from the fact that the maximal condition is preserved under forming extensions.   ∎

Thus finitely generated nilpotent groups satisfy the maximal condition, and it follows at once by refining the lower central series suitably that a finitely generated nilpotent group has a central series whose factors are cyclic groups of prime or infinite order. This structure, incidentally, makes it transparent that *a finitely generated nilpotent torsion group is finite.* We shall see in 1.3.5 that this result extends to finitely generated soluble groups.

As a further application of 1.2.12 we record a very important criterion for a group extension to be nilpotent due to P. Hall (1958).

1.2.17 *Let N be a normal subgroup of a group G such that N and G/N' are nilpotent. Then G is nilpotent.*

The crucial point to notice here is that we require G/N', not merely G/N, to be nilpotent. Indeed the result would be false in that case, as a glance at S_3 shows.

Proof of 1.2.17 The basic idea is that the nilpotency of G/N' tells us something about the action of G by conjugation on N_{ab}, and hence about its action on subsequent lower central factors of N. Regard N_{ab} as a G-operator group, the elements of G acting by conjugation. Since G/N' is nilpotent, a central series for G/N' can be intersected with N_{ab} to yield a G-series in N_{ab} with the property that each of its factors is trivial as a G-module, that is, each element of G acts as the identity automorphism on such a factor. We call a module M having a series with G-trivial factors *G-polytrivial.*

Set $F_i = \gamma_i(N)/\gamma_{i+1}(N)$: thus $N_{ab} = F_1$ is G-polytrivial. If F_i is G-polytrivial, then F_{i+1}, as an image of $F_i \otimes_{\mathbb{Z}} N_{ab}$, will be G-polytrivial provided that $F_i \otimes_{\mathbb{Z}} N_{ab}$ is G-polytrivial. This will be proved in 1.2.18 below.

Now we have that every lower central factor of N is a polytrivial G-module. It is very easy to see from this fact, and the nilpotency of G/N', how to construct a central series for G, thus proving G to be nilpotent.   ∎

To complete the proof, we need to establish:

1.2.18 *If A and B are polytrivial G-modules for any group G, then $A \otimes_{\mathbb{Z}} B$ is a polytrivial G-module.*

Proof By hypothesis there exist two series of G-submodules $0 = A_0 < A_1 < \cdots < A_r = A$ and $0 = B_0 < \cdots < B_s = B$ such that A_i/A_{i+1} and B_j/B_{j+1} are trivial G-modules.

Define a series in $T = A \otimes_{\mathbb{Z}} B$ as follows: let T_k be generated by all tensors $a \otimes b$ such that $a \in A_i$ and $b \in B_j$, where $i + j \leq k$. Thus

$$0 = T_0 = T_1 \leq T_2 \leq \cdots \leq T_{r+s} = T.$$

Suppose that $g \in G$, $a \in A_{i+1}$ and $b \in B_{j+1}$. Then, using additive notation for modules, we have $ag = a + a'$ and $bg = b + b'$, where $a' \in A_i$ and $b' \in B_j$. Therefore

$$(a \otimes b)g = (a + a') \otimes (b + b') = a \otimes b + a \otimes b' + a' \otimes b + a' \otimes b',$$

and consequently

$$(a \otimes b)g \equiv a \otimes b \mod T_{i+j+1}.$$

Therefore, each T_k is a G-module and each T_{k+1}/T_k is G-trivial. Hence, T is G-polytrivial, as required. ∎

Careful examination of the proof of 1.2.17 reveals that if N and G/N' are nilpotent of classes c and d, respectively, then G is nilpotent of class at most

$$\binom{c+1}{2} d - \binom{c}{2}.$$

However the best possible bound was shown by Stewart (1966) to be $cd + (c - 1)(d - 1)$.

The upper central series of a group exhibits certain features which are dual to the inheritance properties of the lower central series. Indeed 1.2.11 has the following analogue.

1.2.19 *Let G be an Ω-operator group such that each $Z_i(G)$ is Ω-invariant. If F_i denotes $Z_i(G)/Z_{i-1}(G)$, then for each $i > 0$ there is an injective Ω-homomorphism*

$$\theta : F_{i+1} \to \mathrm{Hom}(G_{ab}, F_i)$$

given by

$$(zZ_i(G))^\theta : gG' \mapsto [z, g]Z_{i-1}(G),$$

where $z \in Z_{i+1}(G), g \in G$.

Here $\mathrm{Hom}(G_{ab}, F_i)$ is an Ω-module via *diagonal action*, that is, if $\omega \in \Omega$ and $\theta \in \mathrm{Hom}(G_{ab}, F_i)$, then θ^ω is defined by

$$(gG')^{\theta^\omega} = \left(\left((gG')^{\omega^{-1}} \right)^\theta \right)^\omega,$$

where $g \in G$. The proof is a straightforward verification.

From this result we can see the strong influence of the centre of a nilpotent group on the structure of the group. Further evidence of this influence is furnished by:

1.2.20 *Let G be any group and i any positive integer.*

(i) *If $Z(G)$ has no non-trivial π-elements, then $Z_i(G)/Z_{i-1}(G)$ has no such elements.*

(ii) *If $Z(G)$ has finite exponent dividing e, then so does $Z_i(G)/Z_{i-1}(G)$.*

This follows easily from 1.2.19. Thus, in particular, *a nilpotent group with torsion-free centre is torsion-free.* Also, if a nilpotent group G has class c and $Z(G)$ has exponent dividing e, then G has exponent dividing e^c.

In addition there is the following result.

1.2.21 *A finitely generated nilpotent group whose centre is a π-group is itself a finite π-group, where π is any set of primes.*

Proof Since the group satisfies max, its centre is finitely generated and therefore finite. Hence the centre has finite exponent, which must be a π-number. Therefore all the factors of the upper central series have π-exponent by 1.2.20. But these factors are also finitely generated and so they are finite π-groups. Thus the group is a finite π-group. ∎

We note that by contrast the lower central series of a finitely generated torsion-free nilpotent group need not be torsion-free. For example, suppose $n > 1$ and set

$$G = \left\{ \begin{pmatrix} 1 & an & b \\ 0 & 1 & c \\ 0 & 0 & 1 \end{pmatrix} \mid a, b, c \in \mathbb{Z} \right\}.$$

Thus G is a torsion-free nilpotent subgroup of $U_n(\mathbb{Z})$. However,

$$G' = \left\{ \begin{pmatrix} 1 & 0 & dn \\ 0 & 1 & 0 \\ 0 & 0 & 1 \end{pmatrix} \mid d \in \mathbb{Z} \right\},$$

so that $G_{ab} \cong \mathbb{Z} \oplus \mathbb{Z} \oplus \mathbb{Z}_n$.

1.3   Polycyclic groups

The simplest type of abelian group is of course a cyclic group. So a natural class of soluble groups to study consists of groups that have a cyclic series, that is, a series with cyclic factors. Such groups are called *polycyclic groups* and the major initial impulse to their systematic study was given by K. A. Hirsch in a series of five papers, starting in 1938. Indeed this work of Hirsch really marks the beginning of the theory of infinite soluble groups.

It is clear that all finite soluble groups are polycyclic. In fact polycyclic groups can be characterized by the following fundamental property.

1.3.1 *The polycyclic groups are precisely the soluble groups which satisfy the maximal condition on subgroups. Also polycyclic groups are finitely presented.*

Proof Cyclic groups clearly satisfy the maximal condition: that polycyclic groups also satisfy this condition, follows from the observation that the maximal condition on subgroups is preserved under forming extensions.

Conversely, if G is a soluble group with the maximal condition, all subgroups of G are finitely generated and hence all factors of the derived series of G are finitely generated abelian groups. The derived series can therefore be refined to a cyclic series. Hence G is polycyclic.

The final statement follows from the finite presentability of cyclic groups and the fact, due to P. Hall, that finitely presented groups form an extension closed class—see Robinson (1996: 2.2.4). ∎

Thus all finitely generated nilpotent groups are polycyclic: as has been observed, they are characterized by the existence of a *cyclic central series*. It is therefore natural to consider also the groups which have a *normal cyclic series*: these are called *supersoluble groups*. The group A_4, which has no non-trivial cyclic normal subgroups, is the smallest polycyclic group which is not supersoluble. On the other hand, the infinite dihedral group is a supersoluble group which is not nilpotent. Thus supersoluble groups fall strictly between finitely generated nilpotent groups and polycyclic groups.

It is natural to enquire how far supersoluble groups are from being nilpotent. In fact they are *virtually nilpotent*: here we follow the convention that if $\mathbf{X}$ is a property of groups, a group is called *virtually* $\mathbf{X}$ if it has a normal $\mathbf{X}$-subgroup of finite index.

1.3.2 *Let G be a supersoluble group. Then*

(i) Fit(G) *is nilpotent and $G/\text{Fit}(G)$ is a finite abelian group. In particular, G' is nilpotent.*

(ii) *If A is a maximal abelian normal subgroup of G, then $A = C_G(A)$.*

Proof

(i) Let $F = \text{Fit}(G)$. By 1.3.1 the group G satisfies the maximal condition and so F is a product of *finitely many* nilpotent normal subgroups. Therefore F is nilpotent by Fitting's Theorem (1.2.9).

Next suppose that $1 = G_0 < G_1 < \cdots < G_n = G$ is a normal cyclic series of G and set $F_i = G_{i+1}/G_i$. Since F_i is cyclic, its automorphism group is finite and abelian: if we write

$$C = \bigcap_{i=0}^{n-1} C_G(F_i),$$

then it follows that G/C is finite and abelian. Moreover $[G_{i+1} \cap C, C] \leq G_i \cap C$ and so the $G_i \cap C$ form a central series of C. Hence C is a nilpotent subgroup and, since C is normal in G, we have $C \leq F$. Therefore G/F is finite and abelian.

(ii) Let $A < C = C_G(A)$. Since C/A is a non-trivial normal subgroup of the supersoluble group G/A, it contains a non-trivial cyclic normal subgroup $\langle x, A \rangle / A$. But then $\langle x, A \rangle$ is an abelian normal subgroup strictly larger than A. ∎

It is fact that the torsion elements of a supersoluble group are subject to some restrictions. Indeed Zappa (1941a) proved that a supersoluble group G has a normal series

$$1 = G_0 < G_1 < \cdots < G_n = G$$

in which each factor is cyclic of prime or infinite order and the factors appear in the following order from the left: first the odd order factors in descending order of magnitude, then the infinite factors, and finally the factors of order 2, (see Robinson 1996).

It follows at once from this result that *the elements of odd order in a supersoluble group form a fully invariant subgroup.* That the elements of even order do not behave in this way is shown by the infinite dihedral group.

We now describe an important invariant associated with polycyclic groups.

1.3.3 *The number of infinite factors in a cyclic series of a polycyclic group G is an invariant $h(G)$, called the Hirsch number of G.*

Proof Consider a cyclic series of G. Any refinement of the series will itself be cyclic and will have the same number of infinite cyclic factors. To see this, suppose that H/K is an infinite cyclic factor and that $K \leq K^* < H^* \leq H$. Then clearly exactly one of K^*/K, H^*/K^*, or H/H^* is infinite cyclic and the others are finite. By the Refinement Theorem any two cyclic series of G have isomorphic refinements, and consequently they must have the same number of infinite cyclic factors. ∎

A group is said to be *poly-infinite cyclic* if it has a series with infinite cyclic factors. Such a group is clearly polycyclic and torsion-free. However not every torsion-free polycyclic group is poly-infinite cyclic, as the following example of Hirsch (1952) shows:

$$G = \langle\, x, y, z \mid x^z = x^{-1},\ y^z = y^{-1},\ [x,y] = z^{4m} \,\rangle,$$

where $m \geq 1$.

In a certain sense polycyclic groups are close to being poly-infinite cyclic.

1.3.4 (Hirsch 1946) *Polycyclic groups are virtually poly-infinite cyclic*

In order to prove this result we will need a number of auxiliary lemmas concerning finitely generated groups which are of independent interest.

1.3.5 *A finitely generated soluble torsion group is finite.*

Proof We argue by induction on the derived length d of such a group G. If $d = 0$, there is nothing to prove. So assume $d > 0$ and put $A = G^{(d-1)}$. Thus we may suppose that G/A is finite, so that A is finitely generated since G is. But A is an abelian torsion group, so it is finite. Therefore G is finite. ∎

1.3.6 *Suppose that G is a finitely generated group and F is a finite group. Then the number of normal subgroups N of G such that $G/N \cong F$ is finite.*

Proof Each such normal subgroup is the kernel of a homomorphism from G onto F. Since G is finitely generated and since a homomorphism is determined by its effect on a set of generators of G, there are only finitely many such homomorphisms. ∎

It is a corollary of this result which we need for 1.3.4.

1.3.7 *Suppose that H is a finitely generated normal subgroup of a group G and that K is a normal subgroup of finite index in H. Then the normal core K_G of K in G has finite index in H.*

Proof If $g \in G$, then $K^g \lhd H^g = H$ and $H/K^g \cong H/K$. By 1.3.6 there can only be a finite number of K^g's. The intersection of these is the core K_G, which is therefore of finite index in H. ∎

Proof of 1.3.4 Suppose that $1 = G_0 \lhd G_1 \cdots \lhd G_n = G$ is a cyclic series in the polycyclic group G. If $n \leq 1$, then G is cyclic and the result is obvious. We may therefore assume $n > 1$ and that $H = G_{n-1}$ is virtually poly-infinite cyclic. Thus there exists a poly-infinite cyclic normal subgroup K of finite index in H.

By 1.3.7 $|H : K_G|$ is finite. Also K_G, like K, is poly-infinite cyclic. This means that we may assume $K \lhd G$. If G/H is finite, then so is G/K, and we are finished. So we may assume G/H to be infinite cyclic, with generator xH say. Since H/K is finite, x^r centralizes H/K for some $r > 0$. Setting $J = \langle x^r, K \rangle$, we have $J \lhd \langle x, H \rangle = G$, since $[x^r, H] \leq K$. Moreover, $G/J = (\langle x, J \rangle / J) \cdot (HJ/J)$ is finite. Since H contains no positive power of x, the group J/K is infinite cyclic, and thus J is poly-infinite cyclic. ∎

As a corollary to 1.3.4 we have another result showing the importance of poly-infinite cyclic groups.

1.3.8 *Every polycyclic group G is an image of a poly-infinite cyclic group $\bar{G}$.*

Proof By 1.3.4 there is a poly-infinite cyclic normal subgroup N such that G/N is finite. By induction on $|G : N|$, we may assume that there is a normal subgroup M of G such that $N \leq M$, M is an image of a poly-infinite cyclic group $\bar{M}$ and $Q = G/M$ is finite cyclic. Now it is well known that G embeds in the standard wreath product $M\,wr\,Q$—see H. Neumann (1967)—and there is an obvious surjective homomorphism $\bar{M}\,wr\,Q \twoheadrightarrow M\,wr\,Q$. Thus we may replace G by $\bar{M}\,wr\,Q$. Now we write

$$G = Q \ltimes L,$$

where Q is finite cyclic and L is the base group of the wreath product; thus L is poly-infinite cyclic. Let $Q = \langle q \rangle$ and let $X = \langle x \rangle$ be an infinite cyclic group. Make X act as an automorphism group on L via $x \mapsto q$ and put $\bar{G} = X \ltimes L$. Then we have a surjective homomorphism $\bar{G} \twoheadrightarrow G = Q \ltimes L$ with $x \mapsto q$, and of course $\bar{G}$ is poly-infinite cyclic. ∎

An open question here is: what is the least possible value of $h(\bar{G})$? Could it be the length of a shortest cyclic series of G?

As a further corollary to 1.3.4 we have the useful:

1.3.9 *An infinite polycyclic group contains a non-trivial free abelian normal subgroup.*

Proof By 1.3.4 such a group G has a poly-infinite cyclic normal subgroup H of finite index. The smallest non-trivial term of the derived series of H is torsion-free and hence free abelian: further, being characteristic in H, it is normal in G. ■

In the study of infinite soluble groups, and indeed of infinite groups in general, a central role is played by finiteness conditions, that is, group theoretical properties which are possessed by all finite groups. Among the most important finiteness conditions is residual finiteness. If **X** is a class of groups, we say that a group G is *residually* **X** if

$$\bigcap_{G/N \in \mathbf{X}} N = 1$$

or, equivalently, G is a subdirect product of **X**-groups. To assert that a group is residually **X** is a precise way of saying that **X**-images of G are plentiful.

As a first application of 1.3.9 we will show that *virtually polycyclic groups are residually finite* (Hirsch 1954). This result follows at once from a stronger theorem of Mal'cev (1958). The proof given here is due to J. S. Wilson.

1.3.10 *Suppose that H is a subgroup of a virtually polycyclic group G. Then H coincides with the intersection of all the subgroups of finite index in G which contain H.*

Notice that this is equivalent to saying that *every subgroup of G is closed in the profinite topology on G*. Groups with this property are often called *subgroup separable*. The residual finiteness of polycyclic groups follows on taking H to be 1.

Proof of 1.3.10 If G is abelian, then $H \triangleleft G$ and G/H is finitely generated and abelian. The result in this case follows at once from the Structure Theorem for finitely generated abelian groups. Suppose now that ℓ is the Hirsch length of G. If $\ell = 0$, then G is finite and there is nothing to prove. We assume, therefore, that $\ell > 0$ and proceed by induction on ℓ.

By 1.3.9 there exists a non-trivial torsion-free abelian normal subgroup A of G. Clearly $h(G/A) < \ell$ and so we may assume the theorem holds in G/A. Now the result is equivalent to showing that if $g \notin H$, there exists a subgroup K of finite index in G such that $K \geq H$ and $g \notin K$. If $g \notin HA$, then the existence of such a K follows at once from the truth of the result for G/A. Thus we may assume that $g \in HA$. Then $g = ha$ for some $h \in H$ and $a \in A$, where $a \notin H \cap A$ since $g \notin H$. By the abelian case, there is a subgroup B of finite index in A such that $H \cap A \leq B$ and $a \notin B$. Then $A^m \leq B$ for some positive integer m and $|A : A^m| < \infty$. By the induction hypothesis the theorem holds for G/A^m, so as before we may assume that $g \in HA^m$, and say $g = h_1 a_1$ where $h_1 \in H$, $a_1 \in A^m$. But then $ha = h_1 a_1$, so that $h_1^{-1} h = a_1 a^{-1} \in H \cap A \leq B$. However $a_1 \in A^m \leq B$, so $a \in B$, a contradiction. ∎

A group G is said to be *conjugacy separable* if two elements are conjugate in G whenever their images in every finite quotient of G are conjugate.

1.3.11 (Remeslennikov 1969a; Formanek 1976) *Every virtually polycyclic group is conjugacy separable.*

The proof of this result involves significant algebraic number theory: for details the reader may consult (Segal 1983).

The fact that polycyclic groups are so richly endowed with finite images has important implications for the study of algorithmic properties of these groups, as will be seen in Chapter 9. It also motivates a wide variety of questions as to which properties of a polycyclic group can be recognized from its finite images. For example, Hirsch (1946) showed that nilpotency is just such a property.

1.3.12 *Suppose that G is a polycyclic group which is not nilpotent. Then G has a finite image which is not nilpotent.*

Proof Suppose that G is a counterexample of minimal Hirsch length. Then G is infinite and by 1.3.9 it has a non-trivial torsion-free abelian normal subgroup A. Thus A is finitely generated and free abelian of rank r, say. If p is any prime, $h(G/A^p) < h(G)$ and so G/A^p is nilpotent, by the minimality assumption on the Hirsch length. Now A/A^p is elementary abelian of order p^r and hence it is contained in $Z_r(G/A^p)$, by 1.2.8(i) and a simple induction on r. It follows that

$$B = [A, \underbrace{G, \ldots, G}_{r}] \leq A^p \quad \text{and hence} \quad B \leq \bigcap_p A^p = 1$$

since A is free abelian. Thus $A \leq Z_r(G)$. But G/A is nilpotent, so it follows that G is nilpotent, a contradiction. ∎

Recently a significant generalization of this last result has been found by Endimioni (1998), which relates the nilpotent length of a polycyclic group to that of its finite quotients.

1.3.13 *If every finite quotient of a polycyclic group G has nilpotent length at most l, then G has nilpotent length at most l.*

Of course Hirsch's result is the case $l = 1$. For further versions of Hirsch's Theorem, valid in wider classes of infinite soluble groups, see 5.3.12 below and its sequel.

There is an analogue of 1.3.12 for supersolubility, which lies somewhat deeper.

1.3.14 (Baer 1959) *Suppose that G is a polycyclic group which is not super-soluble. Then G has a finite image which is not supersoluble.*

Proof Assume that G is not supersoluble. Then by max, there is a largest normal subgroup of G whose quotient group is not supersoluble. Since we may replace G by this quotient group, nothing is lost in assuming that all proper quotients of G are supersoluble.

Next G' must be infinite. For otherwise, by the residual finiteness of G, there is a normal subgroup L of G with finite index such that $G' \cap L = 1$. Then there is a G-isomorphism $G' \simeq^G G'L/L$, which shows that there is a G-invariant series in G' with cyclic factors. But this clearly implies that G is supersoluble.

Now we argue that G' is nilpotent. Supposing this to be false, we may apply 1.3.12 to produce a normal subgroup N of G' such that G'/N is finite and non-nilpotent. Let M be the core of N in G. Then G'/M is a torsion group, so it is finite and hence $M \neq 1$. Thus G/M is supersoluble and 1.3.2 shows that G'/M, and hence G'/N, is nilpotent. By this contradiction G' is nilpotent.

By 1.2.20 the subgroup $Z(G')$ is infinite, whence it contains a non-trivial free abelian normal subgroup of G. Evidently we may choose such a subgroup A which is *rationally irreducible* with respect to G: this means that $A \otimes_{\mathbb{Z}} \mathbb{Q}$ is a simple $\mathbb{Q}G$ module. Then $[A, G'] = 1$. Any $g \in G$ induces an automorphism in A, which may be represented by an element $\bar{g}$ of $GL_r(\mathbb{Z})$, where r is the rank of A. Let p be any prime; then G/A^p is supersoluble. It follows that there is a $\langle g \rangle$-invariant series in A/A^p with cyclic factors, and this in turn implies that the characteristic polynomial of $\bar{g}$ is a product of linear factors modulo p.

At this point algebraic number theory enters the arena. By a result of Heilbronn (1967: 229), we conclude that $\bar{g}$ has a characteristic root d in $\mathbb{Z}$: clearly $d = \pm 1$. It follows that

$$A_g = \{a \in A \mid a^g = a^d\}$$

is non-trivial. Since $A \leq Z(G')$, we see that $A_g \triangleleft G$. Hence A/A_g is finite because A is G-rationally irreducible. Notice that every subgroup of A_g is $\langle g \rangle$-invariant.

Let $g_1, g_2, \ldots, g_n$ be generators of G. Then $I = \bigcap_{i=1}^n A_{g_i}$ is non-trivial since each $|A : A_{g_i}|$ is finite. Also every subgroup of I is normalized by each g_i, and hence is normal in G. Therefore, there exists a non-trivial cyclic normal subgroup of G and, finally, we obtain the contradiction that G is supersoluble.   ∎

Another indication of the important role of the finite quotients of a polycyclic group is the following theorem of Kegel (1966), which gives a criterion for subnormality.

1.3.15 *Let H be a subgroup of a virtually polycyclic group G. Then H is subnormal in G if and only if its image in every finite quotient is subnormal.*

Proof Of course only the sufficiency of the condition requires comment. Assume that the image of H in every finite quotient of G is subnormal. We show by induction on $h(G) > 0$ that H is subnormal in G. By 1.3.9 there is a non-trivial free abelian normal subgroup A of G.

Since $A/(H \cap A)$ is finitely generated abelian, there is a $k > 0$ such that $(H \cap A)A^k/(H \cap A)$ is torsion-free. Also

$$(H \cap A)A^k/(H \cap A) \simeq A^k/(H \cap A^k),$$

so we may replace A by A^k, that is, we may assume that $A/(H \cap A)$ is free abelian: let r be its rank.

Now let p be any prime; then $h(G/A^p) < h(G)$, so by the induction hypothesis HA^p is subnormal in G. Hence

$$[A, {}_n H] \leq (HA^p) \cap A = (H \cap A)A^p,$$

for some $n > 0$. Since $(H \cap A)A^p \lhd HA$ and $A/(H \cap A)A^p$ is elementary abelian of order p^r, we may take n to be r in the above inclusion *for all primes* p. Consequently

$$[A, {}_r H] \leq \bigcap_p (H \cap A)A^p = H \cap A,$$

since $A/(H \cap A)$ is free abelian. It follows that H is subnormal in HA, and since HA is subnormal in G, we reach the desired conclusion. ∎

Finitely generated nilpotent groups have even better residual finiteness properties than polycyclic groups.

1.3.16 *Let G be a finitely generated nilpotent group and let p be a prime. If G has no p'-elements, then G is a residually finite p-group.*

Proof If G is abelian, it is the direct product of a finite p-group and a free abelian group, so the statement is obviously true. Thus we may assume that G has nilpotent class $c > 1$. If $Z = Z(G)$, then G/Z has no p'-elements by 1.2.20, so this group is residually finite-p, by induction on c.

Let $1 \neq g \in G$: we have to 'exclude' g from some normal subgroup with index a power of p. For this purpose we may suppose that $g \in Z$. Hence $g \notin Z^{p^i}$ for some $i > 0$. Now G/Z^{p^i} is residually finite by 1.3.10. Hence $g \notin L$ where $Z^{p^i} \leq L \lhd G$ and G/L is finite.

Next G/L is the direct product of a finite p-group and a p'-group. Thus we may conclude that gL is a p'-element and $g^m \in L$ for some positive p'-number m. However $g^{p^i} \in Z^{p^i} \leq L$ and therefore $g \in L$, a contradiction. ∎

An immediate corollary of 1.3.16 is:

1.3.17 *If G is a finitely generated nilpotent group and p is any prime, then $\bigcap_{i=1,2,\ldots} G^{p^i}$ is a finite p'-group.*

In particular we obtain the well-known theorem of Gruenberg (1957): *a finitely generated torsion-free nilpotent group is residually finite-p for all primes p.*

It is a remarkable fact that there is strong converse to Gruenberg's theorem.

1.3.18 (Seksenbaev 1965) *If G is a polycyclic group which is residually finite-p for infinitely many primes p, then G is torsion-free and nilpotent.*

Proof Assume that G is not nilpotent. There exists an i such that $\gamma_i(G)/\gamma_{i+1}(G)$ is finite since the alternative is that $h(\gamma_j(G)) > h(\gamma_{j+1}(G))$ for all j: put $L = \gamma_i(G)$. If π is the set of primes dividing $|L/[L,G]|$, then each $L/[L,_j G]$ is a π-group by the tensor product argument—see 1.2.11.

Now choose a prime p not in π for which G is residually finite-p. Then there is a descending central series of G,

$$G = G_0 \geq G_1 \geq \cdots G_j \geq G_{j+1} \cdots ,$$

such that each G/G_j is a finite p-group and $\bigcap_{j=1,2,\dots} G_j = 1$. Hence $L \not\leq G_j$ for some j and $L/L \cap G_j$ is a non-trivial finite p-group. But $[L,_j G] \leq L \cap G_j$. Since $L/[L,_j G]$ is a π-group, we reach the contradiction that $L \cap G_j = L$.

Finally, any group which is residually finite-p for even two different primes p is clearly torsion-free. ∎

Another interesting property of polycyclic groups is indicated by the following result of Rhemtulla (1967).

1.3.19 *Let G be a virtually polycyclic group and let H be a subgroup of G. Then any intersection of conjugates of H is the intersection of finitely many of them. Conversely, any finitely generated soluble group with this intersection property is polycyclic.*

In particular it follows that *the normal core of a subgroup of a virtually polycyclic group is the intersection of finitely many of its conjugates.* This turns out to be an important result in the algorithmic theory of polycyclic groups since it provides an effective means of constructing the core of an arbitrary subgroup, as will be seen in Chapter 9.

Frattini subgroups

Recall that the *Frattini subgroup* Frat(G) of a group G is defined to be the intersection of all the maximal subgroups of G, provided that G has such subgroups. If not, we define Frat(G) to be G. This important subgroup was introduced by G. Frattini in 1885. He proved that if G is a finite group, then Frat(G) is nilpotent Frattini (1885). Subsequently, Wielandt (1937) showed that if $G' \leq$ Frat(G) in a finite group G, then the group is nilpotent. We will apply 1.3.12 to show that these two properties are shared by polycyclic groups.

1.3.20 *Suppose that G is a virtually polycyclic group. Then*

(i) Fit($G/$ Frat(G)) $=$ Fit(G)/ Frat(G): *in particular* Frat(G) *is nilpotent;*
(ii) G *is nilpotent if and only if* $G' \leq$ Frat(G).

Proof

(i) Set $F =$ Frat(G) and let N/F be a nilpotent normal subgroup of G/F. If N is not nilpotent, it has a finite non-nilpotent image N/M by 1.3.12. By 1.3.7 we may replace M by its core in G and so assume that $M \lhd G$.

Put $\bar{G}=G/M$, etc. Suppose that $\bar{P}$ is a Sylow subgroup of $\bar{N}$: it is enough to show that $\bar{P}\triangleleft\bar{G}$. Write $\bar{K}=\bar{P}\bar{F}$ and note that $\bar{K}\triangleleft\bar{G}$. Then the 'Frattini argument'—see Robinson (1996)—shows that $\bar{G}=\bar{K}N_{\bar{G}}(\bar{P})=\bar{F}N_{\bar{G}}(\bar{P})$. Hence $N_{\bar{G}}(\bar{P})=\bar{G}$ and $\bar{P}\triangleleft\bar{G}$, by the non-generator property of the Frattini subgroup (Robinson 1996).

(ii) If G is nilpotent, then every maximal subgroup is normal in G and has prime index, so that $G'\leq\mathrm{Frat}(G)$. Conversely, assume that $G'\leq\mathrm{Frat}(G)$. If G is not nilpotent, it has a finite non-nilpotent image H by 1.3.12. But the hypothesis implies that $H'\leq\mathrm{Frat}(H)$ and so H is nilpotent by Wielandt's Theorem, a contradiction. $\blacksquare$

Examples of polycyclic groups

There is a close link between polycyclic groups and algebraic number theory, arising from a famous theorem of Dirichlet: *if O is the ring of integers of an algebraic number field,*[2] *then its group of units O is finitely generated.* Thus, if we put $A=O^+$, the additive group of O, and assume that H is a subgroup of O which generates O as a ring, the group $G=H\ltimes A$ is polycyclic.

We illustrate the construction by means of a specific example. Let $f=t^2-t-1\in\mathbb{Z}[t]$; this is irreducible over $\mathbb{Q}$ and its splitting field is $\mathbb{Q}(\sqrt{5})$ since the roots of f are $z=(1+\sqrt{5})/2$ and $-z^{-1}=(1-\sqrt{5})/2$. The ring of algebraic integers in $\mathbb{Q}(\sqrt{5})$ is $O=\mathbb{Z}\oplus\mathbb{Z}z$ and the group of units of O is $\langle-1\rangle\times\langle z\rangle$. We may choose H to be $\langle z\rangle$ since H generates O as a ring; put $A=O^+\simeq\mathbb{Z}\oplus\mathbb{Z}$. Then

$$G=H\ltimes A,$$

which is a polycyclic group with $h(G)=3$.

Notice that the group G has a faithful two-dimensional representation over $\mathbb{Z}(\sqrt{5})$, in which

$$(z^r,a)\mapsto\begin{bmatrix}z^r & 0\\ a & 1\end{bmatrix}.$$

Now z can be identified with the integral matrix $\begin{bmatrix}0 & 1\\ 1 & 1\end{bmatrix}$, that is, with the companion matrix of f. Therefore G has a faithful three-dimensional representation over $\mathbb{Z}$, a result which foreshadows the Auslander–Swan Theorem—see 3.3.1 below.

1.4 Soluble groups with the minimal condition

Having identified the soluble groups with the maximal condition, we are led naturally to inquire about soluble groups with min, the minimal condition for subgroups. Certainly such groups must be periodic since the infinite cyclic group does not satisfy min. Indeed it is well known that the abelian groups with min are precisely the direct products of finitely many finite cyclic groups and *quasicyclic*

---

[2] An *algebraic integer* is a root of a monic polynomial over $\mathbb{Z}$.

groups, that is, groups of Prüfer type p^∞—see Fuchs (1960, 1970). Therefore each factor of an abelian series in a soluble group with min must have this structure. As a consequence, *the soluble groups with min are exactly the poly-(finite cyclic or quasicyclic) groups*. Thus there is a clear analogy with polycyclic groups.

More precise information about soluble groups with min is given by a fundamental theorem of Černikov.

1.4.1 (Černikov 1940) *A soluble group satisfies the minimal condition if and only if it is virtually a direct product of finitely many quasicyclic groups*

A group which is virtually a direct product of finitely many quasicyclic groups is called a *Černikov group*. To establish 1.4.1 we need a result about groups satisfying a weaker minimal condition, the *minimal condition for normal subgroups*, $\min -n$.

1.4.2 *Suppose the group G satisfies* $\min -n$. *Then G has a unique minimal subgroup F of finite index. Furthermore F is characteristic in G.*

The subgroup F is therefore the intersection of all the subgroups of finite index in G, that is, the *finite residual* of G.

Proof of 1.4.2 Since G has $\min -n$, there is a minimal *normal* subgroup F of finite index in G. Suppose that H is any subgroup of finite index in G. Then the core H_G also has finite index, and therefore so has $H_G \cap F$. By minimality of F we have $H_G \cap F = F$, so that $F \leq H$, and the result follows.   ∎

We are now able to prove Černikov's Theorem.

Proof of 1.4.1 Suppose G is a soluble group with min and F is its finite residual; we may assume G to be of infinite order, so that $F \neq 1$ by 1.4.2. Hence F contains a non-trivial abelian normal subgroup A of G. If $1 \neq a \in A$, then conjugates of a have the same order as a. Now A is an abelian group with min and it is readily seen from the structure of such groups that A has only a finite number of elements of each given order. Therefore $C_G(a)$ has finite index in G and $F \leq C_G(a)$. It follows that $a \in Z(F)$ and $Z(F) \neq 1$.

If $Z(F) = F$, then F is abelian with min and it has no proper subgroups of finite index. This implies that F is a direct product of finitely many quasicyclic groups and G is Černikov, as required.

If $Z(F) \neq F$, we may apply the preceding argument to $G/Z(F)$, which has finite residual $F/Z(F)$, to show that $Z(F) \neq Z_2(F)$. Let $z \in Z_2(F) \backslash Z(F)$ and $x \in F$. Then $z^x = zz_1$ where $z_1 = [z, x] \in Z(F)$. Hence the order of z_1 divides the order of z—remember that G is a torsion group. So there are only finitely many possibilities for z_1 in $Z(F)$ and hence for z^x. Consequently $C_F(z)$ has finite index in F and $F = C_F(z)$. This shows that $z \in Z(F)$, a contradiction. Hence $F = Z(F)$ and the result is proved.   ∎

Constructing Černikov groups

Suppose that D is a direct product of finitely many quasicyclic groups and F is any finite group. Then the standard wreath product $W = D \ wr \ F$ is a Černikov group, as is every subgroup of W. In fact every Černikov group arises in this way. For, if G is a Černikov group with finite residual D, then G embeds in the wreath product $D \ wr \ (G/D)$. Thus we have:

1.4.3 *The Černikov groups are precisely the groups that arise as subgroups of standard wreath products $D wr F$ where D is a direct product of finitely many quasicyclic groups and F is a finite group.*

Of course we get all the soluble Černikov groups by taking F to be soluble. Clearly, to get further information about Černikov groups, it is necessary to study modules over finite groups which are direct sums of finitely many quasicyclic groups: these have been called *Černikov modules*. A detailed investigation of such modules has been carried out by Hartley (1977). For example, it is shown that, *if D is a Černikov module over a finite group, there is a finite submodule D_0 such that D/D_0 is the direct sum of modules which are irreducible, in the sense that they have no proper infinite submodules.*

Nilpotent groups with min

Somewhat more can be said of the structure of nilpotent groups with min.

1.4.4 (Baer 1955d) *A nilpotent group G satisfies the minimal condition if and only if $Z(G)$ satisfies the condition and $G/Z(G)$ is finite.*

Clearly such groups satisfy min. In order to establish the converse we need a result about *radicable groups*, that is, groups each of whose elements is an nth power for every positive integer n. The term *divisible* is often used for abelian radicable groups.

1.4.5 *Suppose that A is a radicable normal abelian subgroup of a group G. Let H be a subgroup of G such that $[A, {}_r H] = 1$ for some $r \geq 1$ and H_{ab} is a torsion group. Then $[A, H] = 1$.*

Proof Set $A_0 = A$ and $A_{i+1} = [A_i, H]$ for $i \geq 0$: then $A_r = 1$. If $a \in A$ and $h \in H$, the mapping $(a, hH') \mapsto [a, h]A_2$ is well defined since by 1.2.3, $[A, H'] \leq [A, H, H] = A_2$. Also it is bilinear because A is abelian: therefore it induces a homomorphism from $A \otimes_{\mathbb{Z}} H_{ab}$ onto A_1/A_2. But $A \otimes_{\mathbb{Z}} H_{ab}$ is trivial since A is radicable and H_{ab} is torsion. Hence $A_1 = A_2$ and so $A_1 = A_r = 1$. ∎

Proof of 1.4.4 Suppose that G is a nilpotent group of class c with min. By 1.4.1 G is Černikov and therefore it has a radicable abelian normal subgroup A with finite index. For any $g \in G$, we have $[A, {}_c \langle g \rangle] = 1$ by definition of c. Hence $[A, g] = 1$ by 1.4.5. Thus $A \leq Z(G)$ and $G/Z(G)$ is finite. ∎

Notice that as a consequence of 1.4.4 and Schur's theorem on centre-by-finite groups—see Robinson (1996)—*if G is a nilpotent group with min, then G' is finite.*

1.5 Soluble groups with the minimal condition on normal subgroups

First we remark that in a nilpotent group satisfying min $-n$, the minimal condition on normal subgroups, each factor of a central series satisfies min, so that the group itself has min. Thus for nilpotent groups the properties min and min $-n$ are identical. However the situation is completely different for soluble groups.

1.5.1 (Čarin 1949; Duguid and McLain 1956) *There exist metabelian groups satisfying* min $-n$ *which do not satisfy* min.

Proof Let K be the algebraic closure of the field of p elements $\mathbb{Z}_p$. Then the multiplicative group K^* of K is a direct product of infinitely many quasicyclic groups, one for each prime $q \neq p$. Let X be a subgroup of K^* and F the subfield of K generated by X. Denote the additive group of F by A, so that multiplication by elements of X induces automorphisms in A and thus X embeds in $\mathrm{Aut}(A)$. Accordingly, we set $G = X \ltimes A$. Suppose that B is a non-trivial normal subgroup of G contained in A and choose any $0 \neq b \in B$. Then we may write

$$b^{-1} = \sum_{x \in X} f_x x,$$

where $f_x \in \mathbb{Z}_p$ and all but a finite number of the f_x are zero. Now $bx \in B$ for all $x \in X$, so that B contains

$$\sum_{x \in X} f_x(bx) = bb^{-1} = 1.$$

This implies that $A = B$ and it follows that A is a minimal normal subgroup of G.

If we now take X to be the q-component of K^*, that is, a group of type q^∞ where $q \neq p$, then G satisfies min $-n$ since both A and G/A satisfy min $-G$, the minimal condition for G-invariant subgroups. However G does not satisfy min since A is an infinite elementary p-group. ∎

It is an important theorem due to Baer (1964b) that a soluble group with min $-n$ cannot contain elements of infinite order.

1.5.2 *A soluble group with* min $-n$ *is locally finite.*

The proof is based on a result about automorphism groups. If G is a group of automorphisms of a group A, then G is called *irreducible* if A has no proper non-trivial G-invariant subgroups.

1.5.3 *Let G be an irreducible group of automorphisms of an abelian group A. If G is locally finite, then A is an elementary abelian p-group for some prime p.*

Proof (J. E. Roseblade) Assume that A is not elementary abelian p for any p. Since A is characteristically simple, i.e., it has no proper non-trivial characteristic subgroups, it must be a direct product of copies of $\mathbb{Q}$ and hence is radicable.

Clearly G is not trivial. Hence there exists $a \in A$ such that $[a, G] \neq 1$. Now $[a, G]$ is G-admissible, so $A = [a, G]$. Since A is radicable, there exists $b \in A$ such that $a = b^2$. Also $b \in [a, G]$, so $b \in [a, F]$ where F is some finitely generated, and hence finite, subgroup of G. Thus $b \in [b^2, F] = [b, F]^2$ and $[b, F] = [b, F]^2$. But $[b, F]$ is a finitely generated torsion-free abelian group, so it follows that $[b, F] = 1$ and $[a, F] = 1$, which yields the contradiction $a = 1$. ∎

Proof of 1.5.2 Suppose that G is a soluble group with min $-n$. If G is abelian, it has min and the result is certainly true. So we suppose G is not abelian and proceed by induction on the derived length of G. Thus we may assume that if D is the last non-trivial term of the derived series of G, then G/D is locally finite. Since G has min $-n$, the subgroup D satisfies min $-G$, and so it has an ascending G-series whose factors are simple $\mathbb{Z}(G/D)$-modules. Let E be any factor of this series. If $C = C_G(E)$, then $D \leq C$ and so G/C is locally finite. Also G/C is isomorphic with an irreducible group of automorphisms of the abelian group E. From 1.5.3 we deduce that E is elementary abelian. Hence D is a torsion group. Therefore G is torsion and, being soluble, it is locally finite. ∎

Finally, we use Baer's theorem to prove a result about the Fitting subgroup of a metanilpotent group with min $-n$. Here a group is termed *metanilpotent* if it is an extension of a nilpotent group by a nilpotent group.

1.5.4 *If G is a metanilpotent group satisfying* min $-n$*, then* $\mathrm{Fit}(G)$ *is nilpotent.*

Proof Let $F = \mathrm{Fit}(G)$. Also let $A \triangleleft G$ where A and G/A are nilpotent: thus $A \leq F$. If F/A' is nilpotent, then F is nilpotent by 1.2.17. Therefore we may assume that A is abelian.

Since G satisfies min $-n$, there is an $i > 0$ such that $[A, \, _iF] = [A, \, _{i+1}F] = B$, say. If $B = 1$, then F is nilpotent since F/A is. So we may assume that $B \neq 1$. By min $-n$ there is an ascending G-composition series $0 = B_0 < B_1 < \cdots < B_\beta = B$. (Here, of course, β is an ordinal.) If N is nilpotent and normal in G, so is NA and hence N centralizes all $B_{\alpha+1}/B_\alpha$. It follows that the B_α form an ascending F-central series of B.

Next put $C_\alpha = C_F(B_\alpha)$. Then $C_\alpha \triangleleft G$, $A \leq C_\alpha$ and $F = C_0 \geq C_1 \geq \cdots$. By min $-n$ there are only finitely many C_α's: let these be $C_{\alpha_0+1}, C_{\alpha_1+1}, \ldots, C_{\alpha_r+1}$, where $\alpha_0 < \alpha_1 < \cdots < \alpha_r$ and

$$C_{\alpha_i+1} = \cdots = C_{\alpha_{i+1}} > C_{\alpha_{i+1}+1},$$

for $i = 0, 1, \ldots, r - 1$. Note that $C_{\alpha_0+1} = F$ and $C_{\alpha_r+1} = C_F(B)$.

Writing j for α_{i+1}, we see that C_j/C_{j+1} is isomorphic with a group of automorphisms of B_{j+1} which acts trivially on B_j and B_{j+1}/B_j. Consequently C_j/C_{j+1} is isomorphic with a subgroup of $\mathrm{Hom}(B_{j+1}/B_j, B_j)$. The latter group has finite exponent since B_{j+1}/B_j is elementary abelian p for some p by 1.5.2. Hence C_j/C_{j+1} has finite exponent for $j = 0, 1, \ldots, r-1$.

It now follows that $F/C_F(B)$ has finite exponent. But $A \le C_F(B)$, so that $F/C_F(B)$ is also a Černikov group and hence this quotient group is finite. In addition $C_F(B)$ is nilpotent since F/B is nilpotent. Therefore F is a product of finitely many nilpotent normal subgroups of G and so F is nilpotent.   ∎

Whether $\mathrm{Fit}(G)$ need be nilpotent for any soluble group G with min $-n$ remains an open question.

The structure of soluble groups with min $-n$ has been investigated in a number of articles by B. Hartley and D. McDougall. In Hartley (1977) it is shown that *if G is a soluble group with* min $-n$ *which has derived length $d \ge 2$, then $|G| \le \aleph_{d-2}$*. In particular *a metabelian group satisfying* min $-n$ *is countable* (McDougall 1970*a*).

On the other hand, Hartley's bound is sharp for $d = 3$, and thus soluble groups with min $-n$ need not be countable. A classification of metabelian groups with min $-n$ which have no proper subgroups of finite index is described in Hartley and McDougall (1971): this depends crucially on the use of injective hulls of modules over nilpotent torsion groups.

Metanilpotent groups with min $-n$ have been investigated by Silcock (1973), especially in the case where there are no proper subgroups with finite index. For example, it is shown that, if G is such a group, there is a radicable abelian subgroup A with min such that $G = G'A$ and $G' \cap A \le Z(G)$. In Silcock (1975) numerous examples are constructed of soluble groups of arbitrary derived length whose lattices of normal subgroups are well ordered: of course all of these groups satisfy min $-n$.

In conclusion we mention that the parallel study of soluble groups satisfying max $-n$, the maximal condition on normal subgroups, calls for very different techniques and involves the theory of finitely generated modules over polycyclic groups. This is developed at length in Chapters 4 and 8.

2

NILPOTENT GROUPS

The topic of this chapter may be said to be the calculus of commutators insofar as it applies to nilpotent groups. Of course nilpotent groups have a great deal of commutativity built into their structure, so it is not surprising that commutators in such groups exhibit special features. The topics to be explored in this chapter are the extraction of roots and Mal'cev completions, the commutator collection process and the theory of isolators in nilpotent groups.

2.1 Extraction of roots in nilpotent groups

One of the most interesting aspects of the theory of nilpotent groups is that concerned with the extraction of roots. Let π be a (non-empty) set of primes. Then a group G is said to be π-*radicable,* or just *radicable* if π is the set of all primes, if for every $g \in G$ and every positive π-number n, the equation $x^n = g$ has a solution x in G. So this means that each element of G must have an nth root in G for all positive π-numbers n.

If a group has unique extraction of π-roots, it is clearly π-*torsion-free,* that is, it does not contain non-trivial π-elements. Thus the radicable groups in which the extraction of roots is unique are those that are torsion-free.

We note that extraction of roots is not always possible in a torsion-free group, as is shown by an infinite cyclic group. However, an infinite cyclic group can be embedded in a torsion-free radicable group, for example, in the additive group of rational numbers $\mathbb{Q}$. This simple fact has a powerful generalization to all torsion-free nilpotent groups, namely the following theorem of Mal'cev, which is the main result in this section.

2.1.1 (Mal'cev 1949) *A torsion-free nilpotent group G can be embedded in a nilpotent group G^* in which the extraction of roots is unique in such a way that every element of G^* has a positive power in G. Moroever the group G^* is unique up to isomorphism.*

A group with the properties of G^* is called a *Mal'cev completion* or *radicable hull* of the group G. If we replace 'torsion-free' by 'π-torsion-free', where π is any set of primes, the proof of 2.2.1 will in fact produce a unique π-radicable nilpotent group G^*_π in which G embeds.

The first thing to notice is that extraction of roots, if possible, is always unique in a torsion-free nilpotent group. This follows from

2.1.2 *Suppose that G is a torsion-free nilpotent group with elements $a, b \in G$ such that $a^n = b^n$ for some $n > 0$. Then $a = b$.*

Proof This goes by induction on the nilpotent class of G, the result being obvious if G is abelian. Since $G/Z(G)$ is torsion-free by 1.2.20, we may assume by induction that $a = bz$ where $z \in Z(G)$. Then $a^n = b^n z^n = a^n z^n$, so that $z^n = 1$. Hence $z = 1$, as required. ∎

Groups in which extraction of roots is unique when it is possible are sometimes called *R-groups*. Thus torsion-free locally nilpotent groups are R-groups.

Embedding in radicable groups—the proof of Mal'cev's Theorem

There are several different approaches to proving Mal'cev's Theorem. One is to show that a finitely generated torsion-free nilpotent group can be represented in a natural way as a unipotent group of rational matrices. This unipotent representation is then embedded in a Lie algebra of matrices by the so-called logarithmic map (e.g. see, Segal 1983: ch. 3). Here we shall follow the approach due to P. Hall (1969), which entails some preliminary results on commutators.

Suppose that $p = y_1 y_2 \cdots y_n$ is any product of n elements of a group. Form a partition of the set $\{y_1, \ldots, y_n\}$ into r subsets $Y_1, \ldots, Y_r$ by labelling each y_i with one of the *labels* in the set $R = \{1, \ldots, r\}$. Now let S be any non-empty subset of these labels and define Y_S to be the set of all commutators whose entries come only from Y_s, where $s \in S$, and which have at least one component in each Y_s, $s \in S$. Thus commutators in Y_S have weight at least $|S|$.

Next we order linearly the non-empty subsets of R, first by increasing size and then lexicographically. With the above notation we prove an expressibility result for p.

2.1.3 *The element p is expressible as $p = \prod_{\emptyset \neq S \subseteq R} q_S$, where q_S is a product of elements of Y_S and the $2^r - 1$ factors q_S occur in the given ordering of the subsets of R.*

Proof This expression for p is achieved by a *collection process*. Suppose that y_t is the first y in the product p with label 1. If $t = 1$, we need do nothing. If $t > 1$, shift y_t in front of y_{t-1} by using $y_{t-1} y_t = y_t y_{t-1} [y_{t-1}, y_t]$. Now $[y_{t-1}, y_t] \in Y_S$, where $S = \{1, \lambda\}$ and λ is the label of y_{t-1}. Repeating this process as often as necessary, we get an expression $p = q_1 p'$, where $q_1 \in Y_1$ and $p' \in Y_2 \cdots Y_r Y_{\{1,2\}} \cdots Y_{\{1,r\}}$. We then collect the factors in p' belonging to Y_2 and so on, until the required result is achieved. The proof of 2.1.3 is now complete. ∎

Suppose that S is any subset of R, say $S = \{i_1, i_2, \ldots, i_m\}$ with $i_1 < i_2 < \cdots < i_m$, and set $p_S = y_{i_1} \cdots y_{i_m}$. Thus p_S may be obtained from p by setting $y_j = 1$ for all $j \notin S$. Making the same substitutions in the expression for p given in 2.1.3, we obtain at once:

2.1.4 *The element p_S is expressible in the form $p_S = \prod_{T \subseteq S} q_T$ where the factors q_T occur in the prescribed order on the subsets of R.*

This result follows from 2.1.3 because any commutator is trivial if it has an entry equal to 1. We are now able to express the q_S in terms of the p_S by a recursive procedure. For example, $q_{\{\alpha\}} = p_{\{\alpha\}}$, and for $\alpha < \beta$ in R we have by 2.1.4

$$p_{\{\alpha,\beta\}} = q_{\{\alpha\}} q_{\{\beta\}} q_{\{\alpha,\beta\}} = p_{\{\alpha\}} p_{\{\beta\}} q_{\{\alpha,\beta\}}.$$

Hence

$$q_{\{\alpha,\beta\}} = p_{\{\beta\}}^{-1} p_{\{\alpha\}}^{-1} p_{\{\alpha,\beta\}},$$

and so on.

Now specialize to the case where $p = x_1^r x_2^r \cdots x_n^r$ and label the factors as follows:

$$x_1^{(1)} x_1^{(2)} \cdots x_1^{(r)} x_2^{(1)} \cdots x_2^{(r)} \cdots x_n^{(1)} \cdots x_n^{(r)},$$

so that $x_j^{(i)} = x_j$ for $1 \le i \le r$. If $S \subseteq R$ with $|S| = w$, then

$$p_S = x_1^w x_2^w \cdots x_n^w,$$

a product which depends only on $|S|$. By 2.1.4 the element q_S depends only on $|S|$, so that we may write $q_S = \tau_w(x_1, x_2, \ldots, x_n) = \tau_w(x)$, where

$$\tau_w(x) = x_1^w x_2^w \cdots x_n^w = \tau_1(x)^w \tau_2(x)^{\binom{w}{2}} \cdots \tau_k(x)^{\binom{w}{k}} \cdots \tau_w(x).$$

We note that $\tau_1(x) = x_1 \cdots x_n$.

Since $\tau_w(x) = q_S$ with $|S| = w$ and q_S is a product of commutators each of which has at least w entries, we immediately obtain:

2.1.5 *If G is any group, then $\tau_w(G) \subseteq \gamma_w(G)$, where $\tau_w(G) = \{\tau_w(x) \mid x_i \in G\}$.*

Using Hall's methods, we shall now prove Mal'cev's Theorem for finitely generated torsion-free nilpotent groups. The extension to arbitrary torsion-free nilpotent, and indeed even to locally nilpotent groups, will then follow by standard procedures, as indicated below.

Suppose that G is a finitely generated torsion-free nilpotent group. Then by 1.2.16 and 1.2.20 there is a central series

$$1 = G_n \triangleleft G_{n-1} \triangleleft \cdots \triangleleft G_0 = G,$$

in which each factor G_{i-1}/G_i is infinite cyclic, say with generator $u_i G_i$, for $i = 1, 2, \ldots, n$. Then each a in G has a unique expression

$$a = u_1^{\alpha_1} u_2^{\alpha_2} \cdots u_n^{\alpha_n},$$

where α_i is an integer. We will write for convenience $a = u^\alpha$, where α denotes the vector $(\alpha_1, \alpha_2, \ldots, \alpha_n)$. The α_i are called the *canonical parameters* of a with respect to the *basis* $\{u_1, u_2, \ldots, u_n\}$ of G.

Next let $b \in G$, where $b = u^\beta$, and write $ab = c = u^\gamma$, say. Here the γ_i are functions of the $2n$ integer variables α_j, β_j. Similarly, if m is any integer, $a^m = u^\omega$, where the ω_i are functions of m and the n integer variables α_j.

Let F be any field of characteristic 0, or more generally a *binomial ring*, that is, a ring R in which the 'binomial coefficients' $r(r-1)\cdots(r-n+1)/n!$ exist for each $r \in R$. Let G^F be the set of all formal products u^α with exponents $\alpha_1, \ldots, \alpha_n$ in F—here confusion with normal closures is unlikely. The idea is to define multiplication and exponentiation in G^F by using the functions γ_i and ω_i in the obvious way. We will then show that G^F *is a radicable nilpotent group in which G embeds as a subgroup.*

The bulk of the proof resides in showing that G^F is a group. Crucial to the argument is the use of the commutator properties just developed to establish:

2.1.6 *The functions γ_i and ω_i are polynomials in the variables α_j, β_j.*

Proof Since the γ_i and ω_i are functions of integer variables, the statement of 2.1.6. means, strictly speaking, that there are polynomials $f_i(x, y)$ of the $2n$ arguments x_j, y_j such that $f_i(\alpha, \beta) = \gamma_i$, with a similar comment about ω_i. (There is no ambiguity in identifying the function γ_i with the polynomial f_i since polynomials are uniquely determined by their values at integer arguments.)

By construction γ_i and ω_i are integer valued polynomials, and we observe that γ_i depends only on $\alpha_1, \ldots, \alpha_i, \beta_1, \ldots, \beta_i$ since $G_i \lhd G$: indeed γ_i is a $\mathbb{Z}$-linear combination of products of powers of these α's and β's. Similarly for the ω_i.

If $n = 1$, then G is infinite cyclic and we see at once that $\gamma_1(\alpha_1, \beta_1) = \alpha_1 + \beta_1$, $\omega_1(\alpha_1) = m\alpha_1$. Let $n > 1$ and proceed by induction on n. For ease of exposition, let Γ_n and Ω_n denote the hypotheses that $\gamma_1, \ldots, \gamma_n$ (respectively $\omega_1, \ldots, \omega_n$) are polynomials.

We assume that Γ_i and Ω_i hold for $i < n$ and aim to show that Γ_n holds. By an easy calculation

$$c = ab = u^\alpha u^\beta = u_1^{\alpha_1 + \beta_1} \prod_{i=2}^{n} (u_1^{-\beta_1} u_i^{-1} u_1^{\beta_1})^{-\alpha_i} u_2^{\beta_2} \cdots u_n^{\beta_n}. \tag{2.1}$$

Now

$$u_1^{-\beta_1} u_i^{-1} u_1^{\beta_1} = u_1^{-\beta_1} (u_i^{-1} u_1 u_i)^{\beta_1} u_i^{-1} \tag{2.2}$$

and $u_i^{-1} u_1 u_i = u_1 u_{i+1}^{c_{i1}} \cdots u_n^{c_{in-i}}$, with constants c_{ij}, since $[G_i, G] \leq G_{i+1}$. The set $\{u_1, u_{i+1}, \ldots, u_n\}$ forms a canonical basis of size $n - i + 1$ for the group it generates. Hence for $i > 1$ we can use the property Ω_{n-i+1} to deduce that $(u_i^{-1} u_1 u_i)^{\beta_1} = u_1^{\beta_1} u_{i+1}^{\varphi_{i1}} \cdots u_n^{\varphi_{in-i}}$ where the φ_{ij} are polynomials in β_1 (and the c_{ij}).

We now apply the property Γ_{n-i+1} and equation (2.2) to give

$$u_1^{-\beta_1} u_i^{-1} u_1^{\beta_1} = u_{i+1}^{\varphi_{i1}} \cdots u_n^{\varphi_{in-i}} u_i^{-1} = u_i^{-1} u_{i+1}^{\psi_{i1}} \cdots u_n^{\psi_{in-i}},$$

where the ψ_{ij} are further polynomials in β_1. By Ω_{n-i+1} again we get

$$(u_1^{-\beta_1} u_i^{-1} u_1^{\beta_1})^{-\alpha_i} = u_i^{\alpha_i} u_{i+1}^{\theta_{i1}} \cdots u_n^{\theta_{in-i}}, \tag{2.3}$$

where the θ_{ij} are polynomials in β_1 and α_i. Now substitute (2.3) in (2.1) and apply Γ_{n-1} $n-1$ times to obtain Γ_n.

Establishing Ω_n is a little more difficult. Let Ω_n^+ denote the statement Ω_n for $m > 0$. At this point we bring into play the words τ_i which were defined earlier. Set $v_i = \tau_i(u_1^{\alpha_1}, \ldots, u_n^{\alpha_n})$, which is of course a word in the arguments, so that repeated applications of Γ_n yield $v_i = u_1^{\lambda_{i1}} \cdots u_n^{\lambda_{in}}$, where the λ_{ij} are polynomials in the α_j.

Recall from 2.1.5 that, if c is the nilpotent class of G, then $v_{c+r} = 1$ for all $r > 0$. Now $v_1 = \tau_1(u_1^{\alpha_1}) = u_1^{\alpha_1} \cdots u_n^{\alpha_n} = a$ and hence, by the recurrence relation defining the τ_i, we have for $m > 0$

$$a^m = u_1^{m\alpha_1} u_2^{m\alpha_2} \cdots u_m^{m\alpha_n} v_c^{-\binom{m}{c}} \cdots v_2^{-\binom{m}{2}}. \tag{2.4}$$

But $v_2, \ldots, v_c \in G' \le G_1$, and so by Ω_{n-1} we obtain that $v_i^{-\binom{m}{i}} = u_2^{\mu_{i2}} \cdots u_n^{\mu_{in}}$ for $i \ge 2$, where the μ_{ij} are polynomials in m and the α_j. We now substitute back into (2.4) and apply Γ_{n-1} repeatedly, to get the statement Ω_n^+ with polynomials $\omega_i = \omega_i(m, \alpha_1, \ldots, \alpha_i)$.

To complete the argument observe that $a^{-1} = u_n^{-\alpha_n} \cdots u_1^{-\alpha_1} = u^\delta$ where the δ_i are polynomials in the α_j. Let $m > 0$: then by Ω_n^+ we obtain $a^{-m} = u^\varepsilon$, where the ε_i are polynomials in m and α_j. Hence, if $m > 0$, $r > 0$, then by Ω_n^+ and Γ_n we have $a^{r-m} = a^\zeta$, where $\zeta_i = \zeta_i(m, r, \alpha)$ is a polynomial in m, r, and α_j. Hence $\omega_i(r, \alpha) = \zeta_i(m, m+r, \alpha)$ for $m > 0$, $r > 0$, and this must be a polynomial identity. Now for any integer r, choose $m > 0$ such that $m + r > 0$. Then

$$a^r = a^{m+r-m} = u_1^{\zeta_1(m,m+r,\alpha)} \cdots u_n^{\zeta_n(m,m+r,\alpha)} = u_1^{\omega_1(r,\alpha)} \cdots u_n^{\omega_n(r,\alpha)},$$

and the proof of 2.1.6 is complete. ∎

We are now in a position to prove that G^F, the set of all formal products of the u^α, is a group. Take the case of the associative law. This of course holds in G and by our analysis the law expresses itself in terms of polynomial identities which must be satisfied by the α_i and ω_i. It follows that the associative law holds for G^F. Similar remarks apply to the other axioms. Therefore, G^F is a group.

Furthermore, if we define G_i^F by

$$G_i^F = \{u^\alpha \mid \alpha_1 = \alpha_2 = \cdots = \alpha_i = 0\},$$

then G_i^F is a subgroup and $G^F = G_0^F > G_1^F > \cdots > G_n^F = 1$ is a central series of G^F. Moreover, G_i^F / G_{i+1}^F is isomorphic with the additive group of F. Also it is obvious that G^F has the same nilpotent class as G.

Finally, let $F = \mathbb{Q}$; then by an induction argument on $n - i$ every a in $G^{\mathbb{Q}}$ has some positive integral power in G. Also, by a further induction, given any $b \in G$ and $m > 0$, there exists $a \in G^{\mathbb{Q}}$ with $a^m = b$, that is, $G^{\mathbb{Q}}$ is radicable. The proof of the existence part of 2.1.1 is now complete.

Uniqueness of the completion

To finish the proof of Mal'cev's Theorem we must address question of uniqueness of the completion. Let $G^{\mathbb{Q}}$ be the completion of the group G just constructed and suppose that $\theta : G \to H$ is an embedding of G in some torsion-free radicable nilpotent group H. Then we may define $\theta^* : G^{\mathbb{Q}} \to H$ by $(u^\alpha)^{\theta^*} = (u^\theta)^\alpha$. Here, of course, if $u^\alpha = u_1^{\alpha_1} \cdots u_n^{\alpha_n}$, $\alpha_i \in \mathbb{Q}$, we mean that $(u^\theta)^\alpha = (u_1^\theta)^{\alpha_1} \cdots (u_n^\theta)^{\alpha_n}$. Then θ^* is a homomorphism and $\mathrm{Ker}\,(\theta^*) \cap G = 1$, so $\mathrm{Ker}(\theta^*) = 1$: for, if $x \in \mathrm{Ker}(\theta^*)$, then $x^k \in G$ for some $k > 0$, and hence $x = 1$.

Applying this with H a completion of G, we get $G^\theta \le G^{\mathbb{Q}\theta^*} \le H$. If $h \in H$, then $h^m \in G^\theta$ for some $m > 0$ and thus $h^m = (g^{\theta^*})^m$, with $g \in G^{\mathbb{Q}}$, which shows that $h = g^\theta$. Therefore $G^{\mathbb{Q}\theta^*} = H$.

Extension to locally nilpotent groups

Suppose that G is a torsion-free locally nilpotent group and that $X \le Y$ are finitely generated subgroups of G; then there is a commutative diagram,

$$
\begin{array}{ccc}
X & \hookrightarrow & Y \\
\downarrow & & \downarrow \\
X^{\mathbb{Q}} & \longrightarrow & Y^{\mathbb{Q}}
\end{array}
$$

where $X^{\mathbb{Q}} \to Y^{\mathbb{Q}}$ arises from $X \to Y^{\mathbb{Q}}$ just as in the above discussion of uniqueness. This allows us to form $\varinjlim X^{\mathbb{Q}}$, which will be a completion of G. Clearly it is locally nilpotent and torsion-free. It follows that 2.1.1 is valid when G is a torsion-free locally nilpotent group.

A general account of the theory of completions has been given by Hilton, Mislin, and Roitberg (1975). In particular they prove Mal'cev's Theorem using cohomological methods.

The preceding argument depends essentially on the functions γ_i and ω_j being polynomials. It is important to note how closely this property is tied in with the nilpotency of the group G. Suppose G is the non-nilpotent metacyclic group with the presentation $\langle u_1, u_2 \mid u_1^{u_2} = u_1^{-1} \rangle$. Then, as before, we may regard $\{u_1, u_2\}$ as a basis for G, each element of G being uniquely expressible in the form $u_1^{\alpha_1} u_2^{\alpha_2}$. We can also define the functions, γ_i in the same way as before. However,

$$
u_1^{\alpha_1} u_2^{\alpha_2} u_1^{\beta_1} u_2^{\beta_2} = u_1^{\alpha_1 + \beta_1} u_2^{(-1)^{\beta_1} \alpha_2 + \beta_2}.
$$

Thus $\gamma_2 = (-1)^{\beta_1} \alpha_2 + \beta_2$, which is not a polynomial and therefore cannot be extended to a function on $\mathbb{Q}$. However, in some cases such an extension is possible: see Magid (1982) in this connection.

There is an extensive literature on the classification of torsion-free nilpotent groups which is surveyed by Grunewald and Segal (1984).

2.2   Basic commutators

If $G = \langle x_1, x_2, \ldots, x_n \rangle$ is a finitely generated abelian group, then every element of G can be expressed in the form $x_1^{\alpha_1} x_2^{\alpha_2} \cdots x_n^{\alpha_n}$, $\alpha_i \in \mathbb{Z}$. We have seen in 2.1 how this idea can be generalized to finitely generated torsion-free nilpotent groups by using terms of the upper central series. In this section we turn to another way of constructing a 'canonical' basis for a finitely generated nilpotent group, this time using properties of the lower central series. The key ideas are due to P. Hall (1969) and M. Hall (1959).

First, a property of the lower central series of a group is needed.

2.2.1 *Suppose that G is an arbitrary group which is generated by a subset X. Then $\gamma_n(G)$ is generated by all conjugates of the commutators $[x_1, x_2, \ldots, x_n]$, with $x_i \in X$. Thus $\gamma_n(G)$ is generated by all such $[x_1, x_2, \ldots, x_n]$ modulo $\gamma_{n+1}(G)$.*

In order to prove this we first note a simple result.

2.2.2 *Let G be a group such that $G = \langle X \rangle$ and suppose H is a subgroup such that $H = Y^H$ for some subset Y. Then $[H, G]$ is generated by the conjugates of $\{[y, x] \mid x \in X, y \in Y\}$ in G.*

Proof Set $K = \langle [x, y]^g \mid x \in X, \, y \in Y, \, g \in G \rangle$. Then K is clearly normal in G and $K \le [H, G]$ since $[H, G] \lhd G$. Also $yK \in Z(G/K)$, where $y \in Y$, so that $H/K \le Z(G/K)$. Hence, $[H, G] \le K$, so $[H, G] = K$ and thus 2.2.2 is proved. ∎

Proof of 2.2.1 We use induction on n. Assuming the result true for n, we deduce from 2.2.2 that $\gamma_{n+1}(G)$ is generated by all conjugates of the commutators $[x_1, \ldots, x_n], x_i \in X$. ∎

From 2.2.1 we deduce that if G is a finitely generated nilpotent group generated by a set $X = \{x_1, \ldots, x_m\}$, then any element $g \in G$ can be expressed as a product $x_1^{\alpha_1} \cdots x_m^{\alpha_m} u_1^{\gamma_1} \cdots u_t^{\gamma_t}$, where $u_1, \ldots, u_t$ are commutators of various lengths in $x_1, \ldots, x_m$. It was to obtain some sort of canonical form for this expression that the notions of basic commutator and the collection process were developed.

The collection process

In order to describe this we need to set up some formal apparatus. We shall consider finite *strings* or *words* in a finite alphabet $x_1, x_2, \ldots, x_m$. Then we introduce the key concepts of *formal commutator* and *weight*. The commutators of weight 1 are to be the $x_i, i = 1, 2, \ldots, m$. Weight will be denoted by w, so that $w(x_i) = 1, i = 1, 2, \ldots, m$. If u, v are commutators, then the weight of the commutator $[u, v] = u^{-1}v^{-1}uv$ is defined by $w([u, v]) = w(u) + w(v)$. There are clearly only finitely many commutators of each given weight.

The commutators $c_1, c_2, \ldots$ are to be ordered by subscript in the following way: $c_1, \ldots, c_m$ will be $x_1, \ldots, x_m$ in that order. Beyond these we list the c_i's in order of weight, assigning an arbitrary ordering to commutators of the same weight. A string of commutators $c_{j_1}, c_{j_2}, \ldots, c_{j_s}$ is said to be in *collected form* if $j_1 \le j_2 \le \cdots \le j_s$ in the ordering just defined.

An arbitrary string of commutators can be expressed as

$$c_{j_1} \cdots c_{j_r} c_{j_{r+1}} \cdots c_{j_s}$$

where $c_{j_1} \cdots c_{j_r}$ is the *collected part* with $j_1 \leq \cdots \leq j_r$ and $c_{j_{r+1}} \cdots c_{j_s}$ is the *uncollected part,* where j_{r+1} is not the least of $j_{r+1}, \ldots, j_s$. We note that the collected part is empty if j_1 is not the least of the suffixes in the given order.

The collection process proceeds as follows. Let j_t be the least of the sub-scripts in the uncollected part. Thus $t \geq r + 1$. In the string replace $c_{j_{t-1}} c_{j_t}$ by $c_{j_t} c_{j_{t-1}} [c_{j_{t-1}}, c_{j_t}]$. The effect of this is to move c_{j_t} one step to the left and to introduce a commutator whose weight indicates that it appears later in the ordering than c_{j_t}. Repeat the process until c_{j_t} has been moved to the $(r + 1)$th position, when it is absorbed into the collected part. Observe that the process will not in general terminate.

If we now think of $x_1, \ldots, x_m$ as the generators of a group G, it is obvious that the collection process does not alter the group element represented as the product of a string of commutators. In addition, it is evident that only certain commutators are generated in the collecting process—for example, $[x_2, x_1]$ may appear, but not $[x_1, x_2]$ since x_1 is collected before x_2. The commutators that arise in the process are called *basic commutators.*

Now suppose that G is a group generated by elements $x_1, \ldots, x_m$. The *basic commutators of weight 1* are $c_i = x_i$, $i = 1, 2, \ldots, m$. The basic commutators of higher weight are defined inductively by weight and are ordered by their subscripts: for a given weight the ordering is chosen arbitrarily. Suppose that all basic commutators of weight less than n have been defined. Then the basic commutators of weight n are of the form $c_t = [c_r, c_s]$, where c_r and c_s are basic, $w(c_r) + w(c_s) = n$, and $c_r > c_s$. Here $\leq$ is the ordering of basic commutators. Also, if $c_r = [c_u, c_v]$, then $c_v \leq c_r$. By careful consideration of the steps involved in the collection process, the following result can be established.

2.2.3 *Suppose that G is a group generated by finitely many elements $x_1, x_2, \ldots, x_m$. Then modulo $\gamma_{r+1}(G)$:*

(i) *any element $g \in G$ can be written in the form $g = c_1^{e_1} c_2^{e_2} \cdots c_n^{e_n}$, where $c_1, c_2, \ldots, c_n$ are basic commutators of weights $1, 2, \ldots, r$;*
(ii) *$\gamma_r(G)$ is generated by the basic commutators of weight r.*

If F is a free group, then it turns out that, modulo $\gamma_{r+1}(F)$, the basic commutators of weight r form a basis for the free abelian group $\gamma_r(F)/\gamma_{r+1}(F)$: see M. Hall (1959: ch. 11) for a detailed account.

The collection formula

Here we are concerned with expressing a power $y = (x_1 x_2 \ldots x_m)^n$ as a product of powers in a group G. We may suppose that G is generated by $x_1, x_2, \ldots, x_m$. The simplest case is where G is nilpotent with class at most 2, when there is the well-known formula

$$(x_1 x_2 \cdots x_m)^n = x_1^n x_2^n \cdots x_m^n \prod_{i>j=1}^{m} [x_i, x_j]^{\binom{n}{2}}.$$

In general the idea is to express y in terms of basic commutators in the x_i's. In order to do so, we need to remove some of the arbitrariness in the ordering of basic commutators described above. This is achieved by insisting in addition that, for basic commutators $[c, d]$ and $[e, f]$ of weight n, the order that prevails is:

$$[c, d] < [e, f] \quad \text{if } d < f \text{ or if } d = f \text{ and } c < e.$$

We then obtain the *collection formula*.

2.2.4 *Let $G = \langle x_1, x_2, \ldots, x_m \rangle$ be a finitely generated group and let $n > 0$. Then the product $(x_1 x_2 \ldots x_m)^n$ can be collected in the form*

$$(x_1 \cdots x_m)^n = x_1^n \cdots x_m^n \; c_{m+1}^{e_{m+1}} \cdots c_j^{e_j} \; d_1 \cdots d_w,$$

where $x_1, \ldots, x_m, c_{m+1}, \ldots, c_j$ are basic commutators in $x_1, \ldots, x_m$ in the correct order and $d_1, \ldots, d_w$ are basic commutators occurring later than the c_j's in the ordering. The exponents e_i can be expressed as linear combinations of the form

$$e_i = b_1 \binom{n}{1} + b_2 \binom{n}{2} + \cdots + b_r \binom{n}{r}$$

where $m + 1 \leq i \leq j$, r is the weight of c_i and the b_k's are non-negative integers which depend on c_i and not on n.

Again we refer to M. Hall (1959: ch. 12) for the details of the proof.

From the collection formula we can deduce a useful result for nilpotent groups.

2.2.5 *If G is a nilpotent group of class c and p is a prime such that $p > c$, then G^p consists of pth powers.*

Proof Let $x_1, \ldots, x_m$ be elements of G. Then for sufficiently large j

$$(x_1 \cdots x_m)^p = x_1^p \cdots x_m^p \; c_{m+1}^{e_{m+1}} \cdots c_j^{e_j} :$$

for by taking j large enough we can ensure that all the d_i's are trivial, since G is nilpotent. Also

$$e_i = b_1 \binom{p}{1} + b_2 \binom{p}{2} + \cdots + b_r \binom{p}{r},$$

where r is the weight of c_i and the b_k's are non-negative integers. Note that $r < p$. Since p is a prime, it follows that p divides e_i, which means that

$$(x_1 \cdots x_m)^p = x_1^p \cdots x_m^p \; y_{m+1}^p \cdots y_j^p,$$

a product of pth powers.

Now we show that the pth powers form a subgroup. Suppose that $a, b \in G$. Then by the above formula $(ab)^p = a^p b^p d_1^p \cdots d_r^p$, for some elements $d_1, \ldots, d_r$ which, it should be noted, are in G'. Hence $a^p b^p = (ab)^p d_r^{-p} \cdots d_1^{-p}$.

Set $H = \langle a, b \rangle$ and $K = \langle ab, d_1, \ldots, d_r \rangle$. Then $d_1, \ldots, d_r \in H'$ and $K \leq \langle ab, H' \rangle$, so we see that the nilpotent class of K is less than that of H. By induction we may assume that a product of pth powers in K is a pth power. But then $a^p b^p$ is a pth power and it follows that $G^p = \{g^p \mid g \in G\}$. ∎

Next we apply this result to prove an interesting theorem of G. Higman.

2.2.6 (Higman 1955) *Suppose that G is a finitely generated nilpotent group and π is any infinite set of primes. Then $\bigcap_{p \in \pi} G^p$ is finite. Therefore $\bigcap_{p \in \pi} G^p = 1$ if G is torsion-free.*

Proof Let $I = \bigcap_{p \in \pi} G^p$. Suppose that G has nilpotent class c. If $c = 1$, the result is an immediate consequence of the structure theorem for finitely generated abelian groups, so suppose $c > 1$. By induction on c we may assume that $\bigcap_{p \in \pi} (G^p \gamma_c(G)/\gamma_c(G))$ is finite and hence $I \gamma_c(G)/\gamma_c(G)$ is finite. Now let $x \in I \cap \gamma_c(G)$ and suppose x has infinite order. Since π is infinite, we may choose $p \in \pi$ such that $p > c$. Since $x \in G^p$ by hypothesis, we have $x = y_p^p$ for some $y_p \in G$ by 2.2.5; also $x \in Z(G)$, so $\langle x \rangle \triangleleft G$. It follows at once that $G/\langle x \rangle$ has elements of infinitely many prime orders, which is impossible in a finitely generated nilpotent group. Hence x has finite order and $I \cap \gamma_c(G)$ is finite. Therefore I is finite, as required. ∎

For results in the other direction see Seksenbaev (1965) and Robinson (1972a). For example, Seksenbaev proved that *if G is a polycyclic group and $\bigcap_p G^p = 1$, where the intersection is over all primes p, then G is torsion-free and nilpotent.*

2.3 The theory of isolators

Let H be a subgroup of a group G and let π be a (non-empty) set of primes. Then the *π-isolator* of H in G, $I_{\pi,G}(H)$ or $I_\pi(H)$, is defined to be the set of all $x \in G$ such that $x^m \in H$ for some positive π-number m. If π is the set of all primes, we write $I_G(H)$ or $I(H)$. Should $I_\pi(H) = H$, we say that H in *π-isolated in G*, or simply isolated, if π is the set of all primes.

In general $I_\pi(H)$ need not be a subgroup, as is shown by the 2-isolator of the identity subgroup in S_3, for example. A group G such that $I_\pi(H)$ is a subgroup for all $H \leq G$ is said to have the *π-isolator property*, or the *isolator property* if π is the set of all primes. If in addition $|I_\pi(H) : H|$ is finite for all subgroups H, then G is said to have the *strong π-isolator property* (or *strong isolator property*). Notice that if G has the isolator property, all isolators in G are isolated. The main thrust of this section is that nilpotent—and even locally nilpotent—groups possess a rich isolator theory. This will play an important role in our study of centrality in Chapter 8.

The main result about isolators is:

2.3.1 *Let G be a locally nilpotent group. Then*

(i) *G has the π-isolator property for all sets of primes π;*
(ii) *if G is finitely generated, then G has the strong π-isolator property.*

In order to establish this we need a result which is essentially due to Mal'cev.

2.3.2 *Let G be a finitely generated nilpotent group, H a subgroup of G and π a set of primes. Let $x_1, \ldots, x_n$ be a set of generators of G and suppose that $x_i^{r_i} \in H$ for some positive π-number r_i and $i = 1, \ldots, n$. Then each element of G has a positive π-power in H and $|G : H|$ is a finite π-number.*

Proof Since the result is obvious for abelian groups, we proceed by induction on the class of G, say c. Thus we may assume that $|G : H\gamma_c(G)|$ is a finite π-number d. Since G is finitely generated and nilpotent, $\gamma_c(G)$ is finitely generated, say by $y_1, \ldots, y_s$. By the induction hypothesis $y_i^{s_i} \in H \cap \gamma_c(G)$ for some positive π-number s_i. It follows that

$$|\gamma_c(G) : H \cap \gamma_c(G)| = |H\gamma_c(G) : H|$$

is a positive π-number, say e. Then $|G : H| \leq de$. Finally, H is subnormal in G, so that each g in G has a positive π-power in H. ∎

Proof of 2.3.1 Recall that G is a locally nilpotent group. Let $H \leq G$ and $x, y \in I_\pi(H)$. Then $x^r, y^s \in H$ where r, s are positive π-numbers. Taking G in 2.3.2 to be $\langle x, y \rangle$, we deduce that $|\langle x, y \rangle : \langle x^r, y^s \rangle|$ is a finite π-number, and that $(xy^{-1})^m \in \langle x^r, y^s \rangle$ for some π-number $m > 0$. Hence $(xy^{-1})^m \in H$ and $xy^{-1} \in I_\pi(H)$, which is therefore a subgroup. Part (ii) follows at once from (i) and 2.3.2. ∎

Next we give what is essentially a much more general version of 2.3.2 due to P. Hall (1969). It has wider applicability and shows how strong the arithmetic properties of a nilpotent group are.

Suppose that $\theta(x_1, x_2, \ldots, x_n)$ is a word in the variables $x_1, x_2, \ldots, x_n$. If $H_1, H_2, \ldots, H_n$ are subgroups of a group G, define

$$\theta(H_1, H_2, \ldots, H_n)$$

to be the subgroup generated by all $\theta(h_1, h_2, \ldots, h_n)$, where $h_i \in H_i$, $i = 1, 2, \ldots, n$. Hall's result is:

2.3.3 *Let G be a finitely generated nilpotent group with subgroups H_i, K_i, $i = 1, 2, \ldots, n$, such that $K_i \leq H_i$ and $|H_i : K_i| = m_i$ is finite. Then $|\theta(H_1, \ldots, H_n) : \theta(K_1, \ldots, K_n)|$ is finite and divides some power of $m_1 m_2 \cdots m_n$ for every n-variable word θ.*

To deduce 2.3.2 from 2.3.3 take $H_i = \langle x_i \rangle$, $K_i = \langle x_i^{r_i} \rangle$, and $\theta = x_1 x_2 \cdots x_n$. Then $\theta(H_1, \ldots, H_n) = G$ and $\theta(K_1, \ldots, K_n) \leq H$.

In order to prove 2.3.3 we need a criterion for a finitely generated nilpotent group to be a p-group.

2.3.4 *Suppose that H and K are subgroups of a finitely generated nilpotent group G such that $K < H$ and $|H : K|$ is either infinite or finite and divisible by a given prime p. Assume in addition that $|NH : NK|$ is a finite p'-number whenever $1 \neq N \triangleleft G$. Then G is a finite p-group.*

Proof Note that since $|NH : NK| \neq |H : K|$, we must have $N \not\leq K$ and $N \cap H \neq 1$. By 1.2.20 it is enough to show that the centre Z of G is a finite p-group. Suppose first that Z has an element z of infinite order and put $N = \langle z \rangle$. Since $N \cap H \neq 1$, we may suppose $N \leq H$. But $N \cap K \triangleleft G$, which implies that $N \cap K = 1$: for otherwise

$$|(N \cap K)H : (N \cap K)K| = |H : K|$$

is a finite p'-number. If $M = \langle z^p \rangle$, then

$$|MH : MK| = |H : NK| \cdot |NK : MK| = |H : NK| \cdot p,$$

a contradiction, which shows that Z is finite.

Next assume that z is an element of Z with prime order and again write $N = \langle z \rangle$. Then $N \leq H$ since $N \cap H \neq 1$ Now

$$|H : K| = |H : NK| \cdot |NK : K| = |NH : NK| \cdot |N|,$$

which is finite and therefore divisible by p. Also

$$|NH : NK| = |H : (H \cap N)K|,$$

which is a p'-number. Since p divides $|H : K|$, it follows that $|N| = p$, and the result is proven. ∎

Proof of 2.3.3 Suppose that H_i, K_i satisfies the hypotheses of 2.3.3, but not the conclusion. Put $H = \theta(H_1, \ldots, H_n)$ and $K = \theta(K_1, \ldots, K_n)$. Since G, satisfies the maximal condition, we may choose M to be a normal subgroup of G, which is maximal with respect to the condition that $|MH : MK|$ is infinite or finite with a prime divisor p that does not divide $m = m_1 \cdots m_n$. Write $G^* = G/M$, etc.: then G^* is a finite p-group. This means that

$$|H_i^* : K_i^*| = |H_i : K_i(M \cap H_i)|$$

divides m_i, so $H_i^* = K_i^*$ for all i since p does not divide m. But this implies that

$$H^* = \theta(H_1^*, \ldots, H_n^*) = \theta(K_1^*, \ldots, K_n^*) = K^*,$$

that is, $MH = MK$, a contradiction. ∎

In order to exploit this result further we introduce a definition. If H, K are subgroups of a group G and π is a set of primes, we shall say that H and K are *π-equivalent*, and write

$$H \underset{\pi}{\simeq} K,$$

if for any $h \in H$ and $k \in K$, there exist positive π-numbers m and n such that $h^m \in K$ and $k^n \in H$.

2.3.5 *Let θ be a word in $x_1, x_2, \ldots, x_n$ and let π be a set of primes. Suppose G is a locally nilpotent group and H_i, K_i are subgroups satisfying $H_i \underset{\pi}{\simeq} K_i$, $i = 1, 2, \ldots, n$. Then $\theta(H_1, \ldots, H_n) \underset{\pi}{\simeq} \theta(K_1, \ldots, K_n)$.*

Proof Let $H = \theta(H_1, \ldots, H_n)$ and $K = \theta(K_1, \ldots, K_n)$. Denoting π-isolators of subgroups by bars, we see from 2.3.1 that $U \underset{\pi}{\simeq} \bar{U}$ for any $U \leq G$, so that we need only prove that $H \leq \bar{K}$. Let $x = \theta(h_1, \ldots, h_n)$, where $h_i \in H_i$. We need to show that $x \in \bar{K}$. Let $P = \langle h_1, \ldots, h_n \rangle$, a finitely generated nilpotent group. Now $h_i^{m_i} \in K_i$ for certain positive π-numbers m_i, $i = 1, 2, \ldots, n$. If we set $P_i = \langle h_i \rangle$ and $Q_i = \langle h_i^{m_i} \rangle$, then $|P_i : Q_i| = m_i$ is a π-number for each i and so we may apply 2.3.3 and deduce that $|\theta(P_1, \ldots, P_n) : \theta(Q_1, \ldots, Q_n)|$ is a π-number. Now $x \in \theta(P_1, \ldots, P_n)$ and $\theta(Q_1, \ldots, Q_n)$ is subnormal in $\theta(P_1, \ldots, P_n)$, from which it follows that $x^r \in \theta(Q_1, \ldots, Q_n) \leq K$ for some positive π-number r. Hence $x \in \bar{K}$, as required. ∎

An important special case of 2.3.5 is where $\theta(x_1, x_2) = [x_1, x_2]$. In this event we get

$$[H_1, H_2] \underset{\pi}{\simeq} [K_1, K_2],$$

wherever $H_i \underset{\pi}{\simeq} K_i$, $i = 1, 2$. To prove this, note that $\theta(L, M) \underset{\pi}{\simeq} \theta(\bar{L}, \bar{M})$ for arbitrary subgroups L and M; thus $[\bar{L}, \bar{M}] \leq \overline{[L, M]}$.

These ideas can be used to establish a very useful isolator property of normalizers. For this we need a preliminary result:

2.3.6 *Let π be a set of primes, G a locally nilpotent group, N a normal subgroup of G and M a subgroup of G containing N. Set $C = C_G(M/N)$. Then, denoting π-isolators by bars, we have:*

(i) $\bar{C} \leq C_G(\bar{M}/\bar{N})$;
(ii) $N = \bar{N}$ *implies that $C = \bar{C}$;*
(iii) $M \leq \bar{N}$ *implies that C is π'-isolated in G;*
(iv) *if $M \triangleleft G$ and M/N is a finite π-group, then G/C is also a finite π-group.*

Proof

(i) By the special case of 2.3.5 mentioned above, $[\bar{C}, \bar{M}] \leq \overline{[C, M]} \leq \bar{N}$, whence $\bar{C}$ centralizes $\bar{M}/\bar{N}$ as claimed.
(ii) If $\bar{N} = N$, then $[\bar{C}, \bar{M}] \leq N$. Hence $[\bar{C}, M] \leq N$ and thus $\bar{C} = C$.
(iii) Suppose that $x^m \in C$ where $m > 0$ is a π'-number, and put $X = \langle x \rangle$ and $Y = \langle x^m \rangle$. Then by 2.3.5 we have $[X, M] \underset{\pi'}{\simeq} [Y, M]$, so that $[X, M] \leq I_{\pi'}(M)$ since $[Y, M] \leq N$. Since $M \leq \bar{N}$, we have $M \underset{\pi}{\simeq} N$ and hence, by 2.3.5 again, $[X, M] \underset{\pi}{\simeq} [X, N]$. Since $N \triangleleft G$, we obtain

$[X, M] \leq I_\pi(N)$. Since $[X, M] \underset{\pi'}{\sim} [Y, M] \leq N$ and $[X, M] \leq [X, \bar{N}] \leq \bar{N}$, it follows that $[X, M] \leq N$ and so $x \in C$, as required.

(iv) Note that G/C is finite since M/N is finite. Clearly $M \leq \bar{N}$ and by (iii) the subgroup C is π'-isolated, so G/C is a finite π-group.   ∎

We are now in a position to establish the following major result on isolators.

2.3.7 *Let π be a set of primes and let H be a subgroup of a locally nilpotent group G. Then*
$$I_\pi(N_G(H)) \leq N_G(I_\pi(H)),$$
with equality if G is finitely generated.

Proof For any subgroup K of G, denote $I_\pi(K)$ by $\bar{K}$. Set $N = N_G(H)$. Then, since $N \underset{\pi}{\sim} \bar{N}$ and $H \underset{\pi}{\sim} \bar{H}$ by 2.3.1, we may deduce from 2.3.5 that $[\bar{N}, \bar{H}] \leq [\overline{N, H}] \leq \bar{H}$ since $[N, H] \leq H$. Hence $\bar{N} \leq N_G(\bar{H})$.

Suppose now that G is finitely generated and let $x \in N_G(\bar{H})$. In order to show that $x \in \bar{N}$, we may assume that $G = \langle \bar{N}, x \rangle$, so that $\bar{H} \triangleleft G$. Now $|\bar{H} : H|$ is a π-number by 2.3.2 and, since H is subnormal in $\bar{H}$, it follows that $\bar{H}^m \leq H$ for some π-number $m > 0$. Also $\bar{H}^m \triangleleft G$ since $\bar{H} \triangleleft G$. If $C = C_G(\bar{H}/\bar{H}^m)$, then G/C is a finite π-group by 2.3.6(iv), and so $G = \bar{C}$. But $C \leq N$ since $\bar{H}^m \leq H$ and hence $\bar{N} = G$, so the result is proved.   ∎

One immediate consequence of 2.3.7 is that *in a locally nilpotent group normalizers of isolated subgroups are isolated.* This is due to Gluškov (1952) and Smirnov (1951). It can also be deduced (see G. Baumslag 1971: 21) from the following strong residual finiteness property. *Let G be a finitely generated torsion-free nilpotent group and H an isolated subgroup of G. Then $\bigcap_{i=1,2,\ldots}(G^{p^i} H) = H$ for any prime p.* The case $H = 1$ is the well-known result of Gruenberg that finitely generated torsion-free nilpotent groups are residually finite p-groups for every prime p (see 1.3.17).

Centralizers in locally nilpotent groups also have important isolator properties.

2.3.8 *Suppose that G is a locally nilpotent group which is π-torsion-free. Then*

(i) *$C_G(H)$ is π-isolated for every subgroup H;*
(ii) *$Z_i(G)$ is π-isolated for $i \geq 0$.*

Proof

(i) follows immediately from 2.3.6(ii) with $N = 1$ and $M = H$. To prove
(ii) we use induction on i, the case $i = 0$ being true since $I_\pi(1) = 1$. Assume that $i \geq 1$ and $Z_{i-1}(G)$ is π-isolated. Then $Z_i(G) = C_G(G/Z_{i-1}(G))$ is π-isolated by 2.3.6(ii), which completes the proof.   ∎

If we specialize to the case of torsion-free locally nilpotent groups, the following result can be proved. (Here the upper central series is extended to ordinals with the usual requirement of completeness at limit ordinals.)

2.3.9 *Let H be a subgroup of a torsion-free, locally nilpotent group G. Then the following hold:*

(i) *$G/Z_\alpha(G)$ is torsion-free for all ordinals α;*
(ii) *$Z_\alpha(H) = Z_\alpha(I(H)) \cap H$, for all α;*
(iii) *if H is a hypercentral group, then $I(H)$ is hypercentral with the same upper central length;*
(iv) *$(I(H))^{(n)} = I(H^{(n)})$, for $n = 1, 2, \ldots$;*
(v) *if H is normal in G, then $G/C_G(H)$ is torsion-free;*
(vi) *if $K \lhd H \leq G$ and G centralizes H/K, then G also centralizes $I_G(H)/I_G(K)$.*

Proof

(i) is immediate from 2.3.8(ii).
(ii) We may assume that $G = I(H)$. Suppose that $Z_\alpha(H) = Z_\alpha(G) \cap H$ for some ordinal α. Then

$$[Z_{\alpha+1}(G) \cap H, H] \leq [Z_{\alpha+1}(G), G] \cap H \leq Z_\alpha(G) \cap H = Z_\alpha(H),$$

and hence $Z_{\alpha+1}(G) \cap H \leq Z_{\alpha+1}(H)$. On the other hand, by 2.3.5 we have

$$[Z_{\alpha+1}(H), G] = [Z_{\alpha+1}(H), I(H)] \leq I([Z_{\alpha+1}(H), H]) \leq I(Z_\alpha(H))$$
$$\leq I(Z_\alpha(G)) = Z_\alpha(G).$$

Hence $Z_{\alpha+1}(H) \leq Z_{\alpha+1}(G) \cap H$.
In addition, if λ is a limit ordinal and $Z_\alpha(H) = Z_\alpha(G) \cap H$ for all $\alpha < \lambda$, then

$$Z_\lambda(H) = \bigcup_{\alpha < \lambda} Z_\alpha(H) = \bigcup_{\alpha < \lambda} (Z_\alpha(G) \cap H) = \left(\bigcup_{\alpha < \lambda} Z_\alpha(G) \right) \cap H = Z_\lambda(G) \cap H.$$

This completes the proof of (ii).
(iii) If $H \leq Z_\alpha(H)$, then $H \leq Z_\alpha(I(H))$ by (ii) and so $I(H)/Z_\alpha(I(H))$ is torsion. By (i) we obtain $I(H) = Z_\alpha(I(H))$.
(iv) This is immediate from 2.3.5.
(v) For $C_G(H)$ is normal in G and it is isolated in G by 2.3.8(i), so the result follows.
(vi) Since $[K, G] \leq [H, G] \leq K$, both H and K are normal in G. Also $I_G(K) \leq I_G(H)$. Let $L = I_G(K)$ and put $\bar{G} = G/L$ and $\bar{H} = HL/L$. Since $[H, G] \leq L$, we have $[\bar{H}, \bar{G}] = 1$, so that $\bar{H} \leq Z(\bar{G})$. By (v) the group $\bar{G}/Z(\bar{G})$ is torsion-free and so $I_G(H)/L = I_{\bar{G}}(\bar{H}) \leq Z(\bar{G})$. Hence $[I_G(H), G] \leq L = I_G(K)$, as required. ∎

The theory of isolators was used by Lennox (1976a) to prove the following theorem.

2.3.10 *Let G be a finitely generated nilpotent group, H a subgroup of G and π a set of primes. Then*

(i) *$|G : N_G(H)|$ is a finite π-number if and only if $|H : H_G|$ is a finite π-number;*
(ii) *if $|G : H^G|$ is a finite π-number, then so is $|G : H|$.*

Proof Most of the work lies in proving (i). Assume that $|G : N_G(H)|$ is a finite π-number and let bars denote π-isolators. Since H is subnormal in G, we have $G = \overline{N_G(H)} = N_G(\bar{H})$, so that $\bar{H} \triangleleft G$ and $H \leq H^G \leq \bar{H}$. Hence $|H^G : H|$ is finite. If $g \in G$, then $|H : H \cap H^g| = |HH^g : H^g| \leq |H^G : H|$, so $|H : H \cap H^g|$ is finite. But H has only finitely many conjugates, so it follows that $|H : H_G|$ is finite and thus H^G/H_G is a finite nilpotent group. Also $|H^G : H^g|$ is a π-number for all $g \in G$ since $H^G \leq \bar{H}$. Consequently, $|H^G : H_G|$ is a π-number.

Conversely, assume that $|H : H_G|$ is a finite π-number. Clearly, we can assume $H_G = 1$, so H and hence H^G is a finite π-group. Hence $G/C_G(H^G)$ is a finite π-group since G is nilpotent. Since $N_G(H) \geq C_G(H^G)$, the result follows.

To establish (ii) we assume that that $|G : H^G|$ is finite and argue by induction on the subnormal defect of H, say $d > 1$. Let $H \triangleleft H_1$, where H_1 has subnormal defect $d - 1$. Then $|G : H_1^G|$ is finite, so $|G : H_1|$ is finite by induction. Hence $|G : N_G(H)|$ is finite and by (i) we see that $|H : H_G|$ is finite. Consequently, $|H^G : H_G|$ is finite and thus $|G : H_G|$ is finite.   ∎

We will now give an application of this result. Following Roseblade (1978), we call a subgroup H of a group G *orbital* if H has finitely many conjugates in G. Thus H is orbital in G if and only if $|G : N_G(H)|$ is finite. Further G is said to be *orbitally sound* if H/H_G is finite for every orbital subgroup H of G. As an immediate corollary of 2.3.10(i), we observe that *finitely generated nilpotent groups are orbitally sound*.

Hartley and Rhemtulla (unpublished) have used an isolator theory approach to extend this result to prove the following theorem of Roseblade (1978): *every polycyclic group has an orbitally sound subgroup of finite index.*

We observed at the beginning of this section that in the group S_3 the isolator $I_2(1)$ is not a subgroup, so that for non-nilpotent soluble groups there is in general no isolator theory. However Rhemtulla and Wehrfritz (1984) have proved a weaker result.

2.3.11 *Let G be a finitely generated soluble group. Then G is polycyclic if and only if G has a subgroup of finite index satisfying the strong isolator property.*

The proof uses isolator properties of linear groups, as does the proof of the following partial extension to soluble groups of finite rank, (for the terminology see Section 5.1).

2.3.12 *A reduced soluble group with finite abelian ranks has a subgroup of finite index with the strong isolator property.*

By contrast, Rhemtulla, Weiss, and Youssif (1984) have established:

2.3.13 *Suppose that G is a finitely generated soluble group and π is a finite set of primes. Then G has a subgroup of finite index with the π-isolator property if and only if G is nilpotent-by-finite.*

For finitely generated soluble groups in general little can be said, except for the important case of a subgroup whose isolator is the whole group.

2.3.14 (Lennox 1983) *Suppose that G is a finitely generated virtually soluble group and H is a subgroup of G. Then H has finite index in G if and only if $I(H) = G$.*

Proof If $|G : H|$ is finite, then clearly $I(H) = G$. For the converse suppose S is a soluble normal subgroup of finite index in G. We argue by induction on d, the derived length of S, which may be assumed positive. Let $A = S^{(d-1)}$. By induction $|G : HA|$ is finite, so that HA is finitely generated. Thus we may assume that $G = HA$. Since $H \cap A \lhd G$, we may also assume that $H \cap A = 1$.

The result is true for H by the induction hypothesis. Let $a \in A$ and $h \in H$; then $(ha)^m \in H$ for some $m > 0$ and hence $a^{1+h+\cdots+h^{m-1}} \in H \cap A = 1$. Putting $h = 1$, we see that a has finite order and so A is torsion. Also $a^{\langle h \rangle}$ is finite, so $h^k \in C_H(a)$ for some $k > 0$. Since the result holds for H, we may deduce that $|H : C_H(a)|$ is finite. Finally, since G is finitely generated, A is finitely generated as an H-module and hence A is finite, as required. ∎

It should be pointed out that the Mal'cev completion of a finitely generated torsion-free nilpotent group G may be constructed with the aid of isolators. By 3.3.4 below, the group G embeds in $U_n(\mathbb{Z})$, the group of $n \times n$ integral unitriangular matrices, for some n. Thus G embeds in $U = U_n(\mathbb{Q})$. Now U is clearly a torsion-free, radicable nilpotent group. Define G^* to be the isolator of U in G^*. Then by 2.3.1 we see that G^* is a radicable subgroup of U. It follows by uniqueness that G^* is the Mal'cev completion of the given group G.

Isolator theory also enables us to extend 1.2.17. The following is a result of Möhres (1988: 4.3.1), which will find application in Chapter 12.

2.3.15 *Suppose that G is a locally nilpotent torsion-free group and N is a nilpotent normal subgroup of G such that $G/I_G(N')$ is nilpotent. Then G is nilpotent.*

Proof By 2.3.9(iii) $I_G(N)$ is nilpotent since N is. Also, $I_G(I_G(N)') = I_G(N')$ by 2.3.5, so we deduce that $G/I_G(I_G(N)')$ is nilpotent. Thus we may assume that $N = I_G(N)$.

Suppose that N has nilpotent class c. If $c \leq 1$, the result is clear, so let $c > 1$. Set $M = I_N(\gamma_c(N))$; thus the class of N/M is less than c. Also $I_G(M) = I_N(M) = M$ since $N = I_G(N)$, which shows that G/M is torsion-free.

By induction on c we may assume that G/M is nilpotent, so there is a series

$$M = N_0 \triangleleft N_1 \triangleleft \cdots \triangleleft N_m = N,$$

such that $[G, N_i] \le N_{i-1}$ for $1 \le i \le m$.

For $k = 0, 1, 2, \ldots, 2m$ we define

$$T_k = \langle [N_i, N_j] | 0 \le i, j \le m, i + j \le k \rangle.$$

Thus we have the series

$$1 = T_0 \triangleleft T_1 \triangleleft \cdots \triangleleft T_{2m} = N',$$

since $M \le Z(N)$. Next

$$[N_i, N_j, G] \le [N_j, G, N_i][N_i, G, N_j] \le [N_{j-i}, N_i][N_{i-1}, N_j].$$

Therefore, $[T_k, G] \le T_{k-1}$ and G stabilizes the series of T_k's. By 2.3.9(vi) we see that G will also stabilize the series

$$1 = I_G(T_0) \triangleleft I_G(T_1) \triangleleft \cdots \triangleleft I_G(T_{2m}) = I_G(N').$$

Finally, G is nilpotent since $G/I_G(N')$ is nilpotent.   ∎

3

SOLUBLE LINEAR GROUPS

In this chapter we shall discuss soluble groups which are linear, that is, which admit a faithful representation as a group of matrices over a commutative ring with identity. The rings considered will usually be either a field or the ring of integers. While every finite group is linear, it will emerge that linearity imposes considerable restrictions on the structure of an infinite group. Knowledge of the structure of soluble linear groups will be applied to give information about polycyclic groups, and more generally soluble groups of finite rank. The standard reference for infinite linear groups is still the book by Wehrfritz (1973a).

3.1 Mal'cev's Theorem

If R is a commutative ring with identity, we write, $GL_n(R)$ for the group of all invertible $n \times n$ matrices with entries in R: here invertibility requires that the determinant of the matrix be a unit of the ring R. A group G is said to be *R-linear of degree* n if it is isomorphic with a subgroup of $GL_n(R)$. Equivalently, one could say that G is isomorphic with a subgroup of $GL(F_n)$, the group of all R-automorphisms of a free R-module F_n of rank n. Thus one can always think of G as acting on the module F_n. The cases of greatest interest are where R is a field or the ring of integers $\mathbb{Z}$. To avoid constant repetition, it will be our convention that the term 'linear group', without reference to an underlying ring, will mean *a linear group over a field*.

There has been considerable activity in the theory of linear groups over non-commutative division rings in recent years. Such groups are often called *skew linear*. For an account of these groups see the book by Shirvani and Wehrfritz (1986).

We begin with a well-known fact.

3.1.1 *An arbitrary finite group G is R-linear of degree $|G|$ for any commutative ring R with identity.*

The right regular representation of G allows us to represent elements of G by permutation matrices, and these may be assumed to have their entries in R. Thus one can think of linearity as a type of finiteness condition.

There have been many studies on the question of the linearity of a given group. For example, Mal'cev (1940) has found necessary and sufficient conditions for an abelian group to be linear of given degree over a given field: the conditions pertain to the structure of the primary components of the group.

Concerning linear groups in general, we quote without proof a remarkable theorem of Tits, generally called the *Tits Alternative* (Tits 1972).

3.1.2 *Let G be a linear group over a field F.*

(i) *If $\text{char}(F) = 0$, then either G is virtually soluble or it has a free subgroup of rank 2.*

(ii) *If $\text{char}(F) \neq 0$ and G is finitely generated, the same conclusion holds.*

Of course, if a free subgroup of rank 2 occurs in a linear group, so does a free subgroup of every countably infinite rank. Tit's result underscores how widespread soluble linear groups must be: for example, it follows that *every linear torsion group of characteristic 0 is soluble-by-finite and hence is locally finite.*

Another notable property of linear groups was found by Mal'cev (1940).

3.1.3 *A finitely generated linear group is residually finite.*

This is a further indication of the kind of restriction linearity places on the structure of a group. For a proof of 3.1.3 the reader is referred to Wehrfritz (1973a).

Next we give some examples of infinite soluble groups that are linear.

3.1.4

(i) *The infinite dihedral group $\text{Dih}(\infty)$ is $\mathbb{Z}$-linear of degree 2.*

(ii) *The restricted standard wreath product $\mathbb{Z} \text{ wr } \mathbb{Z}$ is $\mathbb{R}$-linear of degree 2.*

Proof

(i) Let $x = \begin{bmatrix} 1 & 1 \\ 0 & 1 \end{bmatrix}$ and $y = \begin{bmatrix} -1 & 0 \\ 0 & 1 \end{bmatrix}$. Then $\langle x, y \rangle$ is a subgroup of $\text{GL}_n(\mathbb{Z})$. Now $x^y = x^{-1}, y^2 = 1$ and x has infinite order, so $\langle x, y \rangle$ is an infinite dihedral group.

(ii) Let c be a real transcendental number: define $x_0 = \begin{bmatrix} 1 & 1 \\ 0 & 1 \end{bmatrix}$ and $y = \begin{bmatrix} 1 & 0 \\ 0 & c \end{bmatrix}$. Then by a simple matrix calculation

$$y^{-i} x_0 y^i = \begin{bmatrix} 1 & c^i \\ 0 & 1 \end{bmatrix} = x_i,$$

say. Thus $x_i^y = x_{i+1}$ for all $i \in \mathbb{Z}$. Let $G = \langle x_0, y \rangle$ and write $A = \langle x_i \mid i \in \mathbb{Z} \rangle$. Then A is an abelian normal subgroup of G and $G/A \simeq \mathbb{Z}$. Suppose there is a relation $x_0^{\ell_0} x_1^{\ell_1} \cdots x_k^{\ell_k} = 1$, where not all the integers ℓ_i are 0. Then matrix multiplication reveals that $\ell_0 + \ell_1 c + \cdots + \ell_k c^k = 0$. But this is impossible since c is transcendental. It follows that A is freely generated by the x_i and therefore $G \simeq \mathbb{Z} \text{ wr } \mathbb{Z}$. (In fact, $\mathbb{Z} \text{ wr } \mathbb{Z}$ is not $\mathbb{Z}$-linear by 3.2.1 below.) ∎

Triangularizable, unitriangularizable, and diagonalizable groups

Let G be an R-linear group of degree n, where R is a commutative ring with identity. If G is conjugate to a subgroup of $T_n(R)$, the group of (upper) triangular matrices with units on the diagonal, then G is said to be *triangularizable*.

Similarly, G is *unitriangularizable* if it is isomorphic with a subgroup of $U_n(R)$, the group of upper unitriangular matrices in $Gl_n(R)$ (i.e. upper triangular matrices with 1's on the diagonal). Finally, G is *diagonalizable* if it is isomorphic with a group of diagonal matrices in $GL_n(R)$, that is, with a subgroup of $U(R) \times U(R) \times \cdots \times U(R)$, with n factors: here $U(R)$ is the group of units of R.

The next result is simple but revealing.

3.1.5

(i) *A diagonalizable linear group is abelian.*
(ii) *A unitriangularizable linear group is nilpotent.*
(iii) *A triangularizable linear group is nilpotent-by-abelian.*

Proof Of course (i) is obvious. As for (ii) recall from 1.2.6 that $U_n(R)$ is nilpotent of class at most $n - 1$. Finally, the map which assigns to each triangular matrix its diagonal yields a homomorphism from $T_n(R)$ to the abelian group

$$\underbrace{U(R) \times U(R) \times \cdots \times U(R)}_{n},$$

which has kernel precisely $U_n(R)$. ∎

In 1951 Mal'cev, improving on earlier results of Lie and Kolchin, proved a fundamental theorem about soluble linear groups which can be regarded as a partial converse to 3.1.5. First we recall some terminology. Let V be a non-zero finite dimensional vector space over a field F and let G be a subgroup of $GL(V)$. Then G is *irreducible* if V is a simple FG-module. Also G is said to be *imprimitive* if there is a decomposition $V = V_1 \oplus V_2 \oplus \cdots \oplus V_k$, with $k > 1$, such that elements of G permute the subspaces V_i. Of course, G is called *primitive* if it is not imprimitive.

Our immediate objective is to prove the following theorem.

3.1.6 (Mal'cev 1951) *Let V be an n-dimensional vector space over an algebraically closed field F and let G be a soluble subgroup of $GL(V)$.*

(i) *If G is irreducible, then G has a normal diagonalizable subgroup D such that $|G:D| \leq g(n)$ for some function g.*
(ii) *In general G has a normal triangularizable subgroup T such that $|G:T| \leq f(n)$ for some function f.*

Thus, if G is irreducible, it is virtually abelian, and in general it is virtually nilpotent-by-abelian.

The main tool needed in the proof of 3.1.6 is the following special case of that result.

3.1.7 *Let V and G be as in 3.1.6. If G is primitive and irreducible, then there is a scalar subgroup S such that $|G : S| \leq n^2 h(n^2)$, where $h(m)$ is the maximum number of automorphisms of an abelian group of order at most m.*

Here a *scalar subgroup* is a subgroup consisting of scalar linear mappings; it is necessarily central in G. Thus 3.1.7 implies that $|G : Z(G)| \le n^2 h(n^2)$.

Proof of 3.1.7 Let A be any abelian normal subgroup of G. Then V is completely reducible as an FA-module since G is irreducible. By Clifford's Theorem—see Curtis and Reiner (1966: §49)—we have $V = V_1 \oplus V_2 \oplus \cdots \oplus V_k$ where V_i is the sum of all the simple FA-submodules of given isomorphism type. Further, elements of G permute the V_i. Since G is primitive, it follows that $k = 1$, so that V is a direct sum of isomorphic simple FA-modules. Now F is algebraically closed, so a simple FG-module has dimension 1. Consequently, A is scalar, and hence $A \le C = Z(G)$: this holds for all abelian normal subgroups A.

Next let B/C be a maximal abelian normal subgroup of G/C. Our immediate objective is to prove that

$$|B : C| \le n^2 \quad \text{and} \quad C_G(B/C) = B.$$

From this it will follow that

$$|G : B| = |G : C_G(B/C)| \le |\operatorname{Aut}(B/C)| \le f(n^2),$$

and hence that $|G : C| \le n^2 f(n^2)$. Since C is scalar, the desired result follows.

Suppose first that $C_G(B) \ne C$. Then $C_G(B)/C$ contains a non-trivial abelian normal subgroup D/C of G/C. Now BD/C is abelian since $[B, D] = 1$, from which we infer that $D \le B$ by maximality of B/C. Therefore $D \le Z(B)$ and D is abelian. By the first paragraph of the proof $D \le C$, a contradiction. Hence $C_G(B) = C$.

Next choose elements $b_1, b_2, \dots, b_r$ from distinct cosets of C in B and suppose the b_i are linearly dependent in the vector space $\operatorname{End}_F(V)$. Then there is a non-trivial relation of the form $\sum_{i=1}^s f_i b_i = 0$, where $0 \ne f_i \in F$, with shortest length $s \le r$. (Here the b_i may need to be relabelled.) Now $b_1 b_2^{-1} \notin C$, so $[b_1 b_2^{-1}, x] \ne 1$ for some x in B; thus $[b_1, x] \ne [b_2, x]$. Since $[b_i, x]$ is in C, it is scalar, equal to $t_i 1$, say. So $t_1 \ne t_2$ and $x^{-1} b_i x = b_i [b_i, x] = t_i b_i$. It follows that

$$0 = t_1 \left(\sum_{i=1}^s f_i b_i \right) - x^{-1} \left(\sum_{i=1}^s f_i b_i \right) x$$

$$= t_1 \left(\sum_{i=1}^s f_i b_i \right) - \sum_{i=1}^s f_i t_i b_i$$

$$= \sum_{i=2}^s (t_1 - t_i) f_i b_i.$$

By minimality of s we deduce that $(t_1 - t_2) f_2 = 0$ and $f_2 = 0$, a contradiction. It follows that the b_i are linearly independent and thus $r \le \dim_F (\operatorname{End}_F(V)) = n^2$. Hence $|B : C| \le n^2$, as claimed.

It remains to prove that $K = C_G(B/C)$ equals B; now $B \le K$ since B/C is abelian. If $k \in K$, the assignment $bC \mapsto [b, k]$ is a well-defined homomorphism

$\theta_k : B/C \rightarrow C$. Furthermore, $k \mapsto \theta_k$ determines a homomorphism from K to $H = \text{Hom}(B/C, C)$, with kernel $C_K(B) = C$. Thus K/C is isomorphic with a subgroup of H. But C is scalar, so we have $|H| \leq |\text{Hom}(B/C, F^*)| \leq |B/C|$ since finite subgroups of $F^* = U(F)$ are cyclic. Finally, $|K:C| \leq |B:C|$ and $|K| \leq |B|$, so that $K = B$ and the proof is complete. ∎

Proof of 3.1.6

(i) If G is primitive, the result follows directly from 3.1.7, so we may assume that G is imprimitive. Then there is a decomposition $V = V_1 \oplus V_2 \oplus \cdots \oplus V_k$, where $k > 1$ and elements of G permute the subspaces V_i. If $g \in G$, then $V_i g = V_{i'}$, where $g^\pi : i \mapsto i'$ is a permutation of $\{1, 2, \ldots, k\}$. Now plainly $\pi : G \rightarrow S_k$ is a homomorphism and, if its kernel is K, we have $|G:K| \leq k! \leq n!$.

Next K acts on each subspace V_i as a linear group. Since $n_i = \dim(V_i) < n$, the induction hypothesis tells us that $K/C_K(V_i)$ has a normal diagonalizable subgroup $D_i/C_K(V_i)$ such that $|K : D_i| \leq g(n_i)$. Writing $D^* = D_1 \cap D_2 \cap \cdots \cap D_k$, we see that D^* acts diagonally on each V_i, and hence on V. Also

$$|K : D^*| \leq \prod_{i=1}^{k} g(n_i) \leq \bar{g}(n)^n,$$

where $\bar{g}(n) = \max\{g(i) \,|\, i < n\}$. Therefore $|G : D^*| \leq \bar{g}(n)n! = m$, say. Replacing D^* by its normal core, we obtain a normal diagonalizable subgroup D of G with $|G : D| \leq m!$. Define $g(n) = m!$.

(ii) In the general case we form an FG-composition series in the module V, say $0 = V_0 < V_1 < \cdots < V_r = V$, and let $m_i = \dim(V_{i+1}/V_i)$. Since G acts as an irreducible linear group on V_{i+1}/V_i, by (i) there is a diagonalizable normal subgroup $D_i/C_G(V_{i+1}/V_i)$ such that $|G:D_i| \leq g(m_i)$. Put $T = D_1 \cap D_2 \cap \cdots \cap D_r$. Then T is triangularizable and

$$|G : T| \leq \prod_{i=1}^{r} g(m_i) \leq \bar{g}(n)^n,$$

where $\bar{g}(n) = \max\{g(i) \,|\, i < n\}$. To complete the proof define $f(n)$ to be $\bar{g}(n)^n$. ∎

Mal'cev's Theorem has many important consequences, most of which are to be found in his paper (Mal'cev 1951). One of the most useful is:

3.1.8 *Let G be a soluble linear group. Then G is nilpotent-by-abelian-by-finite, and if G is irreducible, it is abelian-by-finite.*

Proof Let G be F-linear and let $\bar{F}$ be the algebraic closure of F; thus G is also $\bar{F}$-linear. Applying Mal'cev's Theorem, and also 3.1.5, we see that G is nilpotent-by-abelian-by-finite.

Now assume that G is irreducible and acts on the F-vector space V. There is a normal subgroup T of finite index that acts triangularly on the $\bar{F}$-space

$\bar{V} = V \otimes_F \bar{F}$. Then T' acts unitriangularly on $\bar{V}$ and hence also on V. But $[V, T']$, that is, the subgroup generated by all $v(-1 + t)$ with $v \in V, t \in T'$, is a G-invariant subspace of V, so by irreducibility $[V, T'] = 0$ and $T' = 1$, which shows that G is abelian-by-finite.   ∎

We remark that 3.1.8 does not hold for soluble skew-linear groups—see Shirvani and Wehrfritz (1986).

Sometimes 3.1.8 can be used to show that a particular type of soluble group is not linear.

3.1.9 *Let A, B, C be non-trivial torsion-free abelian groups. Then the wreath product (A wr B) wr C is a soluble group of derived length 3, which is not linear over any field.*

Proof If the group in question were linear, it would be nilpotent-by-abelian-by-finite, and then $(A^m \ wr \ B^m) \ wr \ C^m$ would be nilpotent-by-abelian for some $m > 0$. But it is easily seen that this is not the case.   ∎

For example, $(\mathbb{Z} \ wr \ \mathbb{Z}) \ wr \ \mathbb{Z}$ is a finitely generated torsion-free soluble group of derived length 3, which is not linear.

Another very important consequence of Mal'cev's Theorem is:

3.1.10 (Zassenhaus 1938) *A soluble linear group of degree n has derived length bounded by some function of n. Thus a locally soluble linear group is soluble.*

This follows directly from 3.1.6 and the fact that $T_n(F)$ has derived length at most $1 + [\log_2(n)]$, (see 1.2.5). The proof of 3.1.6 does give a crude upper bound for the derived length, however much better bounds are known, for example:

3.1.11 (Newman 1972) *Let G be a soluble linear group of degree $n \geq 60$. Then the derived length d of G satisfies*

$$5 \log_9(n - 1) + a \leq d \leq 5 \log_9(n - 2) + a + \tfrac{3}{2},$$

where $a = 17/2 - 15(\log 2)(2 \log 3)^{-1}$.

Thus the derived length is essentially logarithmic in n.

On the other hand, there exist soluble skew-linear groups of degree 1 with arbitrarily large derived length (see Shirvani and Wehrfritz 1986). Of course these groups are subgroups of the multiplicative group of a division ring.

We mention also that Dixon (1967) has shown that *the Fitting subgroup of a completely reducible soluble linear group of degree n has index at most* $2^{n-1} \cdot 3^{(2n-1)/3}$, *and the bound is attained when* $n = 2 \cdot 4^k, k \geq 0$. (Notice that such groups are abelian-by-finite by 3.1.8.) Finally, bounds for the Fitting length of a soluble linear group have been given by Frick and Newman (1972).

Next we will prove a theorem on nilpotent linear groups which is a strengthened form of the second part of 3.1.8.

3.1.12 (Suprunenko 1963) *Let G be a nilpotent linear group of degree n and class c. If G is irreducible, then $G/Z(G)$ is finite with order bounded by a function of n and c.*

Proof Let G be F-linear, where F is a field. We first deal with the case where F is algebraically closed. By 3.1.6 the group G has a diagonalizable normal subgroup D such that $|G : D| \leq g(n)$. Let $x \in Z_2(G)$ and $g \in G$. Then $[x, g]$ belongs to $Z(G)$, which is scalar, so $[x, g] = \lambda 1$ for some $\lambda \in F$. Now $\det([x, g]) = 1$, so that $1 = \det(\lambda 1) = \lambda^n$. Therefore $1 = [x, g]^n = [x^n, g]$ since $x \in Z_2(G)$, and thus $x^n \in Z(G)$. Hence $(Z_2(G)/Z(G))^n = 1$, and it follows via 1.2.20 that $(G/Z(G))^m = 1$, where $m = n^{c-1}$.

Now let $d \in D$. Then d is similar to a diagonal matrix, with diagonal elements $\delta_1, \delta_2, \ldots, \delta_n$ say, ($\delta_i \in F$). Since $d^m \in Z(G)$ by the first paragraph, it follows that $\delta_1^m = \delta_2^m = \cdots = \delta_n^m$. Thus $\delta_i = \delta_1 \phi_i$, where $\phi_i^m = 1$ for $i = 2, 3, \ldots, n$. Here the ϕ_i belong to the torsion subgroup of F, so they generate a finite cyclic group. From this we infer that $DZ(G)/Z(G)$ is isomorphic with a subgroup of a direct product of $n-1$ cyclic groups of order m. Therefore $|DZ(G)/Z(G)|$ divides m^{n-1} and hence $|G : Z(G)| \leq m^{n-1} g(n)$.

Now we are able to tackle the general case. Let $\bar{F}$ denote the algebraic closure of F. We regard G as acting on the n-dimensional F-space V, and hence on the $\bar{F}$-space $\bar{V} = V \otimes_F \bar{F}$. Form an $\bar{F}G$-composition series in $\bar{V}$ and consider a factor of the series, say U. Then $G/C_G(U)$ is an irreducible nilpotent $\bar{F}$-linear group, so that by the algebraically closed case $Z(G/C_G(U)) = H(U)/C_G(U)$ has finite index $\ell(U)$ bounded by a function of n and c. Let $H = \bigcap_U H(U)$. Then $[H, G] \leq \bigcap_U C_G(U)$, so that $[H, G]$ acts unitriangularly on $\bar{V}$ and hence on V. But G is F-irreducible. It follows that $[H, G] = 1$ and $H \leq Z(G)$. Finally,

$$|G : Z(G)| \leq |G : H| \leq \prod_U |G : H(U)| \leq (\max\{\ell(U)\})^n. \qquad \blacksquare$$

Corollary 3.1.13 *If G is an irreducible nilpotent linear group, then G' is finite. Thus if G is torsion-free, it is abelian.*

For by Schur's Theorem, (e.g. see, Robinson 1996: 10.1.4), the finiteness of $G/Z(G)$ implies that of G'.

Another important application of Mal'cev's Theorem is to the structure of polycyclic groups.

3.1.14 (Mal'cev 1951) *A polycyclic group is nilpotent-by-abelian-by-finite.*

Proof Let G be a polycyclic group. Then there is a *normal* series $1 = G_0 \triangleleft G_1 \triangleleft \cdots \triangleleft G_m = G$ such that each infinite factor G_{i+1}/G_i is free abelian with finite rank and G-rationally irreducible, that is, $(G_{i+1}/G_i) \otimes \mathbb{Q}$ is a simple $\mathbb{Q}G$-module. Let $C_i = C_G(G_{i+1}/G_i)$. If G_{i+1}/G_i is infinite, G/C_i may be regarded as an irreducible $\mathbb{Q}$-linear group via conjugation. Hence it has an abelian normal subgroup A_i/G_i with finite index. Of course this is also true if G_{i+1}/G_i is finite, when we take A_i to be G_i. Set $A = \bigcap_{i=1}^m A_i$; then G/A is finite and A' centralizes each G_{i+1}/G_i. It follows that A' is nilpotent and G is nilpotent-by-abelian-by-finite.   ∎

3.2   Soluble ℤ-linear groups

Our main objective here is to show that ℤ-linear groups have the remarkable property that all their soluble subgroups are polycyclic. Thus soluble subgroups of the groups $GL_n(\mathbb{Z})$ are a rich source—and in fact the only source—of polycyclic groups. Our first object is to prove this striking fact.

3.2.1 (Mal'cev 1951) *A soluble ℤ-linear group is polycyclic.*

The proof makes use of a well-known result about algebraic number fields due to Dirichlet.

3.2.2 *The group of units of an algebraic number field is finitely generated.*

For a proof of this see, for example Janusz (1996: 13.12).

Proof of 3.2.1 Let G be a soluble ℤ-linear group acting faithfully on a free abelian group A of rank n. The proof is by induction on n, which we may assume to be greater than 1. Suppose first that G acts rationally reducibly on A. So there is a proper non-zero G-submodule B such that A/B is torsion-free. Set $C = C_G(B) \cap C_G(A/B)$. By the induction hypothesis $G/C_G(B)$ and $G/C_G(A/B)$ are both polycyclic, whence so is G/C. Now let $c \in C$; then the map $a + B \mapsto [a, c]$ is a homomorphism c' from A/B to B, and further $c \mapsto c'$ is an injective homomorphism from C into the finitely generated abelian group $\mathrm{Hom}(A/B, B)$. Therefore, C is finitely generated and G is polycyclic.

Next assume that G is rationally irreducible. By 3.1.8 the group G is abelian-by-finite, and evidently we may assume it is abelian. Since $V = A \otimes \mathbb{Q}$ is a simple $\mathbb{Q}G$-module, Schur's Lemma tells us that $E = \mathrm{End}_{\mathbb{Q}G}(V)$ is a division ring. Since G is abelian, it is contained in the centre of E, which is a field F say. Now F is an algebraic number field because $\dim(E)$ is finite (it equals n^2). Let $g \in G$; then by the Cayley–Hamilton Theorem g satisfies its characteristic polynomial, which is a monic polynomial in $\mathbb{Z}[t]$. Since the same holds for g^{-1}, we conclude that g is a unit of F. From this we infer that G is a subgroup of the group of units of F and hence that G is finitely generated, by 3.2.2.   ∎

Combining 3.2.1 with the Tits Alternative 3.1.2, we obtain:

3.2.3 *A subgroup of $GL_n(\mathbb{Z})$ which has no free subgroups of rank 2 is virtually polycyclic.*

It is possible to extend 3.2.1 to soluble groups of automorphisms of virtually polycyclic groups.

3.2.4 (Mal'cev 1951; Smirnov 1953; Baer 1955*a*) *If G is a virtually polycyclic group, then every locally soluble subgroup of $\mathrm{Aut}(G)$ is polycyclic.*

Before embarking on the proof of this theorem, we note a frequently used result about derivations. Suppose that M is a module over a group G. A *derivation* from G to M is a map $\delta : G \to M$ such that

$$(xy)^{\delta} = (x^{\delta})y + y^{\delta}, \quad (x, y \in G).$$

It is a simple matter to verify that the set of all derivations is an additive abelian group, $\mathrm{Der}(G, M)$, where addition is defined by the rule $x^{\delta_1 + \delta_2} = x^{\delta_1} + x^{\delta_2}$. Derivations have several interpretations in group theory and one of these is indicated in the next result, which is proved as 10.1.12 below. (For a full account of derivations see 10.1.)

3.2.5 *Let N be a normal subgroup of a group G and let $Q = G/N$. Let A denote the group of automorphisms of G which stabilize the series $1 \lhd N \lhd G$. Then $A \simeq \mathrm{Der}(Q, Z(N))$, where the G-module structure of $Z(N)$ arises from conjugation.*

We can now proceed to the proof of 3.2.4.

Proof of 3.2.4 Let A denote a soluble subgroup of $\mathrm{Aut}(G)$. Now there is a characteristic series in G whose infinite factors are free abelian of finite rank, say $1 = G_0 \lhd G_1 \lhd \cdots \lhd G_m = G$. We argue by induction on m, the case $m \leq 1$ being 3.2.1. Let $m > 1$ and put $C = C_A(G_1) \cap C_A(G/G_1)$. Then the induction hypothesis shows that A/C is polycyclic and so it remains to prove that C is polycyclic.

By 3.2.5. $C \simeq D = \mathrm{Der}(G/G_1, G_1)$ and D is abelian. Suppose that G/G_1 can be generated by r elements. Then any derivation δ in D is completely determined by its effect on the generators of G/G_1 and it follows that D is isomorphic with a subgroup of the direct product of r copies of G_1. From this we infer that D, and hence C, is finitely generated, as required. ∎

Torsion groups of automorphisms of Černikov groups

It is natural to enquire if there is an analogue of 3.2.4 for Černikov groups. Since the endomorphism ring of a group of Prüfer type p^{∞} is isomorphic with R_p, the group of p-adic integer units, the automorphism group of a radicable abelian p-group with finite rank n is isomorphic with $GL_n(R_p)$. Thus we must deal with R_p-linear groups. However, $U(R_p)$ is the direct sum of a cyclic group of order $p - 1$ or 2, according as $p > 2$ or $p = 2$, and an uncountable torsion-free abelian group. Thus no precise analogue of 3.2.2 exists. However, when we restrict attention to torsion groups, results that correspond to 3.2.2 and 3.3.5 appear.

3.2.6 (Baer 1955d; Polovickiĭ 1962) *A torsion group of automorphisms of a Černikov group is Černikov.*

The proof rests on the following result.

3.2.7 (Baer 1955d) *Let G be a radicable abelian p-group of finite rank and suppose that α is a non-trivial automorphism which fixes all elements of G of order p, and all elements of order 4 if $p = 2$. Then α has infinite order.*

Proof Argue by induction on the rank of G, that is, the number of p^∞ factors in a direct decomposition. Assuming the result to be false, we may take the order of α to be a prime q. Since $[G, \alpha]$ is a homomorphic image of G, it is radicable and hence is a direct factor of G. Suppose that $[G, \alpha] \neq G$. Then by induction α must act trivially on $[G, \alpha]$ and $[G, \alpha, \alpha] = 1$. From this we deduce that $1 = [G, \alpha^q] = [G, \alpha]^q = [G, \alpha]$ and $\alpha = 1$. By this contradiction $G = [G, \alpha]$, that is, $G = G^{\alpha - 1}$.

Next let θ denote the endomorphism $1 + \alpha + \alpha^2 + \cdots + \alpha^{q-1}$ of G. Then $G^\theta = (G^{\alpha-1})^\theta = G^{\alpha^q - 1} = 1$, that is, $\theta = 0$. Next choose g in G of least order such that $g^\alpha \neq g$. Then $1 = [g^p, \alpha] = [g, \alpha]^p$; thus, if g_1 denotes $[g, \alpha]$, we have $g_1^\alpha = g_1$ and $g^\alpha = g g_1$. Hence $g^{\alpha^i} = g g_1^i$ and therefore

$$1 = g^\theta = g(gg_1)(gg_1^2) \cdots (gg_1^{q-1}) = g^q g_1^{\binom{q}{2}}.$$

If $q \neq p$, then $g \in \langle g_1 \rangle$ and $g^\alpha = g$. Therefore $q = p$. If $p > 2$, then $\binom{p}{2}$ is divisible by p, so that $g^p = g_1^{-\binom{p}{2}} = 1$; but then $g^\alpha = g$. Consequently $p = 2$ and $|g| = 4$ and again we get $g^\alpha = g$. ∎

Corollary 3.2.8 (Černikov 1950) *A torsion subgroup of $GL_n(R_p)$ is finite.*

Proof Let A be a torsion subgroup of $GL_n(R_p)$ acting on a radicable abelian p-group G of rank n. Then $G_1 = \{g \in G \mid g^{p^2} = 1\}$ is finite and characteristic in G, and hence $A/C_A(G_1)$ is finite. But $C_A(G_1)$ is torsion-free by 3.2.7, so it is trivial and A is finite. ∎

Proof of 3.2.6 Let A be a torsion group of automorphisms of a Černikov group G. There is a characteristic radicable abelian subgroup D with finite index in G, and by 3.2.8 each $A/C_A(D_p)$ is finite, whence $A/C_A(D)$ is finite. Therefore A/B is finite, where $B = C_A(D) \cap C_A(G/D)$. According to 3.2.5, the group B is isomorphic with a subgroup of $E = \mathrm{Der}(G/D, D)$. But E embeds in the direct product of finitely many copies of D. From this it follows that E, and therefore B, is Černikov. Hence A is Černikov. ∎

The same argument will establish:

3.2.9 *A torsion group of automorphisms of a nilpotent Černikov group is finite.*

The point to notice here is that by 1.4.4 a nilpotent Černikov group is centre-by-finite, so in the proof of 3.2.6 the subgroup D is central in G: therefore B is isomorphic with a subgroup of $\mathrm{Hom}(G/D, D)$, which is finite.

3.3 The linearity of polycyclic groups

Having seen that soluble subgroups of $GL_n(\mathbb{Z})$ are polycyclic, we may reasonably ask if every polycyclic group arises as a subgroup of some $GL_n(\mathbb{Z})$. That this is true is one of the most striking results in the theory of polycyclic groups.

3.3.1 (Auslander 1967; Swan 1967) *Every polycyclic-by-finite group is $\mathbb{Z}$-linear.*

Special cases of this result had previously been found by P. Hall (see 1969) and Learner (1962). We precede the proof with a lemma on group rings.

3.3.2 *Let G be a finitely generated group and let I be an (two-sided) ideal of the integral group ring $\mathbb{Z}G$ such that $\mathbb{Z}G/I$ is finitely generated as an additive group. Then I is finitely generated as an ideal and $\mathbb{Z}G/I^i$ is a finitely generated group for all $i > 0$.*

Proof Denote by M the subgroup of $\mathbb{Z}G$ generated by the generators of G and their inverses, together with coset representatives of the generators of $\mathbb{Z}G/I$. Thus M is finitely generated abelian and $\mathbb{Z}G = M + I$. The subgroup M^2, which is generated by all xy with $x, y \in M$, is finitely generated, and therefore so is $V = (M + M^2) \cap I$.

Let J be the ideal of $\mathbb{Z}G$ generated by V; then $J \subseteq I$ and of course J is a finitely generated ideal. Since $\mathbb{Z}G = M + I$, we have

$$M^2 \subseteq (M + M^2) \cap (M + I) = M + V \subseteq M + J.$$

This shows that $M + J$ is a subring containing G and hence $\mathbb{Z}G = M + J$. Thus $\mathbb{Z}G/J \simeq M/M \cap J$, which is a finitely generated group. It follows that I/J is a finitely generated group and so I is finitely generated as an ideal.

Next I/I^2 is a finitely generated bimodule over $\mathbb{Z}G/I$, and the latter is a finitely generated abelian group. From this we may infer that I/I^2, and hence $\mathbb{Z}G/I^2$, is a finitely generated group. Clearly, the argument proves that every $\mathbb{Z}G/I^i$ is a finitely generated group. ∎

The next result is a crucial reduction theorem for 3.3.1. Here the situation is that G is a group with subgroups A, N, H such that $H \triangleleft G$, $N \triangleleft G$, $N \leq H$, and $G = A \ltimes H$. We further suppose that H is finitely generated and $[H, G] \leq N$. The result we require is:

3.3.3 *Suppose that A is $\mathbb{Z}$-linear and H has a faithful $\mathbb{Z}$-representation in which elements of N are represented by unitriangular matrices. Then G has a faithful $\mathbb{Z}$-representation of the same type.*

Proof Let $\rho : H \to GL_n(\mathbb{Z})$ be the $\mathbb{Z}$-representation of H specified. Then ρ extends in the obvious way to a ring homomorphism—also denoted by ρ—from $\mathbb{Z}H$ to $M_n(\mathbb{Z})$, the ring of all $n \times n$ integral matrices. Write $K = \mathrm{Ker}(\rho)$, which is an ideal of $\mathbb{Z}H$, and observe that $\mathbb{Z}H/K$ is a finitely generated group, being isomorphic with a subgroup of $M_n(\mathbb{Z})$, which is free abelian of rank n^2.

Let I denote the right ideal of $\mathbb{Z}H$ generated by the elements $x - 1$ where $x \in N$. If $h \in H$, then $h(x-1) = (x^{h^{-1}} - 1)h \in I$ since $N \triangleleft G$. Thus I is a two-sided ideal, (a so-called *relative augmentation ideal*). Now I^n is generated as a right ideal by all products of the form $(x_1 - 1)(x_2 - 1) \cdots (x_n - 1)$, where $x_i \in N$. But $x_i^\rho \in N^\rho \subseteq U_n(\mathbb{Z})$, from which it follows that $((x_1 - 1)(x_2 - 1) \cdots (x_n - 1))^\rho = (x_1^\rho - 1)(x_2^\rho - 1) \cdots (x_n^\rho - 1) = 0$. Therefore, $I^n \subseteq K$.

Writing $J_1 = (I + K)^n$, we have $J_1 \subseteq I^n + K = K$ since K is an ideal. Denote the torsion subgroup of the group $\mathbb{Z}H/J_1$ by J/J_1. Since $\mathbb{Z}H/K$ is a finitely generated abelian group, $\mathbb{Z}H/J$ is free abelian with finite rank, say r. Now elements of H act on $\mathbb{Z}H/J$ by right multiplication and this yields a $\mathbb{Z}$-representation $\sigma : H \to GL_r(\mathbb{Z})$. Notice that H acts faithfully on $\mathbb{Z}H/K$ since $K = \text{Ker}(\rho)$ and $\rho|_H$ is faithful. But $J \subseteq K$ because $\mathbb{Z}H/K$ is torsion-free while J/J_1 is finite. Therefore, σ is a faithful representation of H. Furthermore, σ represents elements of N by unitriangular matrices because $I^n \subseteq J_1 \subseteq J$.

To complete the proof we show how to extend σ to $G = H \ltimes A$. First we cause A to act on $\mathbb{Z}H$ by conjugation: thus

$$\left(\sum_{i=1}^{k} n_i h_i \right) \cdot a = \sum_{i=1}^{k} n_i h_i^a,$$

where $h_i \in H$, $a \in A$, $n_i \in \mathbb{Z}$. For any $h \in H$, $a \in A$, write

$$h \cdot a = h^a = h([h, a] - 1) + h \equiv h \pmod{I}$$

since $[H, G] \leq N$. Hence $(\mathbb{Z}H) \cdot (a - 1) \leq I$. It follows that $J_1 \cdot a = J_1$ and hence $J \cdot a = J$. Thus A acts on $\mathbb{Z}H/J$. This action turns $\mathbb{Z}H/J$ into a $\mathbb{Z}A$-module, and even into a $\mathbb{Z}G$-module since

$$(((h + J) \cdot a^{-1})h) \cdot a = ((h + J)^{a^{-1}} h)^a = (h + J)h^a.$$

To obtain a *faithful* $\mathbb{Z}G$-module, we take a free abelian group U of finite rank on which A acts faithfully—this exists by hypothesis. Form $V = (\mathbb{Z}H/J) \oplus U$, also a free abelian group of finite rank. If we let H act trivially on U, then V becomes a faithful $\mathbb{Z}G$-module.   ∎

We may now apply the preceding result to finitely generated torsion-free nilpotent groups, for which a stronger statement can be made.

3.3.4 (Hall 1969) *A finitely generated torsion-free nilpotent group G is isomorphic with a subgroup of $U_n(\mathbb{Z})$ for some $n > 0$.*

Proof By 1.2.20 there is a central series of G with infinite cyclic factors, $1 = G_0 \triangleleft G_1 \triangleleft \cdots \triangleleft G_r = G$. Argue by induction on $r > 0$. Thus $N = G_{r-1}$ is isomorphic with a subgroup of some $U_m(\mathbb{Z})$. Now G has the form $\langle g \rangle \ltimes N$ with $\langle g \rangle$ infinite cyclic, so we may apply 3.3.3 with $H = N$ and $A = \langle g \rangle$, using the faithful $\mathbb{Z}$-representation of A in which $g \mapsto \left[\begin{smallmatrix} 1 & 1 \\ 0 & 1 \end{smallmatrix} \right]$. Recall that in the proof of 3.3.3 the element g acts on $\mathbb{Z}H/J$ by conjugation; since G is nilpotent, it follows that the action of G on $\mathbb{Z}H/J$ is a nilpotent one. Hence elements of G are represented by unitriangular matrices. ∎

Proof of 3.3.1 Here G is a virtually polycyclic group and we have to prove that it is $\mathbb{Z}$-linear. By 1.3.4 and 3.1.14 there exist normal subgroups G_0, G_1 such that G/G_0 is finite, G_0/G_1 is free abelian and G_1 is torsion-free nilpotent. It is in fact enough to prove the theorem for G_0. For, if V is a free abelian group of finite rank s on which G_0 acts faithfully, we simply form the induced $\mathbb{Z}G$-module $\bar{V} = V \otimes_{\mathbb{Z}G_0} \mathbb{Z}G$. Clearly $\bar{V}$ is free abelian with rank equal to $s \cdot |G : G_0|$ and it is easy to see that $\bar{V}$ is a faithful $\mathbb{Z}G$-module. From now on we assume that $G = G_0$ and put $N = G_1$. Thus N is finitely generated, torsion-free nilpotent, and G/N is free abelian of finite rank.

Let H be a normal subgroup of G such that $N \leq H$ and G/H is infinite cyclic. By induction on the Hirsch length there is a faithful $\mathbb{Z}$-representation of H in which elements of N are represented by unitriangular matrices. Also $G = \langle g \rangle \ltimes H$ with $\langle g \rangle$ infinite cyclic. Now apply 3.3.3 to deduce that G is $\mathbb{Z}$-linear. ∎

Part of the interest of the Auslander–Swan Theorem lies in the consequence that any polycyclic group may be specified by giving a finite set of integral matrices as generators. This is a useful alternative to giving the group by means of a finite presentation. There are algorithmic consequences of this fact which are explored in Chapter 9.

Several generalizations of the Auslander–Swan Theorem are known, particularly for soluble groups of finite rank—for details see the end of Section 5.1.

Also Levič (1969*a*) and Remeslennikov (1969*b*) have proved that every finitely generated torsion-free metabelian group is $\mathbb{C}$-linear. It is still an open question whether every finitely generated metabelian group is linear.

In conclusion, we mention that examples are known of finitely generated centre-by-metabelian groups, which are not residually finite—for details see 5.3.15. Of course, such groups cannot be linear over any field by 3.1.3.

4

THE THEORY OF FINITELY GENERATED
SOLUBLE GROUPS I

In this chapter we begin the study of finitely generated soluble groups, a very natural class of groups, but one which is very much wider than the class of polycyclic groups. As an example of a finitely generated soluble group, which is not polycyclic, we mention the semidirect product $G = X \ltimes A$, where A is the additive group of dyadic rationals $m2^n, m, n \in \mathbb{Z}$, and $X = \langle x \rangle$ is an infinite cyclic group, with x acting on A by multiplication by 2: thus $ax = 2a$. This group is clearly metabelian, it is generated by the two elements 1 and x, but it is not polycyclic since A is not finitely generated. We note also that G has the finite presentation

$$G = \langle a, b \mid a^b = a^2 \rangle.$$

4.1   Embedding in finitely generated soluble groups

Finitely generated groups are clearly countable, so that the objects of our study belong to the class of countable soluble groups. In fact this is all that one can expect to say about the structure of finitely generated soluble groups in general, as is shown by the following result of B. H. Neumann and H. Neumann (1959).

4.1.1 *A countable soluble group of derived length d may be embedded in a 2-generator soluble group of derived length at most $d + 2$.*

This result is an immediate corollary of

4.1.2 *Let G be a countable group in a variety $\mathbf{V}$. Then G embeds in a 2-generator group in the variety $\mathbf{VA}^2$, of $\mathbf{V}$-by-metabelian groups.*

Proof Let $G = \langle g(1), g(2), \dots \rangle$. Let $C = \langle c \rangle$ be infinite cyclic and form the *unrestricted* wreath product

$$H = G \, Wr \, C.$$

Then clearly $H \in \mathbf{VA}$. Each element in the base group of H has the form $(x_n), n \in \mathbb{Z}$, where $x_n \in G$ and $(x^c)_n = x_{n-1}$. Define an element $h(i)$ in the base group by the rule

$$(h(i))_n = g(i)^{-n}, \text{ where } n \in \mathbb{Z}, \quad i = 1, 2, \dots.$$

Then

$$([h(i), c])_n = (h(i)^{-1})_n (h(i)^c)_n = g(i)^n g(i)^{-n+1} = g(i).$$

Hence $[h(i), c]$ is the constant sequence $\widehat{g(i)}$, that is, all its entries are equal. Next let $D = \langle d \rangle$ be another infinite cyclic group and set $K = H \, Wr \, D$. Then $K \in \mathbf{VA}^2$. We now define an element k in the base group of K by $k_0 = c$, $k_{-2i+1} = h(i)$, and $k_j = 1$ for all other j. Now set

$$g(i)^* = [k^{d^{2i-1}}, k],$$

which is an element of the base group of K. We claim that

$$(g(i)^*)_0 = \widehat{g(i)} \quad \text{and} \quad (g(i)^*)_j = 1, \ j \neq 0.$$

For

$$(g(i)^*)_j = ([k^{d^{2i-1}}, k])_j = [(k^{d^{2i-1}})_j, k_j] = [k_{j-2i+1}, k_j].$$

If $j > 0$, this is trivial by definition of k_j. For $j = 0$, it is $[k_{-2i+1}, k_0] = [h(i), c] = \widehat{g(i)}$. If $j < 0$, then $[k_{j-2i+1}, k_j] = 1$ since $j - 2i + 1$ or j is even. It therefore follows that

$$G^* = \langle g(i)^* \mid i = 1, 2, \ldots \rangle \simeq \langle \widehat{g(i)} \mid i = 1, 2, \ldots \rangle \simeq G.$$

Also $g(i)^* \in \ <d, k> \leq K$ and hence $G^* \leq \langle d, k \rangle \in \mathbf{VA}^2$, so the result is proven. ∎

The upshot of this elegant embedding theorem is that the subgroup structure of finitely generated soluble groups is as complicated as that of countable soluble groups. In another discouraging result, P. Hall had shown earlier (1954) that an arbitrary countable abelian group could be embedded in a finitely generated soluble group of derived length 3. More precisely, he proved the following result.

4.1.3 *Suppose that A is any non-trivial countable abelian group. Then there are $2^{\aleph_0}$ non-isomorphic 2-generator groups G such that $A \simeq Z(G) = G''$. Moreover G is centre-by metabelian and has derived length 3.*

This result combines with 4.1.1 to show that even the study of finitely generated soluble groups of derived length 3 is hard.

Proof of 4.1.3 Hall's construction starts with the free nilpotent group Y of class 2 on a set of generators $\{y_i \mid i \in \mathbb{Z}\}$, so that $Y \simeq F/\gamma_3(F)$, where F is a free group of countably infinite rank. Then Y' is a free abelian group with basis $\{c_{ij} = [y_i, y_j] \mid i < j\}$. Form the quotient group $X = Y/L$, where L is generated by the elements $c_{ij}^{-1} c_{i+k \ j+k}$. Thus, if we put $x_i = y_i L$, then X is generated by the x_i, $i \in \mathbb{Z}$, subject to the relations $[x_i, x_j, x_k] = 1$ and, in addition,

$$[x_i, x_j] = [x_{i+k}, x_{j+k}], \quad (i, j, k \in \mathbb{Z}).$$

Hence X is a torsion-free nilpotent group of class 2. We note that $d_r = [x_i, x_{i+r}]$ is independent of i and that $d_1, d_2, \ldots$ form a basis for the free abelian group X'.

Since the assignments $\alpha : x_i \mapsto x_{i+1}$ preserve the above defining relations, they extend to an automorphism α of X with infinite order. We now form the semidirect product

$$H = \langle \alpha \rangle \ltimes X.$$

Then $H = \langle \alpha, x_0 \rangle$ and $d_r^\alpha = [x_0, x_r]^\alpha = [x_1, x_{r+1}] = d_r$, so that $X' \leq Z(H)$. Now $\bar{H} = H/X'$ is clearly metabelian and so $[H'', H] = 1$. It is easy to see that $\bar{H} \simeq \mathbb{Z} \ wr \ \mathbb{Z}$. Hence $Z(\bar{H}) = 1$ and so $Z(H) = Z(X) = X'$.

We now bring the non-trivial countable abelian group A into the picture. Since X' is free abelian with countably infinite rank, $A \simeq X'/K$ for some K, and $K \lhd H$ since $X' = Z(H)$.

Define $G_K = H/K$, so that

$$Z(G_K) = G_K'' = X'/K \simeq A,$$

since $Z(H/X') = 1$.

It is clear that K can be chosen in $2^{\aleph_0}$ different ways: let us suppose that the resulting G_K's fall into only countably many isomorphism classes. Thus, for some K there are uncountably many isomorphisms $\theta_\lambda : G_{K_\lambda} \to G_K$. If $\alpha_\lambda : H \to G_{K_\lambda}$ is the natural homomorphism from H with kernel K_λ, then $\alpha_\lambda \theta_\lambda$ is a homomorphism from H to G_K with kernel K_λ. Thus we have an uncountable set of homomorphisms from the 2-generator group H to the countable group G_K, which is impossible. It follows that there are $2^{\aleph_0}$ non-isomorphic groups G_K and the proof is complete. ∎

4.2   The maximal condition on normal subgroups

We shall now show that soluble groups satisfying the *maximal condition on normal subgroups*, max $-n$, are intermediate between polycyclic groups and finitely generated soluble groups.

4.2.1 *Soluble groups with* max $-n$ *are finitely generated.*

Proof Suppose G is a soluble group with max $-n$ and derived length d. If $d \leq 1$, then G is abelian and the result is clear. Let $d > 1$ and set $A = G^{(d-1)}$. Then A is abelian and, by induction on d, there is a finite set of generators $x_1 A, \ldots, x_m A$ for G/A. Moreover, since G satisfies max $-n$, the normal subgroup A satisfies max $-G$, the maximal condition on G-invariant subgroups, where G acts on A by conjugation. Hence $A = a_1^G \ldots a_n^G$, for some finite set of elements $a_1, \ldots, a_n$ in A. But A is abelian and so $a_i^G = a_i^{\langle x_1, \ldots, x_m \rangle}$, which shows that $G = \langle a_1, \ldots, a_n, x_1, \ldots, x_m \rangle$, a finitely generated group. ∎

Examples of finitely generated soluble groups of derived length 3 which do not satisfy max $-n$ are given by 4.1.3. For, if A is any countable abelian group which is not finitely generated, the group G given by the theorem has an infinitely generated centre and therefore cannot satisfy max $-n$. However, there is a very important subclass of finitely generated soluble groups which do satisfy max $-n$, as given by the following result of P. Hall, which is foundational for the theory of soluble groups.

4.2.2 (P. Hall 1954) *A finitely generated abelian-by-polycyclic-by-finite group satisfies* max $-n$. *Thus, in particular, finitely generated metabelian groups satisfy* max $-n$.

In particular, the groups $\langle a, b \mid a^b = a^2 \rangle$ and $\mathbb{Z} \, wr \, \mathbb{Z}$ satisfy max $-n$, but they are not polycyclic.

Suppose now that G is a finitely generated abelian-by-polycyclic group. Then there is an abelian normal subgroup A such that $G/A = H$ is polycyclic. Now G acts on A by conjugation: for if $g \in G$, the mapping $a \mapsto a^g$, $a \in A$, is an automorphism of A. Since A is abelian, this G-action may be regarded as an H-action: if $h = gA \in H$, define $a^h = a^g$. Thus A becomes a module for H. Indeed, A can be regarded as a (multiplicatively written) module over $\mathbb{Z}H$, the integral group ring of H. For, if $w = \sum_{i=1}^{m} n_i h_i \in \mathbb{Z}H$, $n_i \in \mathbb{Z}$, $h_i \in H$, we may define

$$a^w = (a^{h_1})^{n_1} \cdots (a^{h_m})^{n_m}.$$

Notice that the $\mathbb{Z}H$-submodules of A are just the normal subgroups of G contained in A.

An easy argument, using the fact that G/A satisfies max, shows that *G satisfies* max $-n$ *if and only if* A *satisfies* max $-\mathbb{Z}H$. Thus, in module theoretic language, G satisfies max $-n$ if and only if A is a noetherian R-module, where $R = \mathbb{Z}H$. Now by 1.3.1 the group $H = G/A$ is finitely presented since it is polycyclic. It follows from this that A is the normal closure in G of a finite subset $\{a_1, \ldots, a_n\}$, (see 11.1.1 below), so that, in additive notation, A is the sum of finitely many cyclic R-modules

$$A = a_1 R + \cdots + a_n R.$$

Thus A is a finitely generated R-module and it will be noetherian if R is a right noetherian ring. That this is true follows from the slightly more general:

4.2.3 *Suppose that G is a virtually polycyclic group. Then $R = \mathbb{Z}G$ is noetherian as a (right) R-module.*

Thus we have reduced 4.2.2 to a result about the integral group rings of polycyclic groups. To see what is involved in 4.2.3 let us look at the special case where $G = \langle t \rangle$ is infinite cyclic. Then $\mathbb{Z}G$ is the polynomial ring $\mathbb{Z}[t, t^{-1}]$. The fact that this ring is noetherian is an immediate consequence of *Hilbert's Basis Theorem*: if J is a commutative noetherian ring with identity, then the polynomial ring $J[x]$ is noetherian—see Atiyah-Macdonald (1969: 81). This link between commutative algebra and group theory, which was initiated by P. Hall, has proved very fruitful, leading to some deep structural properties of finitely generated abelian-by-polycyclic groups.

We will deduce 4.2.3 by an induction argument from a generalization of the Hilbert Basis Theorem due to P. Hall.

4.2.4 *Suppose G is a group and H is a normal subgroup of G such that G/H is finite or cyclic. Suppose further that R is a ring with identity, M is an RG-module and N an RH-submodule of M. If N generates M as an RG-module and N is RH-noetherian, then M is RG-noetherian.*

In order to deduce 4.2.3 from 4.2.4, suppose G is a polycyclic-by-finite group. Then G has a series $1 = G_0 \lhd G_1 \lhd \cdots \lhd G_n = G$ whose factors are finite or cyclic. If $n = 0$, then $G = 1$ and $RG = R$ and the result is true by hypothesis. Let $n > 0$ and put $H = G_{n-1}$. By induction we may assume that RH is noetherian. Now apply 4.2.4 with $M = RG$ and $N = RH$ to get the result that M is RG-noetherian.

Proof of 4.2.4 Suppose first that G/H is finite and let $\{t_1, \ldots, t_m\}$ be a transversal to H in G. Thus $G = \bigcup_{i=1}^m H t_i$ and, since $M = N(RG)$, we have $M = Nt_1 + \cdots + Nt_m$. Let $a \in N$ and $x \in H$. Then $(at_i)x = (ax^{t_i^{-1}})t_i \in Nt_i$, from which we deduce that Nt_i is an RH-submodule of M and that the mapping $a \mapsto at_i$ is an R-isomorphism which maps RH-submodules of N to RH-submodules of Nt_i. Hence Nt_i has $\max - RH$ since N does. Since the property $\max - RH$ is closed under taking extensions, it follows that M has $\max - RH$, and a fortiori, it satisfies $\max - RG$.

The case where G/H is infinite cyclic is dealt with by generalizing the familiar proof of Hilbert's Basis Theorem. Let $G/H = \langle Hx \rangle$; then each element $f \in M$ can be written in the form $f = \sum_{i=r}^s c_i x^i$, where $c_i \in N$ and $r \leq s$. This representation is, of course, not necessarily unique. If $c_s \neq 0$, we call $c_s x^s$ a *leading term* and c_s a *leading coefficient* of F.

Showing that M is RG-noetherian is equivalent to proving that an arbitrary RG-submodule M_0 of M is finitely generated as an RG-module. To do this, form the set N_0 of all leading coefficients of elements of M_0, together with 0. We claim that N_0 is an RH-submodule of N. Indeed, if f and g have leading terms $c_s x^s$ and $d_t x^t$, then $f \pm gx^{s-t}$ certainly belongs to M_0 and has leading coefficient $c_s \pm d_t$, provided this does not vanish. In any event, $c_s \pm d_t \in N_0$. Moreover, if $u \in RH$, then $f(x^{-s}ux^s)$ in M_0 has leading term $(c_s u)x^s$, unless $c_s u = 0$: hence $c_s u \in N_0$ and our claim is established. By hypothesis N satisfies $\max - RH$, so N_0 is finitely generated, say by elements $a_1, \ldots, a_\ell$. By definition there exists an element $f_i \in M_0$ which has a_i as leading coefficient. Without any loss of generality we may assume that all powers of x involved in f_i are positive and that $a_i x^m$ is the leading term of f_i for each i, where m does not depend on i.

Define M_1 to be the RG-submodule generated by $f_1, \ldots, f_\ell$ and set

$$N_1 = M_0 \cap (N + Nx + \cdots + Nx^{m-1}).$$

Note that $N + Nx + \cdots + Nx^{m-1}$ has $\max - RH$, so that its RH-submodule N_1 is finitely generated. Hence the RG-module $M_2 = M_1 + N_1 RG$ is finitely generated. Clearly $M_2 \leq M_0$: we complete the proof by showing $M_2 = M_0$.

Suppose that $f \in M_0 \backslash M_2$. Without loss of generality we may assume that f involves no negative powers of x. We choose such an element f whose leading term is cx^p, with p minimal. If $p < m$, then $f \in N_1 \leq M_2$, a contradiction, so $p \geq m$. Since $c \in N_0$, we may write $c = \sum_{i=1}^\ell a_i u_i$, where $u_i \in RH$. Now the element

$$f' = \sum_{i=1}^{\ell} f_i \cdot (x^{-m} u_i x^p)$$

belongs to M_2, involves no negative powers of x and has leading term

$$\left(\sum_{i=1}^{\ell} a_i u_i \right) x^p = c x^p.$$

Thus f and f' have the same leading term and their difference $f - f'$ belongs to $M_0 \backslash M_2$ and yet involves no powers of x higher than x^{p-1}, contradicting the minimality of p. The result is therefore proved. ∎

We have in fact proved that *finite extensions of finitely generated abelian-by-polycyclic groups satisfy* max $-n$. As a corollary to this result, we have, in contrast to 4.1.3:

4.2.5 (P. Hall 1954) *There are only countably many non-isomorphic finitely generated abelian-by-polycyclic-by-finite groups.*

Proof Suppose that G is a finitely generated abelian-by-polycyclic-by-finite group. Then $G \simeq F/N$, where F is a finitely generated free group. Suppose that A is an abelian normal subgroup of G with G/A virtually polycyclic. Then A corresponds to some normal subgroup M/N of F/N in the above isomorphism, where $F/M \simeq G/A$ and so F/M is virtually polycyclic and hence finitely presented. Since F is finitely generated, this implies that M is the normal closure in F of finitely many of its elements, say $M = \langle m_1, \ldots, m_r \rangle^F$. Consequently, there are only countably many possibilities for M in a given F. Furthermore, M/N is abelian, so $M' \leq N$ and F/M' satisfies max $-n$ by 4.2.2. Hence $N = \langle M', b_1, \ldots, b_n \rangle^F$ for some $b_1, \ldots, b_n \in M$. Therefore there are only countably many choices for N. ∎

In particular, there are only countably many finitely generated metabelian groups, whereas by 4.1.3 there are uncountably many finitely generated centre-by-metabelian groups.

4.3 Residual finiteness

In 1.3.10 we saw that polycyclic groups are residually finite. A great deal of effort has gone into trying to extend this result to other classes of infinite soluble groups. The main objective of this section is to prove:

4.3.1 (P. Hall 1959) *Finitely generated abelian-by-nilpotent-by-finite groups are residually finite.*

Some 15 years later this result was extended to finitely generated abelian-by-polycyclic-by-finite groups by Roseblade and Jategaonkar. We shall discuss their generalizations of 4.3.1 in Chapter 7.

In order to pinpoint the central issues involved in proving such results, we first of all consider some special cases. Suppose that H is a virtually polycyclic group and A is a simple module for $\mathbb{Z}H$. Writing A multiplicatively, we form $G = H \ltimes A$. Then G is a finitely generated abelian-by-polycyclic-by-finite group. Now assume that G is residually finite. Then, since $A \neq 1$, there is a normal subgroup N of finite index in G which does not contain A. Now $N \cap A$ is a proper $\mathbb{Z}H$-submodule of A, so $N \cap A = 1$ since A is a simple module. Therefore, $A \simeq AN/N$ and A is finite.

The implication of the preceding argument is that in order to prove that finitely generated abelian-by-polycyclic groups are residually finite, it will be necessary to show that a *simple module over the integral group ring of a virtually polycyclic group is finite.*

Suppose now that we want to prove this fact in the special case where H is an infinite cyclic group with generator x and let A be a simple $\mathbb{Z}H$-module. Then A is a cyclic module for $\mathbb{Z}H$ and so it is isomorphic with $\mathbb{Z}H/I$, where I is an ideal of $\mathbb{Z}H$. Because A is simple, I must be a maximal ideal, and thus $\mathbb{Z}H/I$ is a field. Since $\mathbb{Z}H = \mathbb{Z}[x, x^{-1}]$, the finiteness of $\mathbb{Z}H/I$ follows at once from *Hilbert's Weak Nullstellensatz*—see Atiyah and Macdonald (1969: 7.9). This asserts that any simple module over a finitely generated algebra over a field is finite dimensional. Indeed, using the argument just given, the Nullstellensatz is easily seen to dispose of the case, where H is finitely generated abelian. Thus, *simple modules over finitely generated abelian groups are finite.*

In 1959 P. Hall proved a version of the Nullstellensatz which covers the case where H is finitely generated nilpotent, while J. E. Roseblade settled the case H virtually polycyclic in 1973.

The following structural property of finitely generated modules over polycyclic groups plays a key role in all that follows: *such modules possess free submodules with torsion quotient modules involving only a finite set of primes.* In order to make this concept precise we introduce a special class of modules.

Suppose that J is a principal ideal domain and π is a set of non-associate primes (i.e. irreducible elements) in J. We say that a J-module M belongs to the class, $\mathcal{M}(J, \pi)$, if it has a free J-submodule F such that M/F is a π-torsion module. This means that if $a \in M$, there is a product x of primes from π, such that $ax \in F$.

For example, take $J = \mathbb{Z}$ and $M = \{m/2^n \mid m, n \in \mathbb{Z}\}$ to be the additive group (i.e. $\mathbb{Z}$-module) of dyadic rationals. Then $M \in \mathcal{M}(\mathbb{Z}, 2)$ since $M/\mathbb{Z}$ is a 2-group.

The basic properties of the class $\mathcal{M}(J, \pi)$ are given in:

4.3.2 *Let J be a principal ideal domain and π be a set of primes in J. Then:*

(i) *The field of fractions of J belongs to $\mathcal{M}(J, \pi)$ if and only if π is a complete set of primes of J;*

(ii) *$\mathcal{M}(J, \pi)$ is closed under the operation of taking submodules.*

(iii) *A module which has an ascending chain of submodules each factor of which is in $\mathcal{M}(J,\pi)$ is itself in $\mathcal{M}(J,\pi)$.*

We leave the easy proofs of these facts as an exercise. It is the property (i) which plays the crucial role in what follows.

4.3.3 *Let J be a principal ideal domain, G a virtually polycyclic group and M is a finitely generated JG-module. Then $M \in \mathcal{M}(J,\pi)$ for some finite set of primes π.*

In fact 4.3.3 is a generalization of the generic flatness lemma of commutative algebra—see Eisenbud (1995). For suppose that $\pi = \{p_1, \ldots, p_r\}$ and set $\lambda = p_1 \cdots p_r$. Let M_0 be a free submodule of a finitely generated JG-module M and let M/M_0 be π-*torsion*. Then M/M_0 is a λ-torsion module. Conversely, λ-torsion implies π-torsion and so the conclusion of the result is that $M \otimes_J J[1/\lambda] \simeq M_0 \otimes_J J[1/\lambda]$, which is a free $J[1/\lambda]$-module. In the case of G abelian this is the content of the generic flatness lemma.

Proof of 4.3.3 Since G is virtually polycyclic, it has a series $1 = G_0 \lhd G_1 \lhd \cdots \lhd G_h = G$, where G_{i+1}/G_i finite or cyclic. If $h = 0$, then G is trivial and M is a finitely generated J-module. The result follows from the structure theorem for finitely generated modules over principal ideal domains. Assume, then, that $h > 0$ and set $N = G_{h-1}$, $R = JG$, and $S = JN$. We split the proof into two cases.

Case 1: G/N is finite. Let T be a transversal to N in G. Since M is finitely generated as a JG-module, there exist $a_1, \ldots, a_k \in M$ such that $M = \sum_{i=1}^{k} a_i R$. Thus M is the finite sum $\sum_{i=1}^{k} \sum_{t \in T} a_i t S$ of cyclic S-modules $a_i t S$. Hence, M is a finitely generated S-module and so $M \in \mathcal{M}(J,\pi)$ for some finite set of primes π, by induction on h.

Case 2: G/N is infinite cyclic. Let $G = \langle t, N \rangle$ and set $H = \langle t \rangle$, so that $G = H \ltimes N$. Since M is finitely generated as an R-module, there is a finitely generated S-module L, such that $M = LR$, so that $M = \sum_{i \in \mathbb{Z}} Lt^i$. Note that each Lt^i is an S-submodule of M since $N \lhd G$.

We now define two sequences of S-modules for each $n > 0$,

$$L_n^+ = \sum_{i=1}^{n} Lt^i \quad \text{and} \quad L_n^- = \sum_{i=-n}^{-1} Lt^i.$$

Put $V = \bigcup_{n=1}^{\infty} L_n^-$, which is the sum of all the Lt^i for negative i. Therefore $V \leq Vt \leq Vt^2 \leq \cdots$ and $\sum_{i=0}^{\infty} Vt^i = M$. The S-modules Vt/V and Vt^{n+1}/Vt^n are J-isomorphic by the mapping $at + V \mapsto at^{n+1} + Vt^n$. Moreover $Vt = V + L$, so $Vt/V \overset{S}{\simeq} L/V \cap L$, which belongs to $\mathcal{M}(J,\pi_1)$ for some finite set of primes π_1, by induction on h. Hence $M/V \in \mathcal{M}(J, \pi_1)$ by 4.3.2(iii), and so we need only show that $V \in \mathcal{M}(J, \pi_2)$ for some finite π_2.

Now V is the union of the chain $0 = L_0^- \leq L_1^- \leq \cdots \leq L_n^- \leq \cdots$. Since $L_{n+1}^- = Lt^{-(n+1)} + L_n^-$, we have

$$L_{n+1}^-/L_n^- \overset{S}{\simeq} Lt^{-(n+1)}/Lt^{-(n+1)} \cap L_n^-,$$

and this is J-isomorphic with $L/L \cap L_n^- t^{n+1} = L/L \cap L_n^+$. Clearly $L \cap L_1^+ \leq L \cap L_2^+ \leq \cdots \leq L \cap L_n^+ \leq \cdots$ is an ascending chain of S-submodules of L. Now S is right noetherian by 4.2.3—note that J is noetherian since it is a principal ideal domain. Hence $L \cap L_m^+ = L \cap L_{m+1}^+ = \cdots$ for some integer m. By the induction hypothesis each $L/L \cap L_n^+$ belongs to $\mathcal{M}(J, \pi_2)$ for some finite set of primes π_2. Therefore $L_{n+1}^-/L_n^- \in \mathcal{M}(J, \pi_2)$ and so, by 4.3.1(iii), we have $V \in \mathcal{M}(J, \pi_2)$. Finally, $M \in \mathcal{M}(J, \pi)$, where $\pi = \pi_1 \cup \pi_2$.   ∎

We are now in a position to prove Hall's version of Hilbert's Weak Nullstellensatz.

4.3.4 *If M is a simple module over the integral group ring of a finitely generated virtually nilpotent group G, then M is finite.*

Proof Let N be a nilpotent normal subgroup with finite index in G. It is easy to see that a simple G-module has a $\mathbb{Z}N$-composition series with finite length. Thus we may assume that $N = G$. Also we may clearly assume that G acts faithfully on M, so that G can be regarded as a group of automorphisms of M. Since M is simple, it is cyclic and so is finitely generated as a module. Therefore $M \in \mathcal{M}(\mathbb{Z}, \pi)$ for some finite set π of primes, by 4.3.3.

The simplicity of M as a module implies that M is characteristically simple as an abelian group. Thus M is either an elementary abelian p-group for some prime p or a vector space over $\mathbb{Q}$. However—and here we use the crucial 4.3.1(i) for the first time—the latter is impossible since $\mathbb{Q}$ cannot belong to $\mathcal{M}(\mathbb{Z}, \pi)$ for any finite π. We may therefore conclude that M *is an elementary abelian p-group for some prime p* and $Mp = 0$. (Notice that the argument so far only requires G to be virtually polycyclic.)

Let z be any element of $Z(G)$. We now use J to denote $F\langle t \rangle$, the group algebra of an infinite cyclic group $\langle t \rangle$ over the field of p elements F. Then J is a principal ideal domain and we may regard M as a JG-module by defining at to be az, for $a \in M$. This definition succeeds because z is central in G. Again M is certainly finitely generated as a JG-module and hence by 4.3.3 we have $M \in \mathcal{M}(J, \pi)$ for some finite set of primes π.

Since M is a simple JG-module, Schur's Lemma shows that the ring of JG-endomorphisms of M is a division ring. Its centre C is therefore a field, clearly of characteristic p, in which F can be embedded by identifying $u \in F$ with $u1$ in C. Now let S be the subring of C generated by F and z, so that $t \mapsto z$ determines a ring homomorphism α from $J = F\langle t \rangle$ onto S. Hence $S \simeq J/I$ for some ideal I of J.

If $I = 0$, then α is a monomorphism, which extends to a monomorphism $\alpha : K \to C$, where K is the field of fractions of J. Let $0 \neq a \in M$ and define $\theta : K \to M$ by $x\theta = (a)(x\alpha)$, $x \in K$. Then θ is a J-monomorphism embedding K as a J-submodule of M. We now apply 4.3.2(i) a second time to show that this is impossible.

It follows that $I \neq 0$ and $f(z) = 0$ for some $0 \neq f \in J$. Therefore $\langle z \rangle$ is finite and $Z(G)$ is torsion. Since G is a finitely generated nilpotent group, $Z(G)$ is finite. Hence G itself is finite by 1.2.21. We conclude that M is finite since $M = \langle ag \mid g \in G \rangle$, where $0 \neq a \in M$. ∎

As an important corollary to 4.3.4 we have:

4.3.5 *The chief factors of a finitely generated virtually metanilpotent group G are finite and its maximal subgroups are of finite index.*

Proof For the first part it is clearly sufficient to prove that a minimal normal subgroup N of such a group G is finite. Let $F = \mathrm{Fit}(G)$. Then F is nilpotent and G/F is virtually nilpotent. Since G is virtually soluble, we may assume N is abelian and so $N \leq F$. If $[N, F] \neq 1$, then $[N, F] = N$. But then $N = [N_r, F] \leq \gamma_{r+1}(F) = 1$ for some r, a contradiction. So $[N, F] = 1$ and N is a module for G/F. Hence N is a simple $\mathbb{Z}(G/F)$-module and so it is finite by 4.3.4.

For the second part let M a maximal subgroup of G and let F be as before. If $MF \neq G$, then $F \leq M$ and M/F is a maximal subgroup of the virtually nilpotent group G/F. Hence $|G : M| < \infty$. So we may assume that $MF = G$. Let r be least such that $MF^{(r)} = G$. Then $F^{(r+1)} \leq M$ and we may assume $F^{(r+1)} = 1$ and hence $A = F^{(r)}$ is abelian. Also $M \cap A \lhd G$ and if $M \cap A < X < A$ and $X \lhd G$, then $M < XM < G$, a contradiction. Consequently $A/M \cap A$ is a chief factor of G and hence it is finite. Thus $|G : M| = |A : M \cap A|$ is finite, as required. ∎

The Artin–Rees Lemma

In order to complete the proof of 4.3.1 we need a generalization of another famous result from commutative algebra, the *Artin–Rees Lemma*.

Suppose R is a ring with max $-r$, the maximal condition for right ideals, and M is a finitely generated R-module. Let I be an ideal of R and N a submodule of M. The Strong Artin–Rees Lemma says that *if R is commutative, there exists a positive integer m such that*

$$MI^n \cap N = (MI^m \cap N)I^{n-m} \text{ for all } n \geq m.$$

Putting $n = m + 1$, we deduce that there exists a positive integer n such that

$$MI^n \cap N \leq NI.$$

This statement is known as the Weak Artin–Rees Lemma.

What we need is a version of the Artin–Rees Lemma valid for $\mathbb{Z}G$, where G is a finitely generated nilpotent group. In order to explain this we must first develop some auxiliary ideas from ring theory.

Suppose R is a ring with identity. An element r is said to be *central* in R if $xr = rx$ for all $x \in R$. An ideal I of R is called *central* if it can be generated by central elements, so that $I = \sum_\lambda r_\lambda R = \sum_\lambda R r_\lambda$, where each r_λ is central in R. Clearly a central ideal commutes with every other ideal of R. More generally, an ideal I of R is said to be *polycentral* if there is a finite series of ideals of R

$$0 = I_0 < I_1 < \cdots < I_s = I$$

such that I_{r+1}/I_r is a central ideal of R/I_r for $r = 0, 1, \ldots, s - 1$. The least such s is called the *height* of the ideal I. We refer to such a series as an *R-central series in I*.

It is not surprising to find that polycentral ideals are plentiful in integral group rings of nilpotent groups. Indeed, suppose G is any group and that H a normal subgroup of G contained in $Z_c(G)$ for some c. Set $H_i = H \cap Z_i(G)$ and let $I_i = I_{H_i}$ denote the right ideal of $\mathbb{Z}G$ generated by all elements $x - 1$, $x \in H_i$.

Now each H_i is normal in G and it is not hard to see that the assignment $g \mapsto gH_i$, $g \in G$, determines a surjective ring homomorphism from $\mathbb{Z}G$ to $\mathbb{Z}(G/H_i)$ with kernel I_i. Thus I_i is a two-sided ideal of $\mathbb{Z}G$. Furthermore, the series

$$0 = I_0 \leq I_1 \leq \cdots \leq I_c = I_H$$

is a $\mathbb{Z}G$-central series in I_H, so that I_H is a polycentral ideal of $\mathbb{Z}G$. To see this, we need to show that I_{i+1}/I_i is a central ideal of $\mathbb{Z}G/I_i$ for $i = 0, \ldots, c - 1$. Suppose that $x \in H_{i+1}$ and $g \in G$. Then in $\mathbb{Z}G$ we have

$$(x - 1)g - g(x - 1) = xg - gx = gx([x, g] - 1).$$

Now $[x, g] \in H_i$, so that $[x, g] - 1 \in I_i$, and hence $(x - 1)g \equiv g(x - 1) \bmod I_i$, which gives us what we need.

It follows that, *if G is a nilpotent group and H is any subgroup of G, then I_H is polycentral in $\mathbb{Z}G$.* To complete the proof of 4.3.1 we need two important properties of polycentral ideals.

4.3.6 *Suppose that M is a right noetherian R-module, where R is a ring with identity. Let J be a sum of polycentral ideals of R each of which has some power annihilating M. Then $MJ^n = 0$ for some $n > 0$.*

Proof Since M is noetherian, $MJ = MI$, where I is a *finite* sum of polycentral ideals, say $I = I(1) + \cdots + I(s)$, and $MI(j)^r = 0$ for some $r > 0$ not dependent on j. Now set $m = s(r - 1) + 1$. Then I^m is the sum of all products of m $I(j)'s$, and in each product at least one $I(j)$ will occur at least r times, so that $MI^m = 0$. Now I, as a sum of finitely many polycentral ideals, is itself polycentral, and so there is a central series

$$0 = I_0 < I_1 < \cdots < I_h = I$$

of ideals of R. If $MI = 0$, then $MJ = 0$, and we are done, so assume $MI \neq 0$. Then, for some $k < h$, we must have

$$0 = MI_k < MI_{k+1}.$$

Now I/I_{k+1} is a polycentral ideal of R/I_{k+1} with height $h - k - 1 < h$. Hence the natural induction hypothesis on height, applied to the R/I_{k+1}-module M/MI_{k+1}, yields the existence of a positive integer t such that $MJ^t \leq MI_{k+1}$. Now I_{k+1}/I_k is central in R/I_k, so $I_{k+1}J \leq JI_{k+1} + I_k$. But $MI_k = 0$; therefore $UI_{k+1}J \leq UJI_{k+1}$ for every submodule U of M. Repeated application of this fact leads to the inclusion $MJ^{\ell t} \leq MI_{k+1}^\ell$ for all $\ell > 0$. Setting $\ell = m$ and recalling that $MI^m = 0$, we obtain $MJ^{mt} = 0$, so that we may take $n = mt$. ∎

In order to state the second and key property of polycentral ideals, we define for a right R-module M and a subset X of R the subset

$$^*X = \{a \mid a \in M, aX = 0\},$$

that is the set of all elements of M which are annihilated by X. The result we need is essentially another version of the Artin–Rees Lemma.

4.3.7 (Robinson 1974) *Let R be a ring with identity and M a right noetherian R-module. Then there is a positive integer $n > 0$ such that $MI^n \cap {}^*I = 0$ for every polycentral ideal I of R.*

Proof We establish the result in two stages, first proving the weaker assertion where n may depend on I. Suppose this result is false and the pair (M, I) provides a counterexample where I has least (polycentral) height h. The fact that M is noetherian allows us to make the further assumption that the pair $(M/U, I)$ is not a counterexample for any non-zero submodule U of M. Thus for such a U there exists $n > 0$, such that $MI^m \cap^* I \leq U$. From this it follows at once that non-zero submodules of M must intersect non-trivially.

Let $0 = I_0 < I_1 < \cdots < I_h = I$ be a central series for I and choose a non-zero central element $x \in I_1$. Then $0 \neq {}^*I \leq {}^*x$, and so $^*x \neq 0$. Since x is central, the mapping $a \mapsto ax^n$ is an R-endomorphism of M and its kernel $^*(x^n)$ is a submodule. Now M is noetherian, so the sequence $^*x \leq {}^*(x^2) \leq \cdots \leq {}^*(x^n) \leq \cdots$ is eventually stable and there is an $m > 0$ such that $^*(x^m) = {}^*(x^{m+1}) = \cdots$. Suppose that $a \in Mx^m \cap^* x$. Then $a = bx^m$ for some $b \in M$ and $0 = ax = bx^{m+1}$. Hence $b \in {}^*(x^{m+1}) = {}^*(x^m)$ and $a = bx^m = 0$. Therefore $Mx^m \cap^* x = 0$. But Mx^m and $^*x$ are submodules of M and $^*x \neq 0$. It follows that $Mx^m = 0$ and so $M(Rx)^m = 0$: for $(Rx)^m = Rx^m$ because x is central. Now I_1, as a central ideal, is a sum of ideals of the form Rx with x central. We may therefore apply 4.3.6 and deduce that $MI_1^\ell = 0$ for some ℓ.

Suppose we have shown that $MI^r I_1^{s+1} = 0$ for some integers r, s. Then $MI^r I_1^s$ is an R/I_1-module and I/I_1 is a polycentral ideal of R/I_1 with height $h - 1$. By minimality of h there is a $t > 0$ such that

$$0 = (MI^r I_1^s)I^t \cap^* I = MI^{r+t} I_1^s \cap^* I,$$

since I_1 is central. But $^*I$ is a non-zero submodule and so $MI^{r+t} I_1^s = 0$. Now we already know that $MI_1^\ell = 0$, so we can repeat the above argument to show that $MI^n = 0$, for some n. This contradiction completes the first step in the proof.

We finish the argument by showing that an n can be found which is independent of I. Suppose this is false for M but true for every proper image of M. Then, just as above, non-zero submodules of M must intersect non-trivially. Let I be any polycentral ideal. Then, as we have just shown, there exists an m, depending on I, such that $MI^m \cap {}^*I = 0$. This means that either $MI^m = 0$ or ${}^*I = 0$. Denote by J the sum of all the polycentral ideals of R which have a power annihilating M. By 4.3.6 $MJ^n = 0$ for some $n > 0$. It is then immediate that $MI^n \cap {}^*I = 0$ for *all* polycentral ideals I. ∎

(For a more general version of the Artin–Rees property see Section 7.2).
We now apply 4.3.7 to finitely generated abelian-by-nilpotent groups.

4.3.8 *Suppose G is a finitely generated group with a normal abelian subgroup A such that G/A is nilpotent. Then there is a positive integer m such that $\gamma_m(H) \cap Z(H) = 1$, and $Z_{m-1}(H) = Z_m(H)$ for every subgroup H such that $HA \triangleleft G$.*

Proof Set $R = \mathbb{Z}(G/A)$ and $I = I_{HA/A}$. Then I is polycentral in R since $HA/A \triangleleft G/A$ and G/A is nilpotent. Also G/A is polycyclic, so R is noetherian by 4.2.3. By 4.3.7 $AI^n \cap {}^*I = 0$ for some n, which translates into multiplicative notation as $[A, {}_nH] \cap C_A(H) = 1$.

Suppose G/A has nilpotent class c. Then $\gamma_{c+1}(H) \leq A$ and hence $\gamma_{c+1+n}(H) \cap Z(H) = 1$. Set $m = c + 1 + n$ to obtain the first part. Next

$$[Z_m(H), {}_{m-1}H] \leq \gamma_m(H) \cap Z(H) = 1,$$

so that $Z_m(H) = Z_{m-1}(H)$. ∎

The second part of 4.3.8 yields a bound on the upper central heights of certain subgroups of finitely generated abelian-by-nilpotent groups, and in the case of finitely generated metabelian groups on all subgroups. We shall develop this aspect of the theory much further in Chapter 8.

The first part of 4.3.8 now enables us to complete the proof of 4.3.1.

Proof of 4.3.1 Suppose G is a finitely generated abelian-by-nilpotent-by-finite group and let $1 \neq x \in G$. By Zorn's Lemma (or because G satisfies max $-n$ by 4.2.2) there is a normal subgroup K of G which is maximal with respect to not containing x. It suffices to prove that G/K is finite. In order to do this we may assume that $K = 1$, so that every non-trivial normal subgroup of G contains x and thus G has a unique non-trivial minimal normal subgroup, M say.

By hypothesis G has an abelian normal subgroup A such that G/A is virtually nilpotent. We may assume that $A \neq 1$: for otherwise G is nilpotent and hence residually finite. Therefore $M \leq A$ and M is a simple $\mathbb{Z}(G/A)$-module and so it is finite by 4.3.4. Hence $C = C_G(M)$ has finite index in G and thus C is finitely generated.

By 4.3.8 we see that $\gamma_m(C) \cap Z(C) = 1$ for some m. Since $M \leq Z(C)$, we have $\gamma_m(C) \cap M = 1$, and hence $\gamma_m(C) = 1$ by definition of M. It follows that C is a finitely generated nilpotent group: let its centre be Z. Then some positive power Z^n of Z is torsion-free, yet $Z^n \cap M = 1$ since M is finite. In consequence $Z^n = 1$, which implies that C, and hence G, is finite, a contradiction. ∎

In 1974, Segal proved the following more precise result about the residual finiteness of finitely generated abelian-by-nilpotent groups G.

4.3.9 *Let G be a finitely generated abelian-by-nilpotent group.*

 (i) *If G is torsion-free-by-finite, then for all but a finite number of primes p the group G is a finite extension of a residually finite p-group.*
(ii) *If G is an extension of an abelian q-group by a nilpotent group where q is a prime, then G is a finite extension of a residually finite q-group.*

From this one may easily deduce:

4.3.10 *Let G be a finitely generated group with an abelian normal subgroup A such that G/A is virtually nilpotent. Let π be the set of primes dividing orders of elements of A. Then there exists a finite set of primes σ such that for every prime $p \notin \sigma$ the group G is a finite extension of a residually (finite nilpotent $\pi \cup \{p\}$) group.*

Segal also showed that this result extends to the holomorphs of such groups. For an application to centralizer properties see 8.3 below. We remark that Segal deduces his theorem from the following module theoretic result, which again depends on certain Artin–Rees type properties of noetherian modules over finitely generated nilpotent groups.

4.3.11 *Suppose that Γ is a finitely generated nilpotent group and A is a noetherian $\mathbb{Z}_p\Gamma$-module where p is prime. Then there is a normal subgroup Δ of finite index in Γ such that*

$$\bigcap_{i=0}^{\infty} [A, {}_i\Delta] = 0.$$

We do not give the proofs here.

4.4   The Fitting and Frattini subgroups in finitely generated soluble groups

In his paper of 1961, P. Hall turned his attention to the Frattini subgroups of finitely generated soluble groups. The most natural question is whether the Frattini subgroup is nilpotent, but a classical result of Gaschütz on finite groups extends this question. Let G be any group; then the subgroup $\mathrm{FFrat}(G)$ is defined by the equation

$$\mathrm{FFrat}(G)/\mathrm{Frat}(G) = \mathrm{Fit}(G)/\mathrm{Frat}(G).$$

Gaschütz (1953) proved that if G is finite, then $\mathrm{FFrat}(G)$ is nilpotent, and therefore coincides with the Fitting subgroup, $\mathrm{Fit}(G)$. (See also 1.3.20).

Hall was also interested in the role of $HP(G)$, the *Hirsch–Plotkin radical* of G, which is defined to be the product of all the locally nilpotent normal subgroups of G. By the Hirsch–Plotkin Theorem the subgroup $HP(G)$ is locally nilpotent in any group G. As the culmination of his work, Hall proved the following theorem, which is our main objective in this section.

4.4.1 (P. Hall 1961) *Suppose that G is a finite extension of a finitely generated metanilpotent group. Then* $\mathrm{Fit}(G) = HP(G) = \mathrm{FFrat}(G) = Cch(G)$ *is nilpotent.*

Here, $Cch(G)$ is defined to be the intersection of the centralizers of the chief factors of G: the heart of the proof of 4.4.1 consists in showing that this subgroup is nilpotent.

Consider, for a moment, the case where G is a finitely generated abelian-by-nilpotent group and A is an abelian normal subgroup of G with $\Gamma = G/A$ finitely generated and nilpotent. Then any chief factor U/V of G with $U \leq A$ is clearly a simple $\mathbb{Z}\Gamma$-module, and is therefore finite.

Armed with the residual finiteness of G, it is not hard to prove that $\mathrm{Frat}(G)$ *centralizes all chief factors U/V of G*, and hence $\mathrm{Frat}\,(G) \leq \mathrm{Cch}\,(G)$. For G/V is residually finite and, since U/V is finite, there exists a normal subgroup N of finite index in G such that $N \cap U = V$. Set $F = \mathrm{Frat}(G)$. Then $FN/N \leq \mathrm{Frat}(G/N)$ and so, by Gaschütz's Theorem, F/N centralizes the chief factor UN/N of the finite group G/N. Hence $[F,U] \leq N \cap U = V$ and F centralizes U/V, as claimed.

The harder part is to show that $C = \mathrm{Cch}\,(G)$ is nilpotent. Let Δ be the subgroup of Γ, which centralizes each chief factor of G inside A, so that in particular Δ centralizes every simple image of A. Since Γ is nilpotent, C is the preimage of Δ under the natural map $G \to \Gamma$. The nilpotency of C amounts to saying that some $[A, {}_n C]$ is trivial, or in module notation, that $A I_C^n = 0$. From these remarks it is clear that the desired result will follow in this case if we can prove that any ideal I of $\mathbb{Z}\Gamma$ which annihilates every simple image of A has some power annihilating A itself.

In order to get a general idea of how this might be proved, let us specialize to the case where Γ is finitely generated abelian and $A = \mathbb{Z}\Gamma$, a ring that we denote by R. Let G be the natural semidirect product of Γ and R, so that G is the wreath product of $\mathbb{Z}$ and Γ.

Suppose that M/N is a chief factor of G with $M \subseteq R$. Then M/N is a simple R-module and hence is R-isomorphic with R/T, where T is a maximal ideal of R. By Hilbert's Weak Nullstellensatz R/T is finite. Conversely, if T is any maximal ideal of R, then R/T is a chief factor of G. Also the centralizer of M/N corresponds to T and the intersection of the centralizers of such chief factors corresponds to the intersection of the maximal ideals of R, that is, to the Jacobson radical $\mathrm{Jac}(R)$ of R.

Now by Hilbert's Strong Nullstellensatz the Jacobson radical of any ring image of $k[x_1, \ldots, x_n]$, with k a field, is nilpotent: it is not difficult to deduce from this that $\mathrm{Jac}(R)$ and hence $\mathrm{Cch}\,(G)$ is nilpotent.

Thus Hall's Theorem 4.4.1 requires an analogue of the Strong Nullstellensatz for the integral group ring R of a finitely generated nilpotent group. In fact

Hall showed that $\operatorname{Jac}(R) \cap Z$ is nilpotent, where Z is the centre of R, or more generally:

4.4.2 *Suppose G is a finitely generated nilpotent group, $R = \mathbb{Z}G$ and M is a finitely generated R-module. Let z be a central element of R which annihilates every simple image of M. Then some power of z annihilates M.*

Proof Since M is noetherian, we may assume that the result is false for M, but true for every proper R-image of M. Then $^*z = \{a \in M \mid az = 0\}$ is a submodule since z is central. If $^*z > 0$, then $Mz^n \leq {}^*z$ for some n, so that $Mz^{n+1} = 0$, a contradiction. So $^*z = 0$ and the R-endomorphism $a \mapsto az$ is injective and $M \overset{R}{\simeq} Mz$.

The next step is to construct an R-module $\bar{M}$ for which the mapping $a \mapsto az$ is an R-automorphism. Let M_1 be an R-module isomorphic with M by means of the assignment $a \mapsto a_1$. Then $a \mapsto a_1 z$ is an injective R-homomorphism from M into M_1 with image $M_1 z$. Thus M can be embedded in M_1 as $M_1 z$. Iteration of the process yields a sequence of R-modules $M = M_0, M_1, \ldots$ and embeddings $M_i \to M_{i+1}$, where the image of M_i is $M_{i+1} z$. Let $\bar{M}$ be the direct limit of this sequence.

On making the appropriate identifications, we identify $\bar{M}$ as the union of an ascending chain

$$M = M_0 \leq M_1 \leq M_2 \leq \cdots.$$

Since $\bar{M}z \geq M_{i+1}z = M_i$, it follows that $\bar{M} = \bar{M}z$, so that $a \mapsto az$, $a \in \bar{M}$, is an automorphism of $\bar{M}$, as required,

Of course, $\bar{M}$ need not be finitely generated as an R-module. However, we can recover a noetherian property by enlarging the group G. If we form the direct product $\bar{G}$ of G with an infinite cyclic group $\langle x \rangle$, then $\bar{M}$ becomes a $\bar{G}$-module on setting $ax = az$, $a \in \bar{M}$. Compatibility with the action of G is ensured by the centrality of z. Now $M_i = M_{i+1}z$, so that $M_i = M_0 x^{-i}$. Since $\bar{M}$ is generated by the M_i, it follows that $\bar{M}$ is finitely generated as an $\bar{R}$-module, where $\bar{R} = \mathbb{Z}\bar{G}$. Since $\bar{G}$ is a finitely generated nilpotent group, $\bar{M}$ is $\bar{R}$-noetherian.

Now let L be a maximal $\bar{R}$-submodule of $\bar{M}$. Then $\bar{M}/L$ is finite by 4.3.4. Now L cannot contain M; for otherwise it would contain $M_i = Mx^{-i}$ and therefore $\bar{M}$. Hence $L \cap M \neq M$ and $M/L \cap M$ is a non-trivial finite abelian group. Moreover, $L \cap M$ is an R-module and is therefore contained in a maximal R-submodule N of M. By hypothesis, $Mz \leq N$, so that the endomorphism action induced by z on $M/L \cap M$ is not surjective. Since $M/L \cap M$ is finite, this endomorphism is not injective and so there exists $b \in M \backslash L$ such that $bz \in L \cap M$. But then $(b + L)z = L$, which is impossible since $a \mapsto az$, as an automorphism of $\bar{M}$, induces an automorphism of $\bar{M}/L$. This contradiction completes the proof. ∎

Proof of 4.4.1 We assume that G is a finitely generated metanilpotent group— the extension of the result to finitely generated metanilpotent-by-finite groups is not difficult and is left to the reader. Let N be a nilpotent normal subgroup with G/N nilpotent. Then $N \leq \operatorname{Fit}(G) = F$, say. Since G/N satisfies max, F is the product of finitely many normal nilpotent subgroups of G, so it is nilpotent, by Fitting's Theorem 1.2.9. The next step is to prove that the Hirsch–Plotkin radical H of G is contained in $\mathrm{Cch}\,(G)$. Clearly $F \leq H$. In order to show that $H \leq F$, we first establish that H centralizes a minimal normal subgroup A of G. This is obvious if $H \cap A = 1$, and so we may assume that $A \leq H$.

Next A is finite by 4.3.5, so there is a chief factor of H of the form A/B. Then A/B is contained in the centre of H/B since H is locally nilpotent, and hence $[A, H] < A$. But $[A, H] \triangleleft G$, so that $[A, H] = 1$ by the minimality of A. Thus $H \leq \mathrm{Cch}\,(G) = X$, say.

Suppose now that X is not nilpotent. Then X/F is a non-trivial normal subgroup of the nilpotent group G/F and therefore X/F contains a non-trivial element zF of $Z(G/F)$. Now G/F' is a finitely generated abelian-by-nilpotent group and so it satisfies $\max -n$ by 4.2.2. Hence $M = F/F'$ is a noetherian $\mathbb{Z}(G/F)$-module. If N is a maximal submodule of M, then $M(z-1) \leq N$, because $z \in X$. By 4.4.2 there is an $n > 0$ such that $M(z - 1)^n = 0$, which, translated into group theoretical language, means that $\langle z, F \rangle/F'$ is nilpotent. By Hall's nilpotency criterion 1.2.17, the subgroup $\langle z, F \rangle$ is nilpotent. But $\langle z, F \rangle \triangleleft G$ since zF is central. Consequently $\langle z, F \rangle = F$, which implies that $z \in F$, a contradiction. It therefore follows that $X = F$ and so X is nilpotent. ∎

So far we have proved that $\mathrm{Fit}(G) = HP(G) = Cch(G)$. To complete the proof of 4.4.1, a simple result is needed.

4.4.3 *If N is a nilpotent normal subgroup of a group G, then $N' \leq \mathrm{Frat}(G)$.*

Proof of 4.4.1 (concluded). It remains to show that $Y = \mathrm{FFrat}(G)$ coincides with $F = \mathrm{Fit}(G)$: obviously $F \leq Y$, so we need to prove the converse inclusion. Suppose this has been done for N abelian. Then by 4.4.3 we have $N' \leq \mathrm{Frat}(G)$, so that $\mathrm{Frat}(G/N') = \mathrm{Frat}(G)/N'$ and $Y/N' \leq \mathrm{FFrat}(G/N') = \mathrm{Fit}(G)/N'$. Hence Y/N' is nilpotent and, since N is nilpotent, Hall's criterion implies that Y nilpotent and $Y \leq F$, as required.

We may therefore assume that N is abelian. Hence G is a finitely generated abelian-by-nilpotent group and thus is residually finite by 4.3.1. Since $\mathrm{Fit}(G) = Cch(G)$, all we need do is show is that Y centralizes an arbitrary minimal normal subgroup L of G. Recall that L is finite by 4.3.4. Since G is residually finite, there is a normal subgroup V of finite index in G such that $L \cap V = 1$. Applying the fact that 4.4.1 holds for finite groups to G/V, we conclude that $YV/V \leq \mathrm{Fit}(G/V)$. Also $L \overset{G}{\simeq} LV/V$, so LV/V is minimal normal in G/V. By the finite case again, YV/V centralizes LV/V, so that $[Y, L] \leq L \cap V = 1$.

Finally, we give the brief proof of 4.4.3. Suppose that M is a maximal subgroup of G which does not contain N'. Then $G = MN'$, which is easily seen to imply that $G = M\gamma_{i+1}(N)$ for all $i \geq 0$. Choosing i to be the nilpotent class of N, we obtain the contradiction $G = M$. ∎

We remark that it is not hard to extend the proof of 4.4.3 to deal with the case where N is hypercentral (see P. Hall 1961).

We saw in 1.3.12 that a non-nilpotent polycyclic group has a finite non-nilpotent image. This result can now be extended to all finitely generated soluble groups.

4.4.4 (Robinson 1970) *Suppose that G is a finitely generated soluble group which is not nilpotent. Then G has a finite non-nilpotent image.*

Proof Suppose the result is false and take G to be a counterexample of smallest derived length. Let A be the smallest non-trivial term of the derived series of G. Then A is abelian and G/A is nilpotent, so that G satisfies max $-n$. Passing to a suitable image of G, we may further assume that every proper image of G is nilpotent.

Next Fit$(G) = F$ is nilpotent, and if $F' \neq 1$, we would have G/F' nilpotent, so that G is nilpotent by Hall's criterion. By this contradiction $F' = 1$ and G/F is nilpotent since $F \neq 1$. Let $1 \neq zF$ be an element of the centre of G/F.

Let M be a maximal G-submodule of F. Then F/M is finite by 4.3.4. If $M = 1$, the group G is polycyclic and 1.3.12 yields a contradiction. Hence $M > 1$ and G/M is nilpotent. But F/M is a minimal normal subgroup of G/M, so that we get $[F, G] \leq M$. Then $[F, z] \leq M$ and by 4.4.2 it follows that $[F, {}_nz] = 1$ for some n. This means that $\langle F, z \rangle$ is nilpotent. But $\langle F, z \rangle \lhd G$, so $z \in F$, a contradiction. ∎

This result readily extends to finitely generated hyper-(abelian or finite) groups. It also allows us to extend several nilpotency criteria known to hold for finite groups.

4.4.5

(i) *If G is a group whose Frattini subgroup is finitely generated, then* Frat(G) *is nilpotent if and only if it is soluble.*

(ii) *If G is a finitely generated soluble group and every maximal subgroup of G is normal, then G is nilpotent.*

(iii) *If G is a finitely generated soluble group with a triple nilpotent factorization, that is, $G = AB = BC = AC$, where A,B,C are nilpotent, then G is nilpotent.*

The truth of the theorem follows from 4.4.4 and the corresponding results for finite groups (for which see Robinson 1996, 5.2).

4.5 Counterexamples

We now present some counterexamples which establish limits to the validity of Hall's theorems. By 4.1.3 any countable abelian group A may be embedded as the centre of a finitely generated centre-by-metabelian group G. Thus, if we take A to be a group of type p^∞, then G is a finitely generated soluble group of derived length 3, which is not residually finite. Also G does not satisfy max $-n$. Thus Hall's residual finiteness theorem 4.3.1 does not extend to the class of finitely generated soluble groups of derived length 3. We shall give a further example due to P. Hall (1959).

4.5.1 *There exists a finitely generated torsion-free soluble group G of derived length 3, satisfying* max $-n$, *which is not residually finite. Furthermore, G has a minimal normal subgroup which is a rational vector space of countably infinite dimension. Also G has a maximal subgroup of infinite index.*

Proof Take V to be a vector space of countably infinite dimension over $\mathbb{Q}$ with basis $\{v_j : j \in \mathbb{Z}\}$. For each $j \in \mathbb{Z}$ choose a prime p_j, such that $p_r \neq p_s$ for $r \neq s$ and every prime occurs among the p_j. Define linear operators α, β on V by the rules

$$(v_j)\alpha = v_{j+1} \quad \text{and} \quad (v_j)\beta = p_j v_j, \quad j \in \mathbb{Z}.$$

Clearly α and β are invertible and, if we put $\beta_i = \alpha^{-i}\beta\alpha^i$, then

$$(v_j)\beta_i = p_{j-i}v_j,$$

for all i, j in $\mathbb{Z}$. It follows that β^H is free abelian with the β_j forming a basis, where $H = \langle \alpha, \beta \rangle$. Thus H is metabelian, and in fact H is isomorphic with $\mathbb{Z}\,wr\,\mathbb{Z}$. Now form the semidirect product

$$G = H \ltimes V.$$

Then clearly $G = \langle v_0, \alpha, \beta \rangle$.

Suppose that U is any non-zero H-invariant additive subgroup of V. Since $(v_j)\alpha = v_{j+1}$, the subgroup U contains a non-zero element of the form

$$u = a_0 v_0 + a_1 v_1 + \cdots + a_r v_r,$$

where $a_i \in \mathbb{Q}$, $a_0 \neq 0$, $r \geq 0$. Suppose u is chosen in U with r minimal and assume $r > 0$. Then U contains the element

$$p_r u - (u)\beta = a_0(p_r - p_0)v_0 + \cdots + a_{r-1}(p_r - p_{r-1})v_{r-1},$$

which contradicts the minimality of r, since $p_r \neq p_0$. Hence $u = a_0 v_0$, where $0 \neq a_0 \in \mathbb{Q}$.

We now show $U = V$. Let p be any prime and n any integer. Then $p = p_j$ for some j by hypothesis and U contains

$$(a_0 v_0)\alpha^j \beta^n \alpha^{-j} = a_0(v_j)\beta^n \alpha^{-j} = (a_0 p_j^n v_j)\alpha^{-j} = a_0 p^n v_0.$$

It follows that U contains all $q v_0$ for $q \in \mathbb{Q}$. Since $(v_0)\alpha^j = v_j$, we deduce that $U = V$.

We conclude that V is a minimal normal subgroup of G. Since G/V is metabelian and finitely generated, it is residually finite by 4.3.1. Because V is contained in every subgroup of finite index in G, we see that V is the finite residual of G.

The fact that $G = HV$ and the minimality of V in G is easily seen to imply that H is a maximal subgroup of G, and obviously $|G : H|$ is infinite. Finally,

G/V satisfies max $-n$ by 4.2.2 and hence, since V is minimal normal, G itself must satisfy max $-n$. ∎

For examples of finitely generated soluble groups of finite rank which are not residually finite, see 5.3.15 below.

P. Hall (1961) adapted the construction of B. H. Neumann and H. Neumann (described in 4.1.1) for embedding a countable soluble group in a 2-generator soluble group to prove:

4.5.2 *There are finitely generated soluble groups of derived length* 3 *with non-nilpotent Frattini subgroups.*

We shall outline Hall's adaptation of the Neumann–Neumann construction before giving the detailed argument. We begin with an arbitrary countable group $H = \{a_0, a_1, \ldots\}$ and write C for the cartesian power $\mathrm{Cr}H^{\mathbb{Z}}$, which we think of as consisting of all sequences $h = (h_n)$, $h_n \in H$. Multiplication of sequences h and k is defined pointwise, that is, $(hk)_n = h_n k_n$, and the shift operator t is defined by $(t^{-1}ht)_n = h_{n+1}$. Then, in the natural way, we may form the semidirect product S of $\langle t \rangle$ and C.

A particular sequence u defined by

$$u_{2^r} = a_r, \quad r \geq 0, \quad \text{and} \quad u_n = 1,$$

if n is not of the form 2^r. Next define a subgroup G of S by

$$G = \langle u, t \rangle.$$

If we write $B = u^G$, then $G/B \cong \mathbb{Z}$. Furthermore, if $\mathbf{V}$ is any variety of groups containing H, then C, and hence B, is in $\mathbf{V}$, and $G \in \mathbf{VA}$. It turns out that B', which is of course normal in G, consists of all restricted sequences, that is, those with only a finite number of non-trivial entries, in H'. Hence B' is isomorphic with $\mathrm{Dr}H'^{\mathbb{Z}}$, the direct product of countably many copies of H', irrespective of the way in which the generators a_n of H are chosen. It also emerges that H is always a homomorphic image of B.

4.5.3 (P. Hall 1961) *Let $H = \langle a_i \mid i \in \mathbb{Z} \rangle$ and let u, t, G and B be defined as above. Then:*

(i) $B' = \mathrm{Dr}(H')^{\mathbb{Z}}$;

(ii) H *is a homomorphic image of* B.

If in addition we assume that $\langle a_i \rangle \cap H' = 1$ and a_i has finite order d for all $i \geq 0$, then

(iii) $B/B' \cong \mathrm{Dr}(\mathbb{Z}_d)^{\mathbb{Z}}$.

Proof Clearly $B = \langle u^{t^m} \mid m \in \mathbb{Z} \rangle$, so that B' is generated by the commutators $c = [u^{t^\ell}, u^{t^m}]$, $\ell < m$, together with their conjugates in B. Now $(u^{t^\ell})_n = u_{\ell+n}$ by definition of t, so that $c_n = [u_{\ell+n}, u_{m+n}]$. It follows that $c_n = 1$ unless there exist integers r, s, such that $\ell + n = 2^r$ and $m + n = 2^s, (0 \leq r < s)$. Given ℓ and m, these equations have a solution (r, s) for at most one value of n since they imply that $m - \ell = 2^r(2^{s-r} - 1)$. For this unique value of n, if it exists, we get $c_n = [a_r, a_s] \in H'$. Thus all commutators c, and their conjugates in B, are sequences with at most one non-trivial component, and that component belongs to H'. It follows at once that $B' \leq \mathrm{Dr}(H')^{\mathbb{Z}}$.

Now let H'_n denote the nth direct factor of $\mathrm{Dr}(H')^{\mathbb{Z}}$, consisting of all sequences h with $h_n \in H'$ and $a_j = 1$ for $j \neq n$. Given $0 \leq r < s$, we define ℓ and m by $\ell + n = 2^r$ and $m + n = 2^s$. Then, writing $c = [u^{t^\ell}, u^{t^m}]$, we have $c_n = [a_r, a_s]$, with other components of c trivial, so that $c \in H'_n$. Also $c \in B'$. Therefore it follows that $H_n \leq B'$ and hence $B' \geq \mathrm{Dr}(H')^{\mathbb{Z}}$, so (i) is proved.

To establish (ii), we let π_1 be the projection of the cartesian product C onto its 1-component. Then π_1 maps $u^{t^{2^r-1}}$ to a_r, so that $B^{\pi_1} = H$ and (ii) is proven.

Finally, we establish (iii). Assume that $\langle a_i \rangle \cap H' = 1$ and a_i has order d for all $i \geq 0$. Since $B = \langle u^{t^m} \mid m \in \mathbb{Z} \rangle$, each element of B is congruent modulo B' to one of the form

$$y = t^{-m_1} u^{\ell_1} t^{m_1} t^{-m_2} u^{\ell_2} t^{m_2} \cdots t^{-m_k} u^{\ell_k} t^{m_k}, \quad k \geq 0, \qquad (*)$$

where $m_1 < m_2 < \cdots < m_k$ and $0 < \ell_i < d$, for $i = 1, 2, \ldots, k$. We interpret the case $k = 0$ to mean that $y = 1$.

Suppose that $k > 0$. Then the element $t^{-m_1} u^{\ell_1} t^{m_1}$ has infinitely many non-trivial components, namely

$$(t^{-m_1} u^{\ell_1} t^{m_1})_{2^r - m_1} = a_r^{\ell_1}, \quad r = 0, 1, 2, \ldots,$$

since $0 < \ell_1 < d$ and each a_r has order d. Note that these do not belong to H' since $\langle a_i \rangle \cap H' = 1$. By the proof of (i) there are at most $k - 1$ values of r for which the $(2^r - m_1)$th component of one of the other factors in the product $(*)$ differs from 1. Hence y has infinitely many components not in H' and by (i) we get $y \notin B'$. It follows that B/B' is the direct product of the cyclic groups of order d generated by the elements $u^{t^m} B', m \in \mathbb{Z}$, and (iii) is proved. ∎

In order to deduce 4.5.2 from this, we need an elementary result about Frattini subgroups.

4.5.4 (P. Hall 1961) *Suppose that G is any group and A is a normal abelian p-subgroup of G, where p is a prime. Then $A^p \leq \mathrm{Frat}(G)$.*

Proof Assume that M is a maximal subgroup of G not containing A^p and put $N = M \cap A$. Then $G = MA$ and hence $N \lhd G$. Thus the p-group A/N is a minimal normal subgroup of G/N, from which it follows that $A^p \leq N \leq M$ and the result is proven. ∎

Proof of 4.5.2 We apply the construction of 4.5.3, choosing H to be a group with non-nilpotent Frattini subgroup as follows. Let p, q be primes such that $q \equiv 1 \pmod{p^2}$, so that there is a primitive q-adic p^2th root of unity, say ω. Let

Q be a group of type q^∞: then the mapping $x \mapsto x^\omega$ is an automorphism α of Q and we can form the semidirect product H of $\langle \alpha \rangle$ and Q.

It is easy to prove that every maximal subgroup of H has finite index: therefore $\mathrm{Frat}(H) = \langle a^p, Q \rangle$. Since $\omega^p - 1$ is a unit in the ring of q-adic integers, we get $[Q_n, a^p] = Q_n$ for all n, so $\mathrm{Frat}(H)$ is not even locally nilpotent. We also see that H is generated by α and all of its conjugates, say $\alpha = \alpha_0, \alpha_1, \alpha_2, \ldots$, and these are all of order p^2. Since $[Q, a^p] = Q$, we see that $H' = Q$. By 4.5.3(iii) we obtain $B/B' \cong \mathrm{Dr}\,(\mathbb{Z}_{p^2})^{\mathbb{Z}}$ and $B' \cong \mathrm{Dr}\,Q^{\mathbb{Z}}$, in the obvious notation. Also G/B is infinite cyclic, so $G^{(3)} = 1$, and in fact the derived length is exactly 3.

We now apply 4.5.4 with B' for A to get $B' = (B')^q \le \mathrm{Frat}(G)$. Hence we have $\mathrm{Frat}(G/B') = \mathrm{Frat}(G)/B'$ and so $u^p \in \mathrm{Frat}(G)$, by 4.5.4 since $u \in B$. Hence $\mathrm{Frat}(G) \ge \langle B', u^p \rangle$. But by 4.5.3(i) the projection of $\langle B', u^p \rangle$ onto its first coordinate is $\langle H', a^p \rangle = \langle Q, a^p \rangle$, which is not locally nilpotent. Hence $\mathrm{Frat}(G)$ is not locally nilpotent. ∎

By varying the group H in 4.5.2, P. Hall (1961) was further able to prove:

4.5.5

(i) *For each $n \ge 1$ there is 2-generator group G with derived length $n + 2$ such that* $\mathrm{Frat}(G)$ *has derived length greater than n.*
(ii) *There exists a 2-generator soluble group of derived length 3, such that $HP(G)$ is not even a hypercentral group.*

4.6 Engel elements in soluble groups

In this final section of the chapter we shall describe some recent work on Engel elements in finitely generated soluble groups, and an important theorem on modules over finitely generated soluble groups on which it depends.

Recall that an element x of a group G is called a *right Engel element* if, for each g in G, there exists an integer $n = n(x, g) > 0$ such that $[x, {}_n g] = 1$. If n can be chosen independent of the element g, then x is called a *bounded right Engel element*. The sets of all right Engel elements and bounded right Engel elements are denoted respectively by, $R(G)$ and $\bar{R}(G)$.

It is easy to see that $R(G)$ contains the hypercentre $\bar{Z}(G)$ and $\bar{R}(G)$ contains the ω-hypercentre $Z_\omega(G)$ in any group G. In general these containments are both strict. For an account of Engel theory up to 1972 see Robinson (1972b).

Gruenberg (1959) showed that if G is a soluble group, then both $R(G)$ and $\bar{R}(G)$ are subgroups, and in a subsequent paper (Gruenberg 1961) he proved that:

4.6.1 *If G is a finitely generated soluble group, then $\bar{R}(G) = Z(G) = \bar{Z}(G)$.*

This result left open the status of the subset $R(G)$ in a finitely generated soluble group G. The situation was finally resolved by Brookes (1986), who proved:

4.6.2 *If G is a finitely generated soluble group, then all of $R(G)$, $\bar{R}(G)$, $\zeta(G)$, $\bar{\zeta}(G)$ coincide.*

Brookes deduced his theorem from a very interesting result about modules, which once again underscores the importance of module theory in the study of finitely generated soluble groups.

4.6.3 *Let G be a finitely generated soluble group and let M be a finitely generated $\mathbb{Z}G$-module. Assume that the cyclic $\mathbb{Z}\langle g \rangle$-module generated by a is finitely generated as a group for all $a \in M$, $g \in G$. Then M is finitely generated as a group.*

This module theoretic result has other applications, of which we mention one.

4.6.4 *Let G be a finitely generated soluble group. Then G is polycyclic if and only if $x^{\langle g \rangle}$ is finitely generated for all $x, g \in G$.*

Proof Only sufficiency of the condition is in doubt, so we assume it holds and proceed by induction on the derived length $d > 0$. Let $A = G^{(d-1)}$. Then G/A is polycyclic by induction hypothesis. Since G is finitely generated, it follows from 4.2.2 that G satisfies $\max -n$ and hence that A is a finitely generated $\mathbb{Z}G$-module. Moreover, by hypothesis $a^{\langle g \rangle}$ is finitely generated as a group for all $a \in A$, $g \in G$. Applying 4.6.3, we conclude that A is finitely generated and hence G is polycyclic. ∎

(Actually the proof only uses 4.6.3 in the case where G is polycyclic and the result is easy to prove directly).

5

SOLUBLE GROUPS OF FINITE RANK

While the term 'rank' has many connotations in algebra, in soluble group theory it refers to the cardinality of a maximal linearly independent subset of some kind. The foundations of the theory of soluble groups of finite rank were laid down in 1951 by Mal'cev in a famous paper (Mal'cev 1951).

We begin by introducing the various ranks of an abelian group and then use them to define the principal classes of soluble groups of finite rank. Thereafter we proceed to develop the properties of these groups.

5.1 The ranks of an abelian group

Let A be an abelian group, written additively for convenience. The *torsion-free rank* of A is defined to be

$$r_0(A) = \dim_{\mathbb{Q}}(A \otimes \mathbb{Q}),$$

the tensor product being formed over $\mathbb{Z}$. Suppose that X is a maximal linearly independent subset of elements of infinite order in A—which exists by Zorn's Lemma. Then $\langle X \rangle$ is a free abelian group and $A/\langle X \rangle$ is periodic. Since the functor $- \otimes \mathbb{Q}$ is left exact and $(A/\langle X \rangle) \otimes \mathbb{Q} = 0$, it follows that $\langle X \rangle \otimes \mathbb{Q} \simeq A \otimes \mathbb{Q}$ and thus

$$r_0(A) = \dim_{\mathbb{Q}}(\langle X \rangle \otimes \mathbb{Q}) = |X|.$$

So $r_0(A)$ is the cardinal of any maximal linearly independent set of elements of infinite order.

Next let p be any prime. Then the *p-rank* of the abelian group A is defined to be

$$r_p(A) = \dim_{\mathbb{Z}_p}(\mathrm{Hom}(\mathbb{Z}_p, A)).$$

Let S be the subgroup $\{a \in A \mid pa = 0\}$. Then $\mathrm{Hom}(\mathbb{Z}_p, A) = \mathrm{Hom}(\mathbb{Z}_p, S)$, and hence

$$r_p(A) = \dim_{\mathbb{Z}_p}(S).$$

These conclusions are summarized in:

5.1.1 *Let A be an abelian group. Then $r_0(A)$ is the cardinality of any maximal linearly independent subset of elements of infinite order, and if p is a prime, $r_p(A)$ is the cardinality of any maximal linearly independent subset of elements of order p.*

As a first step we identify the abelian groups with finite torsion-free rank and those with finite p-rank.

5.1.2 *Let A be an abelian group with torsion subgroup T.*

(i) *$r_0(A)$ is finite if and only if A/T is isomorphic with a subgroup of a finite dimensional rational vector space.*

(ii) *If p is a prime, $r_p(A)$ is finite if and only if the p-component A_p is the direct sum of finitely many cyclic or quasicyclic groups, that is, A_p satisfies min.*

Proof

(i) Since $r_0(A) = r_0(A/T)$, we may as well assume A to be torsion-free. Then A embeds in $A \otimes \mathbb{Q}$, a vector space with $\mathbb{Q}$-dimension $r_0(A)$, via $a \mapsto a \otimes 1$, $(a \in A)$. Conversely, if A is contained in a finite dimensional vector space V, then $A \otimes \mathbb{Q}$ embeds in $V \otimes \mathbb{Q} \simeq V$, and so $r_0(A) \leq \dim(V)$.

(ii) Here we may suppose A to be a p-group. Assume that $r = r_p(A)$ is finite and recall the standard result that if A is non-trivial, then $A = C_1 \oplus A_1$, where C_1 is either a p^∞-group or a non-trivial cyclic group (see Fuchs 1960: 24). Of course, $r_p(A_1) = r - 1$. The same procedure may be applied to A_1 and after r such steps, we will have $A = C_1 \oplus \cdots \oplus C_r \oplus A_r$, with C_i cyclic or quasicyclic; of course $A_r = 1$ since $r_p(A) = r$. The converse is clear. ∎

At this point it is as well to comment that torsion-free abelian groups of finite rank form a very complex class. Many classification schemes have been proposed for them, but none is wholly satisfactory. One indication of the scale of difficulty is the failure of the Krull–Schmidt Theorem, even for groups of rank 2. On the other hand, torsion-free abelian groups of rank 1 are just the non-trivial subgroups of the additive group of rational numbers $\mathbb{Q}$. These are called *rational groups* and they have been completely classified in Baer (1937): see also Fuchs (1960: 42).

Two further invariants of an abelian group, which are often useful are the *total rank*

$$r(A) = r_0(A) + \sum_{p=\text{prime}} r_p(A),$$

and the *reduced rank*

$$\hat{r}(A) = r_0(A) + \max_p \{r_p(A)\}.$$

It is a simple matter to identify the abelian groups for which these ranks are finite.

5.1.3 *Let A be an abelian group.*

(i) *$r(A)$ is finite if and only if A is the direct sum of finitely many cyclic and quasicyclic groups and a torsion-free group of finite rank.*

(ii) *$\hat{r}(A)$ is finite if and only if the torsion subgroup T is a direct sum of a boundedly finite number of cyclic p-groups and p^∞-groups for each prime p, and A/T has finite torsion-free rank.*

Proof Here the only comment necessary is that when $r(A)$ is finite, the torsion subgroup is a direct summand of A since it is the direct sum of a divisible group and a group of finite exponent—see Fuchs (1960: 24).

On the other hand, it should be noted that, even if $\hat{r}(A)$ is finite, the torsion subgroup need not be a direct summand of A. For example, let $C = \mathrm{Cr}_{i=1,2,\dots} \langle a_i \rangle$ and $D = \mathrm{Dr}_{i=1,2,\dots} \langle a_i \rangle$, where a_i has order p_i, with $p_1, p_2, \dots$ the sequence of primes. Obviously D is the torsion-subgroup of C. Now C/D is easily seen to be divisible (i.e. radicable), but plainly C is reduced. Thus C cannot split over D. ■

Classes of soluble groups of finite rank

We shall now survey the main types of soluble group which have finite rank in some sense (see Fig. 5.1)

1. *Groups with finite torsion-free rank.* In general a (not necessarily soluble) group G is said to have *finite torsion-free rank* if it has a series $1 = G_0 \triangleleft G_1 \triangleleft \cdots \triangleleft G_n = G$, such that each non-torsion factor is infinite cyclic. A routine application of the Schreier Refinement Theorem shows that the number of infinite cyclic factors is independent of the series: it is called either the *torsion-free rank* or the *Hirsch number* of G. Plainly, for abelian groups the definition amounts to $r_0(A)$ being finite, and so our terminology is consistent.

2. *Soluble groups with finite abelian ranks.* A soluble group G is said to have *finite abelian ranks*, or to be a *soluble FAR-group*, if there is a series $1 = G_0 \triangleleft G_1 \triangleleft \cdots \triangleleft G_n = G$, such that each factor G_{i+1}/G_i is abelian and $r_p(G_{i+1}/G_i)$ is finite for $p = 0$ or a prime. (Such groups have also been called $\mathcal{S}_0$-groups in Robinson 1968a.) Since the condition is inherited by subgroups and quotients, a soluble group is FAR precisely when each abelian section has finite torsion-free rank and finite p-rank for all primes p. This is the largest class of soluble groups with a finite rank restriction for which one can reasonably expect to be able to describe the group structure in any detail.

3. *Groups with finite Prüfer rank.* A group is said to have *finite Prüfer rank r* if every finitely generated subgroup can be generated by r elements and r is the least such integer. It is easy to show that an abelian group A has finite Prüfer rank precisely when the reduced rank $\hat{r}(A)$ is finite, and then $\hat{r}(A)$ equals the Prüfer rank.

We observe that *a soluble group has finite Prüfer rank if and only if it has a series with abelian factors of finite reduced rank.* To see this, note that the property of having finite Prüfer rank is extension closed: in fact, *if N and G/N have Prüfer ranks r and r' respectively, then G has Prüfer rank at most $r + r'$.* The proof is a simple exercise.

4. *Soluble groups with finite abelian total rank.* A soluble group has finite abelian *total rank*, or is a *soluble FATR-group*, if it has a series in which each factor is abelian of finite total rank. (Other notations that have been used are soluble A_3-groups, and $\mathcal{S}_1$-groups.) Such groups clearly have finite Prüfer rank.

Unlike the groups of 1–3, soluble FATR-groups do not form a quotient closed class, although the class is subgroup and extension closed.

5. *Minimax groups.* A group is called a *minimax group* if it has a series (of finite length) for which each factor satisfies max or min. This important class of groups has received much attention in the soluble case. The simplest example of a minimax group without max or min is the additive group of π-*adic rationals*,

$$\mathbb{Q}_\pi = \left\{ \frac{m}{n} \;\middle|\; m, n \in \mathbb{Z}, n = \text{a positive } \pi\text{-number} \right\},$$

where π is a finite set of primes. In fact these are the only torsion-free abelian minimax groups of rank 1, as may be deduced from the classification of subgroups of $\mathbb{Q}$.

Abelian minimax groups can be more simply described as follows.

5.1.4 *An abelian group A is minimax if and only if it has a finitely generated subgroup X such that A/X satisfies min. Thus an abelian minimax group is max-by-min.*

To see this, choose a finite set of generators for each factor with max in a max–min series and let the preimages of these elements in A generate a subgroup X. Then A/X satisfies min. The converse is clear.

Because of the structure of abelian groups with min, the subgroup X in the preceding argument may be chosen so that A/X is divisible, and hence is a direct sum of finitely many quasicyclic groups. Now suppose that Y is another such subgroup. Then $(X + Y)/X$ and $(X + Y)/Y$ are both finite, so that $A/X \simeq A/(X + Y) \simeq A/Y$. Therefore the finite set of primes $\pi(A/X)$ is independent of X. This set of primes is called the *spectrum* of A: these are the

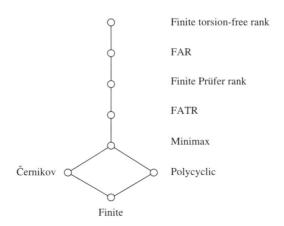

Figure 5.1 Diagram of classes of soluble groups.

primes p for which A has a p^∞-quotient. Also A is called a π-*minimax group* if $\pi(A/X) \subseteq \pi$. For example, $\mathbb{Q}_\pi$ has spectrum π.

The torsion subgroup of an abelian minimax group satisfies min and hence is a direct summand of the group. Thus the classification of abelian minimax groups reduces to the torsion-free case. However torsion-free abelian minimax groups form a complicated class and there are numerous examples of indecomposable, torsion-free, abelian minimax groups of given rank $r \geq 2$.

Example Let $\{v_1, v_2, \ldots, v_r\}$ be a basis for a $\mathbb{Q}$-space V and let $p_1, p_2, \ldots, p_r, q$ be distinct primes. Define A to be the subgroup of V generated by all elements of the following types:

$$\frac{v_1}{p_1^i}, \frac{v_2}{p_2^i}, \ldots, \frac{v_r}{p_r^i}, \frac{v_1 + v_2}{q^i}, \ldots, \frac{v_1 + v_r}{q^i}$$

where $i = 1, 2, \ldots$. Then A is a torsion-free abelian minimax group of rank r, and it can be shown that A is indecomposable (see Fuchs 1970: vol. 2, p. 124).

We shall be interested in *soluble* minimax groups, which evidently form a subgroup, quotient, and extension-closed class. If one forms a series with abelian factors in a soluble minimax group G, it is easy to see that the union of the spectra of the factors is independent of the series. Call this set the *spectrum of* G: it is just the set of primes p for which G has a factor of type p^∞.

There is a further invariant of a soluble minimax group G that is sometimes useful, its minimality. It is clear that G has a series in which every infinite factor is cyclic or quasicyclic. The *minimality* of G, $m(G)$, is defined to be the number of infinite factors in the series: this is independent of the series by the Schreier Refinement Theorem. Thus G is finite if and only if $m(G) = 0$. Minimality can be thought of as an generalization of the Hirsch length of a polycyclic group.

Weak maximal and minimal conditions

Soluble minimax groups can be neatly characterized in terms of weak versions of max and min. A group G is said to satisfy the *weak maximal condition*, wmax, if there are no infinite ascending chains of subgroups $G_1 < G_2 < \cdots$ in which each index $|G_{i+1} : G_i|$ is infinite. Replacing 'ascending' by 'descending', we obtain the dual *weak minimal condition*, wmin.

These finiteness conditions were introduced by Zaĭcev (1968a, 1969a), and also by Baer (1968), who established:

5.1.5 *The following properties of a soluble group G are equivalent:*

(i) G *is a minimax group;*
(ii) G *satisfies wmax;*
(iii) G *satisfies wmin.*

Proof Since wmax and wmin are extension closed properties, it is enough to prove the equivalence of the three properties for an abelian group G. Suppose G

satisfies wmax or wmin. Then G cannot have free abelian subgroups of infinite rank, that is, $r_0(G)$ is finite. Let X be a free abelian subgroup with rank $r_0(G)$. Then G/X is torsion and obviously $\pi(G/X)$ must be finite. Further, G/X cannot have an infinite elementary abelian p-subgroup. Hence $r_p(G/X)$ is finite and by 5.1.2 the group G/X has min. Thus G is minimax.

Conversely, suppose that G is minimax, so that by 5.1.4 it has a finitely generated subgroup X such that G/X has min. If $G_1 \leq G_2 \leq \cdots$ is an ascending chain of subgroups of G, then $G_i \cap X = G_{i+1} \cap X$ and $|G_{i+1}X : G_iX|$ is finite for sufficiently large i, by the structure of G/X. This implies that each $|G_{i+1} : G_i|$ is finite. Therefore G has wmax and by a similar argument it has wmin.   ∎

The following result shows how the various classes of soluble groups under discussion occur naturally as 'poly' classes. (Recall that if **X** is a class of groups, a poly-**X** group is a group with a series of finite length all of whose factors are in **X**.)

5.1.6 *Let G be a soluble group. Then:*

(i) *G has finite Prüfer rank if and only if it is poly-(locally cyclic).*
(ii) *G has FATR if and only if G is poly-(cyclic, quasicyclic, or rational).*
(iii) *G is minimax if and only if it is poly-(cyclic or quasicyclic).*

This follows at once from the definitions.

Some further weak chain conditions

We mention in passing that the weak chain conditions defined above can also be applied to particular types of subgroup, such as normal subgroups or subnormal subgroups. Thus a group G satisfies the *weak maximal or minimal condition for normal subroups*, wmax $-n$ or wmin $-n$, if there are no infinite ascending (respectively descending) chains of normal subgroups in which the index of each term of the series in the next one is infinite. The weak maximal and minimal conditions for subnormal subgroups, wmax $-s$ and wmin $-s$ are defined in a similar fashion manner. These conditions were introduced by Kurdachenko (1979, 1981).

5.1.7

(i) *For soluble groups the conditions wmax $-s$ and wmin $-s$ are equivalent to minimax.*
(ii) *For nilpotent groups the conditions wmax $-n$ and wmin $-n$ are equivalent to minimax.*

Proof If G is soluble with wmax $-s$ or wmin $-s$, then each factor of an abelian series satisfies wmax or wmin respectively. That G is minimax follows at once from 5.1.5. The same argument may be applied to a central series to prove (ii).   ∎

Kurdachenko (1981) has established a generalization of 5.1.7(i) to groups with a subnormal series of general order type.

It is a much harder problem to understand the structure of soluble groups with wmax $-n$ or wmin $-n$, but some results have been obtained. Zaĭcev and Kurdachenko (1982) have shown that if G is metabelian and satisfies wmin $-n$, then G' has a G-invariant series of finite length whose factors satisfy either wmax or wmin for G-invariant subgroups, while G_{ab} is an abelian minimax group. This has stimulated interest in the structure of artinian or noetherian modules over abelian minimax groups—see Kurdachenko and Tushev (1985). Karbe and Kurdachenko (1988) proved that torsion-free residually finite hyperabelian groups with wmin $-n$ are minimax. However, it still seems to open whether a torsion-free soluble group with wmin $-n$ need be a minimax group. Finally, we mention that Karbe (1987) has shown that a soluble torsion group with wmin $-n$ satisfies min $-n$, while metanilpotent torsion groups with wmin $-n$ are countable.

Quite recently, there has been interest in groups which satisfy the maximal condition on subgroups of infinite order or the minimal condition on subgroups of infinite index. Of course, these properties can also be applied to the normal subgroups or subnormal subgroups. They tend to be stronger than the weak-chain conditions discussed above and the consequent effect on the group structure is considerable, especially for soluble groups: for details see Paek (2001, 2002) and the forthcoming article (de Giovanni *et al.* 2004).

The role of linear groups

In any investigation of soluble groups of finite rank the theory of soluble linear groups is likely to be a useful tool, as is easily explained. Let F be an abelian factor in a normal series of a soluble group G and write $\bar{G} = G/C_G(F)$. If F is torsion-free with finite torsion-free rank r, then $\bar{G}$ is $\mathbb{Q}$-linear of degree r since F embeds in $F \otimes_{\mathbb{Z}} \mathbb{Q}$. If F is elementary abelian p of rank r, then of course $\bar{G}$ is $\mathbb{Z}_p$-linear of degree r. Finally, if F is a direct product of r groups of type p^∞, the group $\bar{G}$ is R_p-linear of degree r, where R_p is the ring of p-adic integers: this is because the action of $\bar{G}$ extends to $\mathrm{Hom}(p^\infty, F) \simeq R_p \oplus \cdots \oplus R_p$, ($r$ factors). Thus results about soluble linear groups may be applied to the group $\bar{G}$.

Finally, it should be mentioned that many types of soluble group of finite rank—and even their holomorphs—have been shown to be linear. Of course, the basic result is the Auslander–Swan Theorem (3.3.1), which asserts that polycyclic groups are $\mathbb{Z}$-linear.

For example, Merzljakov (1970) has shown that the holomorph of a polycyclic group is $\mathbb{Z}$-linear. Kopytov (1968), Levič (1969b), and Merzljakov (1968b) have shown independently that soluble FATR-groups with finite torsion subgroups are linear over a field of characteristic 0.

There is a useful sharpened form of this last result for soluble minimax groups due to Wehrfritz (1974).

5.1.8 *Let G be a soluble FATR-group which is torsion-free-by-finite. Then G is $\mathbb{Q}$-linear. In addition, if G is a minimax group with spectrum π, then G is $\mathbb{Q}_\pi$-linear.*

This result has an important application to constructible soluble groups which is described in 11.2. Wehrfritz also showed that the conclusion of 5.1.8 applies to the holomorph of the group G (1973a, 1974). Taking π to be empty in this result, one recovers the theorem of Merzljakov cited above on the holomorph of a polycyclic group.

5.2 Structure theorems for soluble groups of finite rank

In this section we shall establish a number of fundamental theorems which, when taken together, give a good picture of the structure of soluble groups of finite rank. The first of these results is concerned with the widest class of groups under consideration, soluble groups with finite torsion-free rank.

Because the product of two normal torsion subgroups is always a torsion group, in any group G there is a unique largest normal torsion subgroup: it will be denoted here by $\tau(G)$. When G is soluble, the subgroup $\tau(G)$ is locally finite.

5.2.1 (Mal'cev 1951) *Let G be a soluble group with finite torsion-free rank. Then $G/\tau(G)$ has a characteristic series of finite length whose infinite factors are torsion-free abelian of finite rank.*

This theorem may be regarded as saying that the infinite torsion factors can be 'pushed down' to the foot of the series. It follows that *if G is a soluble group with finite torsion-free rank, then G is a soluble FATR-group if and only if $\tau(G)$ is a Černikov group.*

Proof of 5.2.1 Since the statement is certainly true for abelian groups, the derived length d of G may be assumed greater than 1. We write $A = G^{(d-1)}$ and argue by induction on d: there is clearly no loss in assuming that $\tau(G) = 1$, so A is torsion-free abelian of finite rank. Write $S/A = \tau(G/A)$ and $C = C_S(A)$. Then S/C is a $\mathbb{Q}$-linear torsion group. Thus S/C is finite by a well-known theorem of Schur (see Curtis and Reiner 1966). Next C/A is torsion and hence locally finite, while $A \leq Z(C)$. Therefore C' is locally finite, by another well-known theorem of Schur (see Robinson 1996). Because $\tau(G) = 1$, it follows that C is a torsion-free abelian group and clearly it has finite rank. Notice also that S and C are characteristic in G. By the induction hypothesis, the theorem is true for G/S and therefore it is true for G. ∎

Another important structure theorem is obtained when Mal'cev's Theorem on soluble linear groups—see 3.1.6—is applied to the linear groups inherently involved in a soluble group with finite total rank.

5.2.2 (Gruenberg 1961; Mal'cev 1951) *Let G be a soluble group with finite abelian total rank. Then $\mathrm{Fit}(G)$ is nilpotent and $G/\mathrm{Fit}(G)$ is abelian-by-finite. Thus G is nilpotent-by-abelian-by-finite.*

Proof We first show that $G/\mathrm{Fit}(G)$ is abelian-by-finite. From 5.2.1 and the observation that $\tau(G)$ is Černikov, we deduce that there is a normal series of G whose infinite factors are either torsion-free abelian of finite rank or direct products of finitely many quasicyclic p-groups for various primes p. Thus, if F is a typical infinite factor, $G/C_G(F)$ is a linear group over $\mathbb{Q}$ or the field of p-adic numbers. Applying Mal'cev's Theorem to each $G/C_G(F)$, we obtain a normal subgroup N of finite index such that N' acts nilpotently on every factor of the series (whether infinite or not). Therefore N' is nilpotent and $N' \leq \mathrm{Fit}(G)$, which shows that $G/\mathrm{Fit}(G)$ is abelian-by-finite.

To complete the proof we must show that $F = \mathrm{Fit}(G)$ is nilpotent. Now we may refine the series of the first paragraph to one whose factors are abelian and of three possible types: (i) torsion-free and $\mathbb{Q}$-irreducible with respect to G; (ii) an elementary abelian-p chief factor of G; (iii) a direct product of finitely many quasicyclic p-groups with every proper G-invariant subgroup finite: in this last case it is convenient to say that the factor is *p-adically irreducble*. Let A be the smallest non-trivial term in this series. Then $\mathrm{Fit}(G/A)$ is nilpotent, by induction on the length of the series, and so F/A is nilpotent. If N is any normal nilpotent subgroup of G, then $[A, {}_iN] = 1$, for some i. But then we must have $[A, N] = 1$ in each of the three cases, since $[A, N] \vartriangleleft G$. Therefore $[A, F] = 1$ and F is nilpotent. ∎

By a similar argument it may be shown that: *if G is a soluble group with finite Prüfer rank, then $\mathrm{Fit}(G)$ is hypercentral and $G/\mathrm{Fit}(G)$ is abelian-by-finite.* However, $\mathrm{Fit}(G)$ need not be nilpotent under these circumstances. For example, let $p_1, p_2, \ldots$ be the sequence of primes, define

$$G_i = \langle a_i, x_i \mid a_i^{p_i^{i+1}} = 1 = x_i^{p_i^i}, a_i^{x_i} = a_i^{1+p_i} \rangle,$$

and put $G = \underset{i}{\mathrm{D}}_{\mathrm{r}}\, G_i$. Here G is a metabelian torsion group with Prüfer rank 2, but $\mathrm{Fit}(G) = G$ is not nilpotent since G_i has nilpotent class $i + 1$.

Another significant structural feature of soluble FATR-groups is indicated in the next result. Recall here that by 5.2.1 a soluble group of finite torsion-free rank is FATR if and only if $\tau(G)$ is Černikov.

5.2.3 (Robinson 1972b, 10.3) *Let G be a soluble group with finite torsion-free rank such that $\tau(G)$ is finite. Then $G/\mathrm{Fit}(G)$ is polycyclic and abelian-by-finite.*

Proof By 5.2.1 there is a normal series in G whose infinite factors are torsion-free abelian of finite rank and also $\mathbb{Q}$-irreducible with respect to G. Let F be any infinite factor. Then $\bar{G} = G/C_G(F)$ is an irreducible $\mathbb{Q}$-linear group, so it is abelian-by-finite, by 3.1.6. Let $\bar{A} \vartriangleleft \bar{G}$ with $\bar{A}$ abelian and $\bar{G}/\bar{A}$ finite. Now $\bar{F} = F \otimes_{\mathbb{Z}} \mathbb{Q}$ is a completely reducible $\mathbb{Q}\bar{A}$-module. If $\bar{S}$ is any simple $\mathbb{Q}\bar{A}$-direct summand of $\bar{F}$, we have $\bar{S} \overset{\mathbb{Q}\bar{A}}{\simeq} \mathbb{Q}\bar{A}/I$, where I is a maximal ideal of $\mathbb{Q}\bar{A}$. Further the assignment $x \mapsto x + I$ determines an embedding of $\bar{A}_S = \bar{A}/C_{\bar{A}}(\bar{S})$ in the multiplicative group of the algebraic number field $\mathbb{Q}\bar{A}/I$. But the multiplicative group of an algebraic number field is the direct product of a free abelian group and a finite cyclic group (see Fuchs 1960: 76). Since G has finite torsion-free rank, it follows that $\bar{A}_S$ is finitely generated for all S; therefore $\bar{A}$ is finitely generated and $\bar{G}$ is polycyclic, for every factor F.

Finally, let $D = \bigcap_F C_G(F)$, where F is any factor of the series. Then G/D is polycyclic and D is certainly nilpotent, so $D \leq \mathrm{Fit}(G)$. Consequently $G/\mathrm{Fit}(G)$ is polycyclic; by 5.2.2 it is also abelian-by-finite.   ∎

On the other hand, even soluble minimax groups are not in general nilpotent-by-polycyclic, as the following example demonstrates.

Example Let $p > 2$ and q be distinct primes, and let A be a p^∞-group and $X = \mathbb{Q}_q$. Now X is isomorphic with a subgroup of the additive group of p-adic integers consisting of elements that are congruent to $1 \bmod p$ (Hasse 1980); hence X is isomorphic with a subgroup of the *multiplicative* group of p-adic integer units. Thus X acts faithfully on A. We put $G = X \ltimes A$, noting that $\mathrm{Fit}(G) = A$. Thus $G/\mathrm{Fit}(G)$ is not polycyclic, yet G is a metabelian minimax group.

Polyrational groups

A group is said to be *polyrational* if it has a series of finite length whose factors are rational groups, that is, they are isomorphic with subgroups of $\mathbb{Q}$. Polyrational groups are a very natural generalization of poly-infinite cyclic groups.

5.2.4 *A group G is polyrational if and only if it has a characteristic series of finite length whose factors are torsion-free abelian groups of finite rank.*

Proof Only the necessity of the condition is in doubt. Let G be polyrational with a series $1 = G_0 \triangleleft G_1 \triangleleft \cdots \triangleleft G_n = G$ such that G_{i+1}/G_i is rational. Define

$$\bar{G}_{n-1} = \bigcap_{\propto \in \mathrm{Aut}(G)} G_{n-1}$$

and note that $\bar{G}_{n-1}$ is characteristic and $G/\bar{G}_{n-1}$ is torsion-free abelian: also it has finite rank because G does. Now replace G_{n-1} by $\bar{G}_{n-1}$. By repeating this procedure, we obtain a series of the type sought.   ∎

The next theorem shows that at the heart of any non-torsion soluble group with finite torsion-free rank there lies a polyrational group. It is a generalization of 1.3.4.

5.2.5 (Robinson 1968a) *Let G be a soluble group with finite torsion-free rank such that $\tau(G)$ is finite. Then G has a characteristic polyrational subgroup of finite index.*

Proof By 5.2.1 there is a characteristic series whose infinite factors are torsion-free abelian groups of finite rank. There is a useful *factor shifting* technique for moving infinite factors down this series.

Suppose that $X < Y < Z$ are successive terms in the series, with Y/X finite and Z/Y torsion-free abelian of finite rank. Put $C = C_Z(Y/X)$. Then C is characteristic in G: also Z/C is finite and C/X is nilpotent of class ≤ 2 with torsion subgroup $(Y \cap C)/X$. Let ℓ be the exponent of Y/X. We claim that $C^{\ell^2} X/X$ is torsion-free. If $u, v \in C/X$, then $(uv)^{\ell^2} = u^{\ell^2} v^{\ell^2} [v, u]^{\binom{\ell^2}{2}} = u^{\ell^2} v^{\ell^2}$, since ℓ divides $\binom{\ell}{2}$ and $C' \leq Y$. Thus $C^{\ell^2} X/X$ *consists* of powers u^{ℓ^2}, $u \in C/X$. If u^{ℓ^2} has finite order, then $u \in Y/X$ and so $u^{\ell^2} = 1$.

Now put $\bar{Y} = C^{\ell^2} X/X$. Then $\bar{Y}$ is characteristic in G and is torsion-free abelian of finite rank, while $Z/\bar{Y}$ is finite. Replace Y by $\bar{Y}$ in the series. The effect of this procedure is to move an infinite factor to the left past a finite one. By finitely many applications of the technique we can move all the torsion-free factors down the series, thus producing a characteristic polyrational subgroup with finite index in G. ∎

A further comment on the factor shifting technique will prove useful later. With the notation of the proof, suppose that Y/X is a *central* factor of G. Then $C = Z$ and $\bar{Y} = Z^{\ell^2} X/X$. Also $|Z/\bar{Y}|$ is a $|Y/X|$-number. This means that *the factor shifting technique will not introduce new primes if each finite factor of the series is central.* This observation will be used several times in the sequel.

Note that as a consequence 5.2.1 and 5.2.5, *a soluble FATR-group is Černikov-by-polyrational-by-finite.* As another illustration of the importance of polyrational groups, we record a generalization of 1.3.8.

5.2.6 *Let G be a soluble group with finite torsion-free rank and assume that $\tau(G)$ is finite. Then G is isomorphic with a quotient of a polyrational group.*

Proof By 5.2.5 there is a normal polyrational subgroup of finite index in G. Now follow the method of proof of 1.3.8. ∎

Since by 5.2.3 polyrational groups are nilpotent-by-polycyclic, their quotients also have this property. It was noted above that not every soluble minimax group is nilpotent-by-polycyclic. Consequently, *not every soluble minimax group is a quotient of a polyrational group.*

The Prüfer rank of a polyrational group

Suppose that G is a polyrational group with a rational series of length r. Then G has Hirsch length r. Now by a remark on p. 85 the Prüfer rank of G is at most r. It is an interesting fact, due to Zaĭcev (1971a), that the Prüfer rank is exactly r.

5.2.7 *The Prüfer rank of a polyrational group equals its Hirsch length.*

To establish this theorem two preliminary results must first be proved.

5.2.8 (Zaĭcev 1971a; Robinson 1969) *A finitely generated soluble group G with finite abelian total rank is a minimax group.*

Actually the conclusion of 5.2.8 holds true for the wider class of soluble FAR-groups, as will be shown in Chapter 10. But the above form is sufficient for our present purpose.

Proof of 5.2.8 By 5.2.2 there is a nilpotent normal subgroup N such that G/N is abelian-by-finite. Now the tensor product argument for the lower central series—see 1.2.12—shows that N will be minimax if N/N' is. This observation allows us to assume that N is abelian. Therefore G satisfies $\max -n$ by 4.2.2. Then $\tau(G)$, being a Černikov group, must be finite and so we may suppose that $\tau(G) = 1$. Applying 5.2.1, we conclude that there is a normal series of G with finite length whose infinite factors are torsion-free abelian of finite rank and also $\mathbb{Q}$-irreducible with respect to G. Let the smallest non-trivial term of this series be A. Then A must be torsion-free abelian since $\tau(G) = 1$. Further G/A is minimax by induction on the Hirsch length.

Choose a maximal linearly independent subset of A and use it to form a $\mathbb{Q}$-basis of $A \otimes_{\mathbb{Z}} \mathbb{Q}$: this affords a $\mathbb{Q}$-linear representation of G. Now the generators of G and their inverses are represented by rational matrices which involve finitely many primes in the denominators of their entries. Also by $\max -n$ there exist $a_1, a_2, \ldots, a_k$ such that $A = a_1^G a_2^G \cdots a_k^G$. It follows that there is a *finite* set of primes π such that each element of A has a coordinate vector whose entries involve only primes in the set π in their denominators. Therefore A is isomorphic with a subgroup of the direct sum of $r_0(A)$ copies of $\mathbb{Q}_\pi$ and so A is a minimax group, as is G ∎

Next we establish the crucial result needed in the proof of 5.2.7.

5.2.9 (Zaĭcev 1971a) *Let G be a polyrational minimax group of Hirsch length r and let p be a prime which is not in the spectrum of G. Then there exist subgroups $K \triangleleft H \leq G$ such that $|G : H|$ is finite and H/K is elementary abelian of order p^r.*

Proof Let $1 = G_0 \triangleleft G_1 \triangleleft \cdots \triangleleft G_r = G$ be a rational series in G. The proof is by induction on $r > 0$. Writing $N = G_{r-1}$, we know from the induction hypothesis that there are subgroups $M \triangleleft L$ of N such that $|N : L|$ is finite and L/M is elementary abelian of order p^{r-1}. Since $|N : M|$ is finite, there is a $k > 0$ such that $N^k \leq M$. Also N/N^k is finite. Writing C for $C_G(N/N^k)$, we note that G/C is finite and $[N, C] \leq M$. Since CM/M is (finite central)-by-abelian of rank 1, it is abelian and $C^\ell M/M$ is torsion-free of rank 1 for some $\ell > 0$.

Next consider $C^\ell L/M$, which is the direct product $(C^\ell M/M) \times (L/M)$. Now p cannot belong to the spectrum of $C^\ell M/M$ since it is not in the spectrum of G, and consequently $C^\ell M/M \neq (C^\ell M/M)^p$. It follows from the direct product decomposition of $C^\ell L/M$ that $C^\ell L/C^{\ell p}M$ is elementary abelian of order p^r. Finally $|G : C^\ell L|$ is finite. ∎

Proof of 5.2.7 This is now easy. Let G be a polyrational group of Hirsch length r with a series $1 = G_0 \lhd G_1 \lhd \cdots \lhd G_r = G$, each G_{i+1}/G_i being a rational group. Choose $x_i \in G_{i+1} \backslash G_i$ and put $H = \langle x_1, x_2, \ldots, x_r \rangle$. Then $1 = H \cap G_0 \lhd H \cap G_1 \lhd \cdots \lhd H \cap G_r = H$ is a series with rational factors, which shows that H is polyrational with Hirsch length r. Also by 5.2.8 H is a minimax group and so it follows from 5.2.9 that H has an elementary abelian section of order p^r, (for almost all primes p). Certainly this means that H has a subgroup which can be generated by r and no fewer elements. Consequently, the Prüfer rank of G is exactly r. ∎

Before leaving the subject of polyrational groups we note a curious fact.

5.2.10 (Čarin 1954) *Let G be a group with a series of finite length whose factors are non-cyclic rational groups. Then G is nilpotent-by-finite.*

Proof By 5.2.2 and 5.2.3 $\mathrm{Fit}(G)$ is nilpotent and $G/\mathrm{Fit}(G)$ is polycyclic. Furthermore $G/\mathrm{Fit}(G)$ has a series whose factors are quotients of rational groups none of which are infinite cyclic. Hence $G/\mathrm{Fit}(G)$ is finite. ∎

Note the consequence of 5.2.10: *a poly-$\mathbb{Q}$ group is nilpotent*—for a generalization see 5.3.6 below.

The following is another application of 5.2.8.

5.2.11 *If G is a soluble group of finite torsion-free rank and $\tau(G)$ is finite, then $G = X \mathrm{Fit}(G)$, where X is a finitely generated minimax group.*

This follows directly from 5.2.3 and 5.2.8.

Chief factors and maximal subgroups

Many types of infinite soluble groups have the property that their chief factors are finite and their maximal subgroups have finite index. This has already been observed to hold for finitely generated virtually metanilpotent groups (4.3.5), and in fact it is also true for nilpotent-by-polycyclic groups—see 7.1. Soluble groups with FAR also enjoy these properties.

5.2.12 (Robinson 1968a) *Let G be a virtually soluble group with finite abelian ranks. Then chief factors of G are finite and maximal subgroups have finite index.*

Proof To prove the statement for chief factors it is enough to show that any minimal normal subgroup N is finite. If N is infinite, it must be a finite dimensional $\mathbb{Q}$-space and $\bar{G} = G/C_G(N)$ is an irreducible $\mathbb{Q}$-linear group. Now argue as in the first paragraph of the proof of 5.2.3 that $\bar{G}$ is finitely generated. This implies that N must be minimax by the proof of 5.2.8, which is clearly impossible.

Finally, let M be a maximal subgroup of G. There is a largest i such that $A = G^{(i)} \not\leq M$. Thus $G^{(i+1)} \leq M$ and we can assume that $G^{(i+1)} = 1$, that is, A is abelian. Then $G = MA$ and $M \cap A \lhd G$; further $A/M \cap A$ is a chief factor of G, so it is finite and $|G : M|$ is finite. ∎

In conclusion we mention that an extensive theory of soluble minimax groups has been developed by Robinson (1967c). Among the results obtained are: (i) detailed information on the patterns of factors in a series of shortest length with max or min factors in a soluble minimax group: (ii) bounds for the *minimax length*, that is, the length of such a shortest series. The results are strongest for nilpotent groups. For example, the exact value of the minimax length of the group $U_n(\mathbb{Q}_\pi)$ is computed. For details the reader should consult the work cited.

5.3   Residual finiteness of soluble groups of finite rank

A notable property of polycyclic groups is residual finiteness (see 1.3.10), a result which was first proved by Hirsch in 1954. Now there is no chance of proving the residual finiteness of soluble FAR-groups since there is an obvious obstacle: they can have subgroups of type p^∞ or $\mathbb{Q}$. The first main result in this section shows that the presence of such subgroups is the only obstacle to residual finiteness in a soluble FAR-group.

Recall from 2.1 that a group G is said to be *π-radicable*, where π is a set of primes, if each g in G is an mth power of an element of G for every positive π-number m. If G has no non-trivial π-radicable subgroups, G will be called *π-reduced*. (As usual, when π is the set of all primes, we suppress the π in this terminology.)

Our principal objective is to establish:

5.3.1 (Robinson 1968a) *Let G be a soluble group with finite abelian subgroup ranks. Then the finite residual of G coincides with the unique maximum radicable subgroup and this is nilpotent. Furthermore, if G is a minimax group, the finite residual is abelian.*

Since in a radicable group every element of infinite or p power order is contained in a subgroup of type $\mathbb{Q}$ or p^∞ respectively, we may deduce:

Corollary 5.3.2 *A soluble FAR-group G is residually finite if and only if it is reduced or, equivalently, it has no subgroups of type $\mathbb{Q}$ or p^∞ for any prime p.*

Before proving 5.3.1, we must study residual finiteness for abelian groups.

5.3.3 *Let A be an abelian group and π a set of primes. Assume that for each p in π the p-component A_p has finite exponent. Then A is residually finite-π if and only if it is π-reduced.*

Proof Only the sufficiency is in question, so we assume that A is π-reduced and let $I = \bigcap_m A^m$, where m ranges over all positive π-numbers. Note that A/A^m is a residually finite π-group: if we can show that I is π-radicable, it will follow that $I = 1$ and A is residually finite-π.

Let $x \in I$ and $p \in \pi$. Write p^t for the exponent of A_p. Now $x \in A^{p^i}$ for all i, so that $x = a_1^p = a_2^{p^2} = \cdots$, where each a_i is in A^{p^t}. Then $\left(a_i^{-1} a_{i+1}^p\right)^{p^i} = 1$ and yet A^{p^t} has no non-trivial p-elements. Consequently $a_i = a_{i+1}^p$ and $a_1 \in A^{p^i}$ for all i. Next let m be any positive π-number and write $m = p^j n$, where p does not divide n. Then $a_1^p = x \in I \leq A^m$, and, since A^{p^j}/A^m has no elements of order p, it follows that $a_1 \in A^m$ for all m and $a_1 \in I$. Therefore I is π-radicable. ∎

In particular, *a torsion-free abelian group is residually finite if and only if it is reduced.* However, the well-known abelian p-group with generators $a_1, a_2, \ldots$ and relations $a_1^p = 1$ and $a_1 = a_2^p = a_3^{p^2} = \cdots$, is an example of a reduced p-group that is not residually finite. Of course, this group has infinite rank.

Corollary 5.3.4 *Let A be an abelian group such that $r_p(A)$ is finite for all primes p in some set π. Then A is residually finite-π if and only if it is π-reduced.*

This follows directly since the hypotheses imply that the p-component of A is finite.

Torsion-free abelian groups of finite rank have stronger residual finiteness properties, as indicated below.

5.3.5 *Let A be a torsion-free abelian group of finite rank.*

(i) *If A is reduced, then it is residually finite-π for some finite set of primes π.*
(ii) *If A is a minimax group, then it is residually finite-p for all primes p not in the spectrum of A.*

Proof

(i) First recall that by 5.3.4 residually finite-π and π-reduced are the same property for A. Since A is reduced, there is a prime p such that $A \neq A^p$. Put $B = \bigcap_{i=1,2,\ldots} A^{p^i}$ and observe that, because A is torsion-free, A/B is torsion-free and hence B is p-radicable. Then $r_0(B) < r_0(A)$, so that by induction on $r_0(A)$ there is a finite set of primes π_1 such that B is π_1-reduced.

Now put $\pi = \pi_1 \cup \{p\}$ and let R denote the maximum π-radicable subgroup of A. Then $R = R^p$, which implies that $R \leq B$ and hence $R = 1$. Thus A is π-reduced.

(ii) Set $I = \bigcap_{i=1,2,\ldots} A^{p^i}$, where the prime p does not belong to π, the spectrum of A. Now A has a free abelian subgroup X such that A/X is a π-group. Further $A^{p^i} \cap X = X^{p^i}$ since $p \notin \pi$. Thus $I \cap X = \bigcap_{i=1,2,\ldots} X^{p^i} = 1$, so that I is torsion and therefore $I = 1$. Hence A is residually finite-p. ∎

Turning now to soluble FAR-groups, we begin by characterizing the radicable groups among them.

5.3.6 *Let G be a soluble group with finite abelian ranks. Then the following are equivalent:*

(i) *G has no proper subgroups of finite index;*
(ii) *$G = G^m$ for all $m > 0$;*
(iii) *G is radicable and nilpotent.*

Furthermore, if G is a torsion group, the conditions imply that G is abelian.

Proof Since G/G^m is finite, (i) and (ii) are equivalent, while (iii) certainly implies (i). So assume that G satisfies (i). Form a normal series in G such that each factor is either torsion-free abelian and rationally irreducible with respect to G or a direct product of abelian p-groups of finite rank. Let F be a factor of the series: we will argue that F is central in G. Suppose first that F is torsion. For each $m > 0$, the elements of F with order dividing m form a finite G-invariant subgroup and, since G has no proper subgroups of finite index, it must act trivially on F.

Next suppose that F is torsion-free. Then $G/C_G(F)$ is finitely generated and abelian-by-finite, by the proof of 5.2.3. Hence $G/C_G(F)$ is residually finite and $G = C_G(F)$. Thus again G acts trivially on F. It now follows that G is nilpotent.

Next G_{ab} is evidently radicable, having no proper subgroups of finite index. The tensor product argument—see 1.2.12—shows that each lower central factor of G is radicable. The final step is to prove that G is radicable by induction on the nilpotent class $c > 0$. Let $x \in G$ and $m > 0$. By induction hypothesis $x = g^m c_1$ for some $g \in G$ and $c_1 \in \gamma_c(G)$. But $\gamma_c(G)$ is radicable, so $c_1 = c^m$, where $c \in \gamma_c(G)$. Therefore $x = g^m c^m = (gc)^m$ and G is radicable.

Finally, we observe that *if G is any radicable nilpotent group, the elements of finite order belong to the centre.* Indeed, let $g \in Z_2(G)$ have order m. If x is any element of G, we may write $x = y^m$, where $y \in G$; then

$$[x, g] = [y^m, g] = [y, g^m] = 1$$

and thus $g \in Z(G)$. Therefore the torsion subgroup is contained in $Z(G)$ and the final statement follows.   ∎

The structure of radicable hypercentral groups has been described by Černikov (1946, 1948): see also Robinson (1972b: vol. 2).

An essential role in the proof of 5.3.1, and other related results, is played by the following fact, which may be regarded as partial compensation for the failure of reduced soluble FAR-groups to form a quotient closed class.

5.3.7 *Let G be a reduced soluble group with finite abelian subgroup ranks and let A be a maximal abelian normal subgroup of G. Then:*

(i) *G/A is reduced;*
(ii) *if G has finite abelian total rank, then so does G/A.*

Proof

(i) If G/A is not reduced, then 5.3.6 tells us that it has a non-trivial radicable nilpotent normal subgroup, say R_1/A. If the nilpotent class of R_1/A is c, put $R/A = \gamma_c(R_1/A)$; then R/A is radicable and abelian by the tensor product argument. Since G is reduced, A is residually finite by 5.3.4, and R must centralize each $A/A^m, m > 0$, since these quotients are finite. Hence $[A, R] = 1$. Let $r \in R$ be fixed: then the assignment $xA \mapsto [x, r]$ determines a homomorphism from the radicable group R/A to the reduced group A, and this must be trivial. It follows that R is abelian, which contradicts the maximality of A and shows that G/A is reduced.

(ii) Suppose that G/A is not a soluble FATR-group. Then $\tau(G/A)$ is infinite by 5.2.1. Since the Fitting subgroup of a soluble group contains its centralizer (1.2.10), we may deduce that $\mathrm{Fit}(\tau(G/A))$ is also infinite. The latter must be a direct product of infinitely many finite p-groups since G/A is reduced by (i). It follows that G/A has an infinite normal subgroup L/A which is abelian and torsion.

Consider the group A: it is the direct product of a finite group and a torsion-free abelian group of finite rank. It is an easy deduction that a torsion subgroup of $\mathrm{Aut}(A)$ is finite, using Schur's Theorem on the finiteness of torsion subgroups of $GL_n(\mathbb{Q})$. If we write

$$C = C_L(A),$$

then it follows that L/C is finite. In addition $A \le Z(C)$, so that $A = Z(C)$ by maximality of A. Since C/A is abelian, 1.2.20(i) shows that $\pi(C/A) \subseteq \pi(A)$, so that $\pi(C/A)$ is finite. It follows that C/A is finite, hence so is L/A, a final contradiction. ∎

We are now ready to undertake the proof of the main theorem.

Proof of 5.3.1 It is straightforward to show, using Zorn's Lemma, that G contains a unique largest subgroup R which has no proper subgroups of finite index—indeed such a subgroup exists in every group. By 5.3.6 the subgroup R is radicable and nilpotent, and in addition R is obviously contained in the finite residual of G. Moreover, if G is a minimax group, R cannot have subgroups isomorphic with $\mathbb{Q}$. Hence R is a torsion group and by 5.3.6 it is abelian. To complete the proof it is enough to show that G is residually finite under the hypothesis $R = 1$, that is, G is reduced.

Let d denote the derived length of G. Then we may suppose that $d > 1$ by 5.3.3. By Zorn's Lemma there is a maximal abelian normal subgroup A containing $G^{(d-1)}$ and, according to 5.3.7, the group G/A is reduced, so the procedure may be repeated for G/A. By d repetitions of the procedure we obtain a normal series of G,

$$1 = G_0 \lhd G_1 \lhd \cdots \lhd G_d = G,$$

such that each factor is a reduced abelian group. Put $N = G_{d-1}$. By induction on d we may assume that N is residually finite, and of course so is G/N. Now $\bigcap_{m>0} N^m = 1$; therefore it suffices to prove that G/N^m is residually finite for any $m > 0$. Consequently we may suppose that $N^m = 1$ and N is finite. Hence $G/C_G(N)$ is finite and clearly it is enough to show that $C_G(N)$ is residually finite. Therefore we may also suppose that $N \le Z(G)$, a result which implies that G is nilpotent of class 2.

Let $\pi = \pi(N)$; then the π-component G_π of G is finite since G/N is reduced. Also $\tau(G) = G_\pi \times G_{\pi'}$. Now let $1 \ne g \in G$: we have to exclude g from some normal subgroup of finite index in G, and in doing so we may assume that

$g \in N \leq G_\pi$ since G/N is residually finite. Therefore nothing is lost in supposing that $G_{\pi'} = 1$. Now we know that $\tau(G) = G_\pi$ is finite, so that G is a nilpotent FATR-group.

Apply 5.2.5 to obtain a torsion-free normal subgroup L with finite index in G. Then $N \cap L = 1$ and $g \notin L$, so the proof of 5.3.1 is complete. ∎

Some information about the structure of soluble FAR-groups which are torsion can be extracted from 5.3.1. Let G be such a group and let R be the subgroup of the proof of 5.3.1. Thus R is a radicable nilpotent torsion group. By 5.3.6, the group R is actually abelian. Also, since G/R is reduced, its Sylow subgroups are finite.

It follows that *a soluble FAR-group which is a torsion group is an extension of a radicable abelian group by a group with finite Sylow subgroups.* For an account of the structure of locally finite groups with finite Sylow subgroups see M. R. Dixon (1994).

There is a stronger residual finiteness theorem for soluble FATR-groups.

5.3.8 (Robinson 1968a) *Let G be a reduced soluble group with FATR. Then G is residually a finite π-group for some finite set of primes π.*

Proof Let d be the derived length of G. If $d \leq 1$, then G is abelian and the result follows from 5.3.5. Let $d > 1$ and proceed by induction on d. There is a maximal abelian normal subgroup A containing $G^{(d-1)}$. Since G/A is a reduced soluble FATR-group by 5.3.7, induction on d shows that G/A is residually finite-π_1 for some finite set of primes π_1. But by 5.2.5 the quotient group G/A has a normal polyrational subgroup with finite index, so we may assume G/A to be polyrational.

Next A is residually finite-π_2 for some finite set π_2, by 5.3.5. Define π_3 to be the set of all prime divisors of the group orders $|GL_r(p)|$, where $p \in \pi_2$ and r is the total rank of A: clearly π_3 is finite. Writing

$$\pi = \pi_1 \cup \pi_2 \cup \pi_3,$$

which is also finite, we will prove that G is residually finite-π.

Define

$$C = \bigcap_{p \in \pi_2} C_G(A/A^p).$$

Then G/C is a finite π_3-group by definition of π_3. If $1 \neq g \in G$, we may assume that $g \in A$, and hence $g \notin A^m$ for some positive π_2-number m. Now C centralizes A/A^p for all p in π_2, and this is easily seen to imply that A/A^m is contained in the hypercentre of C/A^m. Since C/A is polyrational, 5.2.4 implies that it has a *normal* series whose factors are torsion-free abelian of finite rank.

At this point we apply the factor shifting technique of the proof of 5.2.5 to the group C/A^m, the object being to move the torsion-free factors down the series past the finite hypercentral factor A/A^m. This results in a torsion-free normal subgroup L/A^m whose index in C/A^m is a π_2-number—see the remarks following the proof of 5.2.5. Thus $|G : L|$ is a π-number. Finally, $L \cap A = A^m$, so $g \notin L$ and hence G is residually finite-π. ∎

On the other hand, an abelian FAR-group may fail to be residually finite-π for any finite set of primes π: indeed we need look no further than the group $\bigoplus_p \mathbb{Z}_p$ for an example.

There is a still stronger residual theorem for soluble minimax groups.

5.3.9 *Let G be a reduced soluble minimax group and let p be any prime not in the spectrum of G. Then G has a normal subgroup of finite index, which is residually finite-p.*

Proof We may suppose G to be non-abelian by 5.3.5. Let A be a maximal normal abelian subgroup containing the smallest non-trivial term of the derived series of G. By induction on the derived length there is a normal subgroup L_1 of finite index such that L_1/A is residually finite-p. Note that $A \neq A^p$ since p is not in the spectrum of G. Put $L = C_{L_1}(A/A^p)$; thus G/L is finite. By 5.2.5 we may also assume that L/A is polyrational.

We claim that in fact L is residually finite-p. To this end let $1 \neq x \in L$ and note that we may assume $x \in A$. Hence $x \notin A^{p^i}$ for some i since by 5.3.5 the abelian group A is residually finite-p. Now we apply the factor shifting argument of the proof of 5.2.5 to the group L/A^{p^i}, moving finite p-factors up a suitable series, to produce $M/A^{p^i} \lhd L/A^{p^i}$, with M/A^{p^i} torsion-free and L/M a finite p-group. Finally, $x \notin M$ since $M \cap A \leq A^{p^i}$. ∎

In particular 5.3.9 shows that a polycyclic group has, for every prime p, a normal subgroup of finite index which is residually finite-p, a result of Šmelkin (1968). On the other hand, 5.3.9 is not true for soluble FATR-groups, as is shown by the group $\mathbb{Q}_{2'} \oplus \mathbb{Q}_{3'}$.

We mention also that, if G is a soluble FAR-group, then G/HP(G) is residually finite, where HP(G) is the Hirsch–Plotkin radical. The same conclusion holds for $G/\text{Fit}(G)$ provided that $\text{Fit}(G)$ is nilpotent: for these results see Robinson (1968a).

It is possible to tell if a soluble FAR-group is residually finite by inspecting the centre of its Fitting subgroup.

5.3.10 *Let G be a soluble group with finite abelian ranks. Then G is residually finite if and only if the centre of $\text{Fit}(G)$ is reduced.*

Proof Only the sufficiency is in doubt, so let $F = \text{Fit}(G)$ have reduced centre. Now this implies that every upper central factor of F is reduced. For, if $R/Z_1(F)$ is a radicable subgroup of $Z_2(F)/Z_1(F)$, the map $z \mapsto [z, x]$, where $z \in R$ and $x \in F$ is fixed, is a homomorphism from $R/Z_1(F)$ to $Z_1(F)$. Thus $[R, x]$ is a radical subgroup of $Z_1(F)$ and as such it must be trivial. Hence $[R, F] = 1$ and $R = Z_1(F)$, which establishes our claim.

Finally, by 5.3.1 a radicable abelian subgroup of G must be contained in F and thus is trivial. Therefore G is residually finite by 5.3.1. ∎

In fact it is sufficient for residual finiteness that the centre of the Baer radical be reduced (Robinson 1972*b*). However it does not seem to be known if the Baer radical of a soluble FAR-group can be larger than the Fitting subgroup.

Turning next to nilpotent groups, we record an interesting generalization of a theorem of Gruenberg—see 1.3.17.

5.3.11 *Let G be a torsion-free nilpotent group and let p be a prime. Then G is residually finite-p if and only if $Z(G)$ is p-reduced. If G is a reduced minimax group, then it is residually finite-p for all primes not in the spectrum of G.*

Proof Suppose that $Z(G)$ is p-reduced and let $Z_i = Z_i(G)$. Assuming Z_i to be p-reduced, we argue as in the last proof that each Z_{i+1}/Z_i is p-reduced. Thus induction on the nilpotent class may be used to show that G/Z_1 is residually finite-p. Let $1 \neq x \in Z_1$; then $x \notin Z_1^{p^i}$ for some i, since Z_1 is p-reduced and torsion-free (see 5.3.3). Now we use the factor shifting argument for $G/Z_1^{p^i}$ to get a torsion-free normal subgroup $L/Z_1^{p^i}$ with finite p-power index. Then $x \notin L$ since $L \cap Z_1 = Z_1^{p^i}$. The statement for minimax groups now follows readily from 5.3.5. ∎

We mention in passing that *if G is a torsion-free nilpotent minimax group, then $\bigcap_p G^p = 1$, where p ranges over the complement of the spectrum of G* (Robinson 1968*a*). When G is finitely generated, this fact is due to Higman—see 2.2.6.

Recall that a polycyclic group which is residually finite-p for infinitely many primes p is nilpotent, a result of Seksenbaev (1965), which was proved as 1.3.18. Generalizations of this to soluble groups of finite rank have been given by Robinson (1972*a*).

The importance of the finite quotients of a soluble minimax group is underscored by the next result.

5.3.12 (Robinson 1968*a*) *Let G be a reduced soluble minimax group. If every finite quotient group of G is nilpotent, then G is nilpotent.*

This theorem is reminiscent of previously obtained results for polycyclic groups (1.3.12) and finitely generated soluble groups (4.4.4).

Proof of 5.3.12 There is a normal series of G whose infinite factors are torsion-free abelian and reduced. Let A be the smallest non-trivial term of this series; then G/A is nilpotent by induction on the series length. Suppose first that A is finite. Then by 5.2.5 the group G has a torsion-free normal subgroup F with finite index and thus G embeds in the group $(G/A) \times (G/F)$, which is nilpotent.

Next assume that A is torsion-free and let p be a prime not belonging to the spectrum of A. Then A/A^p is finite and G/A^p is nilpotent by the first paragraph of the proof. If $r = r_0(A)$, we have $[A, {}_rG] \le A^p$ since $|A/A^p| \le p^r$. To complete the proof we show that

$$\bigcap_p A^p = 1.$$

There is a finitely generated subgroup X of A such that A/X is torsion and has no elements of order p, since p is not in the spectrum of A. Hence $X \cap A^p = X^p$, and consequently $\bigcap_p A^p$ meets X trivially. It follows that $\bigcap_p A^p$ is a torsion group and hence must be trivial. This implies that G is nilpotent. ∎

Recently, Detomi (2001) has established a generalized form of 5.3.12.

5.3.13 *Let G be a reduced soluble minimax group and suppose that every finite quotient of G has nilpotent length at most l. Then G has nilpotent length at most l.*

This was proved for polycyclic groups by Endimioni (1998). Since the finite residual of a soluble minimax group is abelian by 5.3.1, we may deduce from 5.3.13 the following result: *if a soluble minimax group has all its finite quotients of nilpotent class at most l, then the nilpotent class of G is at most $l + 1$.*

Frattini subgroups of soluble minimax groups

The result just proved as 5.3.12 has an application to Frattini subgroups of soluble minimax groups.

5.3.14 *Let G be a reduced virtually soluble minimax group. Then $\mathrm{FFrat}(G) = \mathrm{Fit}(G)$, and in particular $\mathrm{Frat}(G)$ is nilpotent.*

The argument here is the same as for 1.3.20, but makes use of 5.3.12. It should be observed that even soluble Černikov groups can have non-nilpotent Frattini subgroups, so it is not redundant that the group in 5.3.13 is assumed to be reduced. This is shown by a standard example, $G = \langle t \rangle \ltimes A$, where A is of type 2^∞, $|t| = 4$ and t acts on A via a 2-adic fourth root of unity. Then $\mathrm{Frat}(G) = \langle t^2, A \rangle$, which is a locally dihedral 2-group.

In addition, 5.3.14 is not valid for reduced soluble FATR-groups. To see this consider the group

$$G = \langle t \rangle \ltimes A,$$

where $A = \mathbb{Q}_{2'} \oplus \mathbb{Q}_{2'}$, $|t| = 4$ and t acts on A according to the matrix

$$\begin{bmatrix} 0 & 1 \\ -1 & 0 \end{bmatrix}.$$

For this group $\mathrm{Frat}(G) = G^2 = \langle t^2, 2A, A(t-1) \rangle$, which is not nilpotent.

We mention that Wehrfritz (1974) has shown 5.3.13 to be valid for the holomorph of a reduced soluble minimax group.

Finally, we give an important example of a finitely generated soluble minimax group which is not residually finite (Robinson 1968a).

5.3.15 *There is a finitely generated centre-by-metabelian minimax group with a central subgroup of type 2^∞.*

Proof Let A and B be two copies of the group $\mathbb{Q}_2$, written multiplicatively,

$$A = \langle a_1, a_2, \ldots \mid a_{i+1}^2 = a_i, i > 0 \rangle \quad \text{and} \quad B = \langle b_1, b_2, \ldots \mid b_{i+1}^2 = b_i, i > 0 \rangle.$$

Form the second nilpotent product N of A and B, that is,

$$N = A * B/[A, B, B][B, A, A].$$

Then N is nilpotent of class 2 and $Z(N) = N' = [A, B]$, which is is generated by all the $[a_i, b_j]$. Observe that

$$[a_i, b_{r-i}] = [a_{i+1}^2, b_{r-i}] = [a_{i+1}, b_{r-(i+1)}].$$

Thus $c_r = [a_i, b_{r-i}]$ depends only on r and the elements c_r generate N'. Also $c_{r+1}^2 = c_r$ and $N' \simeq \mathbb{Q}_2$.

Define an automorphism ξ of N by $a_{i+1}^\xi = a_i$ and $b_j^\xi = b_{j+1}$: thus $c_r^\xi = c_r$. Now write

$$\bar{G} = \langle \xi \rangle \ltimes N$$

and note that $N' \le Z(\bar{G})$. Let X be a cyclic subgroup of N' such that N'/X is of type 2^∞ and put

$$G = \bar{G}/X.$$

This is a 3-generator centre-by-metabelian minimax group with centre N'/X. ∎

Actually it may be shown that *there are uncountably many non-isomorphic finitely generated central extensions of a 2^∞-group by the group G/N'.* One way to see this is to notice that the cohomology group $H^2(G/N', N')$ is uncountable— for the role of H^2 in group theory see Section 10.1.

An interesting feature of the group G of 5.3.15 is that *G is not linear over any integral domain:* this is because of the theorem of Mal'cev (1940) that finitely generated linear groups are residually finite.

6

FINITENESS CONDITIONS ON ABELIAN SUBGROUPS

If a group is soluble, one would expect its abelian subgroups to play a more important role than they do in groups in general. Our object in this chapter is to confirm this expectation by considering the effect on the structure of a soluble group when various finiteness restrictions are imposed on its abelian subgroups. Before we start, it is as well to remember that there are infinite groups in which every proper non-trivial subgroup has prime order: these are the famed *Tarski monsters*—see Ol'šanskiĭ (1991). So in general one cannot expect to prove anything significant.

On the other hand, this phenomenon does not occur for soluble groups. To see why, let G be a soluble group in which every subnormal abelian subgroup is finite, and write $F = \mathrm{Fit}(G)$. If A is a maximal abelian normal subgroup of F, then A is finite, and also $A = C_F(A)$ by 1.2.8. Since $\mathrm{Aut}(A)$ is finite, it follows that F/A, and hence F, is finite. Finally, $C_G(F) \leq F$ by 1.2.10 and $G/C_G(F)$ is finite, whence so is G. So we have shown that *a soluble group all of whose subnormal abelian subgroups are finite must itself be finite.*

This simple result is emblematic of a large body of work in which various finiteness restrictions are placed on sets of abelian subgroups in soluble groups. The first results of this kind were obtained by Schmidt (1945) and Mal'cev (1951) for the minimal and maximal conditions, respectively. Subsequently important contributions were made by Baer, Heineken, Kargapolov, and others. The account presented here is based on the simplified general approach in Robinson (1976c).

6.1 Chain conditions on abelian subgroups

Our first objective is to consider groups whose abelian subgroups satisfy max, that is, are finitely generated. Here the main difficulty is to show that the property is inherited by suitable quotients of a group.

6.1.1 *Let A be an abelian normal subgroup of a group G. If each abelian subgroup of G satisfies max, then the same holds for G/A.*

Note that G itself need not be soluble here. From this we can deduce a famous theorem of Mal'cev.

Corollary 6.1.2 (Mal'cev 1951) *If every abelian subgroup of a soluble group satisfies max, the group is polycyclic.*

Once 6.1.1 is established, the corollary follows quickly by induction on the derived length d. For let $d > 1$ and put $A = G^{(d-1)}$. Then G/A is polycyclic by 6.1.1 and the induction hypothesis. Hence G is polycyclic. But notice that in a free group every abelian subgroup is cyclic, so the property that all abelian subgroups satisfy max is not inherited by quotients in general.

In order to establish 6.1.1, two preliminary results are needed, the first being elementary linear algebra.

6.1.3 *Let A be a torsion-free abelian group of finite rank and let $\theta : A \to A$ be an injective endomorphism. Then A/A^θ is finite.*

Proof By the Cayley–Hamilton Theorem, applied to θ regarded as a linear operator on the vector space $A \otimes \mathbb{Q}$, we have $f(\theta) = 0$ for some $f \neq 0$ in $\mathbb{Z}[t]$. Further, there is no loss in supposing that $m = f(0) \neq 0$, since θ is injective. It follows that $A^m \leq A^\theta$ and so A/A^θ is finite. ∎

The critical result needed to prove 6.1.1 and later theorems is a weak splitting theorem, asserting that a group splits over a normal subgroup 'up to finite index'. (See 10.1 for more on such splittings.)

6.1.4 *Let A be an abelian normal subgroup of a group G such that G/A is abelian and A is not central in G. Assume further that A is torsion-free with finite rank and is $\mathbb{Q}$-irreducible with respect to G. Then there is a subgroup X such that $X \cap A = 1$ and $|G : XA|$ is finite.*

Proof There is an element x of G such that $[A, x] \neq 1$. Thus the assignment $a \mapsto [a, x]$ yields a non-trivial G-endomorphism of A. Now $A/\operatorname{Ker}(\theta) \overset{G}{\cong} A^\theta$ and thus $A/\operatorname{Ker}(\theta)$ is torsion-free. Since $\operatorname{Ker}(\theta) \neq A$ and A is $\mathbb{Q}$-irreducible, we have $\operatorname{Ker}(\theta) = 1$ and θ is injective. Therefore A/A^θ is finite, of order m say, by 6.1.3. Put $C = C_G(A/A^\theta)$; then $C \triangleleft G$ and G/C is finite.

Next let $c \in C$. Then $[c, x] \in A$ since G/A is abelian, and thus $[c, x]^m \in A^\theta$. Now $[c^m, x] \equiv [c, x]^m \bmod [C, x, C]$ and

$$[C, x, C] \leq [A, C] \leq A^\theta = [A, x].$$

Hence $[c^m, x] \in [A, x]$ and we may write $[c^m, x]$ as $[a, x]$ for some $a \in A$. Thus $[c^m a^{-1}, x] = 1$. Now put $X = C_G(x)$, so that $c^m a^{-1} \in X$ and $c^m \in XA$. We have just shown that G/XA is a torsion group. Since $X \cap A = \operatorname{Ker}(\theta) = 1$, the proof will be complete if we can show that G/XA is finitely generated.

Choose a transversal to $[A, x]$ in A, say $\{a_1, a_2, \ldots, a_m\}$. For $j = 1, 2, \ldots, m$ choose, if possible, $g_j \in G$ such that $a_j = [g_j, x]$, but if no such element exists put $g_j = 1$. For any g in G we can write $[g, x] = a_i [b_i, x]$ for some i and $b_i \in A$; this implies that $[gb_i^{-1}, x] = a_i = [g_i, x]$. Consequently $gb_i^{-1} g_i^{-1} \in C_G(x) = X$ and $g \in \langle g_1, g_2, \ldots, g_m, XA \rangle$, which shows that G/XA is finitely generated. ∎

Proof of 6.1.1 The argument follows a pattern common to all the theorems of this type. Let B/A be an abelian subgroup of G/A. Suppose first that $[A, B] = 1$, so that B is nilpotent, and let M be a maximal abelian normal subgroup of B. Then $M = C_B(M)$ and $A \leq M$. For any b in B the assignment $xA \mapsto [x, b]$ yields a homomorphism from M/A to A which depends only on bM: let us call it $(bM)^\tau$. Then $\tau : B/M \to H = \mathrm{Hom}(M/A, A)$ is an injective homomorphism, and M/A and A are finitely generated abelian groups. It follows that H is also finitely generated abelian. Therefore B/M is finitely generated, and as M is finitely generated by hypothesis, B is finitely generated. Hence B/A is finitely generated. This part of the proof will be referred to as the *Hom argument*.

Next, consider the case where A is finite. Here $B/C_B(A)$ is finite and $A \leq Z(C_B(A))$, which means that the argument of the previous paragraph may be applied to show that $C_B(A)/A$, and hence B/A, is finitely generated.

We are now ready to tackle the general case. The torsion subgroup of A may be factored out on the basis of our discussion of the finite case, so assume A is torsion-free. Supposing the result to be false, we choose the pair (A, G) so that A is torsion-free with smallest rank subject to the existence of an abelian subgroup B/A of G/A that is not finitely generated. Clearly A must be $\mathbb{Q}$-irreducible with respect to B, by the minimality of rank. The weak splitting result 6.1.4 may now be applied to show that there is a subgroup X such that $|B : XA|$ is finite and $X \cap A = 1$. However X must be abelian and hence finitely generated. It now follows that B, and hence B/A, is finitely generated. The argument of this paragraph will be referred to as the *minimal rank argument*. ∎

We turn next to the corresponding theorem for the minimal condition.

6.1.5 *Let A be an abelian normal subgroup of a group G. If each abelian subgroup of G satisfies min, then the same holds for G/A.*

By induction on the derived length we rapidly deduce from this result:

Corollary 6.1.6 (Schmidt 1945) *A soluble group in which each abelian subgroup satisfies the minimal condition is a Černikov group.*

Like its predecessor, 6.1.5 depends on a weak splitting result: this time the result asserts that a group splits over a normal subgroup 'up to finite intersection'.

6.1.7 *Let A be an abelian normal subgroup of a group G such that G/A is abelian and A is not central in G. Assume further that A is a divisible p-group which is p-adically irreducible with respect to G. Then there is a subgroup X such that $G = XA$ and $|X \cap A|$ is finite.*

(Recall that A is called *p-adically irreducible* if every proper G-invariant subgroup of A is finite.)

Proof of 6.1.7 Choose x in G such that $[A, x] \neq 1$ and let θ be the G-endomorphism of A in which $a \mapsto [a, x]$. Since A^θ is divisible and G-invariant, $A = A^\theta = [A, x]$ by irreducibility. For any $g \in G$ we have $[g, x] \in [A, x]$ and hence $[g, x] = [a, x]$, where $a \in A$. Therefore $ga^{-1} \in X = C_G(x)$ and $g \in XA$.

It follows that $G = XA$. Finally $X \cap A = \mathrm{Ker}(\theta)$, which must be finite because $A/\mathrm{Ker}(\theta) \simeq A$.   ∎

It is now possible to prove 6.1.5, following the pattern established in the proof of 6.1.1: first the Hom argument, then the minimal rank argument.

Proof of 6.1.5 Let B/A be an abelian subgroup of G/A. Suppose first that $[A, B] = 1$ and let M be a maximal abelian normal subgroup of the nilpotent group B. Then, as before, there is an injective homomorphism $\tau : B/M \to H = \mathrm{Hom}(M/A, A)$. Now the hypothesis on G implies that it is a torsion group. Since M satisfies min, it will therefore suffice to prove that any torsion subgroup of H is finite. Suppose that $\alpha \in H$ has finite order. Then $\mathrm{Im}(\alpha)$ has finite exponent, which shows that α must act trivially on D/A, the divisible part of M/A. Now M/D has finite order, say l, since M satisfies min. Also $\mathrm{Im}(\alpha)$ is contained in the subgroup $\{x \in A \mid x^l = 1\}$, which is finite. Consequently, there are only finitely many possibilities for α, which completes the proof of the case where $[A, B] = 1$. The case where A is finite is dealt with quickly, as in the proof of 6.1.1. The Hom argument is now complete.

In the general case we may assume A to be divisible by factoring out a large enough finite G-invariant subgroup of A. Choose the pair (G, A) with A divisible and of smallest total rank subject to the existence of an abelian subgroup B/A without min. Plainly A is a p-group and it must be p-adically irreducible with respect to B. Applying 6.1.7, we obtain a subgroup X such that $B = XA$ and $X \cap A$ is finite. By the finite case already disposed of, $X/X \cap A$ satisfies min, whence so does B/A, and the proof is complete.   ∎

Minimax groups

Our final objective in this section is to investigate groups whose abelian subgroups are minimax groups i.e., satisfy weak chain conditions: see 5.1.5. The main result is:

6.1.8 *Let A be an abelian normal subgroup of a group G and let π be a set of primes. If each abelian subgroup of G is π-minimax, then the same holds for G/A.*

In the usual way we deduce from this result:

Corollary 6.1.9 (Baer 1968) *A soluble group in which each abelian subgroup is π-minimax is itself π-minimax.*

Notice that this result generalizes 6.1.2 and 6.1.6 since we can take π to be the empty set or else G to be a torsion group.

Proof of 6.1.8 In its structure the proof is similar to the proofs of the previous results 6.1.1 and 6.1.5: however the Hom argument needs to be supplemented because of the presence of Prüfer subgroups. Let B/A be an abelian subgroup of G/A.

Consider first the case $[A, B] = 1$, where most of the extra work is involved. Let D denote the divisible part of A. Our first move is to prove that G/D inherits the property on abelian subgroups. Let $\bar{B}/D$ be an abelian subgroup of B/D and consider its torsion subgroup B_0/D. By 6.1.5, we see that B_0 is a Černikov group and hence it is π-minimax. Thus B_0/D is a direct factor of $\bar{B}/D$, which shows that we may assume $\bar{B}/D$ to be torsion-free. Also it is easy to see that $\bar{B}/D$ will be π-minimax if all its countable subgroups are. Therefore it is permissible to assume that $\bar{B}$ is countable, with $\bar{B} = \{b_1, b_2, \ldots\}$ say.

Next D is a π_0-group for some finite subset π_0 of π. Since $\bar{B}/D$ is abelian and $[D, \bar{B}] = 1$, there are positive π_0-numbers m_{ij} such that

$$1 = [b_i, b_j]^{m_{ij}} = [b_i, b_j^{m_{ij}}].$$

Now define $m_1 = 1$ and $m_i = m_{1i}m_{2i}\cdots m_{i-1i}$ for $i > 1$, and observe that $Y = \langle b_1^{m_1}, b_2^{m_2}, \ldots \rangle$ is abelian and hence is π_1-minimax for some finite subset π_1 of π. Then YD is π_2-minimax, where $\pi_2 = \pi_0 \cup \pi_1 \subseteq \pi$. Also $\bar{B}/YD$ is torsion, while $\bar{B}/D$ is torsion-free and YD/D has finite rank, which implies that $\bar{B}/D$ has finite rank. Furthermore $\bar{B}/YD$ is a π_0-group with finite p-rank for all p, from which we deduce that $\bar{B}/YD$ satisfies min. Since YD/D is π-minimax, it follows that $\bar{B}/D$ is π-minimax.

The previous discussion allows us to factor out by D and assume that A has finite torsion subgroup, say T. Let M denote a maximal abelian normal subgroup of B. By the Hom argument it suffices to prove that $\operatorname{Hom}(M/A, A)$ is π-minimax. Put $\bar{M} = M/A$ and apply the functor $\operatorname{Hom}(\bar{M}, -)$ to the exact sequence $1 \to T \to A \to A/T \to 0$ to get the exact sequence.

$$0 \to \operatorname{Hom}(\bar{M}, T) \to \operatorname{Hom}(\bar{M}, A) \to \operatorname{Hom}(\bar{M}, A/T).$$

If t denotes $|T|$, then $\operatorname{Hom}(\bar{M}, T) \simeq \operatorname{Hom}(\bar{M}/\bar{M}^t, T)$, which is finite. Also an element of $\operatorname{Hom}(\bar{M}, A/T)$ is determined by its effect on a finite subset of $\bar{M}$ since $\bar{M}$ has finite abelian ranks and A/T is torsion-free. Consequently $\operatorname{Hom}(\bar{M}, A/T)$ is isomorphic with a subgroup of the direct product of a finite number of copies of A/T, and so it is π-minimax. Hence $\operatorname{Hom}(\bar{M}, A)$ is π-minimax. The Hom argument now goes through as before, so the case $[A, B] = 1$ is complete. As a consequence, we may factor out by a finite normal subgroup and assume that A is either a divisible p-group or torsion-free.

To finish off the proof we employ the minimal rank argument, choosing the pair (A, G), with $A \triangleleft G$ and A either a divisible abelian p-group or torsion-free abelian: further A is to have minimal rank subject to the existence of an abelian subgroup B/A of G/A which is not π-minimax. Clearly A must be either p-adically irreducible or $\mathbb{Q}$-irreducible with respect to G. Thus weak splitting occurs by 6.1.4 or 6.1.7. The remainder of the argument proceeds just as in 6.1.1 and 6.1.5. ∎

6.2 Finite rank conditions on abelian subgroups

In this section, we will consider groups whose abelian subgroups are subject to varying forms of finiteness of rank. First of all, we discuss groups in which each abelian subgroup A has all ranks finite, that is, $r_0(A)$ and $r_p(A)$ are finite for all primes p. Our first object is to establish:

6.2.1 *Let A be an abelian normal subgroup of a group G. If every abelian subgroup of G has all its ranks finite, then the same is true of G/A.*

This leads in the usual way to

Corollary 6.2.2 (Baer and Heineken 1972) *If each abelian subgroup of a soluble group G has all its ranks finite, then G is a soluble group with finite abelian ranks in the sense of Section 5.1.*

What is being asserted in the Corollary is that finiteness of ranks in a soluble group passes from abelian subgroups to abelian sections. The general method of 6.1 can be applied to prove 6.2.1, but it needs to be supplemented by two additional results.

6.2.3 *Let G be a soluble torsion group in which each abelian p-subgroup has finite Prüfer rank. Then p-sections of G have boundedly finite Prüfer rank and hence are Černikov groups.*

Proof The first step is to show that p-subgroups of G have boundedly finite Prüfer rank. If this is false, there is a sequence of finite p-subgroups $P_1, P_2, \ldots$ with unbounded ranks. Let $Q = \langle P_1, P_2, \ldots, P_{i-1} \rangle$, where $i > 0$, and suppose that Q is known to be a p-group. Since G is locally finite, some conjugate of P_i is contained in the same Sylow subgroup of $\langle Q, P_i \rangle$ as Q. Replacing P_i by this conjugate, we see that $\langle Q, P_i \rangle$ is a p-group. This argument demonstrates that we may assume that $P = \langle P_1, P_2, \ldots \rangle$ is a p-group. Now abelian subgroups of P have finite rank and hence satisfy min; thus P is a Černikov group by 6.1.6, a contradiction which establishes the claim.

To conclude the proof, let H/K be a p-section of G and consider a finite subgroup B/K of H/K. Then it is easy to show that $B = XK$ for some finite p-subgroup X. By the first paragraph, X has bounded Prüfer rank, whence so does H/K. ∎

The next result involves a very useful technique due to Mal'cev. (It has already appeared in the proof of 6.1.8 in a special case.)

6.2.4 (Mal'cev 1951) *Let T be an abelian normal torsion subgroup of a group G and assume that each primary component of T has finite rank. If G/T has a free abelian subgroup of countable rank, then G has a free abelian subgroup with the same rank.*

Proof Let $\{x_1 T, x_2 T, \ldots\}$ be a basis of a free abelian subgroup of G/T with countable rank r. Then $\langle x_i, x_j \rangle'$ is generated by conjugates of $[x_i, x_j]$ and, because all p-ranks of T are finite, it follows that $\langle x_i, x_j \rangle'$ has finite order, m say; hence there is a positive integer n such that x_j^n centralizes $\langle x_i, x_j \rangle'$. Consequently $[x_i, x_j^{mn}] = [x_i, x_j^n]^m = 1$. Therefore for each i and j there is an $\ell_{ij} > 0$ such that $[x_i, x_j^{\ell_{ij}}] = 1$.

Now put $\ell_1 = 1$ and $\ell_i = \ell_{1i} \ell_{2i} \cdots \ell_{i-1 i}$ if $i > 1$. Then $\langle x_1^{\ell_1}, x_2^{\ell_2}, \ldots \rangle$ is abelian and it is clearly free abelian of rank r. ∎

Proof of 6.2.1 Let B/A be an abelian subgroup of G/A. Suppose first that A is a torsion group. Then by 6.2.3 each $(B/A)_p$ is Černikov and thus $r_p(B/A)$ is finite for all primes p. If B/A had infinite torsion-free rank, it would have a free abelian subgroup of countably infinite rank, which is impossible by 6.2.4. It follows that B/A has finite torsion-free rank.

It is now enough to deal with the case where A is torsion-free. If $[A, B] = 1$, the Hom argument may be applied in the usual way. If $[A, B] \neq 1$, use the weak splitting theorem 6.1.4 and the minimal rank argument to complete the proof. ∎

Next we consider soluble groups whose abelian subgroups have finite total rank and show that these are just the finite abelian total rank (FATR)-groups in the sense of 5.1, IV.

6.2.5 (Čarin 1960) *Let G be a soluble group whose abelian subgroups have finite total rank. Then G has finite abelian total rank.*

Proof Let T denote the maximum normal torsion subgroup of G. Since each abelian subgroup of T satisfies min, T is a Černikov group by 6.1.6. Now we know that G is a soluble FAR-group by 6.2.2, and also by 5.2.1 the group G/T is a soluble FATR-group. It follows that G is a soluble FATR-group. ∎

There is no analogue of 6.2.1 for groups whose abelian subgroups have finite total rank, as is made clear by the groups $\mathbb{Q}$ and $\mathbb{Q}/\mathbb{Z}$.

Finally we turn to groups in which abelian subgroups have finite Prüfer rank. Here the results lie at a deeper level. The definitive result is:

6.2.6 (Kargapolov 1962) *Let A be an abelian normal subgroup of a group G. If each abelian subgroup of G has finite Prüfer rank, then the same is true of G/A.*

This has the immediate corollary:

Corollary 6.2.7 *If every abelian subgroup of a soluble group G has finite Prüfer rank, then G has finite Prüfer rank. In particular, if the abelian subgroups of a soluble group have finite Prüfer ranks, they have boundedly finite Prüfer ranks.*

Two additional facts are needed in the proof of 6.2.6.

6.2.8 *Let A be an abelian p-group with finite rank r. Then a Sylow p-subgroup of $\mathrm{Aut}(A)$ has Prüfer rank at most $\frac{1}{2}r(5r - 1)$.*

This useful result was first proved by Kargapolov (1962) in the finite case: the present version is due to Baer and Heineken (1972). A proof can also be found in Robinson (1972b: 7.4).

Corollary 6.2.9 *Let G be a soluble torsion group and let p be a prime. Assume that every abelian p-subgroup of G has finite rank. Then the p-ranks of the abelian subgroups have a finite upper bound r and each p-section of G has Prüfer rank at most $\frac{1}{2}r(5r + 1)$.*

Proof In the first place the bound r exists by 6.2.3. Let P be any p-subgroup of G; then P is Černikov by 6.1.6, and there is a maximal abelian normal subgroup M such that P/M is finite. Furthermore $M = C_P(M)$ by maximality of M. Now M has rank at most r and by 6.2.8 the rank of P/M is at most $\frac{1}{2}r(5r-1)$. Therefore P has Prüfer rank not exceeding $r + \frac{1}{2}r(5r-1) = \frac{1}{2}r(5r+1)$. From this it follows quickly that any p-section of G has Prüfer rank at most $\frac{1}{2}r(5r+1)$.   ■

The real key to 6.2.6 is:

6.2.10 *Let A be a normal subgroup of a soluble torsion group G. Assume that G/A is abelian and A is a direct product of p-groups. If every abelian subgroup of G has finite Prüfer rank, then G has finite Prüfer rank.*

Proof By 6.2.1 the group G/A has finite p-rank for all primes p. All we need to do is bound these p-ranks, and for this purpose we can assume that each p-component of G/A is finite. This implies that the Sylow p-subgroups of G are conjugate. Indeed, if P is a Sylow p-subgroup and D is its radicable part, then $D \le A$ and so D is the radicable part of A and $D \triangleleft G$. Conjugacy now follows from Sylow's Theorem.

Suppose it has been shown that $G = TA$, where T is a direct product of p-groups. Let $M(p)$ be an abelian subgroup of T_p; then the $M(p)$'s generate an abelian subgroup which will have finite Prüfer rank, r say. It follows that each $M(p)$ has rank at most r and by 6.2.9 the subgroup T_p has Prüfer rank at most $\frac{1}{2}r(5r+1)$, as must T.

From now on we concentrate on establishing the existence of T. Let $p_1, p_2, \ldots$ be the sequence of primes and denote by S_i/A the p_i-component of G/A. We will construct a chain of subgroups $G = N_0 \ge N_1 \ge N_2 \ge \cdots$ such that N_i normalizes a Sylow subgroup P_i of N_{i-1} and $G = N_iA$. Assume that N_i has been constructed and let P_{i+1} be a Sylow p_{i+1}-subgroup of N_i. Define N_{i+1} to be $N_{N_i}(P_{i+1})$. Since $P_{i+1} \le N_i \cap S_{i+1} \triangleleft N_i$, the Frattini argument may be applied to show that $N_i = N_{i+1}(N_i \cap S_{i+1})$. Now $(N_i \cap S_{i+1})/(N_i \cap A)$ is a finite p_{i+1}-group, so we have $N_i \cap S_{i+1} = P_{i+1}(N_i \cap A)$ and therefore $N_i = N_{i+1}(N_i \cap A)$. From this it follows that $G = N_iA = N_{i+1}A$, which completes the construction of the chain.

Now the chain will be used to construct the subgroup T. Since P_i satisfies min, there is a $j(i) \ge i$ such that $N_{j(i)} \cap P_i = N_{j(i)+1} \cap P_i = \cdots = T_i$ say. If $k \ge j(i)$ and P is any p_i-subgroup of N_k, then P normalizes P_i since $P \le N_k \le N_{j(i)} \le N_i$. Hence $P \le P_i$ and $P \le N_k \cap P_i = T_i$, which shows that T_i is the unique Sylow p_i-subgroup of N_k. Consequently, $[T_i, T_j] = 1$ if $i \ne j$, so that $T = \langle T_1, T_2, \ldots \rangle$ is the direct product of the T_i's. Finally, $S_i = (N_kA) \cap S_i = (N_k \cap S_i)A$, and if $k \ge j(i)$, we also have $N_k \cap S_i \le T_i(N_k \cap A)$, since $(N_k \cap S_i)/(N_k \cap A)$ is a p_i-group. Therefore $S_i = T_iA$ and $G = TA$.   ■

Proof of 6.2.6 This is now straightforward. Let B/A be an abelian subgroup of G/A. In the first place B/A has all its ranks finite by 6.2.1. Suppose A is a torsion group. Then 6.2.10 shows that the torsion subgroup of B/A has finite

Prüfer rank, whence B/A has finite Prüfer rank. Thus we have reduced to the case where A is torsion-free.

If $[A, B] = 1$, the Hom argument may be applied in the usual way. If $[A, B] \neq 1$, the minimal rank argument may be applied to reach the desired conclusion.   ∎

We end the section by pointing out that the main theorems of Sections 6.1 and 6.2 can be extended to *radical groups*, that is, to groups which have an ascending series with locally nilpotent factors. These results are due to Baer (1968), Baer and Heineken (1972), Čarin (1960), and Plotkin (1956). We shall not give the proofs here since our brief is confined to soluble groups, but a simplified account may be found in Robinson (1976c).

6.3   Chain conditions on subnormal or ascendant abelian subgroups

We will now broaden the investigation by considering the effect of imposing chain conditions only on the subnormal or ascendant abelian subgroups. (A subgroup that is a term of an ascending series in a group is said to be *ascendant*.) A natural first aim would be to try to show that a soluble group whose subnormal abelian subgroups satisfy max is polycyclic. This is in fact true, but algebraic number theory is inescapably involved in the proof. To see why, assume the result has been proved and let K be an algebraic number field. Let A denote the subring of algebraic integers and let X be the multiplicative group of algebraic integer units of K. Now form the natural semidirect product

$$G = X \ltimes A.$$

It is well-known that A is a finitely generated free abelian group. Now it is easy to see that every subnormal abelian subgroup of G is contained in A and hence satisfies max. So the theorem stated implies that G is polycyclic and thus X is finitely generated: in short we have deduced the Dirichlet Units Theorem. It is not surprising therefore that the latter result must be used in the proof of the theorem referred to. This is in contrast to the proof of 6.1.2, for which weak splitting was enough.

6.3.1 *Let A be an abelian normal subgroup of a group G. If every subnormal abelian subgroup of G satisfies max, then the same holds for G/A.*

Proof Let B/A be a subnormal abelian subgroup of G/A. If $[A, B] = 1$, the Hom argument may be applied in the usual way and so we reduce to the case where $[A, B] \neq 1$ and A is torsion-free. The minimal rank argument is also applicable and serves to reduce to the situation where A is $\mathbb{Q}$-irreducible with respect to B.

Put $\bar{A} = A \otimes \mathbb{Q}$ and $\bar{B} = B/C_B(A)$. Then $\bar{A}$ is a simple $\mathbb{Q}\bar{B}$-module and thus

$$\bar{A} \overset{\mathbb{Q}\bar{B}}{\cong} K = \mathbb{Q}\bar{B}/I,$$

where I is a maximal ideal of the group algebra $\mathbb{Q}\bar{B}$. Hence K is an algebraic number field. If $\bar{b} \in \bar{B}$, then $\bar{b}$ and $\bar{b}^{-1}$ satisfy their characteristic equations, which are over $\mathbb{Z}$. Hence $\bar{b}$ corresponds to an algebraic integer unit. Now the assignment $\bar{b} \mapsto \bar{b} + I$ determines an injective homomorphism from $\bar{B}$ to the group of algebraic integer units of K, so it follows by the Dirichlet Units Theorem that $\bar{B}$ is finitely generated.

Finally, $C_B(A)/A$ is finitely generated by the case where $[A, B] = 1$ and consequently B/A is finitely generated. ∎

Corollary 6.3.2 *A soluble group each of whose subnormal abelian subgroups satisfies max is polycyclic.*

This follows by induction on the derived length in the usual way. By using additional techniques we obtain detailed information about the subnormal soluble subgroups of a group whose subnormal abelian subgroups satisfy max. In what follows let $\mathrm{Sol}(G)$, denote the subgroup generated by all the soluble normal subgroups of a group G, that is, the *soluble radical* of G. In general, $\mathrm{Sol}(G)$ will not be soluble, although it is hyperabelian and locally soluble. (A group which has an ascending normal series with abelian factors is said to be *hyperabelian*.)

6.3.3 *Let G be a group in which every subnormal abelian subgroup satisfies max. Then $\mathrm{Sol}(G)$ is polycyclic and it contains all the subnormal soluble subgroups of G.*

Proof Write $S = \mathrm{Sol}(G)$ and $F = \mathrm{Fit}(S)$. Our first concern is to show that F is nilpotent. Suppose there is a countably infinite ascending chain of normal nilpotent subgroups of S, say $N_1 < N_2 < \cdots$. If Z_i is the centre of N_i, then clearly $\langle Z_1, Z_2, \ldots \rangle$ is an abelian normal subgroup of S, so it is subnormal in G and therefore satisfies max. Consequently there is an integer k such that $\langle Z_1, Z_2, \ldots, Z_k \rangle$ contains every Z_i and hence $Z_i \leq N_k$ for all i. Thus $Z_i \leq Z_k$ for $i \geq k$. This shows that $1 \neq Z_k \leq Z(U)$, where $U = \bigcup_{i=1,2,\ldots} N_i$. The argument can be repeated for $U/Z(U)$ since by 6.3.1 the hypothesis on subnormal abelian subgroups is inherited by $U/Z(U)$. We may therefore conclude that $Z_1(U) < Z_2(U) < \cdots$.

Next $V = \bigcup_{i=1,2,\ldots} Z_i(U)$ is a normal hypercentral subgroup of G. If A is a maximal abelian normal subgroup of V, then $A = C_V(A)$ and of course A is finitely generated. Now V is locally nilpotent, so we may apply 3.2.4 to conclude that V/A is a finitely generated nilpotent group. Therefore V satisfies max and $Z_i(U) = Z_{i+1}(U)$ for some i. By this contradiction the chain of N_i's is finite.

The argument so far demonstrates the existence of maximal nilpotent normal subgroups of S. Therefore F is nilpotent, and of course it is finitely generated as well. Next S is hyperabelian, so that $C_S(F) \leq F$ by the proof of 1.2.10. Also S is locally soluble, so by 3.2.4 once again $S/C_S(F)$ is polycyclic, which shows that S is polycyclic.

Now let H be a subnormal soluble subgroup of G. By induction on the subnormal defect of H, which may be assumed to be > 1, we have $H \leq \mathrm{Sol}(H^G)$. But $\mathrm{Sol}(H^G)$ is polycyclic, so $\mathrm{Sol}(H^G) \leq \mathrm{Sol}(G)$ and $H \leq \mathrm{Sol}(G)$, as required. ∎

We turn next to the analogue of 6.3.1 for the minimal condition. However a simple example demonstrates that some modification of the statement is necessary.

Let A be a Prüfer 2^∞-group and let t be the automorphism of A in which $a \mapsto a^5$. Consider $G = \langle t \rangle \ltimes A$, the natural semidirect product. Every subnormal abelian subgroup of G is contained in A and thus satisfies min. But G/A is not even a torsion group. This example suggests that it may be necessary to restrict ourselves to torsion groups.

6.3.4 *Let A be an abelian normal subgroup of a torsion group G. If each subnormal abelian subgroup of G satisfies min, then the same is true of G/A.*

Proof Let B/A be a subnormal abelian subgroup of G/A and put $C = C_B(A)$. Then C is nilpotent, so every abelian subgroup of C is subnormal in G and hence satisfies min. Hence C is Černikov by 6.1.6. Also B/C is a torsion group of automorphisms of the Černikov group A. Therefore B/C is Černikov by 3.2.6, and it follows that B, and hence B/A, satisfies min. ∎

Corollary 6.3.5 *Let G be a soluble torsion group. If every subnormal abelian subgroup of G satisfies min, then G is a Černikov group.*

This may be proved by induction on the derived length.

There is an analogue of 6.3.3 for groups whose subnormal abelian subgroups satisfy min.

6.3.6 *Let G be a group in which each subnormal abelian subgroup satisfies min. Then there is a unique maximum soluble normal torsion subgroup of G: furthermore it is Černikov and contains all subnormal soluble torsion subgroups of S.*

Proof This runs along similar lines to the proof of 6.3.3. Let S be the subgroup generated by all the soluble normal *torsion* subgroups of G. Obviously S is torsion and hyperabelian. Put $F = \mathrm{Fit}(S)$. The first step is to show that F is nilpotent.

Consider an ascending chain of normal nilpotent subgroups of S, say $1 \neq N_1 \leq N_2 \leq \cdots N_\alpha \leq \cdots$, and write $Z_i = Z(N_i)$. Let the intersection $I = Z_1 \cap Z_{\alpha(1)} \cap \cdots \cap Z_{\alpha(i)}$ be chosen minimal, subject to $I \neq 1$: this is certainly possible since Z_1 satisfies min. If $\alpha \geq \alpha(i)$, then $1 \neq I \triangleleft N_\alpha$, so that $I \cap Z_\alpha \neq 1$. It follows from the minimality of I that $I \leq Z_\alpha$ for all $\alpha \geq \alpha(i)$ and hence $I \leq Z(U)$, where $U = \bigcup_\alpha N_\alpha$. By 6.3.4 we may apply the same argument to $U/Z(U)$. Thus, if U is not nilpotent, we will find that $Z_1(U) < Z_2(U) < \cdots$. Write $V = \bigcup_{i=1,2,\ldots} Z_i(U)$, which is a hypercentral torsion normal subgroup of G, and let A be a maximal abelian normal subgroup of V: then A is Černikov. Also V is a torsion group.

Since $A = C_V(A)$, we deduce from 3.2.6 that V/A is Černikov and therefore V is Černikov.

Now let D be the divisible part of V. For any α the subgroup DN_α is nilpotent and so it satisfies min by 6.1.6. Hence DN_α is centre-by-finite by 1.4.4, from which we deduce that $[N_\alpha, D] = 1$ for all $\alpha \geq \alpha(i)$, that is, $D \leq Z(U)$ and V/D is finite. It follows that $Z_i(U) = Z_{i+1}(U)$ for some i, a contradiction. We have now proved that the union of an ascending chain of nilpotent normal subgroups of S is nilpotent. It follows easily that F is nilpotent; of course it is also Černikov.

Finally, let H be a subnormal soluble torsion subgroup of G. Induction on the subnormal defect shows that H^G is a soluble torsion group and consequently $H \leq S$. ∎

Although we cannot expect to be able to show that $\mathrm{Sol}(G)$ is Černikov in the circumstances of 6.3.6, it is a fact that this subgroup is soluble.

6.3.7 *Let G be a group in which each subnormal abelian subgroup satisfies min. Then $\mathrm{Sol}(G)$ is soluble and it contains all subnormal soluble subgroups of G.*

Proof Let S and S_0 denote respectively $\mathrm{Sol}(G)$ and the subgroup generated by all the soluble normal torsion subgroups of G. Then $\mathrm{Fit}(S) = \mathrm{Fit}(S_0) = F$ say, since each subnormal nilpotent subgroup of G is a torsion group. Now F is Černikov by 6.3.6; let D be its divisible part. For each prime p the group $S/C_S(D_p)$ is locally soluble and also linear over the field of p-adic numbers. Therefore, $S/C_S(D_p)$ is soluble by 3.1.10, whence so is $S/C_S(D)$. Since F/D is finite, $S/C_S(D) \cap C_S(F/D)$ is soluble. But the group $(C_S(D) \cap C_S(F/D))/C_S(F)$ is abelian—see 3.2.5—and $C_S(F) \leq F$. Consequently, S is soluble. The final statement is proved by induction on the subnormal defect. ∎

We remark that 6.3.2 and 6.3.5 can be extended to *subsoluble groups*, that is, to groups which have an ascending series of subnormal subgroups with all factors abelian. (Equivalently, the upper Baer series reaches the group transfinitely.) Thus *a subsoluble group whose subnormal abelian subgroups satisfy max is polycyclic*. There is a similar result for soluble torsion groups in which each subnormal abelian subgroup has min. The proofs are essentially the same, 6.3.3 and 6.3.6 being invoked to deal with Baer groups—see also Baer (1967c).

Minimax groups

The analogue for minimax groups of results like 6.3.2 is quickly seen to be false. For suppose that A is a group of type p^∞ and let X be the multiplicative group of p-adic integers π satisfying $\pi \equiv 1 \pmod{p}$. Define G to be the natural semidirect product $X \ltimes A$. Then every subnormal abelian subgroup of G is contained in A and therefore satisfies min. However, X is abelian with infinite torsion-free rank. Notice that in this example X is an *ascendant* abelian subgroup, which suggests that restrictions should be placed on the ascendant abelian subgroups. In fact with such hypotheses positive results can be established.

6.3.8 *Let A be an abelian normal subgroup of a group G and let π be a set of primes. If every ascendant abelian subgroup of G is π-minimax, then the same is true of G/A.*

This result has the consequence:

Corollary 6.3.9 (Baer 1968) *A soluble group whose ascendant abelian subgroups are π-minimax is itself π-minimax.*

The proof of 6.3.8 requires some commutative algebra beyond the Dirichlet Units Theorem, specifically:

6.3.10 (Baer 1968) *Let R be an integral domain whose additive group is a torsion-free minimax group. Then the group of units of R is finitely generated.*

(A proof of this result may also be found in Robinson (1972*b*: 10.36).)

6.3.11 *Let A be a torsion-free abelian π-minimax group and let G be an abelian subgroup of $\mathrm{Aut}(A)$. Then G is π-minimax.*

Proof Argue by induction on $r = r_0(A) > 0$. Suppose first that A is $\mathbb{Q}$-irreducible with respect to G and let R be the ring of endomorphisms of A generated by G. Thus R is a commutative ring with identity. If $0 \neq \rho \in R$, then $\mathrm{Ker}(\rho)$ is an R-submodule of A and $A/\mathrm{Ker}(\rho)$ is torsion-free since A is torsion-free. By $\mathbb{Q}$-irreducibility $\mathrm{Ker}(\rho) = 0$ and ρ is injective. It follows that the equation $\rho\sigma = 0$, with $\rho, \sigma \in R$, implies that $\rho = 0$ or $\sigma = 0$; in short R is an integral domain. Next, if a is a fixed element of A, the assignment $\rho \mapsto a^\rho$ determines an injective homomorphism from the additive group of R to A. Therefore the additive group of R is minimax and 6.3.10 shows that the group of units $U(R)$ is finitely generated. Since G is a subgroup of $U(R)$, we deduce that G is finitely generated and therefore π-minimax.

Now suppose that A is $\mathbb{Q}$-reducible with respect to G, so that there is a proper non-trivial G-invariant subgroup B such that A/B is torsion-free. Put $C = C_G(G) \cap C_G(A/B)$; then G/C is π-minimax by induction hypothesis. Also C is isomorphic with a subgroup of $\mathrm{Hom}(A/B, B)$, which is clearly π-minimax. Consequently C is π-minimax, as is G. ∎

Proof of 6.3.8 Let B/A be an ascendant abelian subgroup of G/A. Assume first that A is a torsion group and let T/A be the torsion subgroup of B/A. Then T is torsion and all its subnormal abelian subgroups satisfy min, so that T is Černikov by 6.3.5. The divisible part of T is plainly a π-group, which shows that T/A is π-minimax. Since T/A is a direct factor of B/A, we may assume that B/A is torsion-free. If X_1 denotes an arbitrary abelian subgroup of B, then $X_1 = (X_1 \cap A) \times X$, where X is torsion-free, since $X_1 \cap A$, the torsion subgroup of X_1, has min. The elements with prime order in A generate a finite subgroup, which is centralized by X^m for some $m > 0$. Writing $A_i = \{a \in A \mid a^{p^i} = 1, p = \text{a prime}\}$, we observe that x^m centralizes A_{i+1}/A_i: consequently X^m is ascendant in $X^m A$ and hence in B. Therefore X^m is π-minimax. But $X \simeq X^m$, so X is torsion-free, and hence X_1 is π-minimax. Now apply 6.1.9 to show that B is π-minimax, which takes care of the case where A is a torsion group.

We may now suppose that A is torsion-free. Writing $C = C_B(A)$, we see that B/C is π-minimax by 6.3.11, while C is π-minimax by the Hom argument. Hence B is π-minimax. ∎

Our next objective is to establish analogues of 6.3.3 and 6.3.6 for minimax groups.

6.3.12 *Let G be a group in which every ascendant abelian subgroup is π-minimax. Then $\mathrm{Sol}(G)$ is a soluble π-minimax group which contains all subnormal soluble subgroups of G.*

In proving this result we will need a generalization of 6.3.11 to locally soluble groups of automorphisms.

6.3.13 *Let G be a finite extension of a reduced soluble π-minimax group and let A be a locally soluble group of automorphisms of G. Then A is a soluble π-minimax group.*

Proof By 5.2.5 there is a characteristic subgroup N of finite index in G such that N is torsion-free and soluble. Writing $F = \mathrm{Fit}(N)$, we may assert on the basis of 5.2.3 that G/F is virtually polycyclic. Suppose we have shown $A/C_A(F)$ to be soluble π-minimax. Now $A/C_A(G/F)$ is polycyclic by 3.2.4, whence A/D is π-minimax, where

$$D = C_A(G/F) \cap C_A(F).$$

Now D is isomorphic with a subgroup of $E = \mathrm{Der}(G/F, Z(F))$ by 3.2.5. Further, it is an easy exercise that a derivation is determined by its effect on the generators of the finitely generated group G/F. Consequently E is isomorphic with a subgroup of a *finite* direct power of $Z(F)$, which shows that E is π-minimax, and hence so is A.

By the foregoing discussion we may replace G by F, so that G is torsion-free and nilpotent, with class c say. If $c \leq 1$, then A is $\mathbb{Q}$-linear and locally soluble, so it is soluble by 3.1.10. Also each abelian subgroup of A is π-minimax by 6.3.11, and 6.1.9 shows that A is π-minimax. Now let $c > 1$ and use induction on c. If $Z = Z(G)$, then $A/C_A(G/Z)$ is π-minimax, as is $A/C_A(Z)$. Writing $F = C_A(Z) \cap C_A(G/Z)$, we see that A/F is π-minimax. Finally, F is isomorphic with a subgroup of $\mathrm{Hom}(G/Z, Z)$, and further any $\theta \in \mathrm{Hom}(G/Z, Z)$ is determined by its effect on a maximal linearly independent set of elements of infinite order in $G/G'Z$. Therefore $\mathrm{Hom}(G/Z, Z)$ is π-minimax, as are F and A. ∎

Proof of 6.3.12 By 6.3.6 applied to $\tau(G)$, there is a unique maximum normal soluble torsion subgroup of G: it is Černikov and contains all subnormal soluble torsion subgroups of G. Clearly we may factor out by this subgroup.

The first step is to prove that $S = \mathrm{Sol}(G)$ is a soluble π-minimax group. Let $F = \mathrm{Fit}(S)$: we will show that F is nilpotent. Suppose that L is any nilpotent normal subgroup of S. Then L is π-minimax by 6.3.9, and it is torsion-free by the assumption of the first paragraph. Refine the upper central series of L to an F-invariant series with torsion-free factors that are $\mathbb{Q}$-irreducible relative to F. Since F is generated by nilpotent normal subgroups, L is contained in the hypercentre of F: thus F is hypercentral and hence locally nilpotent. Let M be a maximal abelian normal subgroup of F. By the first paragraph M is torsion-free, and of course it is also π-minimax. Since $M = C_F(M)$, we conclude that F/M, and hence F, is soluble π-minimax by 6.3.13: thus $\mathrm{Fit}(S) = F$ is nilpotent by 5.2.2. Next S is certainly locally soluble and hence $S/C_S(F)$ is soluble π-minimax by 6.3.13. Since $C_S(F) \leq F$, it follows that S is soluble π-minimax.

If H is a subnormal soluble subgroup, then induction on the subnormal defect shows that $H \leq S$. ∎

Whether or not the final part of 6.3.12 is true for ascendant soluble subgroups remains open.

Counterexamples

It is impossible to extend the theorems of this section to wider classes of soluble groups of finite rank, as is shown by a simple example. Let G be the holomorph of $\mathbb{Q}$, the additive group of rational numbers. Every ascendant abelian subgroup of G is contained in $\mathbb{Q}$ and thus has rank ≤ 1. But G has free abelian subgroups of infinite rank, and it is metabelian. So there can be no results like 6.3.9 for soluble FATR-groups, soluble groups with finite Prüfer rank or soluble FAR-groups.

The next two examples show that there are no results like 6.3.2 and 6.3.5 for abelian normal subgroups.

There exists an infinite soluble group in which every abelian normal subgroup is finite. To begin the construction, we let L be the group with generators a_i, $i \in \mathbb{Z}$, and relations

$$a_i^2 = 1, \quad [a_i, a_{i+1}] = c, \quad [a_i, c] = 1,$$

and

$$[a_i, a_j] = 1 \quad \text{if } |i - j| > 1.$$

Thus L is a nilpotent 2-group of class 2. The assignments $a_i \mapsto a_{i+1}$ determine an automorphism t of L. Define G to be the group $\langle t \rangle \ltimes L$. Thus G is a finitely generated centre-by-metabelian group with $G'' = Z(G) = \langle c \rangle$.

Suppose that A is an abelian normal subgroup not contained in $Z(G)$. Clearly $A \leq L$, so A contains an element $u = a_{i_1}^{\ell_1} \cdots a_{i_k}^{\ell_k}$, where $i_1 < i_2 < \cdots < i_k$, $\ell_j \neq 0$ and $k > 0$. Now L also contains $v = u^{t^{i_k - i_1 + 1}}$, and yet $[u, v] = c^{\ell_1 \ell_k} \neq 1$, a contradiction. Hence $A \leq Z(G) = \langle c \rangle$ and A is finite. Notice that G is not even a soluble FAR-group.

The situation is no better for torsion groups. To see why, let $N_i = \langle a_i, b_i \rangle$ be a quaternion group of order 8 for $i = 1, 2, \ldots$, and form the direct product of the N_i with amalgamated centres, say N. Denote by x_i the automorphism of N such that $a_i^{x_i} = b_i$, $b_i^{x_i} = a_i b_i$ and $a_j^{x_i} = a_j$, $b_j^{x_i} = b$, for $j \neq i$. Then $X = \langle x_1, x_2, \ldots \rangle$ is an elementary abelian 3-group. Put $G = X \ltimes N$, a centre-by-metabelian torsion group with $|Z(G)| = 2$.

Suppose that A is an abelian normal subgroup of G and $A \nleq Z(G)$. Now certainly $A \leq N$ and A must contain an element u of N with non-central ith component for some i. But $[u, u^{x_i}]$ has ith component $[u_i, u_i^{x_i}] \neq 1$. Therefore $A \leq Z(G)$ and $|A| \leq 2$.

Further results

A detailed investigation of chain conditions on abelian subgroups which are subnormal or ascendant with bounded defect has been carried out by Robinson (1968c). We quote some of the results without proof.

1. *Let $\alpha > 1$ be an ordinal and let G be a group in which every ascendant abelian subgroup with defect at most α satisfies max. Then $\mathrm{Sol}(G)$ is polycyclic and contains all ascendant soluble subgroups with defect at most α.*

   Let us write

   $$\vartriangleleft^\alpha \mathbf{Ab} - \max$$

   for the group theoretic property that every ascendant abelian subgroup with defect at most α satisfies max. Also let

   $$\max - \vartriangleleft^\alpha \mathbf{Ab}$$

   denote the maximal condition applied to the set of all ascendant abelian subgroups with defect $\leq \alpha$. Then by the result quoted above $\vartriangleleft^\alpha \mathbf{Ab}$-max implies $\max - \vartriangleleft^\alpha \mathbf{Ab}$: also *the converse holds when α is infinite,* but not in fact when $\alpha = 2$. Nevertheless, the following is true.

2. *Let α be a finite ordinal > 1 and let G be a group satisfying $\max - \vartriangleleft^\alpha \mathbf{Ab}$. Then $\mathrm{Sol}(G)$ is polycyclic and it contains all subnormal soluble subgroups of defect at most $\alpha - 1$. From this one can deduce*

3. *A group G satisfies the maximal condition on subnormal (ascendant) abelian subgroups if and only if it satisfies $\max - \vartriangleleft^\alpha \mathbf{Ab}$ for all finite α (respectively for all ordinals α).*

There are similar results for the minimal condition, but as usual it is necessary to restrict attention to ascendant abelian subgroups, which are torsion.

7

THE THEORY OF FINITELY GENERATED
SOLUBLE GROUPS II

Our object in this chapter is to give an account of J. E. Roseblade's extension of P. Hall's theory of finitely generated abelian-by-polycyclic groups, which was described in Chapter 4. The principal aim of the theory is to generalize Hall's Theorem 4.3.1 to the following: *finitely generated abelian-by-polycyclic-by-finite groups are residually finite.* This result was found independently by Jategaonkar (1974) and Roseblade (1976), and its proof will be our main preoccupation in this chapter. Here we shall follow Roseblade's account of this theorem: for another approach see Passman (1984, 1985).

7.1 Simple modules over polycyclic groups

As indicated at the beginning of 4.3, what is called for in proving the residual finiteness theorem, is an analogue of Hilbert's Weak Nullstellensatz. This was established in 1973 by Roseblade in a seminal paper.

7.1.1 (Roseblade 1973a) *Simple modules over integral group rings of virtually polycyclic groups are finite.*

The proof of 7.1.1 will occupy us throughout this section. Now if G is a virtually polycyclic group and M is a simple $\mathbb{Z}G$-module, it follows from arguments in 4.3 that $Mp = 0$ for some rational prime p (see the proof of 4.3.4, second paragraph). Thus M can be regarded as a simple kG-module, where k is the field of p-elements.

In what follows we shall usually take k to be an *absolute field*, that is, each non-zero element of k must be a root of unity, so that k is in fact absolutely algebraic of prime characteristic. Our aim is to prove 7.1.1 in the following stronger form.

7.1.2 *Suppose that k is an absolute field and G is a virtually polycyclic group. Then every simple kG-module has finite k-dimension.*

We recall that in proving that a simple kG-module M over a finitely generated nilpotent group G with k a field was finite, a key step was to choose an element x of infinite order in the centre of G and to study the structure of M as a $k\langle x\rangle$-module and prove that M is not torsion-free. In the case where G is virtually polycyclic, however, there may not even be a normal nilpotent subgroup of finite index and we may not be able to find a non-trivial central element x.

121

What Roseblade did was to show that much of Hall's original argument could be generalized to the polycyclic case by replacing $\langle x \rangle$ by a suitable abelian normal subgroup A of G and analysing the kG-module M in terms of its structure as a kA-module.

This analysis splits into two parts according to whether M is or is not torsion-free as a kA-module, the main objective being to find a suitable A for which M has kA-torsion. An induction argument can then be used to complete the proof. Thus our first major objective is to prove the following result.

7.1.3 *Suppose that k is an absolute field, G a virtually polycyclic group and A an infinite abelian normal subgroup of G. Then a simple kG-module cannot be torsion-free as a kA-module.*

Our discussion will be facilitated by considering the situation where R is a ring with identity generated by a subring S together with a group G which normalizes S: here G is a subgroup of the group of units of R. This notation will be maintained throughout 7.1.

Let U be any S-module and let $x \in G$. Then, if U^x denotes an additive group isomorphic with U_1, via $u \mapsto u^x$, $(u \in U)$, we may regard U^x as an S-module by defining

$$u^x s = (u s^{x^{-1}}) x,$$

where $u \in U$, $s \in S$. Next S-modules U and V are said to be *G-conjugate*, written

$$U \underset{G}{\sim} V,$$

if $V \simeq U^x$ for some $x \in G$. We have at once the following useful fact.

7.1.4 (The Cancellation Law) *Suppose that V and U are S-submodules of an R-module M. Then $U_x/V_x \underset{G}{\sim} U/V$ for any $x \in G$.*

This is true because

$$(V + u)x\, s = (V + u)s^{x^{-1}}x = (V + u)^x s,$$

for all $u \in U$, $s \in S$.

The Cancellation Law will be used to prove the next result, which will lead to a generalization of 4.3.3.

7.1.5 *Suppose that G is a virtually polycyclic group and M a finitely generated R-module. If S satisfies $\max -r$, then there is an ascending series*

$$0 = M_0 \leq M_1 \leq \cdots \leq M_\lambda \leq M_{\lambda+1} \leq \cdots \leq M_\rho = M$$

of S-submodules of M such that the factors $M_{\lambda+1}/M_\lambda$, $0 \leq \lambda < \rho$, are cyclic S-modules which fall into finitely many G-conjugacy classes.

In the interests of concise expression, let us say that, if $\mathbf{X}$ is a class of S-modules, a module M belongs to the class $\acute{P}\mathbf{X}$, if it has an ascending series of S-modules $\{M_\lambda\}$ such that each factor $M_{\lambda+1}/M_\lambda$ is in $\mathbf{X}$. Also denote by

$\mathbf{X}^G$ the class of all S-modules which are G-conjugate to a member of $\mathbf{X}$. Thus the conclusion of 7.1.5 is that $M \in \acute{P}(\mathbf{X}^G)$, where $\mathbf{X}$ is some finite class of cyclic S-modules.

Proof of 7.1.5 Asssume first that $G = \langle x \rangle$ is infinite cyclic, so that by hypothesis $R = \sum_{n=-\infty}^{\infty} Sx^n$. There is a finitely generated S-submodule U of M such that $M = UR$, so $M = \sum_{n=-\infty}^{\infty} Ux^n$. Put

$$U_n^+ = \sum_{i=1}^{n} Ux^i \quad \text{and} \quad U_n^- = \sum_{i=0}^{n} Ux^{-i} \qquad \text{for } n > 0,$$

and set $U_0^+ = U_0^- = 0$. Then $\{U_n^-\}_{n \geq 0}$ is an ascending series which will form half of the series sought.

We further set $V = \bigcup_{n=0}^{\infty} U_n^-$. Then $\{Vx^n\}_{n \geq 0}$ is an ascending series since $U_{n+1}^- x = U + U_n^-$. Clearly $M = \bigcup_{n=0}^{\infty} Vx^n$, so that the series $\{U_n^-\}_{n \geq 0}$ and $\{Vx^n\}_{n \geq 0}$ together yield a series of ordinal type $\omega 2$.

We must now check that the factors of this series have the required properties. For $n \geq 0$ we have $U_{n+1}^- = U_n^- + Ux^{-(n+1)}$. Hence

$$U_{n+1}^-/U_n^- \cong Ux^{-(n+1)}/U_n^- \cap Ux^{-(n+1)},$$

which, by the Cancellation Law (7.1.4), is G-conjugate to $U/(U_n^- x^{n+1} \cap U) = U/(U \cap U_n^+)$. But $0 \leq U \cap U_1^+ \leq \cdots \leq U \cap U_n^+ \leq \cdots$ is an ascending chain of S-submodules of the finitely generated S-module U, and S has max $-r$, so U satisfies the maximal condition on submodules. Hence $U \cap U_m^+ = U \cap U_{m+1}^+$ for some m. Thus U_{n+1}^-/U_n^- is G-conjugate to one of $U/(U \cap U_r^+)$, $0 \leq r \leq m$, a finite number of finitely generated S-modules. So there exists a class of cyclic S-modules $\mathbf{X}_1$ which is *finitary*, that is, it consists of finitely many isomorphism types, such that $U/(U \cap U_r^+) \in P\mathbf{X}_1$ for all r. Hence $V \in \acute{P}(\acute{P}\mathbf{X}_1)^G = \acute{P}(\mathbf{X}_1^G)$. The factors Vx^{n+1}/Vx^n, $n \geq 0$ are all G-conjugate to Vx/V, again by the Cancellation Law. But $Vx = U + V$ and so $Vx/V \cong U/(U \cap V)$, which is a finitely generated S-module and so belongs to $\acute{P}\mathbf{X}_2$ for some finitary class of cyclic S-modules $\mathbf{X}_2$. Hence $M/V \in \acute{P}(\mathbf{X}_2{}^G)$. On setting $\mathbf{X} = \mathbf{X}_1 \cup \mathbf{X}_2$, we see that $M \in \acute{P}(\mathbf{X}^G)$, as required. This completes the proof in the case where G is infinite cyclic.

In the general case there is a series $1 = G_n \lhd G_{n-1} \lhd \cdots \lhd G_1 \lhd G_0 = G$ such that each factor is either finite or cyclic. Let S_1 be the subring of R generated by S and G_1. If G/G_1 is finite, then M is finitely generated as an S_1-module and the natural induction hypothesis yields $M \in \acute{P}(\mathbf{X}^G)$, for a suitable finitary class $\mathbf{X}$ of cyclic S-modules. If G/G_1 is infinite cyclic, we have $G = G_1 C$ for some infinite cyclic subgroup C of G. Clearly, $R = Rg\langle S_1, x \rangle$ and $S_1^x = S_1$. Also S_1 has max $-r$ and hence by the cyclic case, as an S_1-module, M_1 belongs to $\acute{P}(\mathbf{Y}^C)$ for some finitary class $\mathbf{Y}$ of cyclic S_1-submodules. Suppose that $\mathbf{Y} = \{Y_1, \ldots, Y_m\}$. By the induction hypothesis each Y_i belongs to $\acute{P}(\mathbf{X}_i^{G_1})$ for some finitary class $\mathbf{X}_i$ of cyclic S_1-modules. Now put $\mathbf{X} = \bigcup_{i=1}^{m} \mathbf{X}_i$, which is finitary. Then it readily follows that $M \in \acute{P}(\mathbf{X}^G)$ and, since $\mathbf{X}^{G_1} \cup \mathbf{X}^C \subseteq \mathbf{X}^G$, this completes the proof of 7.1.5. ∎

We now use 7.1.5 to gain information about torsion-free modules in the following result. Carrying forward the notation established at the beginning of this discussion, we have:

7.1.6 *Suppose that G is a virtually polycyclic group, S is a noetherian domain and R is a ring generated by S and G, with G normalizing S. Suppose further that M is a finitely generated R-module which is torsion-free as an S-module. Then there exists a free S-submodule F of M and a non-zero ideal Λ of S, such that every finitely generated S-submodule of M/F is annihilated by some product $\Lambda^{x_1} \Lambda^{x_2} \cdots \Lambda^{x_n}$ of G-conjugates of Λ.*

Proof By 7.1.5 there is an ascending series $\{M_\lambda \mid \lambda < \rho\}$ each factor of which is G-conjugate to one of a finite set $\{U_1, \ldots, U_m\}$, say, of cyclic S-modules. We may take U_1 to be M_1, and since M is S-torsion-free, we may also suppose that $U_1 = S$.

Define Φ to be the set of all λ, $(0 \leq \lambda < \rho)$, for which $M_{\lambda+1}/M_\lambda \cong S$. Thus, for each $\lambda \in \Phi$, we have $M_{\lambda+1} = M_\lambda + u_\lambda S$ for some u_λ. Set $F = \sum_{\lambda \in \Phi} u_\lambda S$, so that F is a free S-module. If $m = 1$, then $F = M$ and the result is trivial. So suppose $m > 1$ and define P_i to be the annihilator of U_i in S. If $i \geq 2$, then U_i is not G-isomorphic to $U_1 = S$ and thus $P_i \neq 0$. Put $\Lambda = P_1 P_2 \cdots P_m$, which is non-zero since S is a domain.

Now let U be a finitely generated S-submodule of M. We need to show that $U\Lambda^{x_1} \cdots \Lambda^{x_n} \leq F$ for some $x_1, \ldots, x_n$ in G. Let μ_1 be the least ordinal such that $U \leq F + M_{\mu_1}$. Since U is finitely generated, μ_1 is not a limit ordinal and so, if $U \not\leq F$, we have $\mu_1 = \lambda_1 + 1$.

Now $\lambda_1 \notin \Phi$ since, for any $\lambda \in \Phi$, we have $F + M_\lambda = F + M_{\lambda+1}$. Hence M_{μ_1}/M_{λ_1} is G-conjugate to one of $U_2, \ldots, U_m$, and it is therefore annihilated by a conjugate of one of $P_2, \ldots, P_m$. Consequently there exists $x_1 \in G$ such that $M_{\mu_1} \Lambda^{x_1} \leq M_{\lambda_1}$, and then $U\Lambda^{x_1} \leq F + M_{\lambda_1}$. Since S is noetherian, $U\Lambda^{x_1}$ is a finitely generated S-submodule of M, and if μ_2 is the least ordinal such that $U\Lambda^{x_1} \leq F + M_{\mu_2}$, then $\mu_2 < \mu_1$. The result now follows by induction. ∎

We apply this next to prove:

7.1.7 *Suppose that G, R, S, M, F, Λ are as described in 7.1.6. If M is a simple R-module, then either every maximal ideal of S contains a conjugate of Λ or there exists a maximal ideal of S, whose conjugates intersect in the zero ideal.*

Proof Suppose first of all that M is as defined in 7.1.6, but is not necessarily simple. Suppose that U is a finitely generated S-submodule of M. By 7.1.6 there is an ideal $X = \Lambda^{x_1} \cdots \Lambda^{x_n}$ of S such that $UX \leq F$. Let L be a maximal ideal of S which contains none of the Λ^{x_i}. Then $X \not\leq L$ and hence $S = X + L$, which implies that

$$U = US \leq UX + UL \leq F + UL.$$

Moreover

$$((UL) \cap F)S \leq ULX + FL = UXL + FL \leq FL$$

and thus $(UL) \cap F \leq FL$. Since M is the set theoretic union of all the U, we may infer that

$$M = F + ML \quad \text{and} \quad (ML) \cap F = FL.$$

This means that $M > ML$: for $F > FL$ since F is free.

Now write, $^\circ L$, for the intersection of all the G-conjugates of L. Thus $x(^\circ L) = (^\circ L)x$ for all $x \in G$ and hence $M(^\circ L)$ is an R-submodule of M. But $M(^\circ L) \leq ML < M$ and hence, if M is a simple R-module, we obtain $M(^\circ L) = 0$. However, $^\circ L \leq S$ and M is S-torsion-free. Therefore $^\circ L = 0$, as required. ∎

In order to dismiss the second alternative in 7.1.7 we need:

7.1.8 *Suppose that k is an absolute field, G a virtually polycyclic group and A a normal subgroup of G. Let $S = kA$ and let P be an ideal of S. If S/P has finite k-dimension, then $A^m \leq 1 +^\circ P$ for some $m > 0$ and $S/^\circ P$ has finite k-dimension.*

Proof Consider S/P as an A-module via conjugation. Since $\dim_k(S/P)$ is finite, A acts on S/P as a subgroup of $GL_n(k)$ for some n. Now k is absolute, so it is algebraic of characteristic p for some prime p and hence $GL_n(k)$ is locally finite. Thus the kernel of the action of A on S/P, that is, $(1 + P) \cap A$ has finite index in A, which implies that $A^m \leq 1 + P$ for some $m > 0$. But $A^m \triangleleft G$ and so $A^m \leq 1 + {}^\circ P$. Finally, A/A^m is finite, $k(A/A^m) \simeq S/S(A^m - 1)$ and $S(A^m - 1) \leq {}^\circ P$. It follows that $S/^\circ P$ has finite k-dimension, as required. ∎

Next we prove

7.1.9 *Suppose that k is an absolute field, G a virtually polycyclic group and A an infinite abelian normal subgroup of G. Put $S = kA$ and suppose that M is a simple kG-module which is torsion-free as an S-module. Then every maximal ideal of S contains a conjugate of Λ, where Λ is the ideal of S in 7.1.6.*

Proof Suppose first that L is a maximal ideal of S. Then by the Weak Nullstellensatz $\dim_k(S/L)$ is finite and so $\dim_k(S/^\circ L)$ is finite by 7.1.8. Since A is infinite, $^\circ L \neq 0$. Finally, using the fact that M is a simple kG-module which is S-torsion-free and taking Λ to be the ideal of S given by 7.1.6, we deduce from 7.1.7 that every maximal ideal of S contains a conjugate of Λ. ∎

The next major step in the argument is to show that *the abelian normal subgroup A in 7.1.3 can be chosen in such a way that no maximal ideal of S contains a conjugate of Λ*. This will mean M is not S-torsion-free and induction may be used to complete the proof that simple modules are finite.

An important result of Bergman (1971) plays a key role at this and the next stage of the argument. This result indicates that the abelian normal subgroups of G that we need to look at are what Roseblade has termed the plinths of G.

A *plinth* of an arbitrary group G is a non-trivial finitely generated free abelian normal subgroup A containing no non-trivial subgroup of lower rank which is normal in any subgroup of finite index in G. Thus G and all its subgroups of finite index act rationally irreducibly on A.

Although not every infinite polycyclic group has a plinth, there is a weaker result in this direction:

7.1.10 *Any infinite abelian normal subgroup A of a virtually polycyclic group G contains a plinth of some normal subgroup of finite index in G.*

Proof Since A is infinite, A^m is non-trivial and free abelian for some positive integer m: also of course $A^m \lhd G$. Let B be a non-trivial free abelian subgroup of A^m of least possible rank subject to having only finitely many conjugates in G. Thus the normalizer of B, and hence its core K, has finite index in G. Finally, B is a plinth of K. ∎

The next result shows that plinths are exactly what we need.

7.1.11 *Suppose that G is a virtually polycyclic group, A is a plinth of G, k is an absolute field and $S = kA$. If Λ is any non-zero ideal of S, there is a maximal ideal of S which contains no conjugate of Λ.*

There are three main ingredients in the proof. The first, due to Passman (1985: lemma 3.1), is a group theoretic version of the Theorem of the Primitive Element in field theory.

7.1.12 *Let Γ be a finitely generated abelian group and V a $\mathbb{Q}\Gamma$-module, which is a simple $\mathbb{Q}\Delta$-module for every subgroup Δ of finite index in Γ. Then there is an element $\gamma \in \Gamma$ such that V is a simple $\mathbb{Q}\langle\gamma^m\rangle$-module for all $m \geq 1$.*

Proof Clearly we may assume that Γ does not act trivially on V, so that there is a maximal ideal L of $\mathbb{Q}\Gamma$ such that $V \cong^{\mathbb{Q}\Gamma} F = \mathbb{Q}\Gamma/L$. By the Weak Nullstellensatz, the field F is a finite extension of $\mathbb{Q}$ and so it has only finitely many proper subfields $F_1, F_2, \ldots, F_n$. Set $F_i = X_i/L$, $i = 1, 2, \ldots, n$, and $\Gamma_i = X_i \cap \Gamma$. Denote by Δ_i/Γ_i the torsion subgroup of Γ/Γ_i. Then Γ/Δ_i is torsion-free and Δ_i/Γ_i is finite.

Evidently, F_i corresponds to a proper $\mathbb{Q}\Gamma_i$-submodule of V, so that $|\Gamma : \Gamma_i|$ is infinite by the hypothesis. Hence $|\Gamma : \Delta_i|$ is infinite and we may deduce that $\Gamma \neq \bigcup_{i=1}^{n} \Delta_i$: this is by a theorem of B. H. Neumann (1954) which states that if a group is the set-theoretic union of finitely many cosets of subgroups, then at least one of the subgroups must have finite index. If $\gamma \in \Gamma \backslash \bigcup_{i=1}^{n} \Delta_i$, then γ has no positive power in any Γ_i, and hence in any X_i. Now let $f = L + \gamma$ and $m > 0$. Then f^m belongs to no proper subfield of F. Hence $\mathbb{Q}\Gamma = L + \mathbb{Q}\langle\gamma^m\rangle$ and thus V is a simple $\mathbb{Q}\langle\gamma^m\rangle$-module. ∎

The second of the preliminary results needed is:

7.1.13 *Suppose that A is a free abelian group of finite rank, k is a finite field and $S = kA$. If γ is an automorphism of A, then the maximal ideals of S that are normalized by γ have trivial intersection.*

Proof We may assume that $A \neq 0$ and $|k| = q$. If L is a maximal ideal of S normalized by γ, then γ induces a k-automorphism in the finite extension S/L of k. Hence there exists $n \geq 0$ such that $L + s^\gamma = L + s^{q^n}$ for all $s \in S$. This means that $a^\gamma a^{-q^n} - 1 \in L$ for all $a \in A$.

Set $A_n = \langle a^\gamma a^{-q^n} \mid a \in A \rangle$ and $I_n = \sum_{a \in A_n} (a - 1)S$, that is, the kernel of the natural homomorphism from S to $k(A/A_n)$. Thus $I_n \leq L$. Conversely, if L_1 is a maximal ideal of S containing I_n, then $L_1 = L_1^\gamma$. For, if φ is the Frobenius endomorphism $s \mapsto s^q$ of S, then γ acts like φ^n on A/A_n and thus on S/I_n. In addition, φ stabilizes every ideal of S and so $L_1 = L_1^\gamma$. Clearly γ normalizes the Jacobson radical J of S/I_n, so $J = J^{q^n}$. But J is nilpotent, whence it must be trivial. It follows that the intersection of all the maximal ideals of S which are normalized by γ is precisely $\bigcap_{n=0}^\infty I_n$.

We will show that $\bigcap_{n \in T} I_n = 0$, where T is any infinite set of non-negative integers, and for this purpose it is sufficient to prove that $\bigcap_{n \in T} A_n = 1$. For suppose this is true and $0 \neq \lambda = \sum \lambda_a a \in S$. Then no non-identity element of A belongs to infinitely many A_n's. Now the set $\{ab^{-1} \mid a \neq b, \lambda_a, \lambda_b \neq 0\}$ is finite and hence for some integer m no element of this set belongs to any A_n with $n \geq m$. Therefore $\lambda \notin I_n$ for $n \geq m$.

We now prove that $\bigcap_{n \in T} A_n = 1$ by induction on the rank of A. Suppose first that there is a subgroup B such that $1 < B = B^\gamma < A$ and A/B torsion-free. By induction $\bigcap_{n \in T} (A/B)_n$ is trivial, and so $\bigcap_{n \in T} A_n \leq B$. Also by induction $\bigcap_{n \in T} B_n = 1$, whence $\bigcap_{n \in T} A_n = 1$ since in fact $A_n \cap B = B_n$. To see the latter, suppose that $a^\gamma a^{-q^n} \in B$ for some $a \in A$. Put $c = aB$; then $c^\gamma = c^{q^n}$ in A/B. But γ induces an automorphism in the free abelian group A/B. Therefore $c = 1_{A/B}$ and $a \in B$, as required.

Finally, if no such B exists, write $C = \bigcap_{n \in T} A_n$ and let $\chi = \chi(t)$ denote the characteristic polynomial of γ. Then $|A/A_n| = \chi(q^n)$. Since $\chi(q^n) \to \infty$ with n and T is infinite, the orders $|A/A_n|$ are unbounded for $n \in T$. Hence A/C is infinite and, if B/C is the torsion subgroup of A/C, then A/B is also infinite. Since $C = C^\gamma$, we have $B = B^\gamma$ and hence B and C are trivial. ∎

The final ingredient needed in the proof of 7.1.11 is the following deep theorem of Bergman (1971).

7.1.14 *Let A be a plinth of a group G, k any field and $S = kA$. If P is a nonzero ideal of S normalized by a subgroup of finite index in G, then $\dim_k(S/P)$ is finite.*

We are now in a position to undertake the proof of 7.1.11.

Proof of 7.1.11 Suppose that G is a virtually polycyclic group, A is a plinth of G, k is an absolute field and $S = kA$. Let $0 \neq \lambda \in S$. We have to show that *there is a maximal ideal of S which contains no conjugate of λ*. It is not hard to see, by considering the finite subfield of k generated by the coefficients of λ, that k may be assumed finite.

By Mal'cev's Theorem (3.1.8) $G/C_G(A)$ is abelian-by-finite, so there exists $K \lhd G$ such that $|G : K|$ is finite and $[A, K'] = 1$. By definition A is a plinth of K. Let T be a transversal to K in G and write $\lambda^* = \prod_{t \in T} \lambda^t$. Observe that $\lambda^* \neq 0$ since S is a domain. If L is a maximal ideal of S which contains no conjugate of λ^* under K, then it is easy to see that L contains no conjugate of λ under G. So we may assume that $G = K$.

Now G induces an abelian group Γ of automorphisms in A and we may regard A as a $\mathbb{Z}\Gamma$-module. Put $V = A \otimes_{\mathbb{Z}} \mathbb{Q}\Gamma$. Then the hypotheses of 7.1.12 are satisfied since A is a plinth of G. Thus there exists an element $\gamma \in \Gamma$ such that V is a simple $\mathbb{Q}\langle\gamma^m\rangle$-module for all $m \geq 1$. This means that if x is an element of G inducing γ, then A is a plinth of the group $\langle A, x\rangle$. Moreover by 7.1.13 the element x normalizes an infinite number of maximal ideals of S; let $\mathcal{L}$ be the set of all such ideals. Then $L^g \in \mathcal{L}$ for all $L \in \mathcal{L}$ and $g \in G$ since G induces an abelian group of automorphisms in S. By 7.1.8 an ideal L has only finitely many conjugates, with the result that $\mathcal{L}$ splits up into infinitely many conjugacy classes.

If every member of $\mathcal{L}$ contained a conjugate of λ, then infinitely many members of $\mathcal{L}$ would contain λ, and with it the ideal P generated by all the conjugates of λ under $\langle x\rangle$. But A is a plinth of $\langle A, x\rangle$ and we may now deduce that $\dim_k(S/P)$ is finite from 7.1.14. By this contradiction the proof of 7.1.11 is complete. ∎

It now follows from 7.1.9, 7.1.10, and 7.1.11 that, if k is an absolute field, G a virtually polycyclic group, and A an infinite normal abelian subgroup of G, then a simple kG-module M cannot be torsion-free as a kA-module. Thus 7.1.3 is established.

In order to establish 7.1.2 we have to prove that simple R-modules have finite k-dimension, where $R = kG$. By 7.1.3 it suffices to consider *R-modules which are not S-torsion-free.*

Recall from 4.3 that if X is a subset of a commutative ring S and M is an S-module, then, $^*X$ is the submodule of elements of M which are annihilated by X. Also $\pi_S(M)$ will denote the set of all ideals P of S which are maximal with respect to $^*P > 0$. As is well known, if M is not S-torsion-free and S satisfies the maximal condition on ideals, then $\pi_S(M)$ is non-empty and consists solely of prime ideals of S.

The key property of the set $\pi_s(M)$ is that if $P_1, \ldots, P_n$ are distinct elements of it, then $\sum_{i=1}^{n} {}^*P_i$ is a direct sum. For P_1 is prime and contains none of $P_2, \ldots, P_n$, so that $P_1 < P_1 + P_2 P_3 \cdots P_n$. Now $^*P_1 \cap (^*P_2 + \cdots + {}^*P_n)$ is annihilated by $P_1 + P_2 P_3 \cdots P_n$, so it must be trivial, by the definition of $\pi_S(M)$.

We apply these ideas to establish:

7.1.15 *Let J be any ring with identity, G any group and $R = JG$. Let H be a normal subgroup of G and put $S = JH$. Suppose that M is any R-module. If $P \in \pi_S(M)$ and T is a right transversal to the cosets of $N_G(P)$ in G, then $(^*P)R = \sum_{t \in T}(^*P)t$ and this sum is direct.*

Proof Set $N = N_G(P)$. Obviously $(^*P)x = {}^*(P^x)$, where $x \in G$. Since $G = NT$ and $^*P$ is a JN-submodule, it follows that $(^*P)R = \sum_t(^*P)t = \sum_t {}^*(P^t)$. If $t_1, \ldots, t_n$ are distinct elements of T, then $P^{t_1}, \ldots, P^{t_n}$ are distinct elements of $\pi_S(M)$, so the sum $\sum_{i=1}^n {}^*(P^{t_i})$ is direct. ∎

The importance of this result is that it enables us to see that *if M is a simple R-module, then $^*P$ is simple as a JN-module, where $N = N_G(P)$*. Indeed, if U is any JN-submodule of $^*P$, then $UR = \sum_t Ut$ and this sum is direct since $\sum_t(^*P)t$ is. Hence $U = ((^*P)t \cap UR)t^{-1}$ for any $t \in T$, and so the assignment $U \mapsto UR$ is an injective mapping from the set of JN-submodules of $^*P$ to the set of R-submodules of M. Furthermore, this mapping preserves inclusions, which shows that *if M satisfies $max-R$, then $^*P$ satisfies $max-JN$*.

Proof of 7.1.2 (concluded). Recall that G is a virtually polycyclic group, k is an absolute field and $R = kG$. We need to prove that simple R-modules are finite-dimensional over k.

Of course, if G is finite, every finitely generated R-module has finite dimension over k. Thus we may assume G is infinite and proceed by induction on the Hirsch number $h(G)$. Let M be a simple R-module. Since G is infinite and virtually polycyclic, it has an infinite abelian normal subgroup, and by 7.1.10 this subgroup contains a plinth A of some normal subgroup K of finite index in G. Of course M is finitely generated as a K-module and hence there is a maximal kK-submodule U of M. Consider $\bigcap_{g \in G} Ug$: this is a proper R-submodule of M, so it is zero by simplicity of M. Since there are only finitely many distinct Ug's, it suffices to show that M/U has finite dimension over k. Thus we may assume that $G = K$ and A is a plinth for G. Set $S = kA$.

Next M cannot be torsion-free as an S-module by 7.1.3, so we may choose $0 \neq P \in \pi_S(M)$ and put $N = N_G(P)$. Then $^*P$ is a simple kN-module by the discussion immediately preceding this proof. If $|G : N|$ is infinite, $h(N) < h(G)$ and by the induction hypothesis $\dim_k(^*P)$ is finite. But S/P is faithfully represented by endomorphisms of $^*P$, so $\dim_k(S/P)$ is finite. If, on the other hand, $|G : N|$ is finite, Bergman's Theorem (7.1.14) applies to yield once again the finiteness of $\dim_k(S/P)$.

By 7.1.8 we have $A^m \leq 1 + {}^\circ P$ for some $m > 0$, where ${}^\circ P$ is the intersection of all the G-conjugates of P. But ${}^\circ P$ annihilates $(^*P)x$ for all $x \in G$, so it annihilates M. Hence A^m acts trivially on M and M is a $k(G/A^m)$-module. Since $h(G/A^m) < h(G)$, we conclude that M has finite k-dimension by the induction hypothesis. So the proof of 7.1.2 is, at long last, complete. ∎

7.1.16 *The chief factors of a finitely generated nilpotent-by-polycyclic-by-finite group are finite and its maximal subgroups have finite index.*

This may be deduced from 7.1.2 in exactly the same way as 4.3.5 was from 4.3.4.

7.2   Artin–Rees properties and residual finiteness

Now that it has been established that simple modules over integral group rings of virtually polycyclic groups are finite, we are in a position to prove the main theorem of this chapter, which is:

7.2.1 *Finitely generated abelian-by-polycyclic-by-finite groups are residually finite.*

The final step in the argument was first carried out by Jategaonkar (1974) and another proof was later given by Roseblade (1976). We shall follow Roseblade's exposition here.

Recall that the final step in the proof of the residual finiteness of finitely generated abelian-by-nilpotent groups involved a non-commutative version of the Artin–Rees Lemma (4.3.7). So it is not surprising that a further generalization of that result is crucial here. The standard form of the (Strong) Artin–Rees Lemma in commutative algebra is:

7.2.2 *Suppose that S is a commutative ring, I an ideal of S, M a finitely generated S-module and U a submodule of M. Then there is a positive integer m such that*

$$MI^n \cap U = (MI^m \cap U)I^{n-m}$$

for all $n \geq m$.

Since the version of Artin–Rees needed here is based on an analysis of the proof of 7.2.2, we present the proof for ease of reference.

Proof of 7.2.2 Form the subring S_I of the polynomial ring $S[t]$ corresponding to I, that is,

$$S_I = S + It + I^2 t^2 + \cdots = Rg \langle S, It \rangle.$$

Since S satisfies max $-r$, we have $I = y_1 S + \cdots + y_r S$ for some $y_1, \ldots, y_r$ in S. Therefore, $S_I = S[y_1 t, \ldots, y_r t]$ and S_I satisfies the maximal condition on ideals by Hilbert's Basis Theorem.

The graded module

$$M_I = MS_I = M + MIt + MI^2 t^2 + \cdots$$

is a finitely generated S_I-module and hence it has max $-S_I$. It follows that the S_I-submodule

$$U + (MI \cap U)t + \cdots + (MI^n \cap U)t^n + \cdots$$

is finitely generated and hence it must be the same as

$$(U + (MI \cap U)t + \cdots + (MI^m \cap U)t^m)S_I$$

for some m. The Artin–Rees Lemma follows on equating coefficients of t^n for $n \geq m$. ∎

The key observation in this proof is that the Artin–Rees property follows from Hilbert's Basis Theorem: if S is commutative and noetherian, then so is S_I. In the case where S is non-commutative ring with max $-r$, the maximal condition on right ideals, we can still prove that S_I satisfies max $-r$, provided that I is a central ideal of S. (The proof is left as an exercise for the interested reader.)

It is now time to present the version of the Artin–Rees Lemma to be used here. As in Section 7.1, let R be a ring with identity generated by a subring S and a group G which normalizes S, with G regarded as a subgroup of the group of units of R. The result required is:

7.2.3 *Assume that G is a polycyclic group and S satisfies* max $-r$. *Let U be an R-submodule of a finitely generated R-module M. If I is a central ideal of S normalized by G, then there is a positive integer m such that*

$$MI^n \cap U = (MI^m \cap U)I^{n-m}$$

for all $n \geq m$.

Proof First of all we need to show that R has max $-r$. This is essentially another version of the Hilbert Basis Theorem. In fact the argument of 4.2.4 provides the basis for induction on the length of a cyclic series in G (see Passman 1985: 10.2.7).

Set $K = IR$. Then, since $I = I^G$, we have $K = RI$, which shows that K is an ideal of R. Now

$$R_K = \langle R, Kt \rangle = \langle R, It \rangle = \langle S, G, It \rangle = \langle S_I, G \rangle.$$

Clearly S_I is normalized by G and S_I has max $-r$ since I is central. The version of Hilbert's Basis Theorem mentioned in the first paragraph shows that R_K has max $-r$. Hence the equation in the statement holds with K in place of I. However, if V is any R-module, we have $VK = VI$ since $RI = IR$. The result now follows at once. ∎

Proof of 7.2.1 Since a finite extension of a finitely generated, residually finite group is residually finite, we may restrict attention to finitely generated abelian-by-polycyclic groups. Let G be such a group.

Let $1 \neq x \in G$. By Zorn's Lemma (or max $-n$), there is a normal subgroup K of G which is maximal with respect to not containing x. What has to be proved is that G/K is finite, and clearly we may assume $K = 1$, so that every non-trivial normal subgroup of G contains x. Thus there exists a unique minimal normal subgroup L of G. Let M be an abelian normal subgroup of G such that

$\Gamma = G/M$ is polycyclic. Then M is an R-module in the usual way, where $R = \mathbb{Z}\Gamma$. Also $L \leq M$, so L is an R-module.

Now L is a simple R-module, whence it is a finite p-group by 7.1.1. Thus $C = C_G(L)$ has finite index in G. Set $H = C/M$ and let A be any characteristic abelian subgroup of H. Then A is G-invariant and $S = \mathbb{Z}A$ is noetherian. Clearly, $R = \langle S, \Gamma \rangle$ and $S^\Gamma = S$.

The augmentation ideal $J = \mathrm{Ker}(\mathbb{Z}A \to \mathbb{Z})$ is a central ideal of S since A is abelian, and $J = J^\Gamma$. Put $I = pS + J$, which is also a central ideal of S. Now apply 7.2.3 to get $MI^n \cap L \leq LI$ for some $n > 0$. But $LI = 0$ since $A \leq C/M$, and so $MI^n \cap L = 0$. But MI^n is an R-submodule of M, so we obtain $MI^n = 0$.

Taking A to be trivial, we obtain $Mp^m = 0$ for some m. Returning to an arbitrary A, we have $M(a-1)^n = 0$ for all $a \in A$. Consider the image of $\mathbb{Z}\langle a \rangle$ in the endomorphism ring of M. Since p^m and $(a-1)^n$ both map M to zero, it follows that this image is finite, and because A acts faithfully, $\langle a \rangle$ is finite. But A is finitely generated, so it is actually finite.

We conclude that $H/C_H(M)$ has all its abelian characteristic subgroups finite. Now $H/C_H(M)$ is polycyclic and it is easy to show that it must be finite (cf. 1.3.9). But this implies that M is finitely generated as an abelian group and hence it is finite. Therefore G is polycyclic and consequently it is residually finite. Finally, G must be finite since it has the unique minimal normal subgroup L. The proof of 7.2.1 is now complete. ∎

By proving yet another version of the Artin–Rees Lemma based on 7.2.3, Segal (1975*b*) was able to sharpen the Roseblade–Jategaonkar Theorem as follows.

7.2.4 *Suppose that G is a finitely generated group with an abelian normal subgroup A such that G/A is polycyclic.*

(i) *If A is a p-group for some prime p, then G is virtually a residually finite p-group.*

(ii) *If A is torsion-free, then G is virtually a residually finite p-group for almost all primes p.*

(iii) *If $\pi = \pi(A)$, then G is virtually a residually finite nilpotent $\pi \cup \{p\}$-group for almost all primes p.*

A corollary of this result, also due to Segal (1976), is that *a torsion group of automorphisms of a finitely generated nilpotent-by-polycyclic group is locally finite.*

7.3 Frattini properties of finitely generated abelian-by-polycyclic groups

In 4.4.1, it was shown that the Frattini subgroup of a finitely generated virtually metanilpotent group is nilpotent, a result due to P. Hall. Subsequently,

Roseblade (1973*a*) extended this result to:

7.3.1 *The Frattini subgroup of a finitely generated nilpotent-by-polycyclic-by-finite group is nilpotent.*

The key result needed in the proof of 4.4.1 was this: if G is a finitely generated nilpotent group, $R = \mathbb{Z}G$ and M is a finitely generated R-module, then a central element x of R which annihilates every simple quotient of M has a power annihilating M itself—see 4.4.2. What is needed here is an extension of this result.

7.3.2 *Suppose that G is a virtually polycyclic group, $R = \mathbb{Z}G$ and M is a finitely generated R-module. If H is a nilpotent normal subgroup of G such that the augmentation ideal I_H annihilates every simple quotient of M, then some power of I_H annihilates M.*

This result is actually true for a wider class of coefficient rings than $\mathbb{Z}$—see Roseblade (1976). In order to prove the theorem we shall need an important corollary of 7.1.6.

7.3.3 *Let G be a virtually polycyclic group, let A be a subgroup of $Z(G)$ and write $R = \mathbb{Z}G$ and $S = \mathbb{Z}A$. Suppose that M is a finitely generated R-module which is prime[1] as an S-module. Then $\bigcap_L ML = 0$, where L ranges over all the maximal ideals of S.*

If we specialize to the case where A is the identity subgroup, we deduce that, in multiplicative notation, $\bigcap_p M^p = 1$, where p ranges over all primes. This is essentially Lemma 12 of P. Hall (1959).

Proof of 7.3.3 Let P denote the annihilator of M in S, set $S_1 = S/P$ and let R_1 be the group ring of G over S_1. Now R/PR is an image of R_1, so M can be treated as a finitely generated R_1-module which has no S_1-torsion, since it is prime for S. By 7.1.6 and the proof of 7.1.7, there exists a free S_1-submodule F of M and a non-zero ideal Λ of S_1 such that if L is any maximal ideal not containing Λ, then $(ML) \cap F = FL$.

Now let $\mathcal{L}$ be the set of all maximal ideals of S that do not contain Λ and let Δ denote their intersection. Then $\Lambda\Delta$ is contained in every maximal ideal of S_1 and is therefore 0. Hence $\Delta = 0$. If $X = \bigcap_{L \in \mathcal{L}} ML$, then $X \cap F = \bigcap_{L \in \mathcal{L}} FL$ since $(ML) \cap F = FL$. Also, since F is free, $\bigcap_{L \in \mathcal{L}} FL = F\Delta = 0$, which shows that $X \simeq (F + X)/F$. But M/F is a torsion S_1-module by 7.1.6 and X is torsion-free. Therefore, $X = 0$, as required.

Suppose that M is any finitely generated non-zero R-module: we claim that there is a series of R-submodules

$$0 = M_0 < M_1 < \cdots < M_d = M$$

[1] A module is called *prime* if it is non-trivial and each non-trivial submodule has the same annihilator as the module.

such that M_{i+1}/M_i is prime as an S-module for $0 \leq i < d$. For let $P \in \pi_S(M)$: then $^*P$ is prime as an S-module and $^*P$ is also an R-submodule since S is central in R. Put $M_1 = {}^*P$ and repeat the process for M/M_1 etc. Since M satisfies $\max - R$, we obtain ultimately the required series. ∎

If U is an S-module, define

$$T_S(U) = \bigcap_L UL,$$

where L ranges over all the maximal ideals of S. Then it follows from 7.3.3 that $T_S(M_{i+1}) \leq M_i$ for $0 \leq i < d$, so that $T_S^d(M) = 0$, with the obvious notation. Let us use this fact to extend 4.4.2.

7.3.4 *Let G be a virtually polycyclic group, let A be a subgroup of $Z(G)$ and put $R = \mathbb{Z}G$ and $S = \mathbb{Z}A$. Let M be a finitely generated R-module and suppose that some s in S annihilates every simple quotient of M. Then $Ms^d = 0$ for some $d > 0$.*

Proof All we need do is show that $Us \leq T_S(U)$ for any R-submodule U of M. For then $T_S^i(M)s \leq T_S^{i+1}(M)$ and the result follows from the fact that $T_S^d(M) = 0$. Suppose that L is a maximal ideal of S such that $Us \nleq UL$. Then $s \notin L$ and $L + sS = S$, which implies that no simple image of M can be annihilated by L and thus $M = ML$. By the Artin–Rees Lemma (7.2.3) $ML^m \cap U \leq UL$ for some m, which yields $U = UL$, a contradiction. Therefore $Us \leq UL$ and the result follows. ∎

Proof of 7.3.2 Suppose that G is a virtually polycyclic group, $R = \mathbb{Z}G$, M is a finitely generated R-module and H is a nilpotent normal subgroup of G such that I_H annihilates every simple R-quotient of M. We need to prove that M is annihilated by some power of I_H.

We first show that G may be replaced by any normal subgroup K of finite index in G containing H. We need to prove that I_H annihilates every simple $\mathbb{Z}K$-quotient of M. To see this, let M/U be such a quotient and set $V = \bigcap_{x \in G} Ux$. Since K has finite index in G, the module M/V has a finite series of $\mathbb{Z}K$-submodules each factor of which is isomorphic with a G-conjugate of M/U. Also M is $\mathbb{Z}K$-noetherian. Let W be a maximal R-submodule of M containing V and let U_1 be a maximal $\mathbb{Z}K$-submodule of M containing W. Since I_H annihilates M/W, it annihilates M/U_1, and $M/U_1 \simeq^{\mathbb{Z}K} M/Ux$ for some $x \in G$. Since $H \lhd G$, we deduce that $MI_H \leq U$.

The next step in the proof is to reduce to the case where A is a plinth of G, by using induction on $h(G)$ and the nilpotency class of H. If H is finite, set $A = Z_1(H)$ and $K = C_G(A)$. Then K has finite index in G and $\mathbb{Z}A$ is in the centre of $\mathbb{Z}K$. By 7.3.4 each element of I_A acts nilpotently on M and, since I_A is finitely generated, we obtain $MI_A^n = 0$ for some n. Now M/MI_A is a $\mathbb{Z}(G/A)$-module and H/A has smaller nilpotency class than H. By the induction hypothesis $MI_H{}^m \leq MI_A$ for some m. Since A is central in H,

$$MI_A I_H{}^m = MI_H{}^m I_A \leq MI_A{}^2,$$

and so on. Hence $MI_H{}^{mn} = 0$.

We now assume H to be infinite, in which case so is $A = Z_1(H)$ by 1.2.20. Also $A \lhd G$ and so by 7.1.10 it contains a plinth of some normal subgroup G_0 containing H with finite index in G. If we are able to show that $MI_A{}^n = 0$, then, since $h(G/A) < h(G)$, it will follow in essentially the same way as above that $MI_H{}^{mn} = 0$, for some n. Clearly, we may assume $G = G_0$, so that A *is a plinth of G*.

Now fix the notation $R = \mathbb{Z}G$ and $S = \mathbb{Z}A$. Since M satisfies $\max - R$, we may assume that if V is any non-zero submodule of M, then $MI_A{}^r \leq V$ for some $r = r(V)$, that is, we assume the result is true for all proper quotients of M. Therefore we may suppose that if U and V are non-zero submodules of M, then $U \cap V \neq 0$. It follows that either M is torsion-free or it is an elementary abelian p-group for some prime p. For if M is not torsion-free, choose $p \in \pi(M)$ and let M_p be the p-component of M. Then M_p is a submodule, so it has finite exponent by the noetherian property. Hence $p^k M$ is p-free for some k and $p^k M \cap M_p = 0$. Thus $p^k M = 0$ since $M_p \neq 0$. Now pM is a module image of M: thus, if $pM \neq 0$, the result is true for pM and M/pM and we are done. Therefore, we may assume that $pM = 0$. The upshot of this is that M is torsion-free for J where $J = \mathbb{Z}$ or $\mathbb{Z}_p$.

The next step in the proof is to show that if L is a maximal ideal of S not containing I_A, then $M = M(^\circ L)$. For L has finite index in S and the group of automorphisms induced by G in $S/^\circ L$ is finite, as we see from 7.1.8 with k the field of p elements. Hence L has only finitely many G-conjugates, say

$$L_1, L_2, \ldots, L_n.$$

None of these contains I_A, so $S = I_A + L_i$ for $i = 1, 2, \ldots, n$. Multiplying these n equations together, we arrive at $S = I_A + L_1 L_2 \cdots L_n$. But $L_1 \cdots L_n \leq {}^\circ L$, so it follows that $S = I_A + {}^\circ L$. Now $M(^\circ L)$ is a submodule and hence, by the preceding paragraph, $MI_A{}^n \subseteq M(^\circ L)$ for some n. Since $S = I_A + {}^\circ L$, it follows that $M = MS^n \subseteq MI_A{}^n + M(^\circ L) \subseteq M(^\circ L)$. Therefore $M = M(^\circ L)$.

Suppose M is torsion-free as an S-module. Then by 7.1.6 there exists a free S-submodule F of M and a non-zero ideal Λ of S such that every finitely generated S-submodule of M/F is annihilated by some product of G-conjugates of Λ. By the proof of 7.1.7 we deduce that if L is a maximal ideal which contains no conjugates of Λ, then $M = F + ML$ and $(ML) \cap F = FL$. Applying this fact to the maximal ideal L under consideration, where we have $M = ML$ since $M = M(^\circ L)$, we deduce that $FL = F$. But $F > FL$ since F is free. We may therefore conclude that L contains some G-conjugate of Λ.

These considerations demonstrate that every maximal ideal of S contains some conjugate of $I_A \Lambda$. Suppose that P is a maximal ideal of J, so that $P = p\mathbb{Z}$ or $P = 0$ if $J = \mathbb{Z}_p$. We now apply 7.1.11 with $k = J/P$ and deduce that

$I_A \Lambda \leq PS$. Thus

$$MI_A\Lambda \leq \bigcap_P MP,$$

where the intersection is over all maximal ideals P of J. But this intersection is 0 by 7.3.3 with $S = J$, since M is J-torsion-free. Hence $MI_A\Lambda = 0$, which is a contradiction since M is S-torsion-free and both I_A and Λ are non-zero. We may therefore conclude that M is not S-torsion-free and hence there is a non-zero P in $\pi_S(M)$. Note that $P \cap J = 0$ since M is J-torsion-free.

Let $U = {}^*P$ and $V = UR$, and put $N = N_G(P)$. We show that J, A, N, U satisfy the same hypotheses as J, H, G, M, respectively. First of all, U is a finitely generated JN-module—see the argument following the proof of 7.1.15. What we need to show is that any simple JN-image U/U_1 of U is annihilated by I_A. Set $V_1 = U_1R$. By 7.1.15 we have $V > V_1$, so there exists a maximal R-submodule W of V containing V_1. Because V/W is R-simple, it is finite by 7.1.1, and hence it is annihilated by ${}^\circ L$ for some maximal ideal L of S. Since V is non-zero, $MI_A{}^r \leq V$ for some r by hypothesis. If $I_A \leq L$, then $MI_A{}^{r+1} \leq VL \leq W$. On the other hand, if $I_A \not\leq L$, then $M({}^\circ L) = M$, as we saw above, and so

$$MI_A{}^{r+1} = MI_AI_A{}^r \leq M({}^\circ L)I_A{}^r = MI_A{}^r({}^\circ L) \leq W.$$

Thus in either case $MI_A{}^{r+1} \leq W$. Since V/W is R-simple, this implies that $VI_A \leq W$ and hence $UI_A \leq W \cap U = U_1$, as required.

If $|G : N|$ is infinite, the natural induction hypothesis on $h(G)$ gives $UI_A{}^m = 0$, so that $VI_A{}^m = 0$ and hence $MI_A{}^{r+m} = 0$ since $MI_A{}^r \leq V$. So we may assume that $|G : N|$ is finite. If $J = \mathbb{Z}p$, then $\dim_J(S/P)$ is finite by Bergman's Theorem (7.1.14). By 7.1.8 A^ℓ acts trivially on V for some $l > 0$. But $h(G/A^\ell) < h(G)$ and so $VI_A{}^m = 0$ for some $m > 0$ by induction on $h(G)$. Thus once again $MI_A{}^{m+r} = 0$. Hence we may assume $J = \mathbb{Z}$, which means that Bergman's Theorem cannot be used directly. However it can be applied indirectly as follows.

Let $Y = S/P$, where of course $S = \mathbb{Z}A$. By 4.3.3 there is a free abelian subgroup F of Y such that Y/F is a π-torsion group for some finite set of primes π. By Bergman's Theorem $\dim_\mathbb{Q}(\mathbb{Q}A/P\mathbb{Q}A)$ is finite and it follows from this that $r(F) = m$ is finite. Let p be any prime not in π; then $Y = F + pY$ and $pY \cap F = pF$. Hence $Y/pY \cong F/pF$, so that $\dim_{\mathbb{Z}_p}(S/(P + pS)) = m$. The annihilator X_p of U/pU in S clearly contains $P + pS$ and it follows that $\dim_{\mathbb{Z}_p}(S/X_p) \leq m$. Now pV is not zero since M has no $\mathbb{Z}$-torsion. Hence pV contains $MI_A{}^t$ for some $t > 0$, so that $I_A{}^t$ annihilates U/pU. Hence $I_A{}^t \leq X_p$ and this implies that $I_A{}^m \leq X_p$. So $UI_A{}^m \leq \bigcap_{p \notin \pi} pU$, which is 0. Hence $UI_A{}^m = 0$ and therefore $VI_A{}^m = 0$ as before. The proof of 7.3.2 is now complete. ∎

Proof of 7.3.1 (concluded). Suppose first that G is a finitely generated abelian-by-polycyclic-by-finite group and let A be an abelian normal subgroup of G such that $\Gamma = G/A$ is virtually polycyclic. Set $H = \mathrm{Frat}(G)$: then, since $\mathrm{Frat}(G/A)$ is nilpotent, HA/A is a normal nilpotent subgroup of Γ. By essentially the same argument as that given after the statement of 4.4.1, the group H centralizes every simple G-quotient of A. Thus I_H annihilates each such quotient. By 7.3.2 we have $AI_H^n = 0$ for some $n > 0$. Hence A is contained in the nth term of the upper central series of HA and, since HA/A is nilpotent, we deduce that H is nilpotent as well.

Finally, let G be a finitely generated nilpotent-by-polycyclic-by-finite group with $H = \mathrm{Frat}(G)$. Let N be a nilpotent normal subgroup of G such that G/N is virtually polycyclic. Then G/N' is finitely generated abelian-by-polycyclic-by-finite and hence $\mathrm{Frat}(G/N')$ is nilpotent by the first paragraph. Therefore HN'/N' is nilpotent and it follows by Fitting's Theorem that HN/N' is nilpotent. But N is nilpotent, so HN is nilpotent by 1.2.17 and thus H is nilpotent, as required. ∎

Further Frattini properties

Now that 7.3.1 has been proved, it is routine to deduce the exact analogue of 4.4.1.

7.3.5 *Let G be a finitely generated abelian-by-polycyclic-by-finite group. Then $\mathrm{Fit}(G) = HP(G) = \mathrm{FFrat}(G) = Cch(G)$ and this subgroup is nilpotent.*

We mention for comparison a theorem of Segal (1977b).

7.3.6 *The conclusion of 4.4.1 holds for arbitrary subgroups of finitely generated abelian-by-nilpotent-by-finite groups.*

This result was achieved by establishing a further property of noetherian modules over finitely generated nilpotent groups. An R-module M is said to be *residually simple* if the intersection of its maximal submodules is 0. Also M is called *poly-residually simple* if there is a chain $0 = M_0 \leq M_1 \leq \cdots \leq M_n = M$ of submodules of M such that each M_i/M_{i-1} is residually simple.

7.3.7 (Segal 1977b) *Every finitely generated module over the integral group ring of a finitely generated virtually nilpotent group is poly-(residually simple).*

Segal proved this result by exploiting properties of the Krull dimension of a module. Brookes (1988) generalized 7.3.6 by introducing the idea of a calibrated module, establishing:

7.3.8 *Every finitely generated module over the integral group ring of a virtually polycyclic group is poly-(residually simple).*

One corollary of this result is the extension of 7.3.5 to subnormal subgroups of finitely generated abelian-by-polycyclic groups.

7.4 Just non-polycyclic groups

A *just non-polycyclic group* is a group that is not polycyclic but all of whose proper quotients are polycyclic. The importance of these groups stems from the following simple observation.

7.4.1 *A finitely generated group G which is not polycyclic has a just non-polycyclic quotient.*

Proof Let $\{N_\lambda \mid \lambda \in \Lambda\}$ be a chain of normal subgroups such that each G/N_λ is not polycyclic. Then G/N is not polycyclic, where $N = \bigcup_\lambda N_\lambda$. For suppose G/N were polycyclic. Now polycyclic groups are finitely presented by 1.3.1, and, as is well known, this implies that N is the normal closure in G of a finite number of elements—see 11.1.1. Hence $N = N_\lambda$ for some λ, a contradiction. It follows by Zorn's Lemma that there is a normal subgroup M of G which is maximal with respect to G/M not being polycyclic. Then G/M is the required just non-polycyclic quotient. ∎

It follows from 7.4.1 that if **P** is some property of groups which is inherited by quotient groups, then, in order to prove that any finitely generated group with **P** is polycyclic, it suffices to treat the case of just non-polycyclic groups. For this reason just non-polycyclic groups have long been implicit in the literature. Here we restrict ourselves to soluble just non-polycyclic groups.

From the definition it is obvious that soluble just non-polycyclic groups satisfy $\max - n$ and are therefore finitely generated by 4.2.1: also these groups are clearly abelian-by-polycyclic, so they are residually finite. Thus soluble just non-polycyclic groups are a special kind of finitely generated abelian-by-polycyclic group. Soluble just non-polycyclic groups were first studied explicitly by Groves (1978*c*) in the metanilpotent case and subsequently by Robinson and Wilson (1984) in the general case.

A key role in the theory of soluble just non-polycyclic groups is played by the Fitting subgroup.

7.4.2 *Let A be the Fitting subgroup of a soluble just non-polycyclic group G. Then A is abelian and it is either torsion-free or an elementary abelian p-group for some prime p. Furthermore $A = C_G(A)$, so that G/A acts faithfully on A.*

Proof In the first place, A is nilpotent since G satisfies $\max - n$. If $A' \neq 1$, then G/A' is polycyclic since G is just non-polycyclic, and so A/A' is finitely generated. Since A is nilpotent, it is finitely generated by 1.2.16. Hence A, and therefore G, is polycyclic, a contradiction which shows that A is abelian.

Denote the torsion subgroup of A by T. Since G has $\max - n$, we have $T^e = 1$ for some $e > 0$, and hence $A^e \cap T = 1$. Now in a just non-polycyclic group, two non-trivial normal subgroups must intersect non-trivially. So we have either $A^e = 1$ or $T = 1$, that is, either A is of finite exponent or it is torsion-free. In the former case the same argument shows that A must be a p-group for some prime p. If $A^p \neq 1$, then G/A^p is polycyclic and A/A^p is finite. Hence A

is finite, which is impossible. Therefore A is elementary. Finally, $A = C_G(A)$ by 1.2.10. ∎

This result makes it clear that there is a dichotomy in the soluble just non-polycyclic groups, according to whether the Fitting subgroup is torsion-free or elementary abelian p. It is convenient to say that the group has *characteristic zero* in the former case and *characteristic p* in the latter.

The characteristic zero case yields readily to the classical theory of P. Hall which was described in Chapter 4, whereas the characteristic p-case is much more difficult and requires the deep generalization of Hall's work due to Roseblade.

Here is the result which reveals the structure of soluble just non-polycyclic groups of characteristic zero (see Groves 1978c; Robinson and Wilson 1984).

7.4.3 *Suppose that G is a soluble just non-polycyclic group of characteristic zero. Let A be the Fitting subgroup of G and put $Q = G/A$. Then:*

(i) *G is a minimax group and A is torsion-free of finite rank;*
(ii) *Q is virtually abelian, so G is virtually metabelian;*
(iii) *A is rationally irreducible as a Q-module;*
(iv) *G nearly splits over A, that is, there is a free abelian subgroup X of G such that $|G : XA|$ is finite and $X \cap A = 1$. Furthermore, if G actually splits over A, the complements of A fall into finitely many conjugacy classes.*

During the proof of 7.4.3 we will require the following well-known ring theoretic lemma.

7.4.4 *Let R be a commutative ring with identity and let M be an R-module, which has a composition series of finite length. Denote the socle of M by N. If N and M/N have no isomorphic simple submodules, then $M = N$ and M is completely reducible.*

Proof Evidently, we may assume that M/N is simple. Let $N = L_1 \oplus \cdots \oplus L_n$, where each L_i is simple. Denote the annihilators of M/N and L_i by I and I_i, respectively. Then by the Chinese Remainder Theorem there is an r in R such that $r \equiv 0 \pmod{I}$ and $r \equiv 1 \pmod{I_i}$ for $i = 1, 2, \ldots, n$.

Next r induces an endomorphism ξ in M by multiplication. Since $M\xi \leq N < M$ and Fitting's Lemma is valid in M, it follows that $\mathrm{Ker}(\xi) \neq 0$; hence $N \cap \mathrm{Ker}(\xi) \neq 0$ and $\mathrm{Ker}(\xi)$ has a simple submodule isomorphic with some L_i. But then $(L_i)\xi = 0$, which contradicts $r \equiv 1 \pmod{I_i}$. ∎

Proof of 7.4.3

(i) By hypothesis and 7.4.2 the subgroup A is torsion-free and abelian. Also by 4.3.3 there is a finite set of primes π and a free abelian subgroup B of A such that A/B is a π-group. Let q be any prime not in π. Then $A^q \cap B = B^q$ and G/A^q is polycyclic, which shows that A/A^q is finite. Therefore B/B^q is finite and B is finitely generated, which clearly implies that A has finite rank.

Since G/A is polycyclic, G has finite rank. That G is a minimax group is now a consequence of 5.2.8.

(ii) We form a finite series of $\mathbb{Z}Q$-submodules of A in which the factors are $\mathbb{Z}$-torsion-free and rationally irreducible with respect to Q. If F is any factor of this series, $Q/C_Q(F)$ is an irreducible soluble linear group and by 3.1.8 it is abelian-by-finite.

If the intersection of all the $C_Q(F)$'s were non-trivial, it would contain a non-trivial abelian normal subgroup of Q and this would lead to the existence of a nilpotent normal subgroup of G not contained in A. Of course this is impossible since $A = \mathrm{Fit}(G)$. Therefore the intersection of all the $C_Q(F)$'s is trivial and Q is abelian-by-finite.

(iii) Let P be an abelian normal subgroup of finite index in Q. Denote by J the subgroup generated by all the rationally irreducible $\mathbb{Z}Q$-submodules of A and let U/J be the $\mathbb{Z}$-torsion subgroup of A/J. Then A/U is $\mathbb{Z}$-finitely generated. Put $\bar{A} = A \otimes \mathbb{Q}$ and $\bar{U} = U \otimes \mathbb{Q}$ with $\bar{U} \leq \bar{A}$. Now $\bar{U}$ and $\bar{A}/\bar{U}$ cannot have isomorphic simple $\mathbb{Q}P$-submodules; for otherwise U would contain a non-zero finitely generated $\mathbb{Z}P$-submodule and its normal closure in G would be a non-trivial finitely generated abelian normal subgroup of the just non-polycyclic group G, which is clearly impossible.

Since $\dim_{\mathbb{Q}}(\bar{A})$ is finite, we are now able to apply 7.4.4 to the $\mathbb{Q}P$-module $\bar{A}$, concluding that $\bar{A} = \bar{U}$, since, by definition, $\bar{U}$ contains all simple $\mathbb{Q}Q$-submodules of $\bar{A}$. It follows that A/J is a torsion group. Suppose now that B is a proper non-trivial $\mathbb{Z}Q$-submodule of A. If A_0 is a rationally irreducible submodule of A, then $A_0 \cap B \neq 0$ since A/A_0 is finitely generated group. Hence $A_0 B/B$ is torsion, whence JB/B is torsion. It follows that A/B is torsion, so that A is rationally irreducible with respect to Q.

(iv) By (ii) there is an abelian normal subgroup N/A with finite index in G/A. Observe that $C_A(N) = 1$: for otherwise $[A, N] = 1$ since A is rationally irreducible. At this point we apply a theorem on the cohomology of infinite soluble groups—see 10.3.6 below—to show that there is a subgroup X such that $X \cap A = 1$ and $|G : XA|$ is finite. The same result shows that if $G = X \ltimes A$, the complements of A fall into finitely many conjugacy classes. (We remark that it is possible to avoid using the cohomological theorem by giving a direct argument along the lines of 6.1.4.) ∎

The next result demonstrates that there is a link between the characteristic and the chief factors of a soluble just non-polycyclic group.

7.4.5 *Let G be a soluble just non-polycyclic group. Then G has characteristic zero if and only if the chief factors have boundedly finite ranks.*

Proof Let $A = \mathrm{Fit}(G)$. If L/M is a chief factor of G, it is G-isomorphic with either $(L \cap A)/(M \cap A)$ or LA/MA. Since the chief factors of the polycyclic group G/A are finite, it is a question of the ranks of the G-chief factors of A.

If G has characteristic 0, then A has finite rank r by 7.4.3, and so a G-chief factor of A has rank at most r.

Conversely, assume that the chief factors of G have rank at most r and suppose that G has characteristic $p > 0$. Let $m = |GL_r(p)|$ and put $N = G^m$. Thus G/N is finite and N centralizes all the G-chief factors L/M with $L \leq A$.

Now put $H/A = \mathrm{Fit}(G/A)$ and observe that $H \cap N$ centralizes all chief factors of G. Applying Roseblade's Theorem 7.3.5, we may conclude that $H \cap N$ is nilpotent, whence $H \cap N \leq A$ and H/A is finite. But an infinite polycyclic group has infinite Fitting subgroup. Therefore G/A is finite, A is finitely generated, and G is polycyclic, a contradiction which shows that G must have characteristic 0. ∎

Groups of characteristic p

If G is a soluble just non-polycyclic group of characteristic $p > 0$ and G happens to be metanilpotent-by-finite, there is a result corresponding to 7.4.3, which is provable by essentially the same methods. However, if G is not metanilpotent-by-finite, the situation is much more difficult and a detailed analysis of the group structure calls for full use of the Roseblade theory.

Here we content ourselves with a statement of the main result.

7.4.6 (Robinson and Wilson 1984) *Let G_0 be a soluble just non-polycyclic group of prime characteristic p which is not metanilpotent-by-finite. Then G_0 has a just non-polycyclic section $G = H/K$ with the following properties: write $A = \mathrm{Fit}(G)$ and $Q = G/A$. Then:*

(i) *H is subnormal and has finite index in G_0.*
(ii) *$P = \mathrm{Fit}(Q)$ is a plinth of Q and Q is the semidirect product of P and a free abelian subgroup.*
(iii) *If $S = \mathbb{Z}_p P$, then A is a torsion-free S-module with finite S-rank*

This result provides valuable structural information about soluble just non-polycyclic groups.

7.4.7

(i) *Every soluble just non-polycyclic group G_0 is an extension of an elementary abelian p-group by a virtually metabelian group for some prime p.*
(ii) *Let G be a finitely generated soluble group. If G is not polycyclic, then G has a non-polycyclic quotient which is elementary abelian-p-by-virtually metabelian.*

Proof Consider (i); the statement is true if G_0 has characteristic 0 by 7.4.3. Assume that G_0 has characteristic $p > 0$. Then, with the notation of 7.4.6, the index $|G_0 : N_{G_0}(K)|$ is finite and $\bigcap_{g \in G_0} K^g = 1$, since otherwise G_0 would be polycyclic. The assertion now follows quite simply. Finally, we can deduce (ii) by applying 7.4.1. ∎

What 7.4.7(ii) demonstrates is that if a finitely generated soluble group is not polycyclic, the trouble must occur quite 'high up' in the group.

Next we record a notable property of soluble just non-polycyclic groups.

7.4.8 *If G is a soluble just non-polycyclic group, then* $\mathrm{Frat}(G) = 1$.

Proof Let G be any just non-polycyclic group and suppose that $F = \mathrm{Frat}(G)$ is non-trivial. By 7.3.1 the subgroup F is nilpotent and so $F \leq A = \mathrm{Fit}(G)$. Since G/F is polycyclic, A/F is finitely generated. If G has a characteristic prime, it follows at once that A/F is finite. The same is true when G has characteristic 0: for by 7.4.3 the subgroup A is rationally irreducible with respect to G, so that A/F is torsion.

Next $\mathrm{Fit}(G/F) = A/F$ by 7.3.5. Therefore the polycyclic group G/F has finite Fitting subgroup, which implies that it is finite. Hence F is finitely generated abelian and we reach the contradiction that G is polycyclic. ∎

We conclude by giving two applications of the theory of just non-polycyclic groups that are in many ways typical.

7.4.9 (Lennox 1973a) *Let G be a finitely generated soluble group in which every 2-generator subgroup is polycyclic. Then G is polycyclic.*

Proof Suppose the result is false. Noting that the hypothesis is inherited by quotients, we may conclude on the basis of 7.4.1 that there is a just non-polycyclic group G in which every 2-generator subgroup is polycyclic. Let $A = \mathrm{Fit}(G)$. Then A is abelian: let $1 \neq a \in A$. Now for any $g \in G$, the subgroup $\langle a, g \rangle$ is polycyclic, so that $a^{\langle g \rangle}$ is finitely generated. Since G/A is polycyclic, $G = A\langle g_1 \rangle\langle g_2 \rangle \cdots \langle g_n \rangle$ for some $g_i \in G$. It follows that

$$a^G = a^{\langle g_1 \rangle\langle g_2 \rangle \cdots \langle g_n \rangle}$$

is finitely generated and abelian. Since G/a^G is polycyclic, we reach the contradiction that G is polycyclic. ∎

7.4.10 (Lennox 1973b) *Let G be a finitely generated soluble group such that $G/\mathrm{Frat}(G)$ is polycyclic. Then G is polycyclic.*

Proof Suppose the result is false. Then we may assume by 7.4.1 that G is just non-polycyclic. The result now follows at once from 7.4.8. ∎

Finally, we remark that soluble groups which are *just non-supersoluble*, that is, not supersoluble but all their proper quotients are supersoluble, have been studied in detail by Robinson and Wilson (1984). It turns out that these groups have their Fitting subgroups torsion-free abelian of rank 1, and they are in fact precisely the finitely generated non-abelian subgroups of the one dimensional rational affine group, with the single exception of the infinite dihedral group. In addition there is a precise classification and method of construction of these groups: for details see the paper cited.

8

CENTRALITY IN FINITELY GENERATED SOLUBLE GROUPS

The subject of this chapter is certain finiteness conditions relating to centrality, which hold in particular classes of finitely generated soluble groups, notably finitely generated abelian-by-nilpotent groups. These ideas first appeared in a fundamental paper by Lennox and Roseblade (1970) and since then they have been extended to other classes of groups. As is so often the case with deep properties of infinite soluble groups, the underlying techniques come from module theory.

8.1 The centrality theorems

Any group G with $\max -n$ has finite upper central height, that is, the upper central series has finite length. Indeed, if $H \triangleleft G$, then $Z_n(H) \triangleleft G$, and so H too has finite upper central height. Thus normal subgroups of finitely generated abelian-by-polycyclic groups have finite upper central height, since these groups satisfy $\max -n$ by 4.2.2. Furthermore, it follows from 4.3.8 that there is a bound on the upper central heights of the normal subgroups of a finitely generated abelian-by-nilpotent group.

We shall say that a group G is *stunted* if there is a bound on the upper central heights of all of its subgroups and that the *height* of G is the least such bound. In answer to a question of P. Hall, Lennox, and Roseblade (1970) proved the following theorem:

8.1.1 *Finitely generated abelian-by-nilpotent groups are stunted.*

The proof of this result will occupy us in much of the current chapter.

Hall was also interested in $\max -c$, the *maximal condition for centralizers*. Thus a group G has $\max -c$ if and only if any ascending chain of centralizers of subgroups

$$C_G(H_1) \leq C_G(H_2) \leq \cdots \leq C_G(H_n) \leq \cdots$$

terminates. Notice that since $C_G(C_G(C_G(H))) = C_G(H)$ for any $H \leq G$, the property $\max -c$ is equivalent to $\min -c$, the *minimal condition for centralizers*. We shall say that a group G has *finite gap number* if there is a finite upper bound on the number of strict inclusions, or *gaps*, in any ascending chain of centralizers in G. The second main result to be established in the chapter is:

8.1.2 (Lennox and Roseblade 1970) *Finitely generated abelian-by-nilpotent groups have finite gap number.*

This result depends on a further centrality property. A group G is called *eremitic* if there exists a positive integer e such that $C_G(x^n) \leq C_G(x^e)$ for all $n > 0$ and $x \in G$: equivalently, whenever an element of G has some positive power in a centralizer, it has its eth power in it. The least such integer e is called the *eremiticity* of G. Thus the groups of eremiticity 1 are precisely the groups in which all the centralizers are isolated. The third main theorem of the chapter is:

8.1.3 (Lennox and Roseblade 1970) *Finitely generated abelian-by-nilpotent groups are eremitic.*

All the main results 8.1.1, 8.1.2, and 8.1.3 are proved by a common plan. Just as in Chapter 4, what we need are certain properties of modules over finitely generated nilpotent groups. In order to explain this, suppose Γ is any group and A is a Γ-module. Write A multiplicitively and form the split extension $\Gamma \ltimes A$. We define, for $n > 0$ and $H \leq \Gamma$,

$$A_n(H) = A \cap Z_n(HA),$$

so that

$$A_n(H) = \{a \in A \,|\, [a, x_1, \ldots, x_n] = 1, \ \forall \ x_1, x_2, \ldots, x_n \in H\}.$$

Thus $A_1(H)$ is the set of H-fixed points in A.

The pair (Γ, A) is said to be *stunted* if there is a $k > 0$ such that $A_k(H) = A_{k+1}(H)$ for all $H \leq \Gamma$. The pair (Γ, A) is *eremitic* if there is an $e > 0$ such that $A_1(x^n) \leq A_1(x^e)$ for all $n > 0$ and $x \in \Gamma$: that is, $[a, x^n] = 1$ implies $[a, x^e] = 1$, where $a \in A$.

Now suppose that G is a finitely generated abelian-by-nilpotent group and let A be an abelian normal subgroup with $\Gamma = G/A$ nilpotent. Clearly Γ is stunted of height equal to its nilpotent class and an easy induction on the class of Γ shows that Γ is eremitic. We now have the simple result:

8.1.4 *The group G is stunted and eremitic if the pair (Γ, A) is stunted and eremitic.*

Proof It is evident that G is stunted if the pair (Γ, A) is. Suppose, then, that (Γ, A) is eremitic of eccentricity e_1 and Γ has eremiticity e_2. Then G is eremitic of eremiticity dividing de_1e_2, where d is the exponent of the torsion subgroup of A (which is finite since G satisfies max $-n$ by 4.2.2).

To see why this holds, suppose that $[y, x^n] = 1$, where $y, x \in G$. We need to show that $[y, x^{de_1e_2}] = 1$. Let $z = [y, x^{e_1e_2}]$. Since Γ has eremiticity e_2, we have $[y, x^{e_2}] \in A$ and hence $z \in A$. Also $[z, x^n] = 1$. Therefore $[z, x^{e_1}] = 1$, so $[z, x^{e_1e_2}] = 1$ and thus $[y, x^{e_1e_2}, x^{e_1e_2}] = 1$. Since $[y, x^{e_1e_2n}] = 1$, we have $z^n = [y, x^{e_1e_2}]^n = [y, x^{e_1e_2n}] = 1$ and so $z^d = 1$. It follows at once that $[y, x^{de_1e_2}] = 1$, as required. ∎

This result allows us to concentrate on pairs (Γ, A), where A is a Γ-module. The main results will be proved by focusing attention on submodules of a very special kind. We need first of all to define for Γ a finitely generated nilpotent group and A any Γ-module a collection, $\mathbf{X}(\Gamma, A)$ of subgroups of Γ whose members are to be characterized by the following properties. Let $X \in \mathbf{X}(\Gamma, A)$. Then:

(i) $A_1(X) > 1$;
(ii) $N_\Gamma(X)$ is isolated;

further, if $X < H \leq \Gamma$ and $A_1(H) > 1$, then:

(iii) $|H : X|$ is finite;
(iv) $N_\Gamma(H)$ is not isolated.

Since it is not immediately apparent that such subgroups X exist, our first concern is to deal with this problem.

8.1.5 *The set* $\mathbf{X}(\Gamma, A)$ *is non-empty.*

Proof In the proof we will need certain isolator properties of a finitely generated nilpotent group Γ, some of which were developed in Chapter 2.

Suppose that $H \leq \Gamma$. We shall denote the normalizer $N_\Gamma(H)$ by $N(H)$ and the isolator $I_G(H)$ by $I(H)$. By 2.3.1 the subset $I(H)$ is a subgroup, while $I(N(H)) = N(I(H))$ by 2.3.7. We also need one further property: there is a subgroup H_0 of finite index in H such that $N(H_0) = I(N(H))$.

To prove this, put $N = N(H)$, $I = I(H)$ and $K = I(N)$. Then $|K : N|$ is finite by 2.3.1, so H has only finitely many conjugates under K. Now $H \leq I \lhd K$ since $I(N) = N(I)$, and $|I : H|$ is finite, so each of these conjugates has finite index in I. Define $H_0 = \bigcap_{x \in K} H^x$, so that $|H : H_0|$ is finite. Then it follows that $I = I(H_0)$ and hence that $K = N(I(H_0)) = I(N(H_0))$. Therefore $N(H_0) \leq K$. But K normalizes H_0, so $K = N(H_0)$, as required.

Since A is non-trivial and $A = A_1(1)$, there certainly exist subgroups H of Γ such that $A_1(H) > 1$. Let d be the maximum Hirsch number for any $H \leq \Gamma$ such that $A_1(H) > 1$. Let H_0 be a subgroup of finite index in such an H satisfying $N(H_0) = I(N(H))$, which exists by the last paragraph. Clearly $A_1(H_0) \geq A_1(H)$, so that $A_1(H_0) > 1$. Moreover, $h(H_0) = h(H)$ and $N(H_0)$ is isolated. Thus H_0 satisfies the first three criteria for membership in $\mathbf{X}(\Gamma, A)$.

However, in order to ensure that H_0 satisfies the final criterion, we may need to enlarge H_0 a little as follows. If $H_0 \leq L$ and $A_1(L) > 1$, then $|L : H_0|$ is finite by maximality of d. Therefore, $I(H_0) = I(L)$ and so $N(I(L)) = N(I(H_0)) = I(N(H_0)) = N(H_0)$, since $N(H_0) = I(N(H))$ is isolated. Consequently, $I(N(L)) = N(H_0)$ and thus $|N(H_0) : N(L)|$ is finite. Hence $N(L)$ is isolated if and only if $N(L) = N(H_0)$.

Now let X be a normal subgroup of $N(H_0)$ containing H_0 which is maximal with respect to $A_1(X) > 1$. Then $H_0 \leq X \lhd N(H_0)$ and hence $N(X) \geq N(H_0)$. Since $A_1(X) > 1$, the index $|X : H_0|$ is finite and, on replacing L by X in the above argument, we obtain $N(X) \leq N(H_0)$. Hence $N(X) = N(H_0)$ and $N(X)$ is isolated. Thus (i), (ii), and (iii) hold for the subgroup X.

If $H > X$ and $A_1(H) > 1$, then $|H : X|$ is finite and so $I(H) \leq I(X) \leq I(H_0)$. If $N(H)$ is isolated, then $N(H) = N(H_0)$ since $N(H_0)$ is isolated, so that $H \lhd N(H_0)$, in contradiction to the maximality of X. Hence $N(H)$ is not isolated and (iv) holds. The proof of 8.1.5 is now complete.   ∎

We will show in the following section that for $X \in \mathbf{X}(\Gamma, A)$, the Γ-module $A_1(X)^\Gamma$ has very desirable properties.

8.2   The Fan Out Lemma

We now introduce the major tool needed to establish the centrality theorems of 8.1.

8.2.1 (The Fan Out Lemma) *Suppose that Γ is a finitely generated nilpotent group and A any non-trivial Γ-module. If $X \in \mathbf{X}(\Gamma, A)$, then*

(F1) $A_1(X)^\Gamma = \mathrm{Dr}_{t \in T}\, A_1(X)^t$, *where T is a right transversal to $N_\Gamma(X)$ in Γ.*
(F2) *If B is a submodule of A, then $B \cap A_1(X)^\Gamma > 1$ if and only if $B_1(X) > 1$.*

This key result will be proved by induction on $h(\Gamma)$. But first we will need some corollaries of the statements (F1) and (F2). We shall use $X, X_1, X_2, \ldots$ to denote members of the set $\mathbf{X}(\Gamma, A)$: also $U, U_1, U_2, \ldots$ are the corresponding subgroups $A(X), A_1(X_1), A_1(X_2), \ldots$ and $Y, Y^{(1)}, Y^{(2)}, \ldots$ are the submodules $U^\Gamma, U_1^\Gamma, U_2^\Gamma, \ldots$. In addition N will denote the normalizer in Γ of X and T will be a transversal to N in Γ. For $H \leq \Gamma$, define

$$T_H = \{t \mid t \in T, H \leq N^t\}.$$

The three consequences of (F1) and (F2) we need are given in:

8.2.2 *If* (F1) *and* (F2) *hold for all $X \in \mathbf{X}(A, \Gamma)$, then so do the following statements:*

(F3) $Y_n(H) = \mathrm{Dr}_{t \in T_H}\, U_n^t(H)$, *for all $n \geq 0$ and $H \leq \Gamma$.*
(F4) *If $X_\lambda, (\lambda \in \Lambda)$, are mutually Γ-inconjugate members of $\mathbf{X}(A, \Gamma)$, then*

$$\left\langle Y^{(\lambda)} \mid \lambda \in \Lambda \right\rangle = \mathrm{Dr}_{\lambda \in \Lambda}\, Y^{(\lambda)}.$$

(F5) *If $X_1, \ldots, X_r$ are mutually Γ-inconjugate and if B is a submodule of A such that $B_\Gamma \left(Y^{(1)} \cdots Y^{(r)} \right) > 1$, then there exists an integer i such that $1 \leq i \leq r$ and $B_1(X_i) > 1$.*

Proof

1. (F1) *implies* (F3). Suppose (F1) holds and $1 \neq \xi \in Y_n(H)$. Assume also that $n > 0$ and (F3) holds with $n - 1$ in place of n. By (F1) we have

$$Y = \mathrm{Dr}_{t \in T}\, U^t, \tag{8.1}$$

so $\xi = u_1^{t_1} \cdots u_r^{t_r}$, where $u_i \in U \backslash 1$ and the t_i are distinct elements of T. Let $x \in H$. Then $[\xi, x] \in Y_{n-1}(H)$ and so by induction on n

$$u_1^{-t_1} \cdots u_r^{-t_r} u_1^{t_1 x} \cdots u_r^{t_r x} \in \operatorname*{Dr}_{t \in T_H} U_{n-1}^t(H). \tag{8.2}$$

Suppose that $U^{t_i x} \neq U^{t_j}$ for all $1 \leq j \leq r$. Then $u_i^{t_i x}$ must be a component element of the direct product in equation (8.2). Hence $n > 1$ and $u_i^{t_i x} \in U^t$, for some $t \in T_H$. From equation (8.1) we have $U^{t_i x} = U^t$ and so $N t_i x = N t$. But $t \in T_H$ and so $H \leq N^t = N^{t_i x}$. Hence $H \leq N^{t_i}$ and therefore $U^{t_i x} = U^{t_i}$ since $x \in H$, which is a contradiction.

Thus H permutes the subgroups $U^{t_1}, \ldots, U^{t_r}$ by conjugation and there exists a subgroup H_0 of finite index in H normalizing all the U^{t_i}. By equation (8.1) N is the normalizer of U in Γ and hence $H_0 \leq N^{t_i}$ for all i. By the definition of $\mathbf{X}(\Gamma, A)$, the subgroup N is isolated, so $H \leq N^{t_i}$ for all i. Therefore, $t_i \in T_H$ and equation (8.2) now shows that $u_i^{-t_i} u_i^{t_i x} \in U_{n-1}^{t_i}(H)$ for all i. It follows that $u_i \in U_n(H)$ and thus $\xi \in \operatorname{Dr}_{t \in T_H} U_n^t(H)$, as required.

2. (F4) *follows from* (F1) *and* (F2). Assume that (F1) and (F2) hold. Then (F3) holds as well by (1). We show by induction on r that if $X_1, \ldots, X_r$ are mutually inconjugate elements of $\mathbf{X}(A, \Gamma)$, the product $Y^{(1)} \cdots Y^{(r)}$ is direct.

Suppose that $r > 1$ and $B = Y^{(2)} Y^{(3)} \cdots Y^{(r)}$ is direct, but $Y^{(1)} B$ is not. Hence $Y^{(1)} \cap B > 1$ and so $B_1(X_1) > 1$ by (F2). Since B is a direct product, $Y_1^{(i)}(X_1) > 1$ for some i, where $2 \leq i \leq r$. Let $X = X_i$ and $Y = U^\Gamma$, where $U = A_1(X)$. Since $i \geq 2$, we conclude that X and X_1 are not conjugate. Also $Y_1(X_1) > 1$.

By (F3) we have

$$Y_1(X_1) = \operatorname*{Dr}_{t \in T_{X_1}} U_1^t(X_1),$$

and so $Y_1(X_1) > 1$ implies that $U_1^t(X_1) > 1$ for some t. Thus $A_1(X^t) \cap A_1(X_1) > 1$, whence

$$A_1(\langle X^t, X_1 \rangle) > 1. \tag{8.3}$$

By non-conjugacy of X and X_1 and the definition of $\mathbf{X}(\Gamma, A)$, neither of X^t and X_1 contains the other. Therefore, on setting $H = \langle X^t, X_1 \rangle$, we obtain $X^t < H$ and $X_1 < H$.

But X^t and X_1 belong to $\mathbf{X}(\Gamma, A)$ and thus $|H : X^t|$ and $|H : X_1|$ are finite by equation (8.3). Hence $I(H) = I(X^t) = I(X_1)$, so that $N(I(H)) = N(I(X^t)) = N(I(X_1))$. Therefore $I(N(H)) = I(N(X^t)) = I(N(X_1)) = N(X^t) = N(X_1)$, since the last two subgroups are isolated. Consequently $N(H) \leq N(X^t) = N(X_1)$. But $H = \langle X_1, X^t \rangle$, so $N(X_1) = N(X^t)$ normalizes H and hence $N(X_1) = N(X^t) = N(H)$ is isolated. Since $X_1 < H$, we obtain a contradiction to criterion (iv) for membership in $\mathbf{X}(\Gamma, A)$.

3. (F5) *follows from* (F1) *and* (F2). The proof goes by induction on r, the case $r = 1$ being (F2). So let $r > 1$ and assume that $B \cap (Y^{(1)} Y^{(2)} \cdots Y^{(r)}) > 1$: set $C = Y^{(2)} \cdots Y^{(r)}$. Suppose that $B \cap Y^{(1)} = B \cap C = 1$. Since $B \cap (Y^{(1)} C) > 1$, we have $Y^{(1)} \cap (BC) > 1$ and so from (F2) we obtain $(BC)_1(X_1) > 1$, that is,

$(BC) \cap A_1(X_1) > 1$. Since the product BC is direct, either $B_1(X_1) > 1$, which contradicts $B \cap Y^{(1)} = 1$, or else $C_1(X_1) > 1$, in contradiction to $C \cap Y^{(1)} = 1$. This completes the proof of 8.2.2. ∎

To prove the Fan Out Lemma one further property of the set $\mathbf{X}(\Gamma, A)$ is needed.

8.2.3 *Suppose that* $X \in \mathbf{X}(\Gamma, A)$ *and* Δ *is an isolated subgroup of* Γ *containing* $N_\Gamma(X)$. *Then* $X \in \mathbf{X}(\Delta, A)$.

Proof We need to check criteria (i)–(iv) for membership in $\mathbf{X}(\Delta, A)$. Let $N = N_\Delta(X)$. Since $X \le N \le \Delta$ and $A_1(X) > 1$, conditions (i) and (ii) hold. Clearly (iii) holds for subgroups H of Δ with $X < H$ since $X \in \mathbf{X}(\Gamma, A)$. For (iv) we need to show that if $X < H \le \Delta$ and $A_1(H) > 1$, then $N_\Delta(H)$ is not isolated in Δ. By (iii) $|H : X|$ is finite and so $I(N(H)) = N(I(H)) = I(N(X)) = N(X)$. Hence $N(H) \le N(X) \le \Delta$. Therefore $N(H) \le N \le \Delta$, so that $N(H) = N_\Delta(H)$. If $N_\Delta(H)$ were isolated in Δ, then $N(H) = N_\Delta(H)$ would be isolated in Γ since Δ is. But this contradicts the fact that $X \in \mathbf{X}(\Gamma, A)$. ∎

Proof of the Fan Out Lemma (8.2.1) We establish (F1) and (F2) by induction on $h = h(\Gamma)$. If $h = 0$, the group Γ is finite and so, if $X \in \mathbf{X}(\Gamma, A)$, then $N(X) = \Gamma$. Thus $X \lhd \Gamma$ and (F1) and (F2) are trivially true. Suppose, then, that $h > 0$ and the result holds for all pairs (Δ, A) with $h(\Delta) < h$. Let Γ have nilpotent class c. For any $H \le \Gamma$, write $N^1(H)$ for $N_\Gamma(H)$ and, inductively, $N^{i+1}(H) = N(N^i(H))$: thus $N^c(H) = \Gamma$. Denote by $m_\Gamma(H)$ the smallest positive integer m such that $N^m(H) = \Gamma$.

Let $X \in \mathbf{X}(\Gamma, A)$. If $m_\Gamma(X) = 1$, then $X \lhd \Gamma$ and (F1) and (F2) are trivially true. So we may assume $m_\Gamma(x) = m > 1$. Put $\Delta = N^{m-1}(X)$; then $\Delta \lhd \Gamma$. Since $X \in \mathbf{X}(\Gamma, A)$, the subgroup $N_\Gamma(X)$ is isolated and, since the operators I and N commute, it follows that $N^i(X)$ is isolated for all i. Thus Δ is isolated and consequently $h(\Delta) < h$. By the induction hypothesis (F1) and (F2) hold for (Δ, A).

We note that $X^x \in \mathbf{X}(\Gamma, A)$ and $N(X^x) = N(X)^x \le \Delta^x = \Delta$ for any $x \in \Gamma$, so by 8.2.3, we have $X^x \in \mathbf{X}(\Delta, A)$.

Now let R be a right transversal to $N = N(X)$ in Δ. Applying (F1) to (Δ, A), we obtain

$$A_1(X)^\Delta = \underset{r \in R}{\mathrm{Dr}} \, A_1(X)^r. \tag{8.4}$$

Let S be a right transversal to Δ in Γ. If s_1, s_2 are distinct elements of S, then X^{s_1} and X^{s_2} are not conjugate in Δ. For, if $X^{s_1 \delta} = X^{s_2}$ with $\delta \in \Delta$, then $s_1 \delta s_2^{-1} \in N$ and so $s_1 \delta s_2^{-1} \in \Delta$. But $\Delta \lhd \Gamma$, so $s_1 s_2^{-1} \in \Delta$ and hence $s_1 = s_2$, a contradiction.

Since (F1) and (F2) hold for (Δ, A), it follows from 8.2.2 that (F4) holds as well, so that the product

$$\prod_{s \in S} A_1(X^s)^\Delta \tag{8.5}$$

is direct. Now set $T = RS$, so that T is a right transversal to N in Γ. Then

$$A_1(X)^\Gamma = A_1(X)^{\Delta S} = \prod_{s \in S} A_1(X)^{\Delta s} = \prod_{s \in S} A_1(X^s)^\Delta.$$

We observed that this product is direct over S, and equations (8.4) and (8.5) show that

$$A_1(X)^\Gamma = \prod_{s \in S} A_1(X)^{\Delta s} = \operatorname*{Dr}_{\substack{r \in R \\ s \in S}} A_1(X)^{rs}.$$

Finally, $Y = \operatorname{Dr}_{t \in T} U^t$, where $U = A_1(X)$. Since this holds for the transversal T, it holds for all transversals. Thus (F1) is established.

In proving (F2) we may assume by induction and 8.2.2 that (F5) holds for (Δ, A). We now deduce that if B is any Δ-submodule of A and if

$$B \cap \left(\prod_{s \in S} A_1(X^s)^\Delta \right) > 1,$$

then $B_1(X^s) > 1$ for some $s \in S$.

Suppose B is any Γ-submodule of A satisfying $B \cap Y > 1$. By what we have just shown (8.5), there is an $s \in S$ such that $B_1(X^s) > 1$, since B is a Δ-submodule. But $B^{s^{-1}} = B$ and hence $B_1(X) > 1$, as required. ∎

For another version of the Fan Out Lemma see K. A. Brown (1981a).

8.3   Proofs of the main centrality theorems

Our object here is to establish 8.1.1, 8.1.2, and 8.1.3. The strategy will be to prove 8.1.1 and 8.1.3 first: it will then be relatively straightforward to deduce 8.1.2.

Let Γ be any group and A a Γ-module. First we indicate a simple reduction. Suppose that B is a submodule such that (Γ, B) and $(\Gamma, A/B)$ are stunted of respective heights h_1 and h_2. Then it follows easily that (Γ, A) is stunted of height at most $h_1 + h_2$. Furthermore, if (Γ, B) and $(\Gamma, A/B)$ are eremitic with eremiticities e_1 and e_2, and if the torsion subgroup of B has finite exponent d, then an argument very similar to that used in 8.1.4 shows that (Γ, A) is eremitic with eremiticity dividing de_1e_2.

The basic idea of the proof in the case where Γ is a finitely generated nilpotent group and A is a noetherian Γ-module is to construct a series of submodules

$$1 = A^{(0)} < A^{(1)} < \cdots < A^{(r)} = A$$

such that $(\Gamma, A^{(i+1)}/A^{(i)})$ is both stunted and eremitic. Since all the $A^{(k)}$ are noetherian, the torsion subgroup of $(A^{(i+1)}/A^{(i)})$ has finite exponent and it will then follow that (Γ, A) is both stunted and eremitic. The series of submodules is constructed using the set $\mathbf{X}(\Gamma, A)$ in the following way.

First observe that the subgroups in $\mathbf{X}(\Gamma, A)$ fall into finitely many conjugacy classes. To see this simply note that if Λ in (F4) were infinite, we could immediately construct an infinite properly ascending chain of submodules of A. Take $X_1, \ldots, X_r$ to be a complete set of representatives of the conjugacy classes of $\mathbf{X}(\Gamma, A)$ and form the submodule $A^* = Y^{(1)} \cdots Y^{(r)}$, where $Y^{(i)} = A_1(X_i)^{\Gamma}$: we will refer to A^* as the *special submodule* of A. Note that $A^* \neq 1$. Put $A^{(0)} = 1$ and assume that $A^{(i)}$ has been defined and $C_i = A/A^{(i)}$ is non-trivial. Now define $A^{(i+1)}$ by

$$A^{(i+1)}/A^{(i)} = C_i^*,$$

so that $A^{(1)} = A^*$. Since A is noetherian, $A^{(s)} = A$ for some $s > 0$ and each factor of the chain

$$1 = A^{(0)} < A^{(1)} < \cdots < A^{(s)} = A$$

is the special submodule C^* of some non-trivial noetherian Γ-module C.

The preceding discussion shows that (Γ, A) will be stunted and eremitic provided that (Γ, A^*) is. As before let $X_1, \ldots, X_r$ be a complete set of representatives of the conjugacy classes of $\mathbf{X}(\Gamma, A)$. Then (F4) shows that $A^* = \mathrm{Dr}_{i=1}^{r} Y^{(i)}$, where $Y^{(i)} = A_1(X_i)^{\Gamma}$, $1 \leq i \leq r$. It is also clear from the Fan Out Lemma that if $H \leq \Gamma$ and $n \geq 0$, then $A^*{}_n(H) = \mathrm{Dr}_{i=1}^{r} Y_n^{(i)}(H)$ and so (Γ, A^*) is stunted and eremitic, provided that $(\Gamma, Y^{(i)})$ is for $1 \leq i \leq r$.

Let X denote a typical X_i and put $U = A_1(X)$ and $Y = U^{\Gamma}$: consider the pair (Γ, Y). From (F1) we obtain $Y = \mathrm{Dr}_{t \in T} A_1(X)^t$, where T is a right transversal to $N = N_{\Gamma}(X)$ in Γ. Since N normalizes X and A, it normalizes $U = A_1(X)$, so that U is an N-module. In fact we have:

8.3.1 U *is a noetherian N-module.*

Proof   Let $U_1 \leq U_2 \leq \cdots \leq U_n \leq \cdots$ be any ascending sequence of N-submodules of U. Since A is noetherian, the sequence $U_1^{\Gamma} \leq U_2^{\Gamma} \leq \cdots \leq U_n^{\Gamma} \leq \cdots$ breaks off and therefore so does the chain $U_1^{\Gamma} \cap U \leq U_2^{\Gamma} \cap U \leq \cdots \leq U_n^{\Gamma} \cap U \leq \cdots$. However, since $U_i^N = U_i$, it follows that $U_i^{\Gamma} = \prod_{t \in T} U_i^t$, and this product is direct by (F1). Hence $U_i^{\Gamma} \cap U$ is merely the 1-component of U_i^{Γ}, that is, U_i itself. Therefore the chain of U_i's breaks off, as claimed. $\blacksquare$

We now exploit the Fan Out Lemma once more to show that (Γ, Y) is stunted and eremitic if the pair (N, U) has these properties. For suppose (N, U) is stunted of height h and eremitic of eremiticity e. The same is clearly true for each of the pairs (N^t, U^t), where t belongs to a right transversal T to N in Γ. If $H < N$, we have $U_{h+1}^t(H) = U_h^t(H)$. Further, if $x \in N^t$ and $n > 0$, then $U_1^t(x^n) \leq U_1^t(x^e)$. The property (F3) yields $Y_n(H) = \mathrm{Dr}_{t \in T_H} U_n^t(H)$ for $H \leq \Gamma$, where $T_H = \{t \mid t \in T, \ H \leq N^t\}$. We deduce at once that $Y_{h+1}(H) = Y_h(H)$, so that (Γ, Y) is stunted.

If $n > 0$ and $U_1^t(x^n) > 1$, then, by property (iii) of $\mathbf{X}(\Gamma, A)$, we have $x^n \in N^t$. But N^t is isolated since N is isolated, and thus $x \in N^t$. By (F3) once again

$$Y_1(x^n) = \mathop{\mathrm{Dr}}_{t \in T_{\langle x \rangle}} U_1^t(x^n).$$

But $U_1^t(x^n) \leq U_1^t(x^e)$, so $Y_1(x^n) \subseteq Y_1(x^e)$. Hence (Γ, Y) is eremitic.

By our argument so far it remains only to prove that (N, U) *is stunted and eremitic*. Now X centralizes U and so by 8.3.1 the N/X-module U is noetherian. Let $L = N/X$. Then we may replace (N, U) by (L, U). Suppose, further, that $1 < K \leq L$ and $U_1(K) > 1$. Then $K = H/X$ for some $H > X$ and $A_1(H) > 1$. Since $X \in \mathbf{X}(\Gamma, A)$, we see that $|H : X|$ is finite and so K is finite. The statement to be proved is therefore equivalent to:

8.3.2 *Suppose that* Γ *is a finitely generated nilpotent group and* A *is a noetherian* Γ*-module such that* $A_1(K) = 1$ *for all infinite subgroups* K *of* Γ. *Then* (Γ, A) *is stunted and eremitic.*

Proof Let P denote the torsion subgroup of Γ. Then P is finite and, since $x \in \Gamma$ and $A_1(x) > 1$ implies $x \in P$ by hypothesis, the pair (Γ, A) is both stunted and eremitic if (P, A) is.

Now the natural semidirect product $P \ltimes A$ is a finite extension of the abelian group A and the torsion subgroup of A has finite exponent since A is a noetherian Γ-module. Therefore 8.3.2 will follow at once from the next result: ∎

8.3.3 *Suppose that* Γ *is a finite group of order* r *and* A *is a* Γ*-module whose torsion subgroup has finite exponent.*

(i) *If* $A = A_\omega(\Gamma)$, *then* Γ *acts nilpotently on* A.
(ii) *The pair* (Γ, A) *is eremitic.*

Proof

(i) Let T be the torsion subgroup of A. Let $B = A/T$ and $b \in B_2(\Gamma)$. If $x \in \Gamma$, then $1 = [b, x^r] = [b, x]^r$. Since B is torsion-free, we have $[b, x] = 1$ and so $B_2(\Gamma) = B_1(\Gamma)$. But $A = A_\omega(\Gamma)$, so $B = B_\omega(\Gamma) = B_1(\Gamma)$ and hence $[A, \Gamma] \leq T$. We may therefore assume that A is torsion.

Clearly we may suppose A to be a p-group for some prime p, of exponent p^e say. Certainly p'-elements of Γ act trivially on A, so we may also assume that Γ is a p-group, of order p^s, say. We prove by induction on $|\Gamma|$ that $[A, {}_{p^{es}}\Gamma] = 1$, which is clear for $s = 0$. Let $r \geq 1$ and let Δ be a maximal subgroup of Γ. Then $|\Gamma : \Delta| = p$ and $[A, {}_{p^{e(s-1)}}\Delta] = 1$ by the induction hypothesis. Put $B = A_{i+1}(\Delta)/A_i(\Delta)$. Then Γ acts on B via an element x, where $\Gamma = \langle \Delta, x \rangle$. Since $(x - 1)^p \equiv x^p - 1 \pmod{p}$ and $x^p \in \Delta$, we get $[B, {}_p\Gamma] = [B, {}_p x] \leq B^p$ and hence $[B, {}_{p^e}\Gamma] = 1$. It follows that $[A, {}_{p^{es}}\Gamma] = 1$.

(ii) The eremiticity of (Γ, A) is proved by an argument very similar to that given in the proof of 8.1.4.

This completes the proofs of the centrality theorems 8.1.1 and 8.1.3. ∎

As a further corollary to 8.1.3, we now prove 8.1.2, that finitely generated abelian-by-nilpotent groups have finite gap number. Of course this theorem, like 8.1.1 and 8.1.3, depends on a result about modules.

Suppose that Γ is a finitely generated nilpotent group and A is a noetherian Γ-module. We say that the pair (Γ, A) has *finite gap number* if there exists an integer ℓ such that, in any sequence

$$A_1(H_1) \leq A_1(H_2) \leq \cdots \leq A_1(H_n) \leq \cdots$$

of centralizers in A of subgroups $H_1, H_2, \ldots, H_n, \ldots$ of Γ there are at most ℓ gaps.

8.3.4 *Suppose that Γ is a finitely generated nilpotent group and A a finitely generated Γ-module. Then (Γ, A) has finite gap number.*

Proof We know that (Γ, A) is eremitic, of eremiticity e say. Now $A_1(H) \leq A_1(K)$ implies that $A_1(H) = A_1(\langle H, K \rangle)$ and so, in considering a chain of centralizers $A_1(H_1) \leq A_1(H_2) \leq \cdots$, we may assume that

$$H_1 \geq H_2 \geq \cdots \geq H_n \geq \cdots .$$

Since $h = h(\Gamma)$ is finite, there can be at most h integers n for which $h(H_n) > h(H_{n+1})$. Thus we may suppose all the H_i to have the same Hirsch number: then $|H_1 : H_n|$ is finite for all n. Therefore $H_1^{r_n} \leq H_n$ for some $r_n > 0$, and so $A_1(H_n) \leq A_1(H_1^{r_n}) \leq A_1(H_1^e)$ for some fixed e by eremiticity. Now set $K = H_1^e$, so that

$$A_1(H_n) \leq A_1(K)$$

for $n \geq 1$. If $H_n K = H_{n+1} K$, we would have

$$A_1(H_n) \cap A_1(K) = A_1(H_{n+1}) \cap A_1(K)$$

and hence $A_1(H_n) = A_1(H_{n+1})$ since $A_1(H_n) \leq A_1(K)$. Thus gaps in the chain $H_1 \geq H_2 \geq \cdots$ occur in the same places as in the chain $H_1 = H_1 K \geq H_2 K \geq \cdots \geq H_n K \geq \cdots$. But $|H_1/K| \leq d = e^\ell$, where ℓ is the length of a series in Γ with cyclic factors. Hence there are at most $\log_2 d$ gaps in this chain and (Γ, A) has finite gap number. The proof that finitely generated abelian-by-nilpotent groups have finite gap number is completed by now familiar arguments. ∎

Further results

It is possible to extend 8.1.1 and 8.1.3 to the following result:

8.3.5 (Lennox and Roseblade 1970) *Metanilpotent-by-finite groups satisfying max $-n$ are both stunted and eremitic.*

Another application of the Fan Out Lemma given in Lennox (1984) is:

8.3.6 *Suppose that Γ is a finitely generated nilpotent group and A is a noetherian Γ-module such that $C_A(x) > 1$ for all $x \in \Gamma$. Then there is a subgroup Δ of finite index in Γ such that $C_A(\Delta) > 1$.*

From this it may be shown that *if G is a finitely generated abelian-by-nilpotent group such that each proper finitely generated nilpotent subgroup is contained in a larger one, then G is nilpotent.*

Wehrfritz (1971*b*) established centrality properties for linear groups over a field.

8.3.7

(i) *A finitely generated linear group over a field is stunted.*
(ii) *A finitely generated linear group over a field is eremitic; furthermore, if the field has characteristic* 0, *the eremiticity is* 1.
(iii) *A linear group of degree n has finite gap number at most* $n^2 - 1$.

The second part of this theorem may be compared to the following result of Lennox (1971):

8.3.8 *A finitely generated, torsion-free, abelian-by-nilpotent group has a subgroup of finite index which is eremitic of eremiticity* 1, *that is, all its centralizers are isolated.*

A different approach extending this result arises from work of Wehrfritz (1971*b*). He deduced 8.3.7(ii) from:

8.3.9 *Let* $p \neq q$ *be primes and let* π *be a set of primes. Suppose that* G *is both a residually finite* $\pi \cup \{p\}$-group *and a residually finite* $\pi \cup \{q\}$-group. *Then centralizers in* G *are* π'-isolated.

Using this result, together with residual finiteness properties of finitely generated abelian-by-nilpotent groups and their holomorphs (see 4.3.10 and the subsequent comment), Segal (1974) was able to prove:

8.3.10 *Suppose that* G *is a finitely generated group with an abelian normal subgroup* A *such that* G/A *is virtually nilpotent. Then the holomorph of* G *has a subgroup of finite index in which every centralizer is* $\pi(A)'$-isolated.

Taking $\pi(A)$ to be empty, we obtain a generalization of 8.3.8. In fact it is not hard also to infer the following:

8.3.11 (Segal 1974) *The holomorph of a torsion-free-by-finite, finitely generated abelian-by-nilpotent-by-finite group is eremitic.*

Segal (1977*b*) used a completely different module theoretic approach based on the concept of Krull dimension to establish:

8.3.12 *Let* G *be a virtually metanilpotent group with* $\max -n$. *If* H *and* K *are subgroups of* G *such that* $[K, {}_n H] = 1$ *for some* $n > 0$, *then* $[K, {}_k H] = 1$, *where the integer* k *depends only on* G.

This provides an alternative proof of the fact that virtually metanilpotent groups with $\max -n$ are stunted. Furthermore, Segal's methods yield an explicit estimate for the integer k and also give a new proof of the eremiticity of virtually metanilpotent groups with $\max -n$.

Counterexamples

First of all we show that there exist finitely generated nilpotent-by-abelian groups which are not stunted.

8.3.13 *There exists a 3-generator group G of upper central height ω such that G' is nilpotent of class 2.*

Proof The construction is a variant of that used in 4.1.3. We first define a nilpotent group K of class 2: it is generated by elements $a_n, b_n, (n \in \mathbb{Z})$, subject to the following relations, which are to hold for all i, j, k.

$$[a_i, a_j] = [b_i, b_j] = [a_i, b_j, b_k] = [a_i, b_j, a_k] = 1,$$
$$[a_i, b_j] = [a_j, b_i], [a_i, b_j] = [a_{i+k}, b_{j+k}],$$
$$[a_i, b_j]^2 = [a_{i+1}, b_j], [a_i, b_i]^2 = 1.$$

Set $A = \langle a_n \mid n \in \mathbb{Z} \rangle$ and $B = \langle b_n \mid n \in \mathbb{Z} \rangle$. Then A and B are free abelian and K is nilpotent of class 2. Put $z_n = [a_i, b_{i+n}] = [a_{i+n}, b_i]$, which is independent of i. With this notation the last two defining relations become $z_{n+1}^2 = z_n$, $n \geq 0$, and $z_0^2 = 1$, so that $Z(K) = K'$ is a group of type 2^∞. It is clear that the mapping ξ defined by $a_n^\xi = a_{n+1}$, $b_n^\xi = b_{n+1}$ is an automorphism of K. Also it is clear by symmetry that the mapping τ given by $a_n^\tau = b_n$, $b_n^\tau = a_n$ is an automorphism of K. Then $H = \langle \xi, \tau \rangle$ is abelian.

Next we form the semidirect product $G = H \ltimes K$. Then $A^\tau = B$ and $G = \langle \tau, \xi, a_0 \rangle$. In addition, since H is abelian, we have $G' \leq K$, so that G is (nilpotent of class 2)-by-abelian.

Denoting factors modulo K' by bars, we have $\langle \bar{\xi}, \bar{A} \rangle \simeq \langle \bar{\xi}, \bar{B} \rangle \simeq \mathbb{Z} \, wr \, \mathbb{Z}$, a group which has trivial centre. Since $C_{\bar{K}}(\bar{\xi}) = C_{\bar{A}}(\bar{\xi}) \times C_{\bar{B}}(\bar{\xi}) = 1$, we have $Z(\bar{G}) = 1$. Now by an easy computation $z_n^\tau = z_n^{-1}$, so τ inverts every element of K'. Moreover $z_n^\xi = z_n$, so K' centralizes both K and ξ and thus $K' \leq Z_\omega(G)$. It follows that $K' = Z_\omega(G)$ because $Z(\bar{G}) = 1$. Since K' is of type 2^∞, the group G has central height ω, as required.   ∎

We remark that Lennox, Neumann, and Wiegold (1990) have shown that *there exists a finitely generated soluble group of derived length 4 whose hypercentre is not nilpotent.*

Another example which shows the limitations of the theory is:

8.3.14 *There exists a 3-generator centre-by-metabelian group which is not eremitic and does not satisfy* max $-c$.

Proof Let G be the group of 8.3.13 and set $G_1 = \langle \xi, K \rangle$. Then $G_1 = \langle a_0, b_0, \xi \rangle$ and $G_1'' \leq K' \leq Z(G_1)$ since $z_n^\xi = z_n$. Now K is nilpotent of class 2 and K' is torsion. It is easy to show that, under these circumstances, if K were eremitic of eremiticity e, the exponent of K' would divide e. But this exponent is clearly infinite, so K is not eremitic and neither is G.

Put $C_n = C_K(\langle b_1, \ldots, b_n \rangle)$ for $n = 1, 2, \ldots$. Then $a_0^{2^n} \in C_n \backslash C_{n+1}$ and hence $C_1 > C_2 > \cdots$ is a strictly descending chain of centralizers in K. Therefore G_1, fails to satisfy $\max - c$. ∎

In the preceding example the group G_1 does not satisfy $\max - n$. However it is in fact possible to vary the construction slightly to produce an example of a 3-generator centre-by-metabelian group which has $\max - n$ but not $\max - c$ (see Lennox and Roseblade 1970: theorem G(iii)).

8.4 Bryant's verbal topology

Bryant (1977) gave an ingenious and much shorter approach to the theorem that finitely generated abelian-by-nilpotent groups satisfy the minimal (and therefore the maximal) condition on centralizers. However, his method does not show that there is a bound on the number of gaps in a centralizer chain.

We will describe the main ideas of Bryant's argument. For any group G we form the free product $G * \langle x \rangle$, where $\langle x \rangle$ is an infinite cyclic group. For any $w \in G * \langle x \rangle$ and $g \in G$, define $w(g)$ to be the element of G obtained by substituting g for x in w. The set $\{g \in G \mid w(g) = 1\}$ is called a *primitive solution set* for w. The *verbal topology* on G is defined by taking all the primitive solution sets to form a sub-base for the closed sets of the topology. Thus any closed set is an intersection of finite unions of primitive solution sets and G becomes a topological group.

A *solution set* for G is defined to be any intersection of primitive solution sets and it is therefore closed. For example, if $H \leq G$, then $C_G(H)$ is a solution set because it is the intersection of the primitive solution sets corresponding to the words $[x, h]$, $h \in H$.

We will need the following elementary topological result.

8.4.1 *Let $\mathcal{C}$ be the set of closed subsets of a topological space X and let $\mathcal{S}$ be a sub-base of $\mathcal{C}$ which is closed under finite intersections. If $\mathcal{S}$ satisfies the minimal condition, then so does $\mathcal{C}$.*

Taking $\mathcal{S}$ to be the set of all solution sets for a group G, we deduce that *the topological group G satisfies the minimal condition on closed sets if it satisfies the minimal condition on solution sets.*

Proof of 8.4.1 We may assume $X \in \mathcal{S}$. Let $\mathcal{C}_0$ denote the set of elements of $\mathcal{C}$ which can be expressed as the union of a finite number of elements of $\mathcal{S}$. Then $\mathcal{C}_0$ is closed under finite intersections and every closed set is an intersection of members of $\mathcal{C}_0$. Suppose that $\mathcal{C}$ does not satisfy min, but $\mathcal{S}$ does. Then $\mathcal{C}_0$ contains an infinite properly descending chain $C_1 \supset C_2 \supset \cdots$. Choose $S \in \mathcal{S}$ minimal with respect to containing such a chain. Note that S exists, since $\mathcal{S}$ has min. Suppose that $C_2 = S_1 \cup \cdots \cup S_n$, where $S_i \in \mathcal{S}$. Then, if $1 \leq i \leq n$, we have the chain

$$S \supset S_i \supseteq C_3 \cap S_i \supseteq C_4 \cap S_i \supseteq \cdots,$$

and each term belongs to $\mathcal{C}_0$. By the hypothesis of minimality, the chain breaks off for each i. But $C_j = (C_j \cap S_1) \cup \cdots \cup (C_j \cap S_n)$ for $j \geq 2$, so the sequence $C_1 \supset C_2 \supset \cdots$ breaks off, a contradiction.   ∎

Next we need

8.4.2 *Let G be a finitely generated group such that every finitely generated group in the variety generated by G satisfies* max $-n$. *Then G satisfies the minimal condition on subgroups which are closed in the verbal topology.*

Proof If $g \in G$, the assignment $w \mapsto w(g)$ determines a homomorphism φ_g from $G * \langle x \rangle$ to G. Put $N = \bigcap_{g \in G} \text{Ker}(\varphi_g)$. If S is a solution set for G, then

$$S = \{g \in G \mid w(g) = 1, \forall\, w \in W\},$$

where $W = \bigcap_{g \in S} \text{Ker}(\varphi_g)$. Thus $W/N \triangleleft (G * \langle x \rangle)/N$, and the latter group belongs to the variety generated by G and hence has max $-n$ by hypothesis. In fact this implies that G satisfies the minimal condition for closed subgroups. For, if $S_1 \supset S_2 \supset \cdots$ is an infinite descending chain of solution sets for G, then $W_1 < W_2 < \cdots$, where $W_i = \bigcap_{g \in S_i} \text{Ker}(\varphi_g) \triangleleft G * \langle x \rangle$. Therefore by 8.4.1 the group G satisfies the minimal condition on closed subgroups.   ∎

Since a finitely generated abelian-by-nilpotent group G satisfies the hypothesis of 8.4.2, we have the immediate corollary:

8.4.3 *Finitely generated abelian-by-nilpotent-by-finite groups satisfy the minimal condition on subgroups closed in the verbal topology. Consequently such groups satisfy* min $-c$ *and therefore* max $-c$.

Since by 1.3.2 supersoluble groups are nilpotent-by-finite, we deduce from the last result that *finitely generated abelian-by-supersoluble-by-finite groups satisfy* max $-c$. However it is still not known if finitely generated abelian-by-polycyclic groups have max $-c$.

8.5   Centrality in finitely generated abelian-by-polycyclic groups

It remains an open question whether finitely generated abelian-by-polycyclic groups have the three centrality properties, stuntedness, eremiticity, and finite gap number. In its module theoretic form one of the main unanswered questions is this: suppose Γ is a polycyclic group and A is a finitely generated Γ-module. Is there a non-zero submodule Y such that (Γ, Y) is stunted and eremitic? Some progress has been made on this question: Lennox and Roseblade (1970) used 8.1.1 and 8.1.3 to establish the following:

8.5.1 *Let G be a finitely generated abelian-by-polycyclic group. Then:*

(i) *there is a bound on the upper central heights of the subnormal subgroups of G;*

(ii) *there is an integer $e = e(G) \geq 0$ such that $C_G(H^n) \leq C_G(H^e)$ for all subnormal subgroups H of G;*

(iii) *the group G satisfies the maximal condition for centralizers of subnormal subgroups.*

It is not known whether there is a bound on the number of gaps in a chain of centralizers of subnormal subgroups in 8.5.1. The proof of this result shows that in order to prove that finitely generated abelian-by-polycyclic groups are stunted and eremitic, it is sufficient to show that the pair (Γ, A) is stunted and eremitic when Γ is a polycyclic group and A is a finitely generated Γ-module such that $C_A(H) = 1$ for all $H \leq \mathrm{Fit}(\Gamma)$.

Segal (1975*b*) proved that analogues of 8.3.10 and 8.3.11 hold for finitely generated abelian-by-polycyclic groups, from which he was able to deduce:

8.5.2 *A finitely generated torsion-free abelian-by-polycyclic group is eremitic.*

In another direction, but also based on the Roseblade–Jategaonkar Theorem described in Chapter 7, Nazzal and Rhemtulla (1991) developed further versions of the Fan Out Lemma and were able to prove another special case of the conjecture:

8.5.3 *Let G be a finitely generated group with an abelian normal subgroup A such that G/A is polycyclic. Assume that either G/A is abelian-by-cyclic or else it has a normal subgroup which is a plinth with abelian quotient group. Then G is stunted, eremitic, and has finite gap number.*

(For the definition of a plinth see 7.1.) On the basis of 8.5.3 and results of Robinson and Wilson on just non-polycyclic groups, we may state:

8.5.4 *A just non-polycyclic soluble group is stunted, eremitic, and has finite gap number.*

For this result the relevant theorems on just non-polycyclic groups are 7.4.3 and 7.4.6 above.

9

ALGORITHMIC THEORIES OF FINITELY GENERATED SOLUBLE GROUPS

The aim of the algorithmic theory of a class of groups $\mathbf{X}$ may be described informally as the collection of information about $\mathbf{X}$-groups, their elements and subgroups that can, in principle at least, be obtained by machine computation. This statement is of course vague, but it does give an intuitive idea of the aims of the chapter.

In order to be precise we shall assume the reader to be familiar with the notion of a *Turing machine*, which is essentially a theoretical model of a computer. The information to be obtained by our algorithms must be capable of being encoded as the output of a Turing machine. This means that our algorithms are algorithms in the classical sense and thus we will not be concerned with their practicality. It is the question of the existence of an algorithm to perform a specific task that is of interest to us here.

Our concern is with the classes of finitely generated soluble groups for which a good algorithmic theory can be developed. The classes with the richest theory are polycyclic groups, finitely generated metabelian groups and, to a lesser extent, finitely generated soluble groups with finite rank. The success of these theories is due in no small degree to the existence of a large body of constructive methods in linear algebra and commutative algebra.

Algorithmic techniques for finitely generated soluble groups have been widely studied over the last 40 years, and the present chapter is intended as an introduction to this developing theory.

9.1 The classical decision problems of group theory

Before introducing the decision problems of the title, we must review some terminology from recursion function theory. A subset S of the set of natural numbers $\mathbb{N}$ is called *recursively enumerable* if its elements are the output of some Turing machine: if S is viewed as the image of a function $\alpha : \mathbb{N} \to S$, then α is said to be *computable*. If, in addition, the complement of S in $\mathbb{Z}$ is recursively enumerable, then S is said to be *recursive*. Thus membership of an integer in a recursive set can be decided with the aid of two Turing machines. It is a fundamental result of recursion theory—due to Kleene (1936)—that there exist recursively enumerable, non-recursive sets: this result underlies the negative solution of all the classical decision problems in group theory. The terms recursively enumerable and recursive can also be applied to a subset of any set whose elements are labelled by natural numbers. In our case the subset will consist of words in some countable set of generators. (For background in recursion theory see, for example, Mal'cev 1970 or Mendelson 1997.)

We now introduce the four fundamental decision problems.

1. *The word problem.* Is there an algorithm which, when given a word w in the generators of a group G, decides if $w = 1$ in G?
2. *The membership problem* (also called the *generalized word problem*). Is there an algorithm which, when given words $w, w_1, w_2, \ldots, w_n$ in generators of a group G, decides if w belongs to the subgroup $\langle w_1, w_2, \ldots, w_n \rangle$ of G?
3. *The conjugacy problem.* Is there an algorithm to decide if two given words w_1, w_2 in the generators of a group G are conjugate?
4. *The isomorphism problem.* Is there an algorithm which can decide if two given groups are isomorphic?

With the exception of the second one, these problems were first raised in 1910 by Max Dehn in the course of his work in algebraic topology. The decision problems are usually stated for finitely presented groups, but they make sense in other contexts, for example, when a group is specified by a finite set of generators which are matrices over $\mathbb{Z}$.

Recursive presentations

A presentation in countably many generators is called *recursive* if the defining relations constitute a recursively enumerable set. Under these circumstances one can write down a presentation of the group for which the set of defining relators is actually a *recursive* set, by attaching to each defining relator a coded subword of the form $(xx^{-1})^i$, $i = 1, 2, \ldots$, where x is a generator.

A group G is said to be *recursively presented* if it has at least one recursive presentation. It is however possible for G to have other presentations that are not recursive. On the other hand, it is easy to show that *if the word problem is soluble for a presentation of G with finitely many generators, then it is soluble for all such presentations of G.* So there is no dependence on the presentation in this case. We mention in passing the celebrated theorem of G. Higman (1961): *every finitely generated, recursively presented group is isomorphic with a subgroup of a finitely presented group.*

Suppose now that G is a finitely generated recursively presented group. Then, if we form the set of *all* relators, that is, words in the generators which are consequences of the defining relators, it is intuitively clear (and not hard to prove) that this is a recursively enumerable set. Hence there is a Turing machine which will tell us if a given word is a relator of G. However, Kleene's Theorem mentioned above suggests that the set of all relators might not be recursive, that is, the set of non-relators might not be recursively enumerable. If this is the case, then no Turing machine will be able to confirm that a given word is *not* a relator, and the word problem for G will be insoluble.

In fact, as is well known, in the 1950s W. W. Boone (1955) and P. S. Novikov (1955) were able to show that the word problem is insoluble for finitely presented groups: this implies that the membership problem and the conjugacy problem are also insoluble. Shortly afterwards, the isomorphism problem was shown to be insoluble for finitely presented groups by Adyan (1957) and Rabin (1958).

On the other hand, there are important classes of infinite groups for which all four decision problems are soluble, a prime example being the class of finitely generated abelian groups. For when such a group is given by means of a finite presentation, its structure can be determined by standard operations with unimodular matrices. Another well-known source of groups with soluble decision problems is the groups that are residually finite and finitely presented in some variety.

Let $\mathbf{V}$ be a variety of groups. If G is a finitely generated group in $\mathbf{V}$, then $G \simeq F/R$ where F is a finitely generated free group in *the variety* $\mathbf{V}$ and R is some normal subgroup. Should R be the normal closure in F of a finite subset, then we have a finite $\mathbf{V}$-presentation of G. The variety $\mathbf{V}$ is said to be *finitely based* if it is determined by a finite set of laws. (For the theory of group varieties see H. Neumann 1967.)

9.1.1 *Let* $\mathbf{V}$ *be a finitely based variety of groups and let* G *be a* $\mathbf{V}$-*group which is finitely presented in* $\mathbf{V}$.

(i) *If* G *is residually finite, the word problem is soluble in* G.
(ii) *If* G *is subgroup separable, the membership problem is soluble in* G.
(iii) *If* G *is conjugacy separable, the conjugacy problem is soluble in* G.

Recall here that a group G is *subgroup separable* if each proper subgroup is the intersection of subgroups with finite index in G. Also G is *conjugacy separable* if two elements are conjugate in G whenever they are conjugate modulo every normal subgroup of finite index. These properties are strong forms of residual finiteness.

Proof of 9.1.1 All three parts are established by a common argument, which we illustrate by proving (i).

Let w be a given word in the generators of the presentation. Two procedures are set in motion which will tell us if w equals 1 in G. The first simply enumerates the words of the form cl, where c is a consequence of the given defining relators and l is a consequence of the laws of $\mathbf{V}$. These words are compared with the word w. If w appears, the procedure stops: then $w = 1$ in G.

The second procedure enumerates the finite $\mathbf{V}$-groups in increasing order $F_1, F_2, \ldots$, for example by constructing their multiplication tables and verifying the laws of the variety $\mathbf{V}$. It then checks each mapping from the set of generators of G to a finite group F_i to see if it is a homomorphism, that is, if the defining relators are equal to the identity in F_i. If this is the case, a finite $\mathbf{V}$-quotient of G has been found, and clearly all quotients arise in this manner. Finally, check to see if $w \neq 1$ in some finite **V**-quotient. If so, the procedure stops and $w \neq 1$ in G. By residual finiteness one of the two procedures will stop, allowing us to decide if $w = 1$ in G. ■

Corollary 9.1.2 *The word problem, the membership problem, and the conjugacy problem are all soluble for virtually polycyclic groups.*

This is because virtually polycyclic groups are subgroup separable and conjugacy separable—see 1.3.10 and 1.3.11. We remark that Segal (1990a) has proved that the *isomorphism problem is soluble for virtually polycyclic groups.* This result lies deeper and rests ultimately on methods from arithmetic group theory.

Negative results for finitely presented soluble groups

The status of the four decision problems for finitely presented soluble groups was open for many years after the appearance of Boone–Novikov Theorem. Finally, it was shown by Kharlampovich (1981) that the word problem is insoluble for finitely presented soluble groups of derived length 3. Subsequently, this was proved in a different way by Baumslag, Gildenhuys, and Strebel (1985, 1986): in both works the authors use results of M. Minsky from recursion theory. In the latter article the following theorem is proved.

9.1.3 *There is a finitely presented soluble group U of derived length 3 and a recursive set of words $w_1, w_2, \ldots$ in the generators of U such that $w_i^p = 1$ with p a prime and $w_i \in Z(U)$ for which there is no algorithm to decide if a given w_i equals the identity in U.*

Here the group U is a semidirect product $H \ltimes A$, where H is torsion-free abelian and A is nilpotent with exponent p^2. It follows that the word problem, the membership problem, and the conjugacy problem are all insoluble for finitely presented nilpotent-by-abelian groups of derived length 3.

The group U can also be used to show that *the isomorphism problem is insoluble for finitely presented soluble groups of derived length 3.* To see how, take a cyclic group $\langle x \rangle$ of order p^3 and define groups G_i, $i = 1, 2, \ldots$, by

$$G_i = (U \times \langle x \rangle)/\left\langle x^{p^2} w_i^{-1} \right\rangle.$$

The point to notice here is that $w_i = 1$ in G if and only if G_i contains no elements of order p^3, that is, G_i is isomorphic with $G^* = U \times \mathbb{Z}_{p^2}$. Since there is no algorithm to decide if $w_i = 1$, there is none that can decide if $G_i \simeq G^*$.

Despite these negative results, there is considerable scope for developing good algorithmic theories for particular types of finitely generated soluble group. For example, Remeslennikov and Romanovskiĭ (1980) showed that the word problem is soluble for free soluble groups of finite rank.

Other likely types of infinite soluble groups that come to mind here are: polycyclic groups, minimax groups, and abelian-by-polycyclic groups. The algorithmic theory of these groups will be developed in succeeding sections. What the theory demonstrates is that the algebraic structure of a group can greatly affect the algorithmic properties of the group. On the other hand, the effect of the algorithmic properties of a group on its algebraic structure is much harder to gauge.

9.2   Algorithms for polycyclic groups

Polycyclic groups form a natural class for which to construct an algorithmic theory. Indeed, as has already been observed, the word problem, the membership problem and the conjugacy problem all have positive solutions for this class (9.1.2), which allows us to perform many basic operations within the group. Another reason to concentrate on polycyclic groups is the existence of convenient finite presentations, the so-called *polycyclic presentations*, which reflect the special structure of the group. Our account of the algorithmic theory of polycyclic-by-finite groups is based on a paper of Baumslag *et al.* (1991).

First of all some terminology. A virtually polycyclic group, that is, a polycyclic-by-finite group, will be called a *PF-group* in the interests of brevity. We will say that a PF-group G is *given* if a finite presentation of it has been specified. A subgroup H of G is *given* if a finite set of generators of H has been provided.

We begin with a very simple algorithm, but one which is used constantly.

9.2.1 (The Normal Closure Lemma) *There is an algorithm which, when given a PF-group G and finite subsets X and Y of G, finds the normal closure of X in $\langle X, Y \rangle$.*

(Of course, in saying that the algorithm 'finds' the normal closure, we mean that it furnishes a finite set of generators for it.)

Proof of 9.2.1 Of course the elements of X and Y are assumed to be given as words in the generators of G. Let $Y = \{y_1, y_2, \ldots, y_n\}$. A sequence of finite subsets X_i is defined recursively by the rules

$$X_0 = X \quad \text{and} \quad X_{i+1} = X_i \cup \left(\bigcup_{j=1}^{n} X_i^{y_j} \cup X_i^{y_j^{-1}} \right).$$

Writing $H_i = \langle X_i \rangle$, we have $H_0 \le H_1 \le H_2 \le \cdots$. Now by the maximal condition, there is an i such that $H_i = H_{i+1}$, and clearly the subgroup H_i is $\langle Y \rangle$-invariant. Thus $H_i = X^{\langle X, Y \rangle} = X^{\langle Y \rangle}$ is the normal closure sought. Furthermore, $H_i = H_{i+1}$ holds if and only if $X_{i+1} \subseteq H_i$, which is decidable since the membership problem is soluble in G. Consequently, the normal closure can be found. ∎

Corollary 9.2.2 *There are algorithms which, when given a PF-group G and a finite subset X of G, find the terms of the derived series and the lower central series of $\langle X \rangle$.*

Proof Put $H = \langle X \rangle$ and suppose we have found the ith term of the derived series $H^{(i)}$: let it have generators $y_1, y_2, \ldots, y_m$. Then $H^{(i+1)}$ is the normal closure in G of the subset $\{[y_i, y_j] \mid i < j = 1, 2, \ldots, m\}$. By 9.2.1 we can find $H^{(i+1)}$. The argument is similar for the terms of the lower central series. ∎

It follows that we can compute the derived length of a given polycyclic group using 9.2.2 and the positive solution of the word problem.

The normal subgroups of finite index in a PF-group are always of interest, so the following fact is useful.

9.2.3 *There is an algorithm which, when a PF-group G is given, lists the normal subgroups of finite index.*

Proof The first step is to enumerate the homomorphisms from G into finite groups. This is accomplished by enumerating the finite groups F and determining which maps from the set of generators of G to F are homomorphisms, as in the proof of 9.1.1. By means of this procedure we construct a list of the finite images of G. For each of these images a finite presentation can be found: then the defining relators form a finite subset whose normal closure in G is the kernel of the relevant homomorphism (see 11.1.1). This normal closure may be found by using 9.2.1, and obviously every normal subgroup of finite index in G is obtainable in this manner. ∎

Corollary 9.2.4 *There is an algorithm to find the soluble radical of a given PF-group G.*

Proof Using 9.2.3 enumerate all pairs (N, i), where N is a normal subgroup with finite index in G and i is a natural number. For each pair find $N^{(i)}$ using 9.2.2 and test to see if $N^{(i)} = 1$, by using the solubility of the word problem in G. Since G is known to be a PF-group, a pair (N, i) such that $N^{(i)} = 1$ will eventually appear. The soluble radical of G/N may be found by enumerating all its subgroups and the preimage of this soluble radical under the canonical homomorphism $G \to G/N$ is the soluble radical of G. ∎

A given PF-group can be tested for solubility, supersolubility and nilpotence.

9.2.5 *There are algorithms which, when a PF-group G is given, decide if the group is soluble, supersoluble, or nilpotent.*

Proof By 9.2.4 the soluble radical S of G may be found. Now decide if $S = G$. This deals with solubility.

As for supersolubility, one has to remember the result of Baer that a PF-group whose finite quotients are supersoluble is itself supersoluble—see 1.3.14. Two procedures are now set in motion. The first enumerates all finite images of G using 9.2.3, and tests each one for non-supersolubility, by enumerating its subgroups. If it finds a non-supersoluble image, the procedure stops: the group G is not supersoluble. The second procedure searches for non-trivial cyclic normal subgroups of G by enumerating non-trivial elements of G and testing the cyclic subgroups they generate for normality by forming normal closures—see 9.2.1. If a cyclic normal subgroup $C_1 \neq 1$ appears and $C_1 \neq G$, the procedure is repeated for G/G_1, for which a finite presentation is at hand. The procedure continues until G is reached, when it stops. By Baer's Theorem one of the two procedures is bound to halt and furnish the answer.

For nilpotence the argument is similar, but it uses Hirsch's Theorem that a PF-group whose quotients are nilpotent is itself nilpotent (1.3.12): it also uses 9.2.2 to test for nilpotence.   ∎

Another simple algorithm tests whether a given subgroup of a PF-group is subnormal.

9.2.6 *There is an algorithm which, when given a PF-group G and a finite subset X, decides if $\langle X \rangle$ is subnormal in G.*

Proof Let $H = \langle X \rangle$. By a theorem of Kegel (1.3.15), the subgroup H is subnormal in G if and only if H^φ is subnormal in G^φ for all homomorphisms φ from G to a finite group. Our first procedure enumerates the finite images of G and checks to see if the image of H is not subnormal; this allows us to detect if H is not subnormal. It remains to show how subnormality of H can be detected.

A second procedure uses 9.2.1 to construct the sequence of successive normal closures of H in G, that is, the sequence of subgroups H_i defined by $H_0 = G$ and $H_{i+1} = H^{H_i}$. Recall that H is subnormal in G if and only if $H = H_i$ for some i. By the Normal Closure Lemma we can find the successive H_i's. The solution of the membership problem enables us to tell if $H_i = H$. Thus subnormality of H can be detected.   ∎

Polycyclic presentations

So far polycyclic groups have been studied using arbitrary finite presentations and no attempt has been made to exploit the special structural features of these groups. In order to construct more sophisticated algorithms we must employ presentations which are directly related to the group structure.

A *polycyclic presentation*[1] of a group G is a finite presentation π in generators $x_1, x_2, \ldots, x_n$ with defining relations of the following types:

(i) $x_i^{x_j} = v_{ij}(x_1, \ldots, x_{j-1})$ and $x_i^{x_j^{-1}} = v'_{ij}(x_1, \ldots, x_{j-1})$, where $1 \leq i < j \leq n$;
(ii) $x_i^{e_i} = u_i(x_1, \ldots, x_{i-1})$, where $1 < e_i \leq \infty$ and $1 \leq i \leq n$. (If $e_i = \infty$, the relation is understood to be vacuous.)

Clearly these relations reflect the existence of a series in G with cyclic factors. Indeed, if $G_j = \langle x_1, x_2, \ldots, x_j \rangle$, then G has the series $1 = G_0 \lhd G_1 \lhd \cdots \lhd G_n = G$ and each factor G_j/G_{j-1} is cyclic. Hence a group with a polycyclic presentation is polycyclic. In fact the converse is true.

---

[1] Closely related to polycyclic presentations are *power-commutator presentations*, which are widely used in computational group theory.

9.2.7 *A group G has a polycyclic presentation if and only if it is polycyclic.*

Proof Let G be a polycyclic group with a series $1 = G_0 \triangleleft G_1 \triangleleft \cdots \triangleleft G_n = G$ in which the factors are cyclic. We can assume that $n > 0$ and $N = G_{n-1}$ has a polycyclic presentation in generators $x_1, x_2, \ldots, x_{n-1}$. Put $G/N = \langle xN \rangle$. Then there are relations of the form $x_i^x = v_i(x_1, \ldots, x_{n-1})$ and $x_i^{x^{-1}} = v_i'(x_1, \ldots, x_{n-1})$, and also $x^e = u(x_1, \ldots, x_{n-1})$ if $e = |G : N|$ is finite. Add the generator x and these relators to the polycyclic presentation of N to get a polycyclic presentation of a group $\bar{G}$.

Now by Von Dyck's Theorem there is a surjective homomorphism $\theta : \bar{G} \to G$. If $\bar{N}$ is the subgroup of $\bar{G}$ generated by $x_1, \ldots, x_{n-1}$, then $\bar{N} \triangleleft \bar{G}$ and evidently the restriction of θ to $\bar{N}$ is an isomorphism from $\bar{N}$ to N. Thus $\mathrm{Ker}(\theta) \cap \bar{N} = 1$. Also θ induces a surjective homomorphism from $\bar{G}/\bar{N}$ to G/N and clearly $|\bar{G}/\bar{N}| = e = |G/N|$. Hence θ induces an isomorphism from $\bar{G}/\bar{N}$ to G/N, so that $\mathrm{Ker}(\theta) = 1$ and $\bar{G} \simeq G$. It follows that G has a polycyclic presentation in $x_1, \ldots, x_{n-1}, x$. ∎

Consistent polycyclic presentations

Let G be a polycyclic group with the presentation π specified in the definition. By omitting the generators $x_{i+1}, \ldots, x_n$ and all relations involving them, we obtain a subpresentation π_i of π, presenting a group H_i in generators $x_1, x_2, \ldots, x_i$. If we write $G_i = \langle x_1, x_2, \ldots, x_i \rangle$, which is a subgroup of G, there is a natural surjective homomorphism $H_i \to G_i$ and also a natural homomorphism $H_i \to H_{i+1}$. These homomorphisms fit together to make the following diagram commute: here the vertical maps are surjective.

$$
\begin{array}{ccccccccc}
1 = H_0 & \to & H_1 & \to & H_2 & \to \cdots \to & H_{n-1} & \to & H_n = G \\
\downarrow & & \downarrow & & \downarrow & & \downarrow & & \downarrow \\
1 = G_o & \hookrightarrow & G_1 & \hookrightarrow & G_2 & \hookrightarrow \cdots \hookrightarrow & G_{n-1} & \hookrightarrow & G_n = G.
\end{array}
$$

Next we establish an easy but basic lemma.

9.2.8 *The following statements about the presentation π are equivalent:*

(i) *the maps $H_i \to H_{i+1}$ are injective;*
(ii) *the maps $H_i \to G_i$ are isomorphisms;*
(iii) *$H_i \simeq G_i$ for each i.*

Proof That (i) implies (ii) is proved by induction on $n - i$ and an easy diagram chase. Obviously (ii) implies (iii). Suppose that (iii) is true. Now PF-groups are *hopfian*, that is, a PF-group cannot be isomorphic with a proper quotient group: hence the given maps $H_i \to G_i$ are isomorphisms. Induction on i and another easy diagram chase show that the $H_i \to H_{i+1}$ are injective. ∎

A polycyclic presentation π satisfying the equivalent conditions of 9.2.8 is called *consistent*. If the presentation is consistent, then $|G_i : G_{i-1}| = |H_i : H_{i-1}| = e_i$, so that no 'collapse' of factors occurs. Thus a consistent polycyclic presentation has the advantage that from it we can read off the orders of the factors in the corresponding series.

Referring to the proof of 9.2.7, we see at once that the factors of the series in $\bar{G}$ have the same orders as those in G. In short a consistent polycyclic presentation has been constructed. Hence *every polycyclic group has a consistent polycyclic presentation*.

It is clearly of importance to be able to recognize consistent presentations of polycyclic groups.

9.2.9 *There is an algorithm which, when given a polycyclic presentation, determines if it is consistent.*

Proof Let π be the given polycyclic presentation and suppose it presents a group G. We employ the notation established above, writing π_i for the sub-presentations of π, where $i = 1, \ldots, n$. If $n \leq 1$, then G is cyclic and its order is e_1, so that the presentation is consistent. Let $n > 1$ and assume that π_{n-1} has been recognized as consistent. It remains to decide if $|G : G_{n-1}| = e_n$. The point to observe is that π will be consistent, if and only if for $i = 1, \ldots, n-1$, the assignments $x_i \mapsto v_{in}(x_1, \ldots, x_{n-1})$, $x_i \mapsto v'_{in}(x_1, \ldots, x_{n-1})$ determine mutually inverse automorphisms of H_{n-1} which fix $u_n(x_1, \ldots, x_{n-1})$: for if this is the case, then $G \simeq (\langle x_n \rangle \ltimes G_{n-1})/L$, where $L = \langle x_n^{-e_n} u_n(x_1, \ldots, x_{n-1}) \rangle^G$, and $|G : G_{n-1}| = e_n$.

Finally, to test if these assignments determine suitable automorphisms, we must verify that they map the defining relators of G_{n-1} to relators and that the resulting endomorphisms are mutually inverse: this calls for several applications of the positive solution of the word problem for G_{n-1}. ∎

The next task is to show how to construct consistent polycyclic presentations.

9.2.10 *There is an algorithm which, when a polycyclic group is given by an arbitrary finite presentation, finds a consistent polycyclic presentation of it.*

Proof First recall that it is possible to pass from one finite presentation of a group to another by a suitable finite sequence of Tietze transformations. Starting with the given presentation, we apply to it finite sequences of Tietze transformations, (which form a recursively enumerable set). Eventually a polycyclic presentation will appear. Then use 9.2.9 to test it for consistency. A consistent polycyclic presentation is bound to appear at some point. ∎

The importance of consistent polycyclic presentations arises from the fact that they allow us to get our hands on the subgroups of a given PF-group.

9.2.11 *There is an algorithm which, when given a PF-group G and a finite subset X, finds a finite presentation of $\langle X \rangle$ in the set of generators X.*

Proof Put $H = \langle X \rangle$. By 9.2.4 we can find the soluble radical S of G. Since $|G : S|$ is finite, the Reidemeister–Schreier method may be applied to give a finite presentation of S. Now use 9.2.10 to obtain a consistent polycyclic presentation of S, with generators $s_1, s_2, \ldots, s_n$ say. Here we assume the generators to be labelled so that, if S_i is the subgroup $\langle s_1, s_2, \ldots, s_i \rangle$, then $1 = S_0 \lhd S_1 \lhd \cdots \lhd S_n = S$.

Since G/S is finite, we can certainly obtain a finite presentation of HS/S in the generators xS, $x \in X$. If the defining relators are $r_1, r_2, \ldots, r_\ell$, then $H \cap S$ is the normal closure of $\{r_1, r_2, \ldots, r_\ell\}$ in H. Now apply the Normal Closure Lemma to find a finite set of generators for $H_0 = H \cap S$.

Because the presentation of S is consistent, a finite presentation of the cyclic group S/S_{n-1} is at hand, and from it we find a finite presentation of $H_0/H_0 \cap S_{n-1}$. Use of the Normal Closure Lemma provides us with finite set of generators for $H_0 \cap S_{n-1}$, and by induction on n a finite presentation of this group can be found. (By consistency we already have a presentation of S_{n-1}.) The presentations of $H/H_0 \cap S_{n-1}$ and $H_0 \cap S_{n-1}$ may be combined in the standard manner to yield a finite presentation of H (detection of the defining relators requires the solution of the word problem). Finally, by expressing the generators of the presentation as the words in X, we obtain a finite presentation of H in X. ∎

There are many useful consequences of 9.2.11.

Corollary 9.2.12 *There is an algorithm which, when given a PF-group G and a finite subset X, finds the Hirsch number of the subgroup $\langle X \rangle$.*

For using 9.2.11 and 9.2.10 we may find a consistent polycyclic presentation of $\langle X \rangle$: then $h(\langle X \rangle)$ is the number of infinite exponents occurring in the presentation.

Corollary 9.2.13 *There is an algorithm which, when given a PF-group G, decides if G is infinite and if so, finds a non-trivial free abelian normal subgroup of G.*

Proof The group G is infinite precisely when $h(G) > 0$ and this is decidable by 9.2.12. Assuming G to be infinite, we enumerate finite subsets of G, form the subgroups they generate, and test each one for commutativity and normality: this is possible by solubility of the word problem and the membership problem. If both properties hold, we determine if the Hirsch number is positive. By these procedures an infinite abelian normal subgroup A will eventually be found. Find a presentation of A using 9.2.11 and determine the structure of A by the standard matrix method. An appropriate power of A will serve as our subgroup. ∎

Corollary 9.2.14 *There is an algorithm which, when there are given PF-groups G and H and also a homomorphism $\theta : G \to H$, by means of its effect on the generators of G, finds $\operatorname{Ker}(\theta)$.*

Proof By 9.2.11 a finite presentation can be found for $\operatorname{Im}(\theta)$, and hence for $G/\operatorname{Ker}(\theta)$. Using the Normal Closure Lemma we may then find a finite set of generators for $\operatorname{Ker}(\theta)$. ∎

9.2.15 *There is an algorithm which, when given a PF-group G, finds its maximum finite normal subgroup.*

Proof Let T denote the maximum finite normal subgroup of G. Find $h(G)$ using 9.2.12: if $h(G) = 0$, then $T = G$, so assume $h(G) > 0$. By 9.2.13 we may find a non-trivial free abelian normal subgroup A of G. A finite presentation of G/A is at hand and $h(G/A) < h(G)$, so by induction it is possible to find the maximum finite normal subgroup of G/A, say T_0/A. Clearly $T \le C_{T_0}(A) = C$, say. By 9.2.11 a finite presentation of the finite group T_0/A may be found and, by checking which of its elements centralize each generator of A, we can find C/A and hence C. Now $C/Z(C)$ is finite since $[A, C] = 1$, so by Schur's Theorem C' is finite and $C' \le T$. Observe that C' may be found by using 9.2.2. Finally, T/C' is the torsion-subgroup of C/C' and it may be found from a finite presentation of C/C'. Thus we have found T.   ∎

Virtually polycyclic subgroups of $GL_n(\mathbb{Z})$

Recall from 3.2.1 and 3.3.1 that the polycyclic groups are exactly the soluble subgroups of the groups $GL_n(\mathbb{Z})$. Thus every PF-subgroup occurs as a subgroup of some $GL_n(\mathbb{Z})$, which raises the question of whether one can find algorithms applicable to PF-subgroups of $GL_n(\mathbb{Z})$ which are more efficient than those for PF-groups given by finite presentations.

Notice that *the word problem is soluble for $GL_n(\mathbb{Z})$* since this is a finitely presented residually finite group. It will be important to know that *the membership problem for PF-subgroups of $GL_n(\mathbb{Z})$ is soluble*. This is because PF-subgroups are known to be closed in the Zariski topology—see Segal (1983: ch. 4]. However, the membership problem has a negative solution for arbitrary subgroups of $GL_n(\mathbb{Z})$ if $n \ge 4$, since it has a negative solution for the direct product of two free groups of rank 2—see Mihailova (1966).

Our next main objective is to establish:

9.2.16 *There is an algorithm which, when a finite subset X of $GL_n(\mathbb{Z})$ is specified, decides if $G = \langle X \rangle$ is a PF-group, and, if G is polycyclic, finds a finite presentation of it.*

This theorem is to be contrasted with results of Rabin (1958): for example, there is no algorithm to decide if a given finitely presented group is polycyclic.

The proof of 9.2.16 depends on a basic result from constructive algebraic number theory.

9.2.17 (Borevič and Šafarevič 1966) *There is an algorithm which, when an algebraic number field K is given, finds a finite presentation for the group of units of K.*

Keep in mind here that K will have the form $\mathbb{Q}[x]/(f)$, where f is some monic irreducible polynomial in $\mathbb{Q}[x]$, so that K is specified by f. Also the group of units of K is a finitely generated abelian group by Dirichlet's Theorem. Two preliminary results must be established before we can undertake the proof of 9.2.16.

9.2.18 *There is an algorithm which, when given a finite subset X of $GL_n(\mathbb{Z})$ such that $G = \langle X \rangle$ is a PF-group, finds the following:*

(i) *a triangularizable normal subgroup T over some algebraic number field K with finite index in G;*
(ii) *a finite presentation of T/U, where U is the unipotent part of T;*
(iii) *a finite presentation of U.*

Proof The first step is to enumerate finite subsets S of G, consisting of words in the generators, and positive integers m. Check to see if $T = \langle S \rangle$ is normal in G, using solubility of the membership problem for the polycyclic subgroup T. Then enumerate up to m cosets of T in G and determine whether their union V satisfies $Vx = V$ for all x in X. This requires the solubility of the membership problem once again. If this is true for V, then $V = G$. Eventually we will find T and m such that $T \triangleleft G$ and $|G : T| \leq m$. Clearly every normal subgroup of finite index in G is obtainable in this way.

At the same time we enumerate algebraic number fields K (via their monic irreducible polynomials) and elements α of $GL_n(K)$, and test each generator t of T to see if $\alpha^{-1} t \alpha$ is upper triangular over K. Mal'cev's Theorem (3.1.6) guarantees that a triple (T, K, α) with $\alpha^{-1} T \alpha$ upper triangular will eventually appear. Observe that the diagonal entries of $\alpha^{-1} T \alpha$ will be units of the number field K.

Next by 9.2.17 it is possible to find a finite presentation for D, the group of units of K. Let $\pi : T \to D \times D \times \cdots \times D$ (with n factors) be the map from an element of T to the diagonal of $\alpha^{-1} T \alpha$. Then π is a homomorphism whose kernel is the unipotent subgroup U of T. The images of the generators of T under π are known. Therefore a finite presentation of $\mathrm{Im}(\pi) \simeq T/U$ can be found, and from this we obtain a finite subset whose normal closure in T is U. Next observe that the proof of the Normal Closure Lemma (9.2.1) can be applied here since it merely calls for solubility of the membership problem. Therefore a finite set of generators for U can be found.

Finally $\alpha^{-1} U \alpha \leq U_n(K)$, the group of all $n \times n$ upper unitriangular matrices. But it is easy to write down a finite presentation of $U_n(K)$ in terms of generating transvections and elements of K. Thus 9.2.11 may be applied to produce a finite presentation of U. ∎

The next result is another tool needed in the proof of 9.2.16.

9.2.19 *Let G be a PF-subgroup of $GL_n(\mathbb{Z})$ with soluble radical S. Then $|G : S| \leq f(n)$, where $f : \mathbb{P} \to \mathbb{P}$ is an increasing computable function, $\mathbb{P}$ being the set of positive integers.*

Proof Let G act on a free abelian group M of rank n. If $n=1$, then $S=G$ and we may take $f(1)=1$: let $m>1$ and proceed by induction on m.

Suppose first that M is rationally reducible with respect to G. Then there is a non-trivial $\mathbb{Z}G$-submodule B of M such that M/B is non-trivial. Let $n_1=h(B)$ and $n_2=h(M/B)$, so that $n_i>0$ and $n=n_1+n_2$. Let $S_1/C_G(B)$ and $S_2/C_G(M/B)$ be the respective soluble radicals of $G/C_G(B)$ and $G/C_G(A/B)$. Writing $N=S_1\cap S_2$, we have $|G:N|\leq f(n_1)f(n_2)$ by induction on n. Also it is easy to see that $N^{(k)}$ is abelian for some k, so N is soluble. If S is the soluble radical of G, then $N\leq S$ and so we have $|G:S|\leq b_1$, where $b_1=\max\{f(n_1)f(n_2)\mid n_i>0, n_1+n_2=n\}$.

Next assume that M is rationally irreducible with respect to G, so that by 3.1.6 there is a free abelian normal subgroup A with finite index in G. Write $C=C_G(A)$ and observe that $C/Z(C)$ is finite, whence C' is finite by Schur's Theorem. Consequently, the elements of finite order in C constitute a finite normal subgroup T of G. Next recall that, by a theorem of Minkowski, finite subgroups of $GL_n(\mathbb{Z})$ have order at most $d(n)$, where d is an increasing function, which can be written down explicitly. Hence $|T| \leq d(n)$ and $|C:D|\leq d(n)!$, where $D=C_C(T)$.

Next, A is in essence a completely reducible subgroup of $GL_n(\mathbb{Q})$, and any irreducible abelian subgroup of $GL_m(\mathbb{Q})$ is isomorphic with a subgroup of the group of units of an algebraic number field of degree at most m over $\mathbb{Q}$ and has Hirsch number at most $m-1$ (by the Dirichlet Units Theorem). If the irreducible constituents of A have degrees $m_1,\ldots,m_k$, then we may conclude that

$$h(A) \leq \sum_{i=1}^{k}(m_i-1) \leq n-1.$$

Consequently, $|G/C| \leq d(n-1)$ and $|G:D| \leq d(n-1)(d(n)!)=b_2$, say. Since D' is finite, $D'\leq T$ and $[D',D]=1$, that is, D is nilpotent. If S is the soluble radical of G, it follows that $D\leq S$ and $|G:S|\leq b_2$. Finally, put $f(n)=\max\{b_1,b_2\}$. ∎

We are now ready to undertake the proof of 9.2.16.

Proof of 9.2.16 The first step is to show how to decide if G is a PF-group. Consider, for an arbitrary prime p, the homomorphism $\theta_p:G\to GL_n(p)$ given by $x\mapsto x(\bmod\, p)$. If G is a PF-group, then $\mathrm{Im}(\theta_p)$ is soluble-by-(finite of order $\leq f(n)$) by 9.2.19. Conversely, suppose each $\mathrm{Im}(\theta_p)$ has this property, that is, $G/\mathrm{Ker}(\theta_p)$ does. Next G is finitely generated, so there is a unique smallest normal subgroup N such that $|G/N| \leq f(n)$. Now $N\,\mathrm{Ker}(\theta_p)/\mathrm{Ker}(\theta_p)$ is soluble and its derived length is bounded by a function of n, by 3.1.10. Since $\bigcap_p \mathrm{Ker}(\theta_p)=1$, it follows that N is soluble and thus G is a PF-group since soluble subgroups of $GL_n(\mathbb{Z})$ are polycyclic.

At this point two procedures are set in motion. The first enumerates the groups $\mathrm{Im}(\theta_p)$, which have explicit sets of generators for successive primes p, and determines if the soluble radical has index at most $f(n)$. If not, it stops.

The second procedure attempts to construct a polycyclic subgroup with finite index in G. It enumerates m-tuples $(g_1,g_2,\ldots,g_m)$ of elements of G, for

$m = 1, 2, \ldots$, and checks to see if $g_i^{g_j^{\pm 1}} \in \langle g_1, \ldots, g_{j-1} \rangle$ for all $1 \leq i < j \leq m$, which is possible by enumerating the elements of $\langle g_1, \ldots, g_{j-1} \rangle$ and searching for $g_i^{g_j^{\pm 1}}$, the word problem being soluble in $GL_n(\mathbb{Z})$. If the m-tuple passes this test, its entries will generate a polycyclic subgroup H. For $\ell = 1, 2, \ldots$ enumerate sequences of ℓ cosets of H, say $y_1 H, \ldots, y_\ell H$, and let U be their union. For each generator x of G, we decide if $xU = U$, that is, if $y_j^{-1} x y_i \in H$ for some $i, j = 1, 2, \ldots, \ell$. This is decidable by the solubility of the membership problem for H. If this is true for each generator x, then $U = G$ and $|G : H| \leq \ell$; the procedure now stops.

If the first procedure stops, G is not a PF-group, while if the second one stops, G is a PF-group and then, by a standard procedure, a finite presentation of G is obtained. It is guaranteed that one procedure will halt. ∎

Corollary 9.2.20 *There is an algorithm which, when given a PF-group G, a finitely generated abelian group M and an explicit G-module structure for M, finds $C_G(M)$.*

Proof Find the order t of the torsion subgroup of M; then $M_0 = tM$ is free abelian and we may find a basis for M_0. Put $C_0 = C_G(M_0)$. The elements of G/C_0 are represented with respect to the basis by elements of $GL_n(\mathbb{Z})$, where $n = h(M_0)$.

Thus by 9.2.16 we may find a finite presentation of G/C_0, and then, using the Normal Closure Lemma, we obtain a finite set of generators for C_0 and hence a finite presentation of C_0. We now enumerate normal subgroups of finite index in C_0 until we find one which centralizes M/M_0. Hence we find $C_{C_0}(M/M_0)$, which is clearly just $C_G(M)$. ∎

As our first major application of 9.2.16, let us consider the problem of constructing the Fitting subgroup.

9.2.21 *There is an algorithm which constructs the Fitting subgroup of a given PF-group G.*

Proof The proof is by induction on $h(G)$, which can be assumed positive; recall that $h(G)$ can be found by 9.2.12. Now apply 9.2.13 to construct a non-trivial free abelian normal subgroup A. By induction on $h(G)$ we can find $F_1/A = \mathrm{Fit}(G/A)$: of course $F = \mathrm{Fit}(G) \leq F_1$. By finding a basis for A we may regard $\bar{F}_1 = F_1/C_{F_1}(A)$ as a subgroup of $GL_n(\mathbb{Z})$, where $n = h(A)$. Now apply 9.2.18 to construct a triangularizable normal subgroup $\bar{T}$ with finite index in $\bar{F}_1$, and its unipotent part $\bar{U}$. A finite presentation of $\bar{F}_1/\bar{U}$ is also available.

It is clear that $\bar{F} \cap \bar{T} = \bar{U}$, where $\bar{F} = F/C_{F_1}(A)$: note that $C_{F_1}(A) \leq F$. Thus $\bar{F}/\bar{U}$ is a finite normal subgroup of $\bar{F}_1/\bar{U}$. Find the maximum finite normal subgroup of $\bar{F}_1/\bar{U}$, using 9.2.15, and test its elements for nilpotent action on A. Hence we may find $\bar{F}$. Finally, $C_{F_1}(A)$ can be obtained by 9.2.20, and thus F is found. ∎

Our next objective is to find the centre.

9.2.22 *There is an algorithm to find the centre of a given PF-group G.*

Proof Consider first the case where G is nilpotent. Clearly we may assume G is infinite. Here our strategy is to get hold of a self-centralizing abelian normal subgroup. Start the procedure by using 9.2.13 to obtain an infinite abelian normal subgroup A_1. Using 9.2.20 we find $C_1 = C_G(A_1)$ and decide if $A_1 = C_1$. Assuming this to be false, we observe that $(C_1/A_1) \cap Z(G/A_1) \neq 1$, and by induction on $h(G)$ we may find $Z(G/A_1)$. Hence by enumeration an element a_2 may be obtained such that $1 \neq a_2 A_1 \in (C_1/A_1) \cap Z(G/A_1)$. Then $A_2 = \langle a_2, A_1 \rangle$ is an abelian normal subgroup strictly larger than A_1. By repetition of this procedure a self-centralizing abelian normal subgroup A is eventually obtained: clearly $Z(G) = C_A(G)$.

Let $x_1, x_2, \ldots, x_n$ be generators of G. Then $Z(G) = C_A(x_1) \cap C_A(x_2) \cap \cdots \cap C_A(x_n)$. Therefore $Z(G)$ can be found by solving a finite set of linear equations over $\mathbb{Z}$.

The general case is now easy. First use 9.2.21 to determine $F = \mathrm{Fit}(G)$, and then find a finite presentation of F. By the first part of the proof, $Z(F)$ may be found. Finally, $Z(G) = C_{Z(F)}(G)$ may be found by solving another finite set of $\mathbb{Z}$-linear equations. ∎

Of course as a consequence of 9.2.22 it is possible to construct the terms of the upper central series and the hypercentre of a PF-group.

The next algorithm plays a key role in the construction of intersections and centralizers.

9.2.23 *There is an algorithm which, when given a PF-group G, a finitely generated abelian group M with an explicit $\mathbb{Z}G$-module structure and an element a of M, finds $C_G(a)$.*

Proof Consider first the case where G is known to be abelian-by-finite. Then an abelian normal subgroup N with finite index in G may be found by enumerating normal subgroups of finite index. Find finite presentations of N and $B = a(\mathbb{Z}N)$; then use 9.2.20 to find $C_N(B)$, which is equal to $C_N(a)$ since N is abelian. Next find a transversal $\{t_1, t_2, \ldots, t_k\}$ to N in G and observe that $C_G(a) \cap Nt_i$ is non-empty if and only if $a = a \cdot xt_i$, that is, $a \cdot t_i^{-1} = a \cdot x$ for some $x \in N$. This is decidable by the positive solution of the conjugacy problem for the PF-group $N \ltimes M$. For each i for which $C_G(a) \cap Nt_i \neq 0$, find an $x_i \in N$ such that $a = a \cdot x_i t_i$ by enumeration. Then $C_G(a)$ is generated by these x_i's and $C_N(a)$.

Now let us return to the general case, arguing by induction on $h(M)$. Suppose first that $h(M) = 0$, so that M and $G/C_G(M)$ are finite. Note that $C_G(M)$ can be found by 9.2.20: now we simply test each element of $G/C_G(M)$ and pick out those which fix a. Hence $C_G(a)$ has been found. Now let $h(M) > 0$.

Using 9.2.21, we find $\mathrm{Fit}(G \ltimes M)$ and then obtain its projection L on G: thus $\mathrm{Fit}(G \ltimes M) = LM$. Here L is the largest nilpotent normal subgroup of G which acts nilpotently on M. Obtain a finite presentation of L and use it to find $D = C_M(L)$. This can be done since $D = M \cap Z(L \ltimes M)$ and $Z(L \ltimes M)$ can be found by 9.2.22. If D is finite, M will be finite since L acts nilpotently on M. Therefore D must be infinite and by the induction hypothesis $H = C_G(a + D)$ can be found.

Next put $A = D + \langle a \rangle$, which is a ZH-submodule of M. Using 9.2.20 we find $K = C_H(D)$. If $k \in K$, the mapping $a + D \mapsto a(k-1)$ is a homomorphism k^θ from A/D to D; further $\theta : K \to U = \mathrm{Hom}(A/D, D)$ is a homomorphism with kernel $C_K(a)$. Since we are able to find finite presentations of K and of $\mathrm{Hom}(A/D, D)$, as well as an explicit description of θ, it is possible to find $C_K(a)$ by using 9.2.14.

Let $P = [a, K]$. Suppose first that P is finite, which is certainly decidable. Then $C_H(A) = C_K(a)$ has finite index in K and thus $K/C_H(A)$ is finite. Also G/L is abelian-by-finite because every polycyclic group is abelian-by-finite modulo its Fitting subgroup. Now $[D, L] = 0$, so $H \cap L \leq K$ and consequently H/K is abelian-by-finite. Thus $H/C_H(A)$ is finite-by-abelian-by-finite, as well as being finitely generated; therefore $H/C_H(A)$ is abelian-by-finite. Since $C_H(A)$ can be found, we are now in the abelian-by-finite case and thus $C_H(a) = C_G(a)$ can be found by the first part of the proof.

Finally, suppose P is infinite. By induction $R = C_H(a + P)$ can be found. We claim that $R = C_H(a)K$. For $C_H(a)K$ is clearly contained in R, while if $h \in H$, then $h \in R$ if and only if $[a, h] = [a, k]$ for some $k \in K$, which is equivalent to $h \in C_H(a)K$.

We now know that $C_H(a)$ satisfies $C_H(a) \cap K = C_K(a)$ and $C_H(a)K = R$. To find $C_H(a)$ enumerate finitely generated subgroups S of H until one is found such that $R = SK$ and $as = a$ for each generator s of S. Then $C_H(a) = C_H(a) \cap SK = S\,C_K(a)$, which means that $C_H(a)$ has been found. ∎

Let G be a group and M a $\mathbb{Z}G$-module. We shall find it expedient to be able to get hold of the kernel $\mathrm{Ker}(\delta)$ of a derivation $\delta : G \to M$, that is, the set of x in G such that $x^\delta = 0$; clearly $\mathrm{Ker}(\delta)$ is a subgroup of G. (For derivations see 3.2.5 and Section 10.1.)

9.2.24 *There is an algorithm which, when given a PF-group G, a finitely generated abelian group M with an explicit $\mathbb{Z}G$-module structure, and a derivation $\delta : G \to M$, finds $\mathrm{Ker}(\delta)$.*

Proof Notice that δ is to be given by its effect on the generators of G: it is clear that this determines δ. The trick here is to define a G-module structure on $L = M \oplus \langle a \rangle$, where $\langle a \rangle$ is infinite cyclic: this is accomplished by the rule

$$(m, an) \cdot g = (mg + g^\delta n, an),$$

where $m \in M$ and $n \in \mathbb{Z}$. Observe that $\mathrm{Ker}(\delta) = C_G(a)$, which can be found by 9.2.23. ∎

This result may be used to compute intersections of subgroups.

9.2.25 *There is an algorithm which, when given a PF-group G together with finite subsets X and Y, finds $\langle X \rangle \cap \langle Y \rangle$.*

Proof Put $H = \langle X \rangle$ and $K = \langle Y \rangle$. The proof is by induction on $h(G) > 0$. First find an infinite abelian normal subgroup A of G. We decide if $A \leq H$. Assuming this to be true, we find $H \cap (KA) = (H \cap K)A$ by the induction hypothesis. Next, find a finite presentation of KA/A and hence a finite set of generators for $K \cap A$. Now enumerate finite subsets U of G and test each to see whether $K \cap A \leq \langle U \rangle \leq H \cap K$ and $\langle U \rangle A = H \cap (KA)$. When such a U appears, it is sure to generate $H \cap K$ since

$$H \cap K = (H \cap K) \cap \langle U \rangle A = \langle U \rangle.$$

Assume now that $A \not\leq H$. Use the method just described to find $H_0 = H \cap (KA)$ and $K_0 = (HA) \cap K$. Also $G_0 = (HA) \cap (KA)$ can be found. Clearly $G_0 = H_0 A = K_0 A$ and $H_0 \cap K_0 = H \cap K$. We may therefore replace (G, H, K) by (G_0, H_0, K_0), that is, we may assume that $G = HA = KA$. Hence $H \cap A \triangleleft G$ and $K \cap A \triangleleft G$. Since $H \cap A$ and $K \cap A$ may be found, we may also assume that $H \cap A = 1 = K \cap A$.

Let $h \in H$; then $h \in KA$, so there is a unique element h^δ of A such that $hh^\delta \in K$. Furthermore, it is routine to verify that $\delta : H \to A$ is a derivation and $\mathrm{Ker}(\delta) = H \cap K$. Hence $H \cap K$ may be found by 9.2.24. ∎

Corollary 9.2.26 *There is an algorithm which computes the normal core of a given subgroup of a PF-group G.*

Proof Let H be a subgroup of a G. Then by a theorem of Rhemtulla (1.3.19) the normal core of H in G is the interection of finitely many conjugates of H. Compute and enumerate intersections of finitely many conjugates of H and check each one for normality in G by conjugating by generators of G and their inverses. Eventually a normal intersection will appear: it will be the normal core of H. ∎

Finally, we are in a position to compute arbitrary centralizers.

9.2.27 *There is an algorithm which, when given a PF-group G and finite subsets X and Y of G, finds $C_{\langle Y \rangle}(X)$.*

Proof Since $C_{\langle Y \rangle}(X) = \bigcap_{x \in X}(C_G(x) \cap \langle Y \rangle)$, it suffices by 9.2.25 to find $C_G(x)$. We argue by induction on $h(G) > 0$. Find a non-trivial free abelian normal subgroup A and determine its G-module structure. If $x \in A$, then $C_G(x)$ can be found by 9.2.23, so suppose that $x \notin A$. By the induction hypothesis $D = C_G(xA)$ can be found. A derivation $\delta : D \to A$ is defined by $d^\delta = [d, x]$. Then $C_G(x) = C_D(x) = \mathrm{Ker}(\delta)$ and $C_G(x)$ can be found by 9.2.24. ∎

Further results

There are many other algorithms known for PF-groups and we list a few of them without going into details of the proofs.

9.2.28 *There are algorithms to perform the following tasks for a given PF-group G.*

(i) *Given finite subsets X, Y of C, find the normalizer $N_{\langle Y \rangle}(\langle X \rangle)$.*
(ii) *Decide if given finite sets of elements and subgroups are simultaneously conjugate in G.*
(iii) *Decide if G is torsion-free.*
(iv) *Find the Frattini subgroup $\varphi(G)$.*

The last of these results is the most difficult. The reader will find these and further algorithms in Baumslag *et al.* (1991). A more recent result asserts that *there is an algorithm to decide if a pair of subgroups of a PF-group permute* (Robinson 2002). It should also be mentioned that Segal's paper (1990*a*) on the solution of the isomorphism problem is a mine of information about algorithms for PF-groups.

After reading the foregoing account the reader may be forgiven for believing that any well-posed algorithmic problem for PF-groups has a positive solution. However, there are some natural problems which are insoluble even for finitely generated nilpotent groups.

All of these rest on the well-known negative solution to Hilbert's Tenth Problem found by Matijasevič (1970). According to this result, there exist integers m and e for which there is no algorithm to decide if a polynomial equation of degree e in m variables over $\mathbb{Z}$ has a solution. On the basis of Matijasevič's Theorem, Remeslennikov (1979) observed that the *epimorphism problem for finitely generated nilpotent groups* has a *negative solution*, that is, there is no algorithm to decide whether there is an epimorphism from one finitely generated nilpotent group to another.

Further negative results were found by Segal (1990*a*). For example, there are no algorithms to perform the following tasks for a finitely generated nilpotent group G:

(i) given $g \in G$ and a finite subset $X \subseteq G$, decide if g belongs to some conjugate of X;
(ii) given finite subsets X, Y of $Z(G)$, decide if there is an epimorphism from $G/\langle Y \rangle$ to $G/\langle X \rangle$.

9.3 Algorithms for finitely generated soluble minimax groups

In this and the following sections we shall discuss algorithms which have been constructed for classes of finitely generated soluble groups that extend beyond polycyclic groups. We need of course to keep in mind the existence of finitely presented soluble groups of derived length 3 with insoluble word problems—see 9.1.3.

In contrast finitely generated soluble minimax groups exhibit better algorithmic behaviour. Here it should be pointed out that for finitely generated groups, the classes of soluble groups with finite abelian rank (FAR), soluble groups with finite Prüfer rank, soluble groups with finite abelian total rank (FATR) and soluble minimax groups all coincide (Robinson 1970)-see 10.5.3. These classes of groups were discussed in detail in Chapter 5.

We begin by considering the word problem. Now it was observed after 5.3.15 that there exist uncountably many non-isomorphic finitely generated centre-by-metabelian minimax groups. These groups cannot all have soluble word problems since there are only countably many algorithms available! Consequently there exist uncountably many groups of this sort with insoluble word problem.

In fact the problem with the groups of 5.3.15 is that they are not even recursively presented: for it is very easy to show that any group with soluble word problem has a recursive presentation. It turns out that this condition is also sufficient in the case of finitely generated soluble minimax groups.

9.3.1 (Cannonito and Robinson 1984) *Let G be a finitely generated virtually soluble minimax group. Then the word problem is soluble for G if and only if it has a recursive presentation.*

Proof Only the sufficiency of the condition is in doubt, so let G have a recursive presentation in a finite set of generators $x_1, x_2, \ldots, x_n$. Clearly, we may assume that G is infinite.

Consider first the case where G is reduced, or equivalently residually finite by 5.3.2. According to 5.3.7 there exists a non-trivial torsion-free abelian normal subgroup A such that G/A is reduced. Evidently we may assume knowledge of a maximal linearly independent subset $\{a_1, \ldots, a_m\}$ of A. If A_0 denotes the subset $\langle a_1, \ldots, a_m \rangle$, then A/A_0 is a torsion group.

Now let w be a given word in the generators x_i. By induction on the torsion-free rank of G, the word problem is soluble for G/A and therefore we may assume that $w \in A$. Hence $w^k = a_1^{r_1} \cdots a_m^{r_m}$ for some positive integer k and integers r_i. Here k and the r_i may be found by enumerating all words of the form $w^{-k} a_1^{r_1} \cdots a_m^{r_m}$ and comparing them with the relators, which form a recursively enumerable set. Clearly $w = 1$ if and only if $w^k = 1$, that is, if $r_1 = \cdots = r_m = 0$, since $a_1, \ldots, a_m$ are linearly independent. Hence it is decidable whether $w = 1$.

Now for the general case. By 5.3.1 there is a radicable abelian normal subgroup D satisfying min such that G/D is residually finite. By the first part of the proof we can tell if the given word w is in D, and so we may assume this to be the case. Let P be the set of all non-trivial elements in D with square-free orders. Thus P is a finite set, which can be assumed known. Observe that $w \neq 1$ if and only if some power of w belongs to P. Enumerate all words of the form $w^{-k} u$ where $k > 0$ and $u \in P$, and compare them with the relators. If some $w^{-k} u$ is a relator, then $w \neq 1$ in G. If w is a relator, then of course $w = 1$ in G. Therefore we are able to tell whether $w = 1$. ∎

Corollary 9.3.2 *The word problem is soluble for any soluble minimax group G which is finitely presented in the variety of soluble groups of derived length at most l, for some $l \geq 0$.*

Proof By hypothesis $G \simeq F/X^F F^{(l)}$, where F is a free group and X is a finite subset of F. Then X^F is recursively enumerable since X is finite, while $F^{(l)}$ is recursively enumerable by induction on l. Therefore G has a recursive presentation and 9.3.1 shows that the word problem is soluble. ∎

The soluble minimax groups for which the word problem is insoluble given in 5.3.15 contain subgroups of Prüfer type, so they are not reduced. It is an interesting fact that finitely generated soluble minimax groups which are reduced always have soluble word problem.

9.3.3 *A finitely generated virtually soluble minimax group G which is reduced has a recursive presentation and hence has soluble word problem.*

Proof By 9.3.1 it is sufficient to show that G has a recursive presentation. One could set about constructing a recursive presentation directly, but it is more economical to appeal to a theorem of Baumslag and Bieri (1976), which asserts that G may be embedded in a finitely presented group and so must have a recursive presentation. (A generalized version of the Baumslag–Bieri Theorem due to Kilsch (1978) is proved below as 11.2.7). ∎

It is not known if the membership problem or the conjugacy problem is soluble for finitely presented soluble minimax groups. However weaker forms of these problems have been shown to be soluble by Robinson (1986a). We state the results without giving proofs.

9.3.4 *Let G be a finitely generated, virtually soluble minimax group with a recursive presentation. Let H be a subgroup of G that is recursively enumerable in terms of the presentation. Then there is an algorithm which decides membership of elements of G in H.*

9.3.5 *Let G be a finitely generated, virtually soluble minimax group with a recursive presentation and let g be an element of G given in terms of the presentation. Then there is an algorithm which decides if any given element of G is conjugate to g.*

The question of making the algorithms in 9.3.4 and 9.3.5 uniform in H and g remains open.

That 9.3.1 is not valid for finitely generated soluble groups of infinite rank is shown by an interesting example of Meskin (1974).

9.3.6 *There is a finitely generated, recursively presented, centre-by-metabelian group which is residually finite but has insoluble word problem.*

Proof We refer to P. Hall's construction of finitely generated centre-by-metabelian groups in 4.1.3. This yields a group $H = \langle t \rangle \ltimes N$, where $N = \langle x_i \mid i \in \mathbb{Z} \rangle$ is nilpotent of class 2 with $[x_i, x_{i+k}] = c_k$ independent of i and $x_i^t = x_{i+1}$. Also $\langle t \rangle$ is infinite cyclic, $H = \langle t, x_0 \rangle$ and $Z(N) = Z(H) = H'' = \langle c_k \mid k = 1, 2, \ldots \rangle$.

Now let $S = \{i_1, i_2, \ldots\}$ be a recursively enumerable, but non-recursive set of positive integers. Add the relations $c_{i_r}^{i_r!} = 1$, $r = 1, 2, \ldots$, to the defining relations of H to form a quotient group G. Evidently, G is a 2-generator, centre-by-metabelian group and it has a recursive presentation. But the word problem for G is insoluble: for if we could decide whether $c_i^{i!}$ equals 1, it would be possible to decide if $i \in S$.

To show that G is residually finite, set $L_k = (t^k)^G$ and note that $x_i L_k = x_{i+k} L_k$, so that $N L_k / L_k$ is finitely generated and G/L_k is polycyclic. Hence G/L_k is residually finite. To complete the proof one shows that $\bigcap_{k=1,2,\ldots} L_k = 1$. To do this observe that L_k is generated by elements of the form t^k, $x_i^{-1} x_{i+k}$ and $c_j^{-1} c_{j+k}$, where $i \in \mathbb{Z}$, $j = 1, 2, \ldots$. Let $1 \neq g \in G$; then g has a unique expression in terms of t, x_i, c_j. By choosing k to exceed any subscript or exponent occurring in this expression for g, it is not hard to show that $g \notin L_k$. ∎

On the other hand, finitely presented centre-by-metabelian groups satisfy max $-n$ by a result of Bieri and Strebel (1980)—see also 11.5.15—and so they have soluble word problem by 9.4.5 below.

9.4 Submodule computability

Our next aim is the creation of an algorithmic theory of finitely generated abelian-by-polycyclic groups. Here we would expect, from our experience in Chapters 4 and 7, that algorithmic properties of finitely generated modules over polycyclic group rings will play an important role. What are needed here are algorithms to decide if elements of such modules are trivial and to find finite presentations of submodules.

For modules over finitely generated commutative rings such algorithms have been known for many years, one of the earliest papers being that of G. Hermann (1926), and they are essentially constructive versions of Hilbert's Basis Theorem: for a modern treatment see Seidenberg (1974, 1978). There has been much recent activity in this area, with the development of Gröbner Basis Theory and efficient practical algorithms. Our task here is to pass from the commutative to the non-commutative case.

A ring R is said to be *computable* if its elements form a computable set (in the sense of 9.1), and if the ring operations are computable (that is, they can be performed by Turing machines). The examples we have in mind are the integral group rings of PF-groups, which are clearly computable.

Suppose that R is a computable ring which satisfies max $- r$, the maximal condition on right ideals: also let M be a finitely generated (right) R-module.

Then M is noetherian and also $M \overset{R}{\simeq} F/K$, where F is the free R-module on a finite set $\{x_1, \ldots, x_k\}$ and K is a submodule. Now F too is noetherian, so K is finitely generated as an R-module, say by $y_1, \ldots, y_m$. Therefore, M has a finite R-presentation $\langle x_1, \ldots, x_n \mid y_1, \ldots, y_k \rangle$. This means that every finitely generated R-module can be specified by a finite R-presentation, which permits the formulation of algorithmic problems for such modules.

Next we come to a crucial definition: this occurs in an important paper of Baumslag, Cannonito and Miller (1981*a*) which extends ideas of Romanovskiĭ (1980).

Let M be a finitely generated R-module, where R is a computable ring with max $-r$, and suppose that M is given by a finite presentation. Then M is called *submodule computable* if there are algorithms which perform the following tasks:

(i) to find a finite R-presentation for the submodule N generated by a given finite subset of M;

(ii) to decide membership in the submodule N of a word w expressed in terms of the generators of M.

Should every finitely generated R-module be submodule computable, the ring R is said to be a *submodule computable ring*.

For example, the ring $\mathbb{Z}$ is submodule computable: indeed there is a standard matrix procedure to find a basis for a given subgroup of a free abelian group and hence obtain a finite presentation of a subgroup of a finitely generated abelian group.

The fundamental theorem for all that follows is:

9.4.1 (Baumslag, Cannonito and Miller 1981*a*) *If G is a virtually polycyclic group, the group ring $\mathbb{Z}G$ is submodule computable.*

We note a useful consequence of this result.

Corollary 9.4.2 *There is an algorithm which, when given a PF-group G, two finitely generated $\mathbb{Z}G$-modules M and N, and a $\mathbb{Z}G$-module homomorphism $\theta : M \to N$, finds the kernel of θ.*

For we can find a finite $\mathbb{Z}G$-presentation of $\mathrm{Im}(\theta)$ and hence a finite set of module generators for $\mathrm{Ker}(\theta)$.

Before undertaking the proof of this major theorem, we give two simple lemmas which will be helpful in the course of the proof. The first of these tells us that it suffices to check submodule computability of free modules.

9.4.3 *Let R be a computable ring satisfying* max $-r$. *If every finitely generated free R-module is submodule computable, then R is submodule computable.*

Proof Let M be a finitely generated R-module; then M has a finite R-presentation, say $\langle a_1, \ldots, a_m \mid w_1, \ldots, w_r \rangle$. Let F be the free R-module on a set $\{f_1, \ldots, f_m\}$: thus the assignment $f_i \mapsto a_i$ determines a surjective

R-homomorphism $\varphi : F \to M$. Take a finite subset $\{b_1, \ldots, b_k\}$ of M and put $N = b_1 R + \cdots + b_k R$. We need to find a finite R-presentation of N. Now the preimage U of N under φ is generated by $\mathrm{Ker}(\varphi)$ and the elements $b_j(f)$, that is, the b_j as words in the free generators $f_1, \ldots, f_m$, and N is generated by the $w_i(f)$ and $b_j(f)$.

Since F is submodule computable, a finite R-presentation of U in the $w_i(f)$, and $b_j(f)$ can be found. In this presentation delete the generators $w_i(f)$ and add them to the relators to get a finite presentation of N in the $b_j(f)$. Finally, for any word v in $f_1, \ldots, f_m$ it is clear that $v \in U$ if and only if $v^\varphi \in N$. By submodule computability of F, membership in N is decidable. Therefore M is submodule computable, whence so is R.   ∎

The second result needed permits the computation of intersections of submodules.

9.4.4 *Let R be a submodule computable ring and M a finitely generated R-module (given by a finite R-presentation). Suppose that L and N are submodules of M given by finite sets of generators. Then there is an algorithm which finds a finite set of generators for $L \cap N$.*

Proof Let $L = u_1 R + \cdots + u_\ell R$ and $N = v_1 R + \cdots + v_n R$. Clearly $L + N$ is generated by all the elements u_i, v_j and submodule computability permits us to find a finite R-presentation of $L + N$ in the u_i, v_j. Suppressing the generators v_i and adding these as relators, we obtain a finite R-presentation of $(L + N)/N$ in the $u_i + N$, and hence of $L/L \cap N$ in the $u_i + (L \cap N)$. The relators in the latter presentation, when expressed on as words in the u_i, form a set of generators for $L \cap N$.   ∎

We are now ready to undertake the proof of the main result on submodule computability.

Proof of 9.4.1 We may assume that a series $1 = G_0 \triangleleft G_1 \triangleleft \cdots \triangleleft G_\ell = G$ in which each G_{i+1}/G_i is cyclic or finite is known. The proof is by induction on ℓ, which can be assumed positive. By the induction hypothesis the theorem is true for $H = G_{\ell-1}$, so the ring $\mathbb{Z}H$ is submodule computable. Let F be a finitely generated free $\mathbb{Z}G$-module with basis $\{f_1, f_2, \ldots, f_q\}$ and let $\{v_1, v_2, \ldots, v_n\}$ be a finite subset of F. By 9.4.3 it is sufficient to show how to find a finite $\mathbb{Z}G$-presentation for the $\mathbb{Z}G$-submodule N generated by v_i, and to decide membership in N.

Case 1: G/H is finite. This is the easy case. Find a finite transversal to H in G, say $\{1 = u_1, u_2, \ldots, u_k\}$. Then F is the free $\mathbb{Z}H$-module on the set $\{f_i u_j \mid i = 1, 2, \ldots, q, \ j = 1, 2, \ldots, k\}$. Since $\mathbb{Z}H$ is submodule computable, we can find a finite $\mathbb{Z}H$-presentation of N in the $f_i u_j$ and also decide membership in N. To obtain a finite $\mathbb{Z}G$-presentation of N, simply omit the $f_i u_j$ for $j > 1$, but retain the defining relators.

Case 2. $G/H = \langle xH \rangle$ is infinite cyclic. An element f of F has a unique expression $f = \sum_{i=r}^{s} c_i x^i$ where $r \le s$ and c_i belongs to the free $\mathbb{Z}H$-module F_0 on $\{f_1, \ldots, f_q\}$. If $c_r \ne 0$ and $c_s \ne 0$, then r and s are called the *lower* and *upper* degrees of f. If $r \ge 0$, then s is the usual degree. Clearly, it may be assumed that no v_i involves negative powers of x. Choose and fix an integer d such that $\deg(v_i) \le d$ for all i and write

$$M = F_0 \oplus F_0 x \oplus \cdots \oplus F_0 x^{d-1}.$$

We will describe how to construct an ascending sequence of finite subsets S_i of F starting with $S_1 = \{v_1, v_2, \ldots, v_n\}$. Suppose that S_i has already been constructed and put $L_i = (S_i)\mathbb{Z}H$. Then, according to 9.4.4, it is possible to find finite sets T_i and T_i^* such that $M \cap L_i = (T_i)\mathbb{Z}H$ and $Mx \cap L_i = (T_i^*)\mathbb{Z}H$. Define the next subset in the sequence to be

$$S_{i+1} = \{fx^j \mid f \in S_i \cup T_i \cup T_i^*, fx^j \in F_0 \oplus F_0 x \oplus \cdots \oplus F_0 x^d\}.$$

Clearly S_{i+1} is finite and $S_i \subseteq S_{i+1}$.

Writing $L_i = (S_i)\mathbb{Z}H$ as before, we note that $L_i = L_{i+1}$ for some i since $S_i \subseteq M$ and M is $\mathbb{Z}H$-noetherian. What is more, we can decide if $L_i = L_{i+1}$ since $\mathbb{Z}H$ is submodule computable: thus i and S_i can be found. Now put $S = S_{i+1}$. We claim that the set S has the following properties:

(i) $N = (S)\mathbb{Z}G$;
(ii) $S \subseteq M$;
(iii) $N \cap M \subseteq (S)\mathbb{Z}H$;
(iv) if $s \in S$ and $\deg(s) < d$, then $sx \in S$ and $sx^{-1} \in S$ provided that sx^{-1} involves no negative powers of x.

In the first place S contains all the v_i's, so (i) is true, while (ii) is true by construction. In addition (iv) is valid: for if $s \in S$ and $\deg(s) < d$, with $s = ux^j, u \in S_i \cup T_i \cup T_i^*$, then $sx = ux^{j+1} \in S_{i+1} = S$ and $sx^{-1} = ux^{j-1} \in S$, provided that negative powers of x do not occur. Thus it remains only to establish (iii).

Let $f \in N \cap M$ and write $f = \sum_{s \in S} sh_s$ where $h_s \in \mathbb{Z}G$, using (i). Let $-\ell$ be the smallest lower degree of a non-zero h_s and assume that $\ell > 0$. Thus $h_s = d_s x^{-\ell} +$ terms of higher degree in x, where $d_s \in \mathbb{Z}H$ and $d_s = 0$ if h_s has lower degree less than $-\ell$. Hence $f = \sum_s sh_s = (\sum_s sd_s) x^{-\ell} + \cdots$. Since f involves no negative powers of x, it must be the case that $\sum_s sd_s \in F_0 x \oplus \cdots \oplus F_0 x^d = Mx$. Therefore $\sum_s sd_s \in Mx \cap (S)\mathbb{Z}H = L_{i+1} = L_i$. It follows from the mode of construction that S contains T_i^*, which by definition generates $Mx \cap (S)\mathbb{Z}H$. Hence $\sum_s sd_s = \sum_s se_s$ where $e_s \in \mathbb{Z}H$ and $e_s = 0$ if s involves no negative powers of x. We now replace d_s by e_s in h_s, obtaining thereby an element $h_s' = e_s x^{-\ell} + \cdots$. Then

$$f = \left(\sum_s sd_s\right) x^{-\ell} + \cdots = \left(\sum_s se_s\right) x^{-\ell} + \cdots = \sum_s sh_s'.$$

If s has lower degree $-\ell$, replace sh'_s in f by $(sx^{-1})xh'_s$, noting that $sx^{-1} \in S$ and xh'_s has lower degree $-\ell + 1$. After at most ℓ repetitions of this procedure we will arrive at a situation in which no h_s involves negative powers of x.

Next let m be the maximum (upper) degree of an h_s with $s \in S$. Thus $h_s = r_s x^m +$ terms of lower degree in x, where $r_s \in \mathbb{Z}H$ and $r_s = 0$ if $\deg(h_s) < m$. Now

$$f = \sum_s sh_s = \left(\sum_s sr_s\right)x^m + \cdots .$$

Therefore $\deg(\sum_s sr_s) \leq \deg(f) - m < d$ since $\deg(f) \leq d$. It follows that $\sum_s sr_s \in M \cap (S)\mathbb{Z}H$. By the construction there exist elements a_s of $\mathbb{Z}H$ such that $\sum_s sr_s = \sum_s sa_s$ and $a_s = 0$ if $\deg(s) = d$. Replace r_s by a_s in $h_s = r_s x^m + \cdots$ to get an element $h'_s = a_s x^m + \cdots$ such that $f = \sum_s sh'_s$. If $\deg(h'_s) = m$, then $a_s \neq 0$ and hence $\deg(s) < d$. Replace the term $sa_s x^m$ by $sx(x^{-1}a_s x^m)$, noting that $sx \in S$. After at most m applications of this procedure, we reduce to the situation where $m = 0$, that is, $h_s \in \mathbb{Z}H$ for all s, and $f \in (S)\mathbb{Z}H$ as required. Thus property (iii) has been verified.

Next we show how to find a finite $\mathbb{Z}G$-presentation of $N = (S)\mathbb{Z}G$ in the generators S. By induction $\mathbb{Z}H$ is submodule computable, so we can obtain a finite $\mathbb{Z}H$-presentation of $(S)\mathbb{Z}H$ in the generators S. Add to this presentation the relations of the form $s' = sx$ where $s, s' \in S$ to produce a finite $\mathbb{Z}G$-presentation of a $\mathbb{Z}G$-module. Our claim is that this presents N.

To establish this, let f be a $\mathbb{Z}G$-word in S such that $f = 0$ in N. We will show that f is a consequence of the defining relators. Without loss of generality we may assume that $f = \sum_s sh_s$, where h_s belongs to $\mathbb{Z}G$ and does not involve negative powers of x. Denote the maximum degree of the h_s's by m, and assume that $m > 0$. Writing $h_s = r_s x^m + \cdots$ lower powers of x, we have $f = (\sum_s sr_s)x^m + \cdots$. Since $f = 0$ in N, it follows that $\deg(\sum_s sr_s) \leq d - 1$ and hence $\sum_s sr_s \in N \cap M \subseteq (S)\mathbb{Z}H$ by property (iii) of the construction. Thus $\sum_s sr_s = \sum_s sa_s$, where $a_s \in \mathbb{Z}H$ and $a_s = 0$ if $\deg(s) = d$. Hence $f = (\sum_s sa_s)x^m + \cdots$. The device of replacing s by sx when $\deg(s) < d$ allows us to reduce the value of m. By repetition of this procedure we reach a situation where $m = 0$ and $f = \sum_s sh_s$ with $h_s \in \mathbb{Z}H$. The relation $f = 0$ is therefore a consequence of the $\mathbb{Z}H$-defining relations for $(S)\mathbb{Z}H$. Finally, convert the presentation of N in the generators S to a presentation of N in $\{v_1, \ldots, v_n\}$.

Finally, let $f \in F$ be a given element. We must show how to decide if $f \in N$ and here it may be assumed that f involves no negative powers of x. Carry out the construction of the set S as above, but choose $d \geq \deg(f)$. Then by property (iii) we have $f \in N$ if and only if $f \in (S)\mathbb{Z}H$. The latter is decidable by submodule computability of $\mathbb{Z}H$, which completes the proof of 9.4.1.   ∎

As our first application of 9.4.1 we establish a result which displays a large class of finitely generated soluble groups with soluble word problem.

9.4.5 (Baumslag, Cannonito, and Miller 1981a) *Let G be a nilpotent-by-PF-group satisfying* $\max - n$. *Then G has a finitely generated recursive presentation for which the word problem is soluble.*

Proof In the first place G is finitely generated since soluble groups with $\max - n$ are finitely generated: also $\max - n$ is inherited by subgroups of finite index. Let N be a nilpotent normal subgroup such that G/N is a PF-group. Of course, G/N is finitely presented: in addition $\gamma_i(N)/\gamma_{i+1}(N)$ is a finitely generated $\mathbb{Z}(G/N)$-module and hence has a finite $\mathbb{Z}(G/N)$-presentation, which may be assumed known. By the usual method of writing down a presentation for a group extension, a finitely generated recursive presentation of G may be written down.

Next let w be a given word in the generators of G. By induction on c, the nilpotent class of N, the word problem is soluble for $G/\gamma_c(N)$, so we may assume that $w \in \gamma_c(N)$. Now by $\max - n$, the module $\gamma_c(N)$ has a finite $\mathbb{Z}(G/N)$-presentation. Therefore by submodule computability of $\mathbb{Z}(G/N)$, we can decide if $w = 1$. ∎

In particular the word problem is soluble for finitely generated abelian-by-polycyclic-by-finite groups since such groups satisfy $\max - n$. An unsolved problem here is whether every finitely presented soluble group with $\max - n$ has soluble word problem. Note that this is the largest quotient closed class of finitely presented soluble groups since a quotient of a finitely presented group with $\max - n$ is finitely presented.

Our next aim is to prove that the membership problem is soluble for finitely generated abelian-by-nilpotent-by-finite groups. For this purpose an additional algorithmic tool is required.

9.4.6 (Baumslag, Cannonito and Miller 1981a) *Let G be a finitely generated virtually nilpotent group. Then there is an algorithm which, when given a finitely generated $\mathbb{Z}G$-module M and finite subsets $\{a_1, \ldots, a_q\}$ and $\{x_1, \ldots, x_r\}$ of M and G, respectively, finds a finite $\mathbb{Z}\langle x_1, \ldots, x_r\rangle$-presentation of the $\mathbb{Z}\langle x_1, \ldots, x_r\rangle$-module generated by $a_1, \ldots, a_q$.*

Proof Write $H = \langle x_1, \ldots, x_r \rangle$. By 9.2.11 we can find a finite presentation of H, and by 9.4.1 a finite presentation of the module $\{a_1, \ldots, a_q\}\mathbb{Z}G$ is obtainable: clearly nothing is lost in taking the latter to be M. By hypothesis there is a nilpotent normal subgroup L with finite index in G (obtainable by enumerating subgroups of finite index). Then $H_0 = H \cap L$ can be found and also a transversal to H_0 in H, say $\{1 = y_1, \ldots, y_d\}$.

Write $N = \{a_1, \ldots, a_q\}\mathbb{Z}H$ and observe that N is generated as a $\mathbb{Z}H_0$-module by the elements $a_i y_j$. Assuming the result is known for H_0, we find a finite $\mathbb{Z}H_0$-presentation of N in the $a_i y_j$. From this we obtain a finite $\mathbb{Z}H$-presentation of N by omitting generators $a_i y_j$ for $j > 1$. This demonstrates that we may replace H by H_0 and assume that H is subnormal in G. By forming successive normal closures and using 9.2.1 and 9.2.11, a series $H = H_0 \lhd H_1 \lhd \cdots \lhd H_\ell = G$ with

H_{i+1}/H_i finite or cyclic may be constructed. Induction on ℓ then reduces to the case where $H \triangleleft G$ and G/H is finite or cyclic.

Suppose first that G/H is finite and let $\{1 = t_1, t_2, \ldots, t_r\}$ be a transversal to H in G. Then $M = \{a_1, \ldots, a_q\}\mathbb{Z}G$ is generated as a $\mathbb{Z}H$-module by the elements $a_i t_j$. A finite $\mathbb{Z}H$-presentation of M in the $a_i t_j$ is obtained from the $\mathbb{Z}G$-presentation of M by replacing $a_i g$ by $(a_i t_j)h$, where $g = t_j h$, $(h \in H)$. Since $\mathbb{Z}H$ is submodule computable, it is possible to find a finite $\mathbb{Z}H$-presentation of $N = \{a_1, \ldots, a_q\}\mathbb{Z}H$.

Now assume that G/H is infinite cyclic, with $G/H = \langle xH \rangle$. Suppose that M has a finite presentation in $a_1, \ldots, a_q$ with defining relators $v_1, \ldots, v_n$. Let F be the free $\mathbb{Z}G$-module on $\{f_1, \ldots, f_q\}$, so that $M \overset{\mathbb{Z}G}{\simeq} F/K$, where K is the submodule generated by $v_1, \ldots, v_n$. Thus $N \overset{\mathbb{Z}G}{\simeq} F_0/F_0 \cap K$, where F_0 is the free $\mathbb{Z}H$-module on $\{f_1, \ldots, f_q\}$. To obtain a finite $\mathbb{Z}H$-presentation of N, it suffices to find a finite set of generators for $F_0 \cap K$. Clearly, we lose nothing by assuming that the v_i involve no negative powers of x.

Apply the method of proof of 9.4.1 to construct a finite subset S of F containing $v_1, \ldots, v_n$, by choosing $d \geq \max\{\deg(v_i) \mid i = 1, \ldots, n\}$. Let $f \in F_0 \cap K$ be written in the form $\sum_{s \in S} s h_s$, where $h_s \in \mathbb{Z}G$. Since $\deg(f) = 0$, the reduction argument given in the proof of 9.4.1 allows us to assume that each h_s belongs to $\mathbb{Z}H$. Hence $F_0 \cap K = F_0 \cap \{v_1, \ldots, v_n\}\mathbb{Z}H$. Finally, F_0 and K are $\mathbb{Z}H$-submodules of $F_0 \oplus F_0 x \oplus \cdots \oplus F_0 x^d$ and submodule computability of $\mathbb{Z}H$ and 9.4.4 allow us to find a finite set of generators for $F_0 \cap K$.   ∎

We are now in a position to establish the solubility of the membership problem for a significant class of finitely generated soluble groups.

9.4.7 (Romanovskiĭ 1980) *The membership problem is soluble for finitely generated abelian-by-nilpotent-by-finite groups.*

Proof Let G be a finitely generated abelian-by-nilpotent-by-finite group. Then G has a finitely generated recursive presentation by 9.4.5. Let A be an abelian normal subgroup of G such that $\bar{G} = G/A$ is nilpotent-by-finite. Then A has a finite $\mathbb{Z}\bar{G}$-module presentation, which we may suppose to be given.

Let $w, w_1, \ldots, w_m$ be words in the generators of G and put $H = \langle w_1, \ldots, w_m \rangle$. The problem is to decide whether $w \in H$ in G. Now the membership problem is soluble for G/A, so we may assume that $w \in HA$. Now express w in the form $w = ha$, with $h \in H$ and $a \in A$, by enumerating such elements ha, comparing them with w and utilizing solubility of the word problem in G. Evidently $w \in H$ if and only if $a \in H \cap A$, so this is what has to be decided.

Since H is finitely generated, a finite presentation of HA/A can be found using 9.2.11. This yields a finite set of generators for $H \cap A$ as a $\mathbb{Z}(HA/A)$-module. Now apply 9.4.6 to get a finite presentation of the $\mathbb{Z}(HA/A)$-submodule generated by $H \cap A$ and a. Finally submodule computability of $\mathbb{Z}(HA/A)$ permits us to decide if $a \in H \cap A$.   ∎

It is still an open question whether the membership problem is soluble for finitely generated abelian-by-polycyclic groups: the challenge here is extend 9.4.6 to polycyclic groups. It is also unknown if the membership problem is soluble for metanilpotent groups satisfying max $-n$.

The methods of the proof of 9.4.5 may be used to establish another useful algorithmic result.

9.4.8 (Baumslag, Cannonito, and Miller 1981*b*) *Let G be a virtually polycyclic group. Then there is an algorithm to decide if a given finitely generated $\mathbb{Z}G$-module is finitely generated as an abelian group.*

Using this it is not difficult to deduce:

9.4.9 *There is an algorithm which, when given a finitely presented soluble group G in the variety of soluble groups with derived length at most k, decides if G is polycyclic and, if so, finds a finite presentation of it.*

9.5 Algorithms for finitely generated metabelian groups

In the remainder of the chapter we shall concentrate on finitely generated metabelian groups, following an article of Baumslag, Cannonito, and Robinson (1994). For such groups one would expect to find a rich algorithmic theory: for, if G is a finitely generated metabelian group, G' is a noetherian module over the finitely generated commutative ring $\mathbb{Z}(G_{ab})$ by 4.2.3, and the methods of constructive commutative algebra are at our disposal.

Another reason for undertaking the study is that a finitely generated metabelian group is finitely presented in the variety of metabelian groups $\mathbf{A}^2$: this is because finitely generated metabelian groups satisfy max $-n$. Thus such groups are finitely expressible in a natural way. In what follows it will be assumed that a finitely generated metabelian group is specified by a finite $\mathbf{A}^2$-presentation.

As one would expect, it is advantageous to use special $\mathbf{A}^2$-presentations which shed light on the group structure. By a *preferred presentation* of a finitely generated metabelian group G we shall mean a finite $\mathbf{A}^2$-presentation of the form

$$G = \langle x_1, x_2, \ldots, x_n \mid R_1 \cup R_2 \rangle,$$

where

(i) R_1 is a finite set of words of the type

$$\prod_{1 \leq i < j \leq n} [x_i, x_j]^{u_{ij}},$$

u_{ij} being a word of the form $x_1^{m_1} \cdots x_n^{m_n}$, $m_i \in \mathbb{Z}$;

(ii) R_2 is a finite set of words of the type

$$x_1^{\ell_1} x_2^{\ell_2} \cdots x_n^{\ell_n} w,$$

where $\ell_i \in \mathbb{Z}$, and w is a word of the type specified in (i).

So the words in R_2 determine a finite presentation of the group G_{ab}, while those in R_1 will in fact serve as defining relators for a finite $\mathbb{Z}G_{ab}$-presentation of G' in the generators $[x_i, x_j]$. First we must show that such presentations can always be found.

9.5.1 *There is an algorithm which, when given a finitely generated metabelian group G by means of a finite $\mathbf{A}^2$-presentation, finds a preferred $\mathbf{A}^2$-presentation.*

Proof Let the given presentation of G be in generators $x_1, x_2, \ldots, x_n$. Now each defining relator can be rewritten in the form $x_1^{\ell_1} \cdots x_n^{\ell_n} w$ where,

$$w = \prod_{1 \le i < j \le n} [x_i, x_j]^{u_{ij}}$$

with u_{ij} a word of the form $x_1^{m_1} \cdots x_n^{m_n}$. This is done by collecting the x_i's in order on the left, introducing commutators $[x_i, x_j]$ as needed and using the fact that commutators commute among themselves. If the ℓ_i are not all zero, this yields a relator to be included in the set R_2; otherwise a relator for the set R_1 is obtained. Thus a preferred presentation of G in the x_i has been found.   ∎

The primary use of a preferred presentation is to find a finite module presentation of the derived subgroup.

9.5.2 *There is an algorithm which, when a finitely generated metabelian group G is given, finds a finite $\mathbb{Z}G_{ab}$-presentation of G'.*

Proof First find a preferred presentation of G with generators $x_1, \ldots, x_n$, using 9.5.1. If F is the free group on $\{x_1, \ldots, x_n\}$, there is a surjective homomorphism $\theta : F \to G$.

Next let M be the $\mathbb{Z}G_{ab}$-module with generators c_{ij}, $i < j = 1, 2, \ldots, n$, and defining relators corresponding to the relators in the $[x_i, x_j]$ occurring in the set R_2 of the preferred presentation. Make M into an F-module via θ and form the semidirect product $F \ltimes M$. Add the relators $[x_i, x_j] = c_{ij}$ and $x_1^{\ell_1} \cdots x_n^{\ell_n} w = 1$, where the latter is a typical relator in the set R_2 of the preferred presentation. This yields a group $\bar{G}$ which is a quotient of $F \ltimes M$. Clearly, $\bar{G}$ is a finitely generated metabelian group and $\bar{G}' = M$. Since the relators of G also hold in $\bar{G}$, there is a surjective of homomorphism from G to $\bar{G}$ and hence a surjective $\mathbb{Z}G$-homomorphism from G' to M in which $[x_i, x_j] \mapsto c_{ij}$. But the module relations defining M also hold in G': thus there is a module homomorphism from M to G' sending c_{ij} to $[x_i, x_j]$. Hence $G' \overset{\mathbb{Z}G}{\simeq} M$.   ∎

A measure of the usefulness of preferred presentations is the extent to which they aid us in getting hold of finitely generated subgroups.

9.5.3 *There is an algorithm which, when given a finitely generated metabelian group G and a finite subset X, find a finite $\mathbf{A}^2$-presentation of $\langle X \rangle$.*

Proof Write $H = \langle X \rangle$. A preferred presentation of G is found by 9.5.1, and from it we obtain a finite presentation of G_{ab}, and hence for $\bar{H} = HG'/G' \simeq H/H \cap G'$. From this we read of a finite set of generators for the $\mathbb{Z}\bar{H}$-module $H \cap G'$. From the preferred presentation of G we can find a finite $\mathbb{Z}G_{ab}$-presentation of G', using 9.5.2. Then use 9.4.6 to obtain a finite $\mathbb{Z}\bar{H}$-presentation of $H \cap G'$ in terms of the generators of $H \cap G'$. Finally, combine the presentations of $H \cap G'$ and $H/H \cap G'$ in the natural way to produce a finite $\mathbf{A}^2$-presentation of H. ∎

The conjugacy problem

Recall that the word problem and the membership problem are soluble for finitely generated metabelian groups by 9.4.7. To solve the conjugacy problem for this class of groups, a further result from constructive commutative algebra is needed, specifically the following.

9.5.4 (Noskov 1982) *There is an algorithm which, when a finitely generated commutative ring R and a finite set X of units of R are given, finds a finite presentation of the group $\langle X \rangle$.*

This result is a generalization of a theorem of Borevič–Šafarevič which was important in the algorithmic theory of polycyclic groups (see 9.2.17). In fact the proof of 9.5.4 is based on that result. Notice that any finitely generated commutative ring with identity is isomorphic with a quotient of some polynomial ring $\mathbb{Z}[x_1, x_2, \ldots, x_n]$, and hence is specifiable by a finite set of polynomials in the x_i.

The first step in solving the conjugacy problem is to solve a module theoretic version of it.

9.5.5 *There is an algorithm which, when given a finitely generated abelian group Q, a finitely generated $\mathbb{Z}Q$-module M, and elements a, b of M, decides if there is an element g of G such that $b = ag$.*

Proof Put $R = \mathbb{Z}Q$. We first decide whether $b \in N = aR$, which is possible by the submodule computability of R (see 9.4.1). Then we may suppose this to be the case and find an $r \in R$ such that $b = ar$. Also a finite presentation of the R-module N may be found by submodule computability: this will have a single generator, so N may be identified with R/I, where I is an ideal with a known finite set of generators. Furthermore a is identified with $1 + I$, so that $b = r + I$. With this notation, we recognize that the problem is to decide if $b \in \bar{Q} = (Q + I)/I$.

First of all decide whether b is a unit of R, or equivalently if $R = rR + I$. This is decidable by submodule computability. Assuming this to be true, we apply 9.5.4 to find a finite presentation of the finitely generated abelian group $\langle b, \bar{Q} \rangle$. From this we can tell if $b \in \bar{G}$. ∎

We are now ready to attack the conjugacy problem.

9.5.6 (Noskov 1982) *There is an algorithm which, when given a finitely generated metabelian group G and elements x and y, decides if x and y are conjugate in G.*

Proof Let $A = G'$ and $Q = G_{ab}$. Obviously, we are able to decide if $xA = yA$, so we may assume this to be true. Observe that x and y are conjugate if and only if they are conjugate modulo $[A, x]$. For if the latter is true, then $y^g = x[x, a] = x^a$

for some $g \in G$ and $a \in A$. Moreover a finite $\mathbb{Z}Q$-presentation of A can be found by 9.5.3. If A is generated as a $\mathbb{Z}Q$-module by $a_1, \ldots, a_m$, the submodule $[A, x]$ is generated by $[a_1, x], \ldots, [a_m, x]$ since A is abelian. Therefore, a finite $\mathbf{A}^2$-presentation of $G/[A, x]$ is at hand. Consequently we may assume that $[A, x] = 1$.

Now put $M = \langle x, A \rangle$, which is an abelian normal subgroup of G containing both x and y, since $xA = yA$. Of course, M is a finitely generated $\mathbb{Z}Q$-module and it is easy to write down a finite $\mathbb{Z}Q$-presentation of M in terms of that of A, after computing conjugates of x by generators of G and their inverses. Finally, apply 9.5.5 to the $\mathbb{Z}Q$-module M and decide whether x and y are conjugate in G. ∎

It is still an open question whether finitely generated abelian-by-nilpotent groups have soluble conjugacy problem.

Another application of Noskov's Theorem (9.5.4) is to find the centre of a finitely generated metabelian group: notice that this subgroup is finitely generated because of the condition max $-n$. The first step is to find maximal abelian normal subgroups.

9.5.7 *There is an algorithm which, when a finitely generated metabelian group G is given, finds a finite $\mathbb{Z}G_{ab}$-module presentation for a maximal abelian normal, subgroup containing G'.*

Proof Write $Q = G_{ab}$. First find a finite presentation of $A = G'$, say in generators $a_1, a_2, \ldots, a_m$. Since $\mathbb{Z}Q$ is submodule computable, a finite presentation in a_i can be found for each a_i^G. Now $a_i^G \overset{\mathbb{Z}G}{\simeq} \mathbb{Z}Q/I_i$, where $I_i = \mathrm{Ann}_{\mathbb{Z}Q}(a_i)$; therefore a finite set of generators of the ideal I_i is available. Next $(Q + I_i)/I_i$ is a finitely generated group of units of the ring $\mathbb{Z}Q/I_i$ and thus 9.5.4 enables us to find a finite presentation of it. The assignment $x \mapsto x + I_i$ yields a surjective homomorphism from Q to $(Q + I_i)/I_i$ with kernel $C_Q(a_i) = C_G(a_i)/A$. Thus we have a finite set of generators for $C_G(a_i)/A$.

Using the finite presentation of Q, we proceed to find $\bigcap_{i=1}^m C_G(a_i)/A$, which equals $C_G(A)/A$. It is decidable whether or not $C_G(A) = A$, and should this be true, A is a maximal abelian normal subgroup of G. Otherwise find b in $C_G(A)\backslash A$ and set $A_1 = \langle A, b_1 \rangle$. This is a $\mathbb{Z}Q$-module and a finite $\mathbb{Z}Q$-presentation of it is obtainable from a finite presentation of A. By repetition of the argument an ascending chain of abelian normal subgroups is generated, $A = A_0 \leq A_1 \leq A_2 \leq \cdots$. Of course this must terminate, and when it does, we will have our maximal abelian normal subgroup. ∎

It is now straightforward to compute the centre.

9.5.8 *There is an algorithm which finds the centre of a finitely generated metabelian group G.*

Proof Write $Q = G_{ab}$ and apply 9.5.7 to obtain a maximal abelian normal subgroup A containing G'. Clearly $Z(G) = \bigcap_{i=1}^n C_A(x_i)$, where $G = \langle x_1, x_2, \ldots, x_n \rangle$. Now $a \mapsto [a, x_i]$, $a \in A$, yields a $\mathbb{Z}Q$-endomorphism of

A with kernel $C_A(x_i)$ and $A/C_A(x_i) \overset{\mathbb{Z}G}{\simeq} [A, x_i]$. A finite $\mathbb{Z}Q$-presentation of $[A, x_i]$ is found by submodule computability and from this a finite set of $\mathbb{Z}Q$-generators for each $C_A(x_i)$ is obtained. This allows us to find finitely many $\mathbb{Z}Q$-module generators of $Z(G)$ and of course they will generate $Z(G)$ as a group. ∎

It is evident that, as a consequence of 9.5.8, the terms of the upper central series and the hypercentre of a finitely generated metabelian group may be computed. Next we treat the Fitting subgroup. Now this subgroup is not in general finitely generated, but by max $-n$ it is the normal closure of a finite subset and so it is finitely specifiable.

9.5.9 *There is an algorithm which, when given a finitely generated metabelian group G, finds a finite subset X such that $\mathrm{Fit}(G) = \langle X \rangle^G$.*

Proof Put $Q = G_{ab}$, $A = G'$ and $R = \mathbb{Z}Q$. It will be convenient for us to write R-modules additively in this proof.

We show how to construct a series of R-submodules $0 = A_0 < A_1 < \cdots < A_\ell = A$ and prime ideals P_i of R such that $A_{i+1}/A_i \overset{R}{\simeq} R/P_i$. Assuming $A \neq 0$, we enumerate non-trivial elements a_1 of A and find a finite R-presentation of $A_1 = a_1 R$. Hence a finite set of generators of the ideal $P_1 = \mathrm{Ann}_R(a_1)$ is at hand since $A_1 \overset{R}{\simeq} R/P_1$. Now there is an algorithm in constructive commutative algebra to test P_1 for primality (see Seidenberg 1978). Eventually, we will find an a_1 for which P_1 is prime since maximal annihilators of non-zero elements of A are prime. Next, decide if $A_1 = A$ and if not repeat the procedure for A/A_1. In this way an ascending chain of submodules is generated, and since A is noetherian, the procedure will terminate finitely at A.

Now let $g \in G$ and put $\bar{g} = gA \in Q$. If $g \in \mathrm{Fit}(G)$, there is a $k > 0$ such that $(\bar{g} - 1)^k$ annihilates each factor A_{i+1}/A_i, and so $(\bar{g} - 1)^k \in P_i$. But P_i is prime, so $\bar{g} - 1 \in P_i$ and $\bar{g} \in 1 + P_i$. From this we deduce that

$$\mathrm{Fit}(G)/A = \bigcap_{i=1}^{\ell} \mathrm{Ker}(\theta_i)$$

where $\theta_i : Q \to (Q + P_i)/P_i$ is the canonical homomorphism defined by $x^{\theta_i} = x + P_i$. Now apply 9.5.4 to get a finite presentation of the group $(Q + P_i)/P_i$, and hence a finite set of generators for the subgroup $\mathrm{Ker}(\theta_i)$ of Q. Therefore, a finite set of generators of $\mathrm{Fit}(G)/A$ is obtainable. By adjoining to this set the R-module generators of A, a finite subset is obtained whose normal closure is $\mathrm{Fit}(G)$. ∎

Further algorithms

We mention without proof some further algorithms which have been constructed for a finitely generated metabelian group G. For details the reader should consult Baumslag, Cannonito, and Robinson (1994).

1. There is an algorithm which constructs the set of all elements of finite order in G in the form $u_1 X_1^G u_2 X_2^G \cdots u_k X_k^G$, where the u_i and X_i are explicitly constructed elements and finite subsets of G respectively.

From this one can deduce:

2. There is an algorithm to determine if G is torsion-free.
3. Given finite subsets X and Y, the centralizer $C_{\langle Y \rangle}(X)$ may be found (in terms of finite subsets of G).
4. There is an algorithm to find the Frattini subgroup $\mathrm{Frat}(G)$.

The last of these is the most challenging algorithm. As might have been predicted, it rests on the possibility of finding the intersection of all the maximal submodules of a finitely generated module over a finitely generated commutative ring.

10

COHOMOLOGICAL METHODS IN INFINITE SOLUBLE GROUP THEORY

It is almost 60 years since cohomology first made its appearance in group theory, by way of the theory of extensions. In a series of ground-breaking works Eilenberg and MacLane (1947), Eckmann (1946), and MacLane (1949) showed that the cohomology groups $H^n(G, M)$ for $n = 0, 1, 2, 3$, occur as natural objects in group theory. Cohomology began to be used to establish purely group theoretic results in the 1950s: perhaps the most famous application is still the theorem of Gaschütz that a non-cyclic finite p-group possesses outer automorphisms of p-power order (Gaschütz 1965, 1966). The lecture notes of Gruenberg (1970), Stammbach (1973), and Bieri (1981) have all been influential in inducing group theorists to acquire the necessary homological techniques.

Our intention in this chapter is to establish results on the cohomology of infinite soluble groups which can be applied to shed light on the structure of these groups. We shall assume a general familiarity with the basics of homological algebra, as found, for example, in MacLane (1963) and Hilton and Stammbach (1997). We begin in 10.1 with a brief review of the concepts and a description of the occurrence of low dimensional cohomology in group theory. After a discussion of soluble groups with finite (co)homological dimension in 10.2, we establish in 10.3 a number of theorems asserting that cohomology groups vanish or are in some sense small. These are then applied in 10.4 and 10.5 to give information about the structure of infinite soluble groups.

10.1 The cohomology groups in group theory

We begin by recalling the definitions of the functors Ext_R^n and Tor_n^R. Let R be a ring with identity and let A and B be right R-modules. Consider a projective resolution $\mathbf{P} \twoheadrightarrow A$ of A, that is, an exact sequence

$$\cdots \to P_{n+1} \overset{\partial_{n+1}}{\to} P_n \overset{\partial_n}{\to} P_{n-1} \to \cdots \to P_1 \overset{\partial_1}{\to} P_0 \to A \to 0$$

in which each P_i is a projective right R-module. Form the complex $\text{Hom}_R(\mathbf{P}, B)$, that is,

$$\cdots \to \text{Hom}_R(P_{n-1}, B) \overset{\delta_n}{\to} \text{Hom}_R(P_n, B) \overset{\delta_{n+1}}{\to} \text{Hom}_R(P_{n+1}, R) \to \cdots,$$

where δ_n is defined by composition in the natural way, and then take its homology $H(\text{Hom}_R(\mathbf{P}, B))$. In this way we obtain the abelian groups

$$\text{Ext}_R^n(A, B) = \text{Ker}(\delta_{n+1})/\text{Im}(\delta_n).$$

It can be shown that up to isomorphism $\mathrm{Ext}^n_R(A, B)$ is independent of the choice of the resolution $\mathbf{P}$. Alternatively, $\mathrm{Ext}^n_R(A, B)$ may be defined by forming an injective resolution $B \rightarrowtail \mathbf{I}$ of B and taking homology of the complex $\mathrm{Hom}_R(A, \mathbf{I})$.

Next we define the functor Tor^R_n. Let A and B be right and left R-modules, respectively. Choose a projective (left) resolution $\mathbf{P} \twoheadrightarrow B$ and then take the homology of the complex $A \otimes_R \mathbf{P}$, that is,

$$\cdots \to A \otimes_R P_{n+1} \overset{\partial'_{n+1}}{\to} A \otimes_R P_n \overset{\partial'_n}{\to} A \otimes_R P_{n-1} \to \cdots,$$

with the obvious induced maps ∂'_n. Then we define $\mathrm{Tor}^R_n(A, B)$ by

$$\mathrm{Tor}^R_n(A, B) = \mathrm{Ker}(\partial'_n)/\mathrm{Im}(\partial'_{n+1}).$$

Again this group is independent of the resolution $\mathbf{P}$. Alternatively, $\mathrm{Tor}^R_n(A, B)$ may be computed from the homology of the complex $\mathbf{P} \otimes_R B$, where $\mathbf{P} \twoheadrightarrow A$ is a projective right resolution of A.

Now let G be a group and let $R = \mathbb{Z}G$. If A is a right $\mathbb{Z}G$-module, we define the *cohomology group of G of dimension n with coefficients in A* to be

$$H^n(G, A) = \mathrm{Ext}^n_{\mathbb{Z}G}(\mathbb{Z}, A),$$

where $\mathbb{Z}$ is a trivial right $\mathbb{Z}G$-module. Dually, if A is a right $\mathbb{Z}G$-module, the *homology groups of G with coefficients in A* are defined by

$$H_n(G, A) = \mathrm{Tor}^{\mathbb{Z}G}_n(A, \mathbb{Z}),$$

where once again $\mathbb{Z}$ is the trivial G-module.

The Gruenberg Resolution

If one wants to compute the (co)homology of a group, it is necessary have a projective resolution of $\mathbb{Z}$ at hand. It was shown by Gruenberg (1960, 1970) that a free resolution can be formed whenever one has a presentation of the group. The fundamental statement is as follows.

10.1.1 (The Gruenberg Resolution) *Let $R \rightarrowtail F \overset{\pi}{\twoheadrightarrow} G$ be a presentation of a group G, where F is a free group. Then the following is a free right $\mathbb{Z}G$-resolution of $\mathbb{Z}$:*

$$\cdots \to \bar{I}^n_R/\bar{I}^{n+1}_R \to I_F\bar{I}^{n-1}_R/I_F\bar{I}^n_R \to \bar{I}^{n-1}_R/\bar{I}^n_R \to$$
$$\cdots \to \bar{I}^2_R/\bar{I}^3_R \to I_F\bar{I}_R/I_F\bar{I}^2_R \to \bar{I}_R/\bar{I}^2_R \to I_F/I_F\bar{I}_R \to \mathbb{Z}G \to \mathbb{Z}.$$

Here $\bar{I}_R$ is the kernel of the natural homomorphism $\pi : \mathbb{Z}F \twoheadrightarrow \mathbb{Z}G$. Also $\mathbb{Z}G \to \mathbb{Z}$ is the augmentation map given by $g \mapsto 1, g \in G$, and $I_F/I_F\bar{I}_R \to \mathbb{Z}G$ is induced from π, while all other maps are the natural ones arising from inclusion.

A simple case is where G is a cyclic group of order n: then we may take F to be an infinite cyclic group $\langle x \rangle$ and $R = \langle x^n \rangle$. In this case the resolution reduces to

$$\cdots \to \mathbb{Z}G \overset{\nu}{\to} \mathbb{Z}G \overset{1-\bar{x}}{\to} \mathbb{Z}G \overset{\nu}{\to} \mathbb{Z}G \to \mathbb{Z} \to 0,$$

where $1 - \bar{x}$ is multiplication by $1 - x$ and ν is multiplication by $1 + x + \cdots + x^{n-1}$. We therefore obtain the important formulas

$$H^{2n+1}(G, M) \simeq \operatorname{Ker} \nu / \operatorname{Im}(1 - \bar{x}),$$

and

$$H^{2n+2}(G, M) \simeq \operatorname{Ker}(1 - \bar{x})/\operatorname{Im}(\nu)$$

for any $\mathbb{Z}G$-module M and $n \geq 0$. Furthermore

$$H_n(G, M) \simeq H^{n+1}(G, M)$$

for $n \geq 1$.

Application of 10.1.1 with $M = \mathbb{Z}$, the trivial $\mathbb{Z}G$-module, and G any group yields explicit formulas for the homology groups. For convenience we shall write

$$H_n(G) = H_n(G, \mathbb{Z}) \quad \text{and} \quad H^n(G) = H^n(G, \mathbb{Z}),$$

where $\mathbb{Z}$ is a trivial $\mathbb{Z}G$-module.

10.1.2 *Let $R \rightarrowtail F \twoheadrightarrow G$ be a presentation of a group G. Then for $n > 0$*

$$H_{2n}(G) \simeq \bar{I}_R^n \cap (I_F \bar{I}_R^{n-1} I_F)/(I_F \bar{I}_R^n + \bar{I}_R^n I_F)$$

and

$$H_{2n-1}(G) \simeq (I_F \bar{I}_R^{n-1}) \cap (\bar{I}_R^{n-1} I_F)/(\bar{I}_R^n + I_F \bar{I}_R^{n-1} I_F).$$

In particular $H_1(G) \simeq I_F / (\bar{I}_R + I_F^2)$, from which it is not hard to prove that

$$H_1(G) \simeq G_{ab}.$$

Also $H_2(G) \simeq (\bar{I}_R \cap I_F^2)/(I_F \bar{I}_R + \bar{I}_R I_F)$, which can be shown to be isomorphic with $(R \cap F')/[R, F]$. Thus we have *Hopf's formula for the Schur multiplier of G*,

$$H_2(G) \simeq (R \cap F')/[R, F]$$

For details of these results the reader is referred to Gruenberg (1970) or MacLane (1963).

Another consequence of 10.1.1 is a useful tool for computing (co)homology over trivial modules.

10.1.3 (The Universal Coefficients Theorem) *Let G be a group and M a trivial $\mathbb{Z}G$-module. Then for $n \geq 1$,*

(i) $H_n(G, M) \simeq (H_n(G) \otimes M) \oplus (\operatorname{Tor}_1^{\mathbb{Z}}(H_{n-1}(G), M))$,
(ii) $H^n(G, M) \simeq \operatorname{Hom}(H_n(G), M) \oplus \operatorname{Ext}_{\mathbb{Z}}^1(H_{n-1}(G), M)$.

Details of the proof may be found in Gruenberg (1970). Notice that by combining 10.1.2 and 10.1.3 one can in principle compute the (co)homology of a group with coefficients in an arbitrary trivial module.

As another consequence of the Gruenberg Resolution, we note

10.1.4 (Baumslag, Dyer, and Miller 1983) *If G is a recursively presented group, the homology groups $H_n(G)$ are recursively presented.*

Proof Let $R \rightarrowtail F \twoheadrightarrow G$ be a recursive presentation of G. Then clearly $\bar{I}_R$ and I_F, together with their powers and products, are recursively enumerable. Since the maps in 10.1.2 are natural, their kernels and images are known and hence $H_n(G)$ is recursively presented. ∎

If G is finitely generated, then $H_1(G) \simeq G_{ab}$ is finitely generated; if G is finitely presented, then $H_2(G) \simeq (F' \cap R)/[F, R]$ is finitely generated since $R/[F, R]$ is. However, finite generation does not extend to higher dimensional homology groups. Indeed Baumslag, Dyer, and Miller (1983) have shown that *if $B_1, B_2, \ldots$ is any sequence of recursively presented torsion-free abelian groups with B_1 and B_2 finitely generated, then there is a finitely presented group G such that $H_n(G) \simeq B_n$ for all n.*

The homology of polycyclic groups

Turning to infinite soluble groups next, we observe the following important result.

10.1.5 (Baumslag, Cannonito and Miller 1981a) *If G is a virtually polycyclic group and M is a finitely generated $\mathbb{Z}G$-module, then each $H_n(G, M)$ is a finitely generated abelian group.*

First of all a general remark. Suppose G is a group for which $\mathbb{Z}G$ is noetherian as a right $\mathbb{Z}G$-module, that is, $\mathbb{Z}G$ satisfies max $-r$. Then there is a finitely generated right free $\mathbb{Z}G$-presentation $\mathbf{F} \twoheadrightarrow \mathbb{Z}$. To see this we start with the augmentation $\varepsilon : \mathbb{Z}G \twoheadrightarrow \mathbb{Z}$: its kernel is the augmentation ideal $K_0 = I_G$, which is finitely generated as a $\mathbb{Z}G$-module. Thus there is a surjective homomorphism $\varphi_1 \colon F_1 \twoheadrightarrow K_0$, where F_1 is a finitely generated free $\mathbb{Z}G$-module. Next let K_1 be the kernel of φ_1, which is a finitely generated $\mathbb{Z}G$-module since F_1 is right $\mathbb{Z}G$-noetherian. Find a surjective homomorphism $\varphi_2 : F_2 \twoheadrightarrow K_1$ with F_2 a finitely generated free module—again the kernel K_2 is finitely generated—and so on. In this way we obtain a finitely generated free $\mathbb{Z}G$-presentation of $\mathbb{Z}$,

$$\cdots \to F_3 \overset{\varphi_3}{\to} F_2 \overset{\varphi_2}{\to} F_1 \overset{\varphi_1}{\to} F_0 = \mathbb{Z}G \overset{\varepsilon}{\twoheadrightarrow} \mathbb{Z}$$
$$\searrow \quad \nearrow \quad \searrow \quad \nearrow \quad \searrow \quad \nearrow$$
$$K_2 \qquad\quad K_1 \qquad\quad K_0$$

Proof of 10.1.5 Since G is virtually polycyclic, $\mathbb{Z}G$ is a right noetherian ring by 4.2.3, and hence there is a finitely generated right free $\mathbb{Z}G$-resolution $\mathbf{F} \to \mathbb{Z}$. Use this resolution to compute the homology groups: thus

$$H_n(G, M) = \mathrm{Tor}_n^{\mathbb{Z}G}(M, \mathbb{Z}) = H_n\left(\mathbf{F} \underset{\mathbb{Z}G}{\otimes} \mathbb{Z}\right).$$

Now

$$F_i \otimes_{\mathbb{Z}G} \mathbb{Z} \simeq F_i/[F_i, G] = (F_i)_G,$$

the trivialization of F_i. So $H_n(G, M) = H_n(\mathbf{F}_G)$, where $\mathbf{F}_G$ is the trivialized $\mathbb{Z}G$-complex

$$\cdots \to (F_{i+1})_G \to (F_i)_G \to \cdots \to (F_1)_G \to (F_0)_G \,.$$

Here the maps are homomorphisms between finitely generated abelian groups and it follows at once that $H_n(G, M)$ is finitely generated. ∎

The reader should note that the construction of the finitely generated free resolution that precedes the proof is an effective one. This remark relies on 9.4.1, which guarantees that we can find a finite presentation for the image of the map $F_i \to F_{i-1}$. Thus we can explicitly find finite presentations of the $(F_i)_G$ involved in the proof. Therefore *there is an algorithm to find $H_n(G, M)$, when G, M and n are given.*

The corresponding result for cohomology is also true, as may be seen by invoking the concept of *Poincaré duality*. A group G is called a *Poincaré duality group of dimension n* if

$$H^k(G, M) \simeq H_{n-k}(G, \hat{\mathbb{Z}} \otimes_{\mathbb{Z}} M)$$

for $0 \leq k \leq n$ and all $\mathbb{Z}G$-modules M. Here $\hat{\mathbb{Z}}$, (which is called the dualizing module), is $\mathbb{Z}$ with some G-action and the tensor product is a $\mathbb{Z}G$-module by the diagonal action. It is known that *all torsion-free polycyclic groups are Poincaré duality groups:* for this and much more about duality groups see Bieri (1981).

10.1.6 (Baumslag, Cannonito, and Miller 1981a) *If G is a virtually polycyclic group and M is a finitely generated $\mathbb{Z}G$-module, each $H^n(G, M)$ is a finitely generated abelian group.*

Proof By 1.3.4 the group G has a normal subgroup H with finite index that is torsion-free. Since H is a Poincaré duality group, $H^k(H, M) \simeq H_{n-k}(H, \hat{\mathbb{Z}} \otimes M)$, which by 10.1.5 is finitely generated because $\hat{\mathbb{Z}} \otimes M$ is clearly a finitely generated G-module.

To show that $H^k(G, M)$ is finitely generated, we apply the Lyndon–Hochschild–Serre spectral sequence for cohomology to the group extension $H \rightarrowtail G \twoheadrightarrow G/H$. (This spectral sequence is discussed in 10.2.) Note that M is finitely generated as a $\mathbb{Z}H$-module since $|G : H|$ is finite, and hence $H^j(H, M)$ is a finitely generated group. Then $H^i(G/H, H^j(H, M))$ is also finitely generated, as may be seen directly from a finitely generated resolution of $\mathbb{Z}$. It now follows from the spectral sequence that $H^k(G, M)$ is finitely generated. ∎

Group theoretic interpretations

The fact that H_0, H^0, H^1, H^2 all have well-established group theoretical interpretations opens up the possibility of applications of cohomology to group theory. But first we recall the connections.

The groups $H_0(G, M)$ and $H^0(G, M)$

To compute $H_0(G, M)$, apply the functor $M \otimes_{\mathbb{Z}G}$ to the *left* version of the Gruenberg Resolution to get

$$\cdots \to M \otimes_{\mathbb{Z}G} \left(I_F / \bar{I}_R I_F\right) \to M \otimes_{\mathbb{Z}G} \mathbb{Z}G \to 0.$$

Now $M \otimes_G \mathbb{Z}G \overset{\mathbb{Z}G}{\simeq} M$ via the mapping $a \otimes g \mapsto ag$. This induces $M \otimes_{\mathbb{Z}G} \left(I_F / \bar{I}_R I_F\right) \mapsto [M, G]$ and hence

$$H_0(G, M) \simeq M/[M, G] = M_G.$$

In a similar way $H^0(G, M)$ may be found by applying $\mathrm{Hom}_{\mathbb{Z}G}(-, M)$ to the *right* Gruenberg resolution to obtain

$$0 \to \mathrm{Hom}_{\mathbb{Z}G}(\mathbb{Z}G, M) \to \mathrm{Hom}_{\mathbb{Z}G}(I_F / I_F \bar{I}_R, M) \to \cdots .$$

Now $\mathrm{Hom}_{\mathbb{Z}G}(\mathbb{Z}G, M) \simeq M$ via $\theta \mapsto (1)\theta$. It follows that the kernel of the map between the Hom's is $\{a \in M \mid ag = a, \forall g \in G\}$, the set of G-fixed points. Thus

$$H^0(G, M) \simeq M^G.$$

Derivations, complements and H^1

Let G be a group and M a right $\mathbb{Z}G$-module: recall from 3.2 that a *derivation* from G to M is a function $\delta : G \to M$ such that

$$(xy)^\delta = (x^\delta)y + y^\delta \quad \text{for } x, y \in G.$$

The derivations clearly form an additive group $\mathrm{Der}(G, M)$, where the group operation is defined by $x^{\delta_1 + \delta_2} = x^{\delta_1} + x^{\delta_2}$. A derivation δ is called *inner* if there exists an a in M such that $x^\delta = a(x - 1)$ for all x in G. The inner derivations form a subgroup of $\mathrm{Der}(G, M)$, denoted by $\mathrm{Inn}(G, M)$.

10.1.7 *If M is a right $\mathbb{Z}G$-module, then $H^1(G, M) \simeq \mathrm{Der}(G, M)/\mathrm{Inn}(G/M)$.*

Proof Application of $\mathrm{Hom}_{\mathbb{Z}G}(-, M)$ to the right Gruenberg Resolution yields

$$\mathrm{Hom}_{\mathbb{Z}G}(\mathbb{Z}G, M) \to \mathrm{Hom}_{\mathbb{Z}G}\left(I_F / I_F \bar{I}_R, M\right) \to \mathrm{Hom}_{\mathbb{Z}G}\left(\bar{I}_R / \bar{I}_R^2, M\right) \to \cdots ,$$

from which the following exact sequence may be derived—for details see Robinson (1996), 11.4:

$$M \to \mathrm{Hom}(I_G, M) \to H^1(G, M) \to 0.$$

Now it can be shown that $\mathrm{Der}(G, M) \simeq \mathrm{Hom}_{\mathbb{Z}G}(I_G, M)$ via the map $\delta \mapsto (g - 1 \mapsto g^\delta)$, where $\delta \in \mathrm{Der}(G, M)$. The image of M under $M \to \mathrm{Der}(G, M)$ is $\mathrm{Inn}(G, M)$, whence the result follows. ∎

We are now in a position to connect derivations with complements in semidirect products. Let M be a $\mathbb{Z}G$-module, written *multiplicatively*, and consider the semidirect product

$$E = G \ltimes M.$$

If X is any complement of M in E, then for each g in G we have $g \in XM$, so that $gg^\delta \in X$ for some unique $g^\delta \in M$. Since $\left(g_1 g_1^\delta\right)\left(g_2 g_2^\delta\right) = g_1 g_2 \left(\left(g_1^\delta\right)^{g_2} g_2^\delta\right)$, the map $\delta : G \to M$ is a derivation. Conversely, if $\delta \in \mathrm{Der}(G, M)$, then $X = \{gg^\delta \mid g \in G\}$ is a complement of M in E.

Now suppose that δ and ε are two derivations from G to M with corresponding complements X and Y. If $a \in M$, observe that $\left(gg^\delta\right)^a = g\left([g, a]g^\delta\right)$, since M is abelian. Hence $Y = X^a$ if and only if $g^\varepsilon = [g, a]g^\delta = g^{-1+a}g^\delta$ for all $g \in G$, that is, $\varepsilon + \mathrm{Inn}(G, M) = \delta + \mathrm{Inn}(G, M)$.

We have therefore proved:

10.1.8 *Let G be a group and M a $\mathbb{Z}G$-module. Then there is a bijection between the set of complements of M in $E = G \ltimes M$ and the group $\mathrm{Der}(G, M)$. Furthermore, two complements are conjugate in E if and only if the corresponding derivations fall in the same coset of $\mathrm{Inn}(G, M)$.*

Since the complement G obviously corresponds to the zero derivation, we deduce:

Corollary 10.1.9 *All the complements of M in $G \ltimes M$ are conjugate if and only if $H^1(G, M) = 0$.*

Near conjugacy of complements

In practice the complements of M in an infinite semidirect product $E = G \ltimes M$ are rarely conjugate. However, the complements may have a weaker property called *near conjugacy*, especially if the group E is soluble with finite rank.

Let M be a given $\mathbb{Z}G$-module and put $E = G \ltimes M$. The complements of M in E are said to be *nearly conjugate* if they fall into finitely many conjugacy classes modulo some finite $\mathbb{Z}G$-submodule of M.

Just what is entailed by near conjugacy when M has finite abelian ranks (FAR) is made clear by the next result. In what follows $M[i]$ denotes the subset $\{a \in M \mid a^i = 1\}$, where $i > 0$.

10.1.10 (Robinson 1976a, b) *Let G be a group and M a $\mathbb{Z}G$-module with finite abelian ranks: put $E = G \ltimes M$. Let $\delta \in \mathrm{Der}(G, M)$ and denote by X_i the complement of M in E corresponding to the derivation $i\delta$.*

(i) *If $\delta + \mathrm{Inn}(G, M) \in H^1(G, M)$ has finite order m, there is a finite set of subgroups $\{G_1, G_2, \dots, G_r\}$ of E, depending only on m, such that each X_i is conjugate modulo $M[m]$ to some G_j and $G_j \cap M = M[m]$. Furthermore, $r \le |M : M^m|$.*

(ii) *Conversely, if the complements X_i fall into a finite number r of conjugacy classes modulo a finite submodule of order s, then $\delta + \mathrm{Inn}(G, M)$ has order at most rs in $H^1(G, M)$.*

Proof

(i) The mapping $ga \mapsto ga^m, (g \in G, a \in M)$, is an endomorphism α of E. Since $m\delta$ is inner, there exists an a in M, such that $(g^m)^\delta = a^{g-1}$ for all g

in G. Write $X = X_1$: then X^α contains $(gg^\delta)^\alpha = g(g^m)^\delta = ga^{g-1} = g^{a^{-1}}$, so that $X^\alpha = G^{a^{-1}}$. Notice that $\ell = |M : M^m|$ is finite. Now choose a transversal to M^m in M, say $\{t_1, t_2, \ldots, t_\ell\}$, and write $a^{-1} = t_j b^m$, $(b \in M, 1 \leq j \leq \ell)$. Then $X^\alpha = G^{a^{-1}} = G^{t_j b^m}$, so that $G^{t_j} = (X^{b^{-1}})^\alpha$. Let G_j denote the preimage of G^{t_j} under α: note that the G_j depend only on m and the choice of transversal, and there are at most ℓ of the G_i. Then $G_j = X^{b^{-1}} \text{Ker}(\alpha) = X^{b^{-1}} M[m]$ and $G_j \cap M = M[m]$. Hence X is conjugate to one of $G_1, \ldots, G_\ell$ modulo $M[m]$. The same argument applies to any $i\delta$, so the conclusion is valid for all the X_i.

(ii) Now suppose that the complements X_i fall into r conjugacy classes modulo some finite submodule F of order s. Then two of $X_1, X_2, \ldots, X_{r+1}$ are necessarily conjugate modulo F, so there exist i and j such that $1 \leq i < j \leq r+1$ and $s(j - i)\delta \in \text{Inn}(G, M)$. It follows that $\delta + \text{Inn}(G, M)$ has order at most rs in $H^1(G, M)$. ∎

Corollary 10.1.11 *The complements of M in $G \ltimes M$ are nearly conjugate if and only if $H^1(G, M)$ is a bounded abelian group.*

It is important to note in 10.1.10(i) that if $mH^1(G, M) = 0$, then each complement is nearly conjugate to one of $G_1, \ldots, G_r$ modulo $M[m]$.

Automorphisms and H^1

Another way in which derivations are used in group theory is in the construction of automorphisms. Let G be a group with a normal subgroup N and let

$$\Gamma = C_{\text{Aut}(G)}(N) \cap C_{\text{Aut}\,G}(G/N),$$

that is, Γ is the subgroup of automorphisms of G that stabilize the series $1 \triangleleft N \triangleleft G$. If $\gamma \in \Gamma$, then $[G, \gamma, N] = 1$ since $[N, G, \gamma] = 1 = [\gamma, N, G]$; thus $[G, \gamma] \leq Z(N) = M$. Notice that M is a $\mathbb{Z}Q$-module via conjugation, where $Q = G/N$.

It is a simple matter to verify that if $\gamma \in \Gamma$, the map $xN \mapsto [x, \gamma]$ is a well-defined derivation $\delta(\gamma)$ from Q to M. Conversely, if $\delta \in \text{Der}(Q, M)$, then $x \mapsto x(xN)^\delta, (x \in G)$, is an automorphism $\gamma(\delta)$ of G belonging to Γ. Furthermore, the maps $\gamma \mapsto \delta(\gamma)$ and $\delta \mapsto \gamma(\delta)$ are mutually inverse isomorphisms, so that

$$\Gamma \simeq \text{Der}(Q, M).$$

Now let us assume that $C_G(N) = M$. Then $\gamma \in \Gamma$ is an inner automorphism of G if and only if it is induced through conjugation by an element a of M, that is, $xN \mapsto [x, \gamma] = x^{-1+\gamma} = x^{-1+a}$, and the corresponding derivation is inner. So in this situation γ corresponds to an element of $\text{Inn}(Q, M)$.

Summing up our conclusions, we have:

10.1.12 *Let G be a group with a normal subgroup N. Let $Q = G/N$ and let Γ be the subgroup of automorphisms stabilizing the series $1 \triangleleft N \triangleleft G$. Then $\Gamma \simeq \text{Der}(Q, Z(N))$. Furthermore, if $C_G(N) = Z(N)$, then $\Gamma \cap \text{Inn}(G)$ maps to $\text{Inn}(Q, Z(N))$ under this isomorphism, so that*

$$\Gamma/(\Gamma \cap \text{Inn}(G)) \simeq H^1(Q, Z(N)).$$

This result can be used to establish the existence of outer automorphisms in infinite soluble groups, as will be shown later in this section.

Group extensions and H^2

The original, and best known, point of entry of cohomology into group theory is the theory of group extensions. Here we will briefly summarize the basic facts and refer to Gruenberg (1970) or Robinson (1996) for a detailed account.

10.1.13 *Let Q be a given group and M a $\mathbb{Z}Q$-module. Then there is a bijection between the set of equivalence classes of group extensions $M \rightarrowtail G \twoheadrightarrow Q$ which induces the given module structure in M by conjugation and the group $H^2(Q, M)$. An extension splits if and only if the corresponding cohomology class is the zero element of $H^2(Q, M)$.*

Corollary 10.1.14 *Every extension of M by Q splits precisely when $H^2(Q, M) = 0$.*

Near splitting

Just as conjugacy of complements is a rare phenomenon, splitting seldom occurs in infinite groups: but a weaker property, called *near splitting*, can be a useful replacement.

Consider a group extension

$$M \overset{\mu}{\rightarrowtail} G \overset{\varepsilon}{\twoheadrightarrow} Q$$

inducing the given module structure in the $\mathbb{Z}G$-module M. The extension is said to be *nearly split* if there is a subgroup X of G such that

$$|G : X\operatorname{Im}(\mu)| < \infty \quad \text{and} \quad |X \cap \operatorname{Im}(\mu)| < \infty.$$

Thus near splitting is splitting 'up to finite index and a finite normal subgroup'. For indeed, $X \cap \operatorname{Im}(\mu) \lhd X\operatorname{Im}(\mu)$, so that $|G : N_G(X \cap \operatorname{Im}(\mu))|$ is finite and hence $(X \cap \operatorname{Im}(\mu))^G$ is a finite normal subgroup of G.

The natural question is: what condition does near splitting impose on the cohomology class of the extension? The answer is furnished by the following result.

10.1.15 (Robinson 1976a, b) *Let Q be a group, M a $\mathbb{Z}Q$-module which has finite abelian ranks and $M \overset{\mu}{\rightarrowtail} G \overset{\varepsilon}{\twoheadrightarrow} Q$ an extension inducing the given $\mathbb{Z}Q$-module structure of M. Let Δ be the cohomology class of the extension. Then the extension is nearly split if and only if Δ has finite order in $H^2(Q, M)$.*

Proof First of all suppose that $m\Delta = 0$, where $m > 0$. Form the push-out of

$$\begin{array}{ccc} M & \overset{\mu}{\to} & G \\ m\downarrow & & \\ M & & \end{array}$$

the vertical map being multiplication by m:

$$
\begin{array}{ccccc}
M & \overset{\mu}{\rightarrowtail} & G & \overset{\varepsilon}{\twoheadrightarrow} & Q \\
m\downarrow & & \theta\downarrow & & \| \\
M & \overset{\bar{\mu}}{\rightarrowtail} & \bar{G} & \overset{\bar{\varepsilon}}{\twoheadrightarrow} & Q
\end{array}
$$

The cohomology class of the extension in the second row is $m\Delta = 0$, that is, the extension splits and there is a subgroup $\bar{X}$ such that $\bar{G} = \bar{X}M^{\bar{\mu}}$ and $\bar{X} \cap M^{\bar{\mu}} = 1$. Writing X_1 for $G^\theta \cap \bar{X}$, we have $G^\theta \cap (\bar{X}M^{\mu\theta}) = X_1 M^{\mu\theta}$. Hence, since $\mu\theta = m\bar{\mu}$,

$$
\begin{aligned}
|G^\theta : X_1 M^{\mu\theta}| &= |G^\theta(\bar{X}M^{\mu\theta}) : \bar{X}M^{\mu\theta}| \\
&\leq |\bar{X}M^{\bar{\mu}} : \bar{X}M^{\mu\theta}| \\
&= |M^{\bar{\mu}} : M^{m\bar{\mu}}| \\
&= |M : M^m|,
\end{aligned}
$$

which is finite. Define X to be the preimage of X_1 under θ. Then we have

$$
|G : XM^\mu| = |G^\theta : X_1 M_1^{\mu\theta}| \leq |M : M^m|.
$$

Also $(X \cap M^\mu)^\theta \leq \bar{X} \cap M^{\bar{\mu}} = 1$, so that $(X \cap M^\mu)^m = 1$ and $X \cap M^\mu$ is contained in $(M[m])^\mu$, which is finite. Hence the extension nearly splits.

Conversely, assume that the extension is nearly split, so there is a subgroup X such that $|X \cap A|$ and $|G : XA|$ are finite, where $A = M^\mu$. Since $X \cap A \lhd XA$ and $|G : XA|$ is finite, the subgroup $A_0 = (X \cap A)^G$ is finite.

Write $\bar{G} = G/A_0$, $\bar{A} = A/A_0$, etc. Then $\bar{X} \cap \bar{A} = 1$ and $|\bar{G} : \bar{X}\bar{A}|$ is finite. Let $\bar{\Delta}$ be the cohomology class of the extension $\bar{A} \rightarrowtail \bar{G} \twoheadrightarrow Q$ and write $\bar{Q}_0$ for the image of $\bar{X}\bar{A}$ under $\bar{G} \rightarrow Q$. Then $m = |\bar{Q} : \bar{Q}_0|$ is finite. It will be sufficient to prove that $m\bar{\Delta} = 0$, since A_0 is finite.

Consider the restriction and corestriction maps

$$
\text{res} : H^2(\bar{Q}, \bar{A}) \rightarrow H^2(\bar{Q}_0, \bar{A}) \quad \text{and} \quad \text{cor} : H^2(\bar{Q}_0, \bar{A}) \rightarrow H^2(\bar{Q}, \bar{A}).
$$

It is a well-known fact that the composite map

$$
\text{res} \circ \text{cor} : H^2(\bar{Q}, \bar{A}) \rightarrow H^2(\bar{Q}, \bar{A})
$$

is simply multiplication by m—see Hilton and Stammbach (1997: 16.4). Since $\bar{A} \rightarrowtail \bar{X}\bar{A} \twoheadrightarrow \bar{Q}_0$ is split, $(\bar{\Delta})\text{res} = 0$ and hence $m\bar{\Delta} = 0$, as claimed. ∎

Corollary 10.1.16 *Let Q be a group and let M be a $\mathbb{Z}Q$-module which has finite ranks as an abelian group. Then every extension of M by Q with the given module structure is nearly split if and only if $H^2(Q, M)$ is a torsion group.*

Other cohomology groups

The group $H^2(Q, M)$ can also be used to describe non-abelian extensions $N \rightarrowtail G \twoheadrightarrow Q$ with given *coupling* $\chi : Q \rightarrow \text{Out}(N)$. Here χ arises from conjugation in G and $M = Z(N)$ is a Q-module by means of χ. The equivalence classes of extensions correspond, in a non-natural way, to elements of $H^2(Q, M)$. Also each homomorphism $\chi : Q \rightarrow \text{Out}(N)$ determines an element of $H^3(Q, M)$, its *obstruction*, which must vanish if an extension is to exist with the given coupling χ. For a detailed account of this theory see Gruenberg (1970) or Robinson (1996).

We mention in passing that group theoretic interpretations of cohomology groups in dimensions greater than 3 are known, although no really convincing applications have yet been given. Thus we will not pursue the matter here. For details see Holt (1979), Huebschmann (1980), and Robinson (1989a).

Some applications

We end this section by presenting three applications of cohomology to infinite soluble groups. Once we have established a series of cohomological vanishing theorems in 10.3, many more applications can be given.

10.1.17 (Bowers and Stonehower 1973; Robinson 1982) *Let G be a group with a series of finite length whose factors are either infinite cyclic or locally finite. Then the maximal torsion subgroups of G fall into finitely many conjugacy classes.*

From this we deduce at once:

10.1.18 (Mal'cev 1951) *The finite subgroups of a virtually polycyclic group fall into finitely many conjugacy classes.*

There is in fact an algorithm that constructs for a given virtually polycyclic group a representative of each conjugacy class of finite subgroups (Baumslag *et al.* 1991).

During the proof of 10.1.17 a standard result on the cohomology of finite groups will be needed. If G is a finite group of order m and M is any $\mathbb{Z}G$-module, then

$$mH^n(G, M) = 0$$

for all $n > 0$: see for example MacLane (1963) or Hilton and Stammbach (1997).

Proof of 10.1.17 Evidently there is nothing to be lost in factoring out by the maximum normal torsion subgroup, so assume that this is trivial. Then by 5.2.1 there is a series of normal subgroups $1 = G_0 \lhd G_1 \lhd \cdots \lhd G_n = G$ in which each infinite factor is torsion-free abelian with finite rank: clearly torsion subgroups of G are finite. Let $n > 0$ and put $A = G_1$. By induction on n the result is true for G/A.

Suppose that $H_1, H_2, \ldots$ are infinitely many non-conjugate finite subgroups of G. We may assume by induction that the H_iA fall into finitely many conjugacy classes. Hence infinitely many of the H_j are contained in a conjugate of some H_iA. Replacing these H_i's by suitable conjugates, we may assume that each H_j is contained in H_iA and

$$K = H_iA = H_{j_1}A = H_{j_2}A = \cdots$$

for infinitely many H_{j_k}.

Next $H_{j_k} \cap A = 1$ since A is torsion-free, so the H_{j_k} are complements of A in K. Now $mH^1(K/A, A) = 0$, where $m = |K/A|$, and it follows from 10.1.11

that the H_{j_k} are nearly conjugate. Since A is torsion-free, this means that the H_{j_k} fall into finitely many conjugacy classes, which is a contradiction.   ∎

The second application is to establish the existence of outer automorphisms of finitely generated nilpotent groups: this result is in the spirit of Gaschütz's classic theorem on outer automorphisms of finite p-groups.

10.1.19 *Let G be a finitely generated torsion-free nilpotent group which is not cyclic. Then* $\mathrm{Out}(G)$ *has an element of infinite order.*

Proof Let A be a maximal normal abelian subgroup of G, so that $A = C_G(A)$. Put $Q = G/A$: if Q is finite, then $[A, G] = 1$ and $G = A$, since G acts nilpotently on A. In this case the statement is obvious, so we may assume that Q is infinite and thus it has an infinite cyclic quotient Q/N. Corresponding to the extension $N \rightarrowtail Q \twoheadrightarrow Q/N$, there is a standard five-term cohomology sequence

$$1 \to H^1(Q/N, A^N) \to H^1(Q, A) \to (H^1(N, A))^Q \to H^2(Q/N, A^N) :$$

for this for this see Hilton and Stammbach (1997: VI, 8). Now $H^1(Q, A)$ is isomorphic with a subgroup of $\mathrm{Out}(G)$ by 10.1.12. If the result is false, $H^1(Q, A)$ must be a torsion group and we conclude from the exact sequence that $H^1(Q/N, A^N)$ is also torsion. But Q/N is infinite cyclic, so that $H^1(Q/N, A^N) = A^N/[A^N, Q]$ and the latter is torsion. Consequently A^N is torsion since Q acts nilpotently on A^N. Therefore $A^N = 1$, which shows that $A = 1 = G$.   ∎

It remains an unanswered question *whether every finitely generated infinite nilpotent group has outer automorphisms*. We remark that Zalesskiĭ (1971b) has shown that *an infinite nilpotent p-group has outer automorphisms*. On the other hand, the same author has given examples of torsion-free nilpotent groups with no outer automorphisms (Zalesskiĭ 1972).

Next recall that a group G is said to be *complete* if the conjugating homomorphism $G \to \mathrm{Aut}(G)$ is an isomorphism. In other words, G is complete if and only if $Z(G) = 1$ and every automorphism of G is inner. Clearly, a nontrivial nilpotent group cannot be complete. More interesting is the following generalization.

10.1.20 (Robinson 1980) *An infinite supersoluble group cannot be complete.*

Proof Suppose that G is an infinite supersoluble group which is complete and let A be a maximal abelian normal subgroup of G. Then $A = C_G(A)$ by 1.3.2. Since $\mathrm{Out}(G) = 1$, it follows from 10.1.12 that $H^1(Q, A) = 1$, where $Q = G/A$. Now consider the exact sequence $A \overset{\alpha}{\rightarrowtail} A \twoheadrightarrow A/A^2$, where $a^\alpha = a^2, (a \in A)$. Note that α is injective here: for an element of order 2 in A would have to belong to the hypercentre of G by supersolubility. Applying the cohomology functor $H(Q, -)$ to the exact sequence $A \overset{\alpha}{\rightarrowtail} A \twoheadrightarrow A/A^2$, we obtain the long exact cohomology sequence

$$1 \to A^Q \overset{\alpha}{\to} A^Q \to (A/A^2)^Q \to H^1(Q, A) \to \cdots :$$

for this see Hilton and Stammbach (1997: VI, 2). Since $A^Q = 1 = H^1(Q, A)$, it follows that $(A/A^2)^Q = 1$, which forces $A = A^2$ by supersolubility of Q. Hence A is finite, as must be G, since $A = C_G(A)$. ∎

On the other hand, it has been shown that *there exist non-trivial torsion-free supersoluble groups with no outer automorphisms* (Menegazzo and Puglisi 2000). Thus the triviality of the centre in 10.1.20 is essential. We mention also that there are numerous finite supersoluble groups which are complete—for a classification of these see Hartley and Robinson (1980).

In Section 10.4 we give a detailed study of complete infinite soluble groups based on vanishing theorems to be established in 10.3.

10.2   Soluble groups with finite (co)homological dimensions

In this section we introduce the homological and cohomological dimensions of a group and, in the soluble case, relate these invariants to the Hirsch number.

If G is a group, the *homological dimension* $hd(G)$ is defined to be the smallest integer $n \geq 0$ such that $H_k(G, M) = 0$ for all $k > n$ and all right $\mathbb{Z}G$-modules M: if no such n exist, $hd(G)$ is taken to be ∞. In homological terms this means that $hd(G)$ is the *flat dimension* of $\mathbb{Z}$ as a trivial $\mathbb{Z}G$-module since $H_k(G, M) = \operatorname{Tor}_k^{\mathbb{Z}G}(M, \mathbb{Z})$.

Dually we define the *cohomological dimension* of G, $cd(G)$, to be the smallest integer $n \geq 0$ such that $H^k(G, M) = 0$ for all $k > n$ and all right $\mathbb{Z}G$-modules M, with $cd(G) = \infty$ if no such n exists. Thus $cd(G)$ is the *projective dimension* of $\mathbb{Z}$ as a trivial $\mathbb{Z}G$-module since $H^k(G, M) = \operatorname{Ext}_{\mathbb{Z}G}^k(\mathbb{Z}, M)$.

For example, a non-trivial free group has homological and cohomological dimensions equal to 1.

The basic properties of these invariants are listed in the following result.

10.2.1 *Let G be an arbitrary group.*

(i) $hd(G) \leq cd(G)$, *and also* $cd(G) \leq hd(G) + 1$ *provided that G is countable.*
(ii) *If S is a subgroup of G, then $hd(S) \leq hd(G)$ and $cd(S) \leq cd(G)$.*
(iii) *If S is a subgroup with finite index in G and G is torsion-free, then $hd(S) = hd(G)$, while $cd(S) = cd(G)$.*

For proofs of these results, together with a detailed account of the homological background, the reader is referred to Bieri (1981).

Corollary 10.2.2 *A group G which has finite homological (or cohomological) dimension is torsion-free.*

Proof Let $hd(G)$ be finite and suppose that S is a cyclic subgroup with finite order $m > 1$. From the formula for the homology of cyclic groups in 10.1, we have $H_{2n+1}(S) \simeq \mathbb{Z}/m\mathbb{Z} \neq 0$ for all $n \geq 0$, which shows that $hd(S) = \infty$. But this contradicts 10.2.1 (ii). Of course, if $cd(G)$ is finite, then so is $hd(G)$ by 10.2.1. ∎

Another consequence of 10.2.1 is that $hd(G) = 0$ if and only if $|G| = 1$: for clearly $hd(\mathbb{Z}) = 1$.

The Lyndon–Hochschild–Serre spectral sequences

The *Lyndon–Hochschild–Serre* (or LHS-) spectral sequences relate the (co)homology of a group to that of a normal subgroup and its associated quotient group. Because of this, these spectral sequences are indispensable tools in the study of the (co)homological dimensions of non-simple groups. Here we can give only the barest account, referring to the standard books Hilton and Stammbach (1997) and MacLane (1963) for details.

Let G be a group with a normal subgroup N and quotient group $Q = G/N$. Let M be any right $\mathbb{Z}G$-module. Elements of G act on $H_j(N, M)$ and $H^j(N, M)$ by the action on the module M and by conjugation in N. These actions combine to make $H_j(N, M)$ and $H^j(N, M)$ into $\mathbb{Z}G$-modules on which the elements of N act trivially. Hence $H_j(N, M)$ and $H^j(N, M)$ become $\mathbb{Z}Q$-modules, and so it is meaningful to form the groups

$$H_i(Q, H_j(N, M)) \quad \text{and} \quad H^i(Q, H^j(N, M)).$$

First we consider homology: set

$$E_{ij}^2 = H_i(Q, H_j(N, M)).$$

These groups, together with certain homomorphisms $d^2 : E_{ij}^2 \to E_{i-2,j+1}^2$ called *differentials,* form the E^2-*page* of the LHS-spectral sequence. Subsequent pages are formed by successively taking the homology of differentials.

What emerges from these pages is a chain of subgroups

$$0 = H_0 \leq H_1 \leq \cdots \leq H_{n+1} = H_n(G, M)$$

such that H_{i+1}/H_i is isomorphic with $E_{i\,n-i}^{\infty}$, a certain section of $E_{i\,n-i}^2$. Thus knowledge of the E_{ij}^2 can tell us a great deal about the structure of the group $H_n(G, M)$. The terminology used is that the LHS-spectral sequence for homology *converges* to the homology groups of G, in symbols

$$E_{ij}^2 = H_i(Q, H_j(N, M)) \underset{i+j=n}{\Rightarrow} H_n(G, M).$$

The E_2-page of the LHS-spectral sequence for cohomology consists of the $E_2^{ij} = H^i(Q, H^j(N, M))$, and certain maps $d_2 : E_2^{ij} \to E_2^{i+2\,j-1}$, the *differentials.* It can be shown that there is a series of subgroups

$$0 = H^0 \leq H^1 \leq \cdots \leq H^{n+1} = H^n(G, M)$$

with H^{i+1}/H^i isomorphic to $E_{\infty}^{n-i\,i}$, a certain section of $E_2^{n-i\,i}$. The cohomology spectral sequence *converges* to the cohomology of G, in symbols

$$E_2^{ij} = H^i(Q, H^j(N, M)) \underset{i+j=n}{\Rightarrow} H^n(G, M).$$

In practice it may be very hard to compute the groups E_{ij}^{∞} and E_{∞}^{ij}. However, if, for example, $H^j(N, M) = 0$ for all $j \geq 0$, it follows at once from the convergence of the spectral sequence that $E_\infty^{n-i\,i} = 0$ and hence that $H^n(G, M) = 0$, with a corresponding conclusion for homology.

Another useful case is where $H^j(N, M) = 0$ for all $j > 1$, which would be true if N were infinite cyclic. In this event $E_2^{ij} = 0$ if $j > 1$ and it can be shown that the spectral sequence 'collapses' to an exact sequence as indicated in 10.2.3 and 10.2.4.

In the next two results N is a normal subgroup of an arbitrary group G, while $Q = G/N$ and M is a $\mathbb{Z}G$-module.

10.2.3 *If $H^j(N, M) = 0$ for $1 < j < m$, then there is an exact sequence*

$$0 \to H^1(Q, M^N) \to H^1(G, M) \to H^1(N, M)^Q \to \cdots \to H^n(Q, M^N)$$
$$\to H^n(G, M) \to H^{n-1}(Q, H^1(N, M)) \to H^{n+1}(Q, M^N) \to H^{n+1}(G, M) \to \cdots$$

for $n < m$.

10.2.4 *If $H_j(N, M) = 0$ for $1 < j < m$, then there is an exact sequence*

$$\cdots \to H_{n+1}(G, M) \to H_{n+1}(Q, M_N) \to H_{n-1}(Q, H_1(N, M)) \to H_n(G, M)$$
$$\to H_n(Q, M_N) \to \cdots \to H_1(N, M)_Q \to H_1(G, M) \to H_1(Q, M_N) \to 0$$

for $n < m$.

We now return to the investigation of groups with finite (co)homological dimension. The next theorem provides some examples of such groups.

10.2.5 *Let G be a torsion-free group with a series whose factors are locally finite or infinite cyclic. Then $hd(G) \le h(G)$ and $cd(G) \le h(G) + 1$.*

Recall here that $h(G)$ is the Hirsch number (or torsion-free rank) of G. We will precede the proof with a comment on the homology of a non-trivial rational group R, that is, a subgroup of $\mathbb{Q}$. Clearly $R = \lim(R_i)$ where $(R_i)_{i \in I}$ is a direct system of infinite cyclic subgroups. Now it is a fact that homology commutes with direct limits (see Bieri 1981). Hence for any $\mathbb{Z}R$-module M and $n > 1$,

$$H_n(R, M) \simeq \lim H_n(R_i, M) = 0,$$

so that $hd(R) = 1$.

Proof of 10.2.5 By 5.2.5 there is a normal polyrational subgroup P with finite index and this implies that G is countable. Hence $hd(G) = hd(P)$ and $cd(G) = cd(P)$ by 10.2.1 (iii). Also of course $h(G) = h(P)$. Therefore we may assume G to be polyrational.

Let $1 = G_0 \triangleleft G_1 \triangleleft \cdots \triangleleft G_n = G$ be a series with non-trivial rational factors, where $n = h(G)$. If $n = 0$, then $G = 1$ and $hd(G) = 0$. If $n = 1$, then G is rational and, as has been observed, $hd(G) = 1$, so $hd(G) = h(G)$. Let $n > 1$ and argue by induction on n. Setting $N = G_{n-1}$, we have $hd(N) \le h(N) = n - 1$.

Let $k > n$ and consider $H_k(G, M)$ for any $\mathbb{Z}G$-module M. To show that $H_k(G, M) = 0$ it will be enough, by the spectral sequence, to prove that

$E_{ij}^2 = H_i(G/N, H_j(N, M)) = 0$ if $i + j = k$. Certainly, $H_j(N, M) = 0$ if $j \geq n$. If $j < n$, then $i = k - j > k - n \geq 1$; hence $H_i(G/N, H_j(N, M)) = 0$ because $hd(G/N) = 1$. Consequently, $H_k(G, M) = 0$ and $hd(G) \leq n$. By 10.2.1 we deduce that $cd(G) \leq hd(G) + 1 \leq n + 1$. $\blacksquare$

In particular, if R is a non-trivial rational group, $hd(R) = 1$ and $cd(R) = 1$ or 2. In fact both cases can occur. Of course, $cd(\mathbb{Z}) = 1$. If R is the group $\mathbb{Q}_p$ of p-adic rationals, there are non-split extensions $\mathbb{Z} \rightarrowtail E \twoheadrightarrow R$ with E abelian—see Fuchs (1960)—so $H^2(R, \mathbb{Z}) \neq 0$ and thus $cd(R) = 2$.

The information just obtained has a nice application to the computation of the (co)homology of Prüfer groups.

10.2.6 (Robinson 1979) *Let P be a Prüfer group of type p^∞ and let M be a trivial $\mathbb{Z}P$-module. Then for $n \geq 1$*

$$H_{2n-1}(P, M) \simeq P \otimes_{\mathbb{Z}} M, \qquad H_{2n}(P, M) \simeq \text{Tor}(P, M),$$

and

$$H^{2n-1}(P, M) \simeq \text{Hom}(P, M), \qquad H^{2n}(P, M) \simeq \text{Ext}(P, M),$$

where $\text{Tor} = \text{Tor}_1^{\mathbb{Z}}$ and $\text{Ext} = \text{Ext}_{\mathbb{Z}}^1$.

Proof It will be enough to show that

$$H_{2n}(P) = 0 \quad \text{and} \quad H_{2n-1}(P) \simeq P$$

for $n \geq 1$: for then we can apply the Universal Coefficients Theorem (10.1.3) to get

$$H_{2n}(P, M) \simeq \text{Tor}(H_{2n-1}(P), M) \simeq \text{Tor}(P, M),$$

$$H_{2n-1}(P, M) \simeq H_{2n-1}(G) \otimes_{\mathbb{Z}} M \simeq P \otimes_{\mathbb{Z}} M,$$

and

$$H^{2n}(P, M) \simeq \text{Ext}(H_{2n-1}(P), M) \simeq \text{Ext}(P, M),$$

$$H^{2n-1}(P, M) \simeq \text{Hom}(H_{2n-1}(P), M) \simeq \text{Hom}(P, M).$$

To establish these facts consider the LHS-spectral sequence for homology, applied to the extension $\mathbb{Z} \rightarrowtail Q \twoheadrightarrow P$, where $Q = \mathbb{Q}_p$, the additive group of p-adic rationals. Since $hd(\mathbb{Z}) = 1$, we obtain from 10.2.4 the exact sequence

$$\cdots \to H_{n+1}(Q) \to H_{n+1}(P) \to H_{n-1}(P) \to H_n(Q) \to H_n(P) \to \cdots$$

$$\cdots \to \mathbb{Z} \to H_1(Q) \to H_1(P) \to 0.$$

But $hd(Q) = 1$, so $H_n(Q) = 0 = H_{n+1}(Q)$ for $n \geq 2$. Hence

$$H_{n+1}(P) \simeq H_{n-1}(P), \quad \text{for} \quad n \geq 2.$$

Thus it remains to compute $H_1(P)$ and $H_2(P)$. We saw after 10.1.2 that $H_1(P) \simeq P$, while $H_2(P) = 0$ because P is a direct limit of cyclic groups P_i and $H_2(P_i, \mathbb{Z}) = 0$, by the formula for the homology of a cyclic group in 10.1. This completes the proof. $\blacksquare$

Returning to the problem of identifying $hd(G)$ when G has finite torsion-free rank, we first find a lower bound in the nilpotent case.

10.2.7 (Stammbach 1970) *Let G be a torsion-free nilpotent group with finite rank. If $n = h(G) > 0$, then $H_n(G)$ is a rational group and it is cyclic if and only if G is finitely generated.*

Proof There is a central series $1 = G_0 \lhd G_1 \lhd \cdots \lhd G_n = G$ with each G_{i+1}/G_i a rational group. If $n = 1$, the result is true since $H_1(G) \simeq G$. Let $n > 1$ and argue by induction on n.

We apply the LHS-spectral sequence for homology to the extension $G_1 \rightarrowtail G \twoheadrightarrow G/G_1$, noting that $H_k(G_1) = 0$ for all $k \geq 2$ since $hd(G_1) = 1$. Furthermore, $H_k(G/G_1) = 0$ for $k \geq n$ since $hd(G/G_1) \leq h(G/G_1) = n - 1$ by 10.2.5. It therefore follows from 10.2.4 that

$$H_n(G) \simeq H_{n-1}(G/G_1, H_1(G_1, \mathbb{Z})).$$

Since $H_1(G_1, \mathbb{Z}) \overset{\mathbb{Z}G}{\simeq} G_1$, we obtain

$$H_n(G) \simeq H_{n-1}(G/G_1, G_1) \simeq H_{n-1}(G/G_1) \otimes_{\mathbb{Z}} G_1$$

by the Universal Coefficients Theorem (10.1.3): note here that $\mathrm{Tor}_1^{\mathbb{Z}}(A, B) = 0$ if A or B is torsion-free (see MacLane 1963: V 6). By induction on n the group $H_{n-1}(G/G_1)$ is rational, whence so is $H_n(G)$.

Finally, suppose that $H_n(G)$ is cyclic. Then it follows from $H_n(G) \simeq H_{n-1}(G/G_1) \otimes_{\mathbb{Z}} G_1$ that both $H_{n-1}(G/G_1)$ and G_1 must be cyclic. By the induction hypothesis G/G_1 is finitely generated, whence so is G. Conversely, if G is finitely generated, G_1 is infinite cyclic, and $H_n(G) \simeq H_{n-1}(G/G_1) \otimes_{\mathbb{Z}} G_1$, which is cyclic by induction. ∎

The next result provides a lower bound for the homological dimension of a soluble group with finite torsion-free rank.

10.2.8 (Stammbach 1970) *Let G be a soluble group with finite torsion-free rank. If $n = h(G) > 0$, there is a $\mathbb{Z}G$-module M such that $M \simeq \mathbb{Q} \simeq H_n(G, M)$. Hence $hd(G) \geq n$.*

Proof Form the derived series $1 = G^{(d)} \lhd G^{(d-1)} \lhd \cdots \lhd G^{(1)} \lhd G^{(0)} = G$ and put $S_i = G^{(i-1)}/G^{(i)}$ and $n_i = h(S_i)$: set

$$L_i = H_{n_i}(S_i, \mathbb{Q}),$$

where $\mathbb{Q}$ is a trivial $\mathbb{Z}S_i$-module. Let T_i be the torsion subgroup of S_i and apply the LHS-spectral sequence for homology to the extension $T_i \rightarrowtail S_i \twoheadrightarrow S_i/T_i$. Notice that $H_j(T_i, \mathbb{Q})$ is the direct limit of the groups $H_j(F, \mathbb{Q})$, where the F's are the finite subgroups of T_i. But $H_j(F, \mathbb{Q}) = 0$ for $j > 0$ since it is a divisible group and yet is annihilated by $|F|$. Therefore $H_j(T_i, \mathbb{Q}) = 0$ for $j > 0$. Thus $E_{ij}^2 = 0$ for $j > 0$ and 10.2.4 implies that $H_k(S_i, \mathbb{Q}) \simeq H_k(S_i/T_i, \mathbb{Q})$. By the Universal Coefficients Theorem

$$H_k(S_i, \mathbb{Q}) \simeq H_k(S_i/T_i) \otimes_{\mathbb{Z}} \mathbb{Q}$$

since $\text{Tor}(S_i/T_i, \mathbb{Q}) = 0$. Also $H_{n_i}(S_i/T_i) \simeq \mathbb{Q}$ by 10.2.7, and therefore

$$L_i \simeq H_{n_i}(S_i/T_i) \otimes_{\mathbb{Z}} \mathbb{Q} \simeq \mathbb{Q}.$$

Now L_i is already a $\mathbb{Z}G$-module via conjugation in S_i/T_i. We make L_i into a different $\mathbb{Z}G$-module L_i^* by defining $a \cdot g = a^{g^{-1}}$, ($a \in L_i^*, g \in G$). This determines a module structure since $\text{Aut}(\mathbb{Q})$ is abelian. We now define a $\mathbb{Z}G$-module

$$M_d = L_1^* \otimes L_2^* \otimes \cdots \otimes L_d^*,$$

where elements of G act diagonally, that is,

$$(a_1 \otimes_{\mathbb{Z}} a_2 \otimes_{\mathbb{Z}} \cdots \otimes_{\mathbb{Z}} a_d) \cdot g = (a_1 \cdot g) \otimes (a_2 \cdot g) \otimes_{\mathbb{Z}} \cdots \otimes_{\mathbb{Z}} (a_d \cdot g).$$

Clearly $M_d \simeq \mathbb{Q}$. To complete the proof we will show that

$$H_n(G, M_d) \simeq \mathbb{Q}.$$

This will be achieved by induction on d. If $d \leq 1$, then G is abelian and, if T is its torsion subgroup, $H_n(T, M_1) = 0$ and $H_n(G, M) \simeq H_n(G/T, M_1) \simeq \mathbb{Q}$ by the spectral sequence and 10.2.7. Now let $d > 1$.

Put $N = G^{(d-1)}$ and apply the LHS-spectral sequence for homology to the exact sequence $N \rightarrowtail G \twoheadrightarrow G/N$. Since $h(N) = n_d$ and $h(G/N) = n - n_d$ by 10.2.5 and induction, the group $E^2_{n-n_d, n_d}$ is the only non-zero E^2-term and all differentials are 0. This means that the spectral sequence collapses and

$$H_n(G, M_d) \simeq H_{n-n_d}(G/N, H_{n_d}(N, M_d)).$$

(This method of proof is often called the *corner argument*.) Since N centralizes each S_i, we see from the definition that M_j is a trivial $\mathbb{Z}N$-module, and by the Universal Coefficients Theorem

$$H_{n_d}(N, M_d) \simeq H_{n_d}(N) \otimes_{\mathbb{Z}} M_d$$
$$\simeq (H_{n_d}(N) \otimes_{\mathbb{Z}} \mathbb{Q}) \otimes_{\mathbb{Z}} M_d$$
$$\simeq H_{n_d}(N, \mathbb{Q}) \otimes_{\mathbb{Z}} M_d$$
$$= L_d \otimes_{\mathbb{Z}} M_d.$$

Furthermore, these are $\mathbb{Z}G$-isomorphisms. But $L_d \otimes_{\mathbb{Z}} L_d^* \simeq \mathbb{Q}$, as $\mathbb{Z}G$-modules, since $(a \otimes b) \cdot g = ar \otimes br^{-1} = a \otimes b$, where $a \in L_d, b \in L_d^*$ and $g \in G$ acts on L_d by multiplication by r. Hence

$$H_{n_d}(N, M_d) \overset{\mathbb{Z}G}{\simeq} L_1^* \otimes_{\mathbb{Z}} \cdots \otimes_{\mathbb{Z}} L_{d-1}^* = M_{d-1},$$

and $H_n(G, M_d) \simeq H_{n-n_d}(G/N, M_{d-1}) \simeq \mathbb{Q}$ by induction on n. $\blacksquare$

We now present the main theorem on soluble groups with finite (co)homological dimension (Bieri 1981: 7.10).

10.2.9 *Let G be a group with a series of finite length whose factors are abelian or locally finite. Then the following are equivalent:*

(i) *$hd(G)$ is finite;*

(ii) $cd(G)$ *is finite;*

(iii) G *is a torsion-free, virtually soluble group with finite abelian total rank (FATR).*

Before proving this theorem, we must establish an auxiliary result.

10.2.10 *Let G be a group with a series whose factors are torsion-free abelian of finite rank or locally finite. Then $G/\tau(G)$ is virtually soluble (where $\tau(G)$ is the largest torsion normal subgroup).*

Proof Let $1 = G_0 \triangleleft G_1 \triangleleft \cdots \triangleleft G_n = G$ be a series of the type indicated. We may assume $n > 1$ and argue by induction on n. Clearly, we may also assume that $\tau(G) = 1$. By the induction hypothesis G_{n-1} is virtually soluble and we may suppose it to be infinite: hence there exists a non-trivial abelian normal subgroup A of G. We may assume that A is maximal with these properties. Write $T/A = \tau(G/A)$ and put $C = C_T(A)$. Now A is torsion-free abelian with finite rank, so by Schur's theorem T/C is finite. In addition $C/Z(C)$ is locally finite, so that C' is also locally finite. Since $\tau(G) = 1$, it follows that C is abelian and $C = A$, by maximality of A. Therefore T/A is finite and, since G/T is virtually soluble by induction on $h(G)$, so is G. ∎

Proof of 10.2.9 (i) implies (iii). Note that G is torsion-free by 10.2.2. Suppose S is a soluble subgroup of G. If S does not have finite torsion-free rank, it will have a subgroup S_0 with finite but unbounded Hirsch number. But $h(S_0) \le hd(S_0) \le hd(G)$ by 10.2.8 and 10.2.1. By this contradiction all soluble subgroups of G have finite torsion-free rank.

Now let $1 = G_0 \triangleleft G_1 \triangleleft \cdots \triangleleft G_\ell = G$ be a series in which each G_{i+1}/G_i is either locally finite or abelian. If $\ell \le 1$, then G is abelian and all is well. So let $\ell > 1$ and proceed by induction. Then $N = G_{\ell-1}$ is a virtually soluble FATR-group. If G/N is locally finite, then 10.2.10 shows that G is virtually soluble. If G/N is abelian, obviously G is virtually soluble, and the result follows at once:

(ii) implies (i). This follows from 10.2.1.

(iii) implies (ii). If G satisfies (iii), then $cd(G)$ is finite by 10.2.5. ∎

Corollary 10.2.11 *A soluble group G has $hd(G)$ or $cd(G)$ finite if and only if it is torsion-free with finite total rank.*

The last result provides a complete characterization of soluble groups with finite homological or cohomological dimension.

Finally, we have the following information about the values of $hd(G)$ and $cd(G)$.

10.2.12 *If G is a group with the properties of 10.2.9, then $hd(G) = h(G)$ and $cd(G) = h(G)$ or $h(G) + 1$.*

These statements follow at once from 10.2.1, 10.2.5, and 10.2.8. It has already been observed that both the possibilities for $cd(G)$ indicated in 10.2.12 can occur: indeed $cd(\mathbb{Z}) = 1 = hd(\mathbb{Z})$ and $cd(\mathbb{Q}_p) = 2 = hd(\mathbb{Q}_p) + 1$. But it turns out to be a very delicate matter to determine the groups for which $cd(G) = h(G)$. The first result on this problem was:

10.2.13 (Gruenberg 1970) *Let G be a torsion-free, nilpotent group of finite rank. Then $cd(G) = h(G)$ if and only if G is finitely generated.*

Proof Assume that $cd(G) = n = h(G)$. If M is any trivial $\mathbb{Z}G$-module, the Universal Coefficients Theorem yields

$$H^{n+1}(G, M) \simeq \mathrm{Ext}(H_n(G), M)$$

since $H_{n+1}(G, M) = 0$. But $cd(G) = n$, so we have $H^{n+1}(G, M) = 0$ and hence $\mathrm{Ext}(H_n(G), M) = 0$. However we may choose M to be the kernel of a surjective homomorphism $F \twoheadrightarrow H_n(G)$, where F is free abelian. Consequently, F must split over K, so that $H_n(G)$ is a free abelian group. On the other hand, $H_n(G)$ is a rational group by 10.2.7, so it is cyclic. By the same result G is finitely generated. ∎

After Gruenberg's Theorem appeared there was a great deal of interest in the problem of identifying the soluble groups G of finite torsion-free rank such that $cd(G) = h(G)$. For example, Gildenhuys and Strebel (1981, 1982) showed that G must be finitely generated and hence a minimax group by 5.2.8. Finally, Kropholler finished off the problem by proving the following remarkable theorem.

10.2.14 (Kropholler 1986b) *The following statements are equivalent for a soluble group G:*

   (i) $hd(G) = cd(G) < \infty$;
 (ii) G *is of type* (FP);
(iii) G *is a duality group*;
(iv) G *is torsion-free and constructible.*

Here a group G is said to be *of type* (FP) if there is a projective $\mathbb{Z}G$-resolution of $\mathbb{Z}$ with finite length with finitely generated modules. *Constructible groups* form the smallest class of groups closed under forming finite extensions, generalized free products and HNN-extensions of its members: these groups are studied in detail in 11.2 below.

Finally a group G is a *duality group of dimension n* if there is a right $\mathbb{Z}G$-module C such that

$$H^k(G, M) \simeq H_{n-k}(G, C \otimes_{\mathbb{Z}} M)$$

for $0 \leq k \leq n$ and all G-modules M.

The proof of Kropholler's theorem is too complex to give here.

10.3   Cohomological vanishing theorems for nilpotent groups

The theme of this section is a tendency for the (co)homology of a nilpotent group with coefficients in a highly non-trivial module to vanish, or at least be small in some sense. As a simple instance of this phenomenon, we recall a well-known result about finite soluble groups: *if G is a finite non-nilpotent group such that the final term A of the lower central series is abelian, then G splits over A and all the complements of A are conjugate.* In cohomological terms the assertion here is that $H^1(Q, A) = 0 = H^2(Q, A)$, whenever Q is a finite nilpotent group and A is a finite $\mathbb{Z}Q$-module satisfying $A = [A, Q]$.

Our aim is to find generalizations of this result to arbitrary nilpotent groups and arbitrary dimensions. Then in 10.4 we shall find numerous applications of this theory to the structure of infinite soluble groups.

Our first two results are (co)homological vanishing theorems which have the following common constituents: a *nilpotent group Q*, a *ring R with identity*, and a *right RQ-module M*. Keep in mind that $H_0(Q, M) = M_Q = M/[M, Q]$ and $H^0(Q, M) = M^Q = C_M(Q)$.

10.3.1 (Robinson 1976a) *If M is RQ-noetherian and $M_Q = 0$, then $H_n(Q, M) = 0 = H^n(Q, M)$ for all $n \geq 0$.*

10.3.2 (Robinson 1976a) *If M is RQ-artinian and $M^Q = 0$, then $H_n(Q, M) = 0 = H^n(Q, M)$ for all $n \geq 0$.*

Notice that the hypotheses of the two theorems are dual, while the conclusion is self-dual, in the duality in which 'noetherian' corresponds to 'artinian', and 'homology' to 'cohomology'. Ultimately the proofs rest on two properties of the functors Tor and Ext.

10.3.3 *Let R be a ring with identity and let r be a central element. Let A and B be R-modules and denote by α and β the respective module endomorphisms of A and B arising from multiplication by r.*

(i) *If A and B are right R-modules, the induced endomorphisms α^* and β_* of $\operatorname{Ext}_R^n(A, B)$ are equal.*

(ii) *If A is a right R-module and B a left R-module, the induced endomorphisms α_* and β_* of $\operatorname{Tor}_n^R(A, B)$ are equal.*

Proof Let us take the case of Ext. Since r is a central element of R, right multiplication by r is a module endomorphism of any right R-module. Let $\mathbf{P} \twoheadrightarrow A$ be a projective right resolution of the R-module A. Then we have the commutative diagram

$$
\begin{array}{ccc}
\mathbf{P} & \twoheadrightarrow & A \\
\downarrow r & & \downarrow r \\
\mathbf{P} & \twoheadrightarrow & A
\end{array}
$$

so that r acts on the complex $\operatorname{Hom}_R(\mathbf{P}, B)$ on the left via the rule $(a)r \cdot \theta = (ar)\theta$, where $a \in P_i$ and $\theta \in \operatorname{Hom}_R(P_i, B)$. Hence $(a)r \cdot \theta = ((a)\theta)r$. Now r also acts on $\operatorname{Hom}_R(\mathbf{P}, B)$ on the right via $(a)\theta \cdot r = ((a)\theta)r$. Therefore the two actions coincide and it follows on taking homology that $\alpha^* = \beta_*$.

The proof for homology is entirely dual and involves forming the complex $\mathbf{P} \otimes_R B$ with $\mathbf{P}$ as before. ∎

Much of the labour involved in proving 10.3.1 and 10.3.2 lies in establishing two technical results which have the following hypotheses in common.

Let Q and X be groups, R a ring with identity, and M a right R-module. Assume that M is made into a right RQ-module and a right RX-module by means of two homomorphisms, written

$$* : Q \to \mathrm{Aut}(M) \quad \text{and} \quad \dagger : X \to \mathrm{Aut}(M).$$

Assume further that X has a central series $1 = X_0 \triangleleft X_1 \triangleleft \cdots \triangleleft X_\ell = X$ with cyclic factors, (so X is a finitely generated nilpotent group). The key results needed are as follows.

10.3.4 *Suppose that M is RQ-noetherian and $M_X = 0$. If in addition*

1.
$$\left[M, \left[X_{i+1}^{\dagger}, Q^* \right] \right] \leq \left[M, X_i^{\dagger} \right]$$

for $0 \leq i < \ell$, then $H_n(X, A) = 0 = H^n(X, A)$ for all $n \geq 0$.

10.3.5 *Suppose that M is RQ-artinian and $M^X = 0$. If in addition*

2.
$$M^{[X_{i+1}^{\dagger}, Q^*]} \geq M^{X_i^{\dagger}}$$

for $0 \leq i < \ell$, then $H_n(X, A) = 0 = H^n(X, A)$ for all $n \geq 0$.

Proof of 10.3.4 We will deal only with cohomology, the proof for homology being dual. Clearly $[M, X_i]$ is an RX-module; it is also an RQ-module because of the condition 1. Let m denote the number of *distinct* submodules $[M, X_i] = [M, X_i^{\dagger}]$. If $m = 1$, then $M = [M, X] = [M, X_0] = 0$ and there is nothing to prove: so assume $m > 1$ and proceed by induction on $\ell + m$.

We first observe that it is sufficient to deal with two special cases: M *torsion-free* and M *elementary* p, as abelian groups. Indeed, since M is RQ-noetherian, its ($\mathbb{Z}$-)torsion subgroup is bounded, say by e; thus $M_1 = eM$ is torsion-free. It follows that there is a series of RQ-, RX-submodules $0 = M_0 \leq M_1 \leq \cdots \leq M_r = M$ in which each M_i with $i > 0$ has the form $e_i M$, ($e_i > 0$), and M_{i+1}/M_i is either torsion-free or elementary abelian p_i for some prime p_i. Use of the long exact cohomology sequence shows that it is enough to prove that $H^n(X, M_{i+1}/M_i) = 0$ for $0 \leq i < r$. Here it is essential to observe that M_{i+1}/M_i is an RQ-, RX-image of M and therefore inherits the properties of M. This justifies our claim.

Next write $X_1 = \langle x \rangle$: then $x \in Z(X)$ and $a \mapsto a(x-1)$ is an RX-endomorphism θ of M. In fact it is also an RQ-endomorphism because $[M, [X_1^{\dagger}, Q^*]] \leq [M, X_0^{\dagger}] = 0$. Two cases must now be distinguished.

Case 1: $\theta = 0$. Here M is a trivial X_1-module, so by the formulas in 10.1 for the cohomology of cyclic groups we have $H^j(X_1, M) \simeq 0$, M or M/dM, where $j \geq 0$ and $d = |X_1|$ if X_1 is finite. This is because M is either torsion-free or an elementary abelian p-group. Furthermore, these isomorphisms are of $\mathbb{Z}X$-modules, since $X_1 \leq Z(X)$. Hence for any $i, j \geq 0$

$$H^i(X/X_1, H^j(X_1, M)) \simeq H^i(X/X_1, A) = 0,$$

where $A = 0$, M or M/dM, by the induction hypothesis applied to the triple $(Q, X/X_1, A)$. Of course, it is necessary to check that the conditions on M are inherited by A, which is easy. Applying the LHS-spectral sequence for cohomology to the exact sequence $X_1 \rightarrowtail X \twoheadrightarrow X/X_1$, we reach the conclusion that $H^n(X, M) = 0$ for all $n \geq 0$.

Case 2: $\theta \neq 0$. Since $\mathrm{Ker}(\theta^i)$ is an RQ-module and M is RQ-noetherian, there is an integer $r > 0$ such that $\mathrm{Ker}(\theta^r) = \mathrm{Ker}(\theta^{r+1})$. The usual argument of Fitting's Lemma shows that $\mathrm{Im}(\theta^r) \cap \mathrm{Ker}(\theta) = 0$. Hence θ induces an injective endomorphism φ in the module $B = \mathrm{Im}(\theta^r)$. Put $C = \mathrm{Im}(\theta^i)/\mathrm{Im}(\theta^{i+1})$ where $i \geq 0$: this is an RQ-, RX-image of M and the number of distinct $[C, X_j]$'s is at most $m - 1$ since $[C, X_1] = 0$, whereas $[M, X_1] \neq 0$. Apply the induction hypothesis to the triple (Q, X, C) to conclude that $H^n(X, C) = 0$ for all $n \geq 0$. This holds for each i. Consequently, the long exact cohomology sequence leads to $H^n(X, M/B) = 0$ and it therefore suffices to prove that $H^n(X, B) = 0$ for $n \geq 0$.

For the final step consider the exact sequence $B \overset{\varphi}{\rightarrowtail} B \twoheadrightarrow \mathrm{Coker}(\varphi)$. The cohomology sequence yields

$$\cdots \to H^n(X, B) \overset{\varphi_*}{\to} H^n(X, B) \to H^n(X, \mathrm{Coker}(\varphi)) \to \cdots .$$

Now $H^n(X, B) = \mathrm{Ext}^n_{\mathbb{Z}X}(\mathbb{Z}, B)$, and by applying 10.3.3 with $R = \mathbb{Z}X$ and $r = x - 1$, we conclude that φ_* is also induced from multiplication by $x - 1$ in $\mathbb{Z}$. But $\mathbb{Z}$ is a trivial X-module, so $\varphi_* = 0$ and $H^n(X, B)$ embeds in $H^n(X, \mathrm{Coker}(\varphi))$. Since $\mathrm{Coker}(\varphi) = B/(B)\varphi = \mathrm{Im}(\theta^r)/\mathrm{Im}(\theta^{r+1})$, we have $H^n(X, \mathrm{Coker}(\varphi)) = 0$ and hence $H^n(X, B) = 0$ for all $n \geq 0$. This completes the proof of 10.3.4. ∎

Proof of 10.3.5 Again we only prove the result for cohomology. Each $M^{X_i} = M^{X_i^\dagger}$ is an RX-, RQ-submodule by condition 2. We argue by induction on $\ell + m$, where m is the number of distinct submodules M^{X_i}. If $m = 1$, then $M = M^{X_0} = M^X = 0$, so we may assume $m > 1$.

By the artinian condition there is an $n > 0$ such that $(n!)M = (n+1)!M = \mathrm{etc.}$ Writing $U = M[n!]$, we see that M/U is divisible. It follows that there is a series of RQ-, RX-submodules $0 = M_r \leq M_{r-1} \leq \cdots \leq M_1 \leq M_0 = M$ in which each M_i with $i > 0$ has the form $M[e_i] = \{a \in M \mid e_i a = 0\}$, for $e_i > 0$, and M_{i+1}/M_i is either divisible or elementary abelian p for some prime p. Notice that the M_{i+1}/M_i inherit the properties of M since $M/M_i \simeq e_i M$ as $RX-$, RQ-modules. Thus the long cohomology sequence allows us to assume that, as an abelian group, M is either divisible or elementary abelian p.

Put $X_1 = \langle x \rangle$ and let θ be the RX-, RQ-endomorphism $a \mapsto a(x - 1)$ of M. Once again two cases must be distinguished.

Case 1: $\theta = 0$. Since M is a trivial $\mathbb{Z}X_1$-module and it is either divisible or elementary abelian p, we have $H^j(X_1, M) \simeq 0$, M or $M[d]$, where $d = |X_1|$ if this is finite. Moreover these are RX-isomorphisms. Hence for any $i, j \geq 0$

$$H^i(X/X_1, H^j(X_1, M)) \simeq H^i(X/X_1, A) = 0$$

by the induction hypothesis applied to $(Q, X/X_1, A)$, where $A = 0$, M or $M[d]$. By the LHS-spectral sequence $H^n(X, M) = 0$ for all $n \geq 0$.

Case 2: $\theta \neq 0$. Since $\mathrm{Im}(\theta^i)$ is an RQ-module and M is RQ-artinian, there is an $r > 0$ such that $\mathrm{Im}(\theta^r) = \mathrm{Im}(\theta^{r+1})$. Hence $M = \mathrm{Im}(\theta) + \mathrm{Ker}(\theta^r)$ and θ induces a surjective endomorphism φ in the module $B = M/\mathrm{Ker}(\theta^r) = \mathrm{Coim}(\theta^r)$. Writing C for $\mathrm{Ker}(\theta^{i+1})/\mathrm{Ker}(\theta^i)$, we observe that C is RQ-, RX-isomorphic with a submodule of M, so it inherits the hypotheses on M. Also $C^{X_1} = C$ and $M^{X_1} \neq M$, so that the number of distinct C^{X_j}'s is at most $m - 1$. Therefore, $H^n(X, C) = 0$ for $n \geq 0$ by the induction hypothesis applied to (Q, X, C). Consequently, $H^n(X, \mathrm{Ker}(\theta^r)) = 0$ and it is enough to prove that $H^n(X, B) = 0$.

In the final step we apply the cohomology functor to the exact sequence $\mathrm{Ker}(\varphi) \rightarrowtail B \overset{\varphi}{\twoheadrightarrow} B$, obtaining thereby the exact sequence

$$\cdots \to H^n(X, \mathrm{Ker}(\varphi)) \to H^n(X, B) \overset{\varphi_*}{\to} H^n(X, B) \to \cdots .$$

Now $H^n(X, B) = \mathrm{Ext}^n_{\mathbb{Z}X}(\mathbb{Z}, B)$ and by 10.3.3 we have $\varphi_* = 0$ since $\mathbb{Z}$ is a trivial $\mathbb{Z}X$-module. Therefore $H^n(X, \mathrm{Ker}(\varphi))$ maps onto $H^n(X, B)$. Since $\mathrm{Ker}(\varphi) = \mathrm{Ker}(\theta^{r+1})/\mathrm{Ker}(\theta^r)$, it follows that $H^n(X, B) = 0$. ∎

We are now in a position to establish the two main theorems 10.3.1 and 10.3.2.

Proof of 10.3.1 The first step is to form an ascending central series of Q with cyclic factors, say $1 = Q_0 \vartriangleleft Q_1 \vartriangleleft \cdots Q_\nu = Q$. Since M is RQ-noetherian, there can be only finitely many distinct submodules $[M, Q_\alpha]$: let the non-zero submodules among these be $[M, Q_{\alpha(i)+1}]$, $i = 1, 2, \ldots, r$, where $\alpha(1) < \alpha(2) < \cdots < \alpha(r)$ and

$$\cdots = [M, Q_{\alpha(i)}] < [M, Q_{\alpha(i)+1}] = \cdots = [M, Q_{\alpha(i+1)}] < [M, Q_{\alpha(i+1)+1}] = \cdots .$$

Let $Q_{\alpha(i)+1} = \langle x_i, Q_{\alpha(i)} \rangle$ and put

$$X = \langle x_1, x_2, \ldots, x_r \rangle.$$

Then

$$[M, Q_{\alpha(i)+1}] = [M, Q_{\alpha(i)}] + [M, x_i] = [M, Q_{\alpha(i-1)+1}] + [M, x_i].$$

Hence

$$[M, Q_{\alpha(i)+1}] = [M, x_1] + [M, x_2] + \cdots + [M, x_i].$$

Put $X_\beta = X \cap Q_\beta$; then $[M, Q_{\alpha(i)+1}] = [M, X_{\alpha(i)+1}]$. If β is an ordinal satisfying $\alpha(i) + 1 \leq \beta \leq \alpha(i+1)$, we have

$$[M, Q_\beta] = [M, Q_{\alpha(i)+1}] = [M, X_{\alpha(i)+1}] \leq [M, X_\beta],$$

so that $[M, Q_\beta] = [M, X_\beta]$ for all β. This shows that $[M, X_\beta]$ is an RQ-module, as well as an RX-module.

Next $[M, X] = [M, Q] = M$. Therefore, if $\alpha(i) + 1 \leq \beta \leq \alpha(i+1)$,

$$[M, [X_{\beta+1}, Q]] \leq [M, [Q_{\alpha(i+1)+1}, Q]] \leq [M, Q_{\alpha(i+1)}] \leq [M, Q_\beta] = [M, X_\beta].$$

Of course $\{X_\beta \mid 0 \leq \beta \leq \nu\}$ is a series of finite length since X is finitely generated and nilpotent. Thus we may apply 10.3.4, where the homomorphisms $*$ and $\dagger$ arise from the Q- and X-module structures of M. It follows that $H^n(X, M) = 0$ for $n \geq 0$.

It is now easy to complete the proof. Since X is subnormal in Q, there is a series $X = Q_0 \lhd Q_1 \lhd \cdots \lhd Q_k = Q$. If $H^n(Q_i, M) = 0$ for some i and all n, it follows from the LHS-spectral sequence that $H^n(Q_{i+1}, M) = 0$. Hence $H^n(Q, M) = 0$ for all $n \geq 0$ by induction on k.

For homology the proof is similar. ∎

Proof of 10.3.2 As before we form an ascending central series in Q with cyclic factors, say $1 = Q_0 \lhd Q_1 \lhd \cdots Q_\nu = Q$. By the artinian condition there are only finitely many RQ-submodules M^{Q_α}. Let the distinct proper submodules among them be $M^{Q_{\alpha(i)+1}}$, $i = 1, 2, \ldots, r$, where $\alpha(1) < \alpha(2) < \cdots < \alpha(r)$ and

$$\cdots = M^{Q_{\alpha(i)}} > M^{Q_{\alpha(i)+1}} = \cdots = M^{Q_{\alpha(i+1)}} > M^{Q_{\alpha(i+1)+1}} = \cdots.$$

Put $Q_{\alpha(i)+1} = \langle x_i, Q_{\alpha(i)} \rangle$ and $X = \langle x_1, x_2, \ldots, x_r \rangle$. Writing $X_\beta = X \cap Q_\beta$, we have $M^{Q_{\alpha(i)+1}} = M^{X_{\alpha(i)+1}}$. Also, if $\alpha(i) + 1 \leq \beta \leq \alpha(i+1)$, then

$$M^{Q_\beta} = M^{Q_{\alpha(i)+1}} = M^{X_{\alpha(i)+1}} \geq M^{X_\beta} \geq M^{Q_\beta},$$

which shows that $M^{Q_\beta} = M^{X_\beta}$ and $M^X = M^Q = 0$. Furthermore,

$$M^{[X_{\beta+1}, Q]} \geq M^{[Q_{\alpha(i+1)+1}, Q]} \geq M^{Q_{\alpha(i+1)}} = M^{Q_\beta} = M^{X_\beta}.$$

Now apply 10.3.5 and deduce that $H^n(X, M) = 0$ for all $n \geq 0$. Since X is subnormal in Q, it follows via the spectral sequence, just as above, that $H^n(Q, M) = 0$ for $n \geq 0$.

Once again the proof is similar for homology. ∎

Several extensions of 10.3.1 and 10.3.2 are known. For example, the proof just given still applies if Q is merely hypercentral, and it is likely that Q can even be locally nilpotent; for partial results in this direction see Robinson (1987b). In addition, there are results for Q a locally supersoluble group in the case that the module has no non-zero R-cyclic RQ-quotients or submodules (Robinson 1987a).

Bounded (co)homology groups

Our next object is to prove theorems asserting that (co)homology groups of nilpotent groups are bounded as abelian groups. Such results can be used to establish near splitting and conjugacy theorems on the basis of 10.1.11 and 10.1.16.

First of all some terminology. Let R be a principal ideal domain and M an R-module. The *torsion-free rank* $r_0(M)$ and the *p-rank* $r_p(M)$ of M as an R-module, where p is irreducible element of R, are defined in the same way as for abelian groups: thus

$$r_0(M) = \dim_F(M \otimes_R F),$$

where F is the field of fractions of R, and

$$r_p(M) = \dim\left(\mathrm{Hom}_{R/pR}(R/pR, M)\right).$$

The *total rank* of M is defined to be

$$r_0(M) + \sum_p r_p(M).$$

Our aim is to establish the following result.

10.3.6 (Lennox and Robinson 1980) *Let Q be a nilpotent group, R a principal ideal domain and M an RQ-module with finite total R-rank. Assume that either M_Q or M^Q is bounded as an R-module. Then $H_n(G, M)$ and $H^n(G, M)$ are bounded R-modules for all $n \geq 0$.*

In applying this theorem one would like to be able to deduce near splitting theorems: for this purpose one needs to have the analogue of 10.1.16 for modules over a principal ideal domain R such that R/rR is finite for any $r \neq 0$ in R. The proof is entirely similar to the case where $R = \mathbb{Z}$.

The most important case of 10.3.6 is where $R = \mathbb{Z}$, and for ease of presentation let us assume we are dealing with this: but the argument is identical in the general case. Another significant choice for R would be the group algebra of an infinite cyclic group over a finite field: we shall find an application for this in the proof of 10.4.5.

Our strategy in proving 10.3.6 is first to deal with the case $n = 0$.

10.3.7 *With the notation of* 10.3.6, *the group M_Q is finite if and only if M^Q is finite.*

Proof We are assuming that $R = \mathbb{Z}$, so M is an abelian group with finite total rank $r(M)$.

Assume that M_Q is finite. By passing to a suitable quotient of M we may assume that the torsion subgroup T of M is divisible. The proof is by induction on $r(M) + c > 0$, where c is the nilpotent class of Q.

Write $C = Z(Q)$ and choose any x in C: let θ denote the $\mathbb{Z}G$-endomorphism $a \mapsto a(x - 1)$ of M. If $\mathrm{Ker}(\theta)$ is finite, then so is M^Q, so we may suppose $\mathrm{Ker}(\theta)$ to be infinite. Hence $r(\mathrm{Im}(\theta)) < r(M)$. Since $\mathrm{Im}(\theta) \overset{\mathbb{Z}G}{\simeq} M/\mathrm{Ker}(\theta)$, the induction hypothesis implies that $(\mathrm{Im}(\theta))^Q$ is finite. If $\mathrm{Im}(\theta)$ is infinite, $r(M/\mathrm{Im}(\theta)) < r(M)$ and therefore $(M/\mathrm{Im}(\theta))^Q$ is finite by induction. It follows that M^Q is finite.

We may therefore suppose that $\mathrm{Im}(\theta)$ is finite. Hence $\mathrm{Im}(\theta) \leq T$ and $(T)\theta$ is both finite and divisible, so that $(T)\theta = 0$. Therefore we have proved that $[M, C] \leq T$ and $[T, C] = 0$.

Next $M_Q = M/[M, Q]$ is finite, of order d say. Thus

$$d[M, C] = [dM, C] \leq [M, Q, C],$$

which implies that $[M, C]/[M, Q, C]$ is finite since M has finite total rank. But $[M, Q, C] = [M, C, Q]$ since $[C, Q] = 1$, so it follows that $[M, C]_Q$ is finite. Since $[M, C, C] = 0$, we recognize that $[M, C]$ is a $\mathbb{Z}(Q/C)$-module satisfying the hypotheses on M. Hence $[M, C]^Q$ is finite by the induction hypothesis. But $(M/[M, C])^Q$ is finite by induction applied to the $\mathbb{Z}(Q/C)$-module $M/[M, C]$, so it follows that M^Q is finite.

Assume that M^Q is finite. We have to prove that M_Q is finite. Replacing M by a suitable multiple, we may suppose the torsion subgroup T to be divisible. The proof is once again by induction on $r(M) + c > 0$.

Let x and θ be as in the first part of the proof above. Then certainly we may assume that $M/\mathrm{Im}(\theta)$ is infinite, so that $r(\mathrm{Im}(\theta)) < r(M)$ and hence $(\mathrm{Im}(\theta))_Q$, that is, $(M/\mathrm{Ker}(\theta))_Q$, is finite by induction. If $\mathrm{Im}(\theta)$ is infinite, then $r(\mathrm{Ker}(\theta)) < r(M)$ and $(\mathrm{Ker}(\theta))_Q$ is finite by induction. This implies that M_Q is finite. Thus we may assume that $\mathrm{Im}(\theta)$ is finite, and hence $\mathrm{Im}(\theta) \leq T$ and $(T)\theta = 0$. As a consequence we have $[M, C] \leq T$ and $[T, C] = 0$.

We show next that $(M/M^C)^Q$ is finite. To see this, let $a + M^C \in (M/M^C)^Q$; then $[a, Q, C] = 0$, so that $[a, C, Q] = 0$ and $[a, C] \leq M^Q$. Writing $m = |M^Q|$, we have $0 = m[a, C] = [ma, C]$, which shows that $ma \in M^C$ and $m(M/M^C)^Q = 0$. Therefore $(M/M^C)^Q$ is finite, as was claimed.

The induction hypothesis may now be applied to M/M^C, which is a $\mathbb{Z}(Q/C)$-module since $[M, C, C] = 0$, and it follows that $(M/M^C)_Q$ is finite. Also $(M^C)_Q$ is finite by the induction hypothesis, from which we may deduce that M_Q is finite, thus completing the proof of 10.3.7. ∎

We are now in a position to undertake the proof of the bounded (co)homology theorem.

Proof of 10.3.6 As usual we deal only with cohomology. Using 10.3.7, we see that M_Q and M^Q are both finite. If T denotes the torsion subgroup of M, then T_Q and T^Q are finite, as are $(M/T)_Q = 0 = (M/T)^Q$ by 10.3.7 again. The cohomology sequence shows that it is enough to prove the theorem for the modules T and M/T, that is, for M torsion and for M torsion-free.

Case 1: M is torsion-free. In this case we have $M^Q = 0$. The proof uses some of the ideas in the proofs of 10.3.2 and 10.3.4. First of all form an ascending central series in Q with cyclic factors,

$$1 = Q_0 \triangleleft Q_1 \triangleleft \cdots Q_\beta = Q.$$

Since $r(M)$ is finite and M/M^{Q_α} is clearly torsion-free, there can be only finitely many submodules M^{Q_α}: let these be $M^{Q_{\alpha(i)+1}}$, $i = 1, 2, \ldots, r$, where $\alpha(1) < \alpha(2) < \cdots < \alpha(r)$ and

$$\cdots = M^{Q_{\alpha(i)}} > M^{Q_{\alpha(i)+1}} = M^{Q_{\alpha(i)+2}} = \cdots = M^{Q_{\alpha(i+1)}} > M^{Q_{\alpha(i+1)+1}} = \cdots.$$

Let $Q_{\alpha(i)+1} = \langle x_i, Q_{\alpha(i)} \rangle$ and put $X = \langle x_1, x_2, \ldots, x_r \rangle$. Since

$$M^{Q_{\alpha(i)+1}} = (M^{Q_{\alpha(i)}})^{\langle x_i \rangle} = (M^{Q_{\alpha(i-1)+1}})^{\langle x_i \rangle},$$

we have $M^X = M^Q = 0$. Since X is subnormal in C, it is sufficient to prove that $H^n(X, M) = 0$ for all n, by the LHS-spectral sequence. Therefore we may assume that $Q = X$ and Q is finitely generated.

Now we know there is a central series in Q with cyclic factors and finite length, say $1 = Q_0 \triangleleft Q_1 \triangleleft \cdots \triangleleft Q_\ell = Q$. The proof proceeds by induction on $r(M) + \ell$. Put $Q_1 = \langle x \rangle$ and let θ be the $\mathbb{Z}Q$-endomorphism of M in which $a \mapsto a(x-1)$. Suppose first that $\theta = 0$. Then $H^j(Q_1, M) \overset{\mathbb{Z}Q}{\simeq} A$, where $A = 0$, M or M/dM in case $d = |Q_1|$ is finite. By the induction hypothesis we conclude that each $H^i(Q/Q_1, H^j(Q_1, M))$ bounded for all i and j; hence $H^n(Q, M)$ is bounded, by the LHS-spectral sequence.

Next let $\theta \neq 0$. If $\mathrm{Ker}(\theta) \neq 0$, the result is true for $\mathrm{Ker}(\theta)$ and $M/\mathrm{Ker}(\theta)$ by the induction hypothesis, from which it is clear that it holds for M. So we may assume $\mathrm{Ker}(\theta) = 0$; it follows from 6.1.3 that $\mathrm{Coker}(\theta) = M/\mathrm{Im}(\theta)$ is finite. Apply the cohomology sequence to the exact sequence $M \overset{\theta}{\rightarrowtail} M \twoheadrightarrow \mathrm{Coker}(\theta)$, noting that $\theta_* = 0$ by 10.3.3. It follows that

$$0 \rightarrow H^n(Q, M) \rightarrow H^n(Q, \mathrm{Coker}(\theta))$$

is exact. Since $\mathrm{Coker}(\theta)$ is finite, $H^n(Q, M)$ is bounded.

Case 2: M is a torsion group. The proof is dual to case 1. By factoring out a suitable finite submodule, we may assume that M is divisible; hence $M_Q = 0$. Form an ascending central series of Q with cyclic factors $1 = Q_0 \triangleleft Q_1 \triangleleft \cdots Q_\beta = Q$. Since M satisfies min, there are only finitely many submodules $[M, Q_\alpha]$, say $[M, Q_{\alpha(i)+1}]$, $i = 1, 2, \ldots, r$, where

$$\cdots = [M, Q_{\alpha(i)}] < [M, Q_{\alpha(i)+1}] = \cdots = [M, Q_{\alpha(i+1)}] < [M, Q_{\alpha(i+1)+1}] = \cdots.$$

Let $Q_{\alpha(i)+1} = \langle x_i, Q_{\alpha(i)} \rangle$ and put $X = \langle x_1, x_2, \ldots, x_r \rangle$. Next we argue that $M_X = M_Q = 0$. We may now replace Q by X, so that there is a finite central series $1 = Q_0 \triangleleft Q_1 \triangleleft \cdots \triangleleft Q_\ell = Q$ with cyclic factors.

Let $Q_1 = \langle x \rangle$ and let θ be the $\mathbb{Z}X$-endomorphism in which $a \mapsto [a, x]$. Suppose first that $\theta = 0$. Now $H^j(Q_1, M)$ is $\mathbb{Z}Q$-isomorphic with $0, M$ or $M[d]$ in case $d = Q_1$ is finite. Therefore, by the induction hypothesis we have $H^i(Q/Q_1, H^j(Q_1, M)) = 0$ for all i, j, whence $H^n(Q, M) = 0$ for $n \geq 0$ by the LHS-spectral sequence.

Now assume that $\theta \neq 0$. If $\mathrm{Im}(\theta) \neq M$, the result is true for $\mathrm{Im}(\theta)$ and for $M/\mathrm{Im}(\theta)$, whence it is true for M. Therefore we may assume that $M = \mathrm{Im}(\theta)$, from which it follows that $\mathrm{Ker}(\theta)$ is finite. From the exact sequence $\mathrm{Ker}(\theta) \rightarrowtail M \overset{\theta}{\twoheadrightarrow} M$ and the cohomology sequence, we obtain an exact sequence

$$H^n(Q, \mathrm{Ker}(\theta)) \rightarrow H^n(Q, M) \rightarrow 0.$$

Since $\mathrm{Ker}(\theta)$ is finite, $H^n(Q, M)$ is bounded, which completes the proof. ∎

We record as consequences of 10.3.6 two vanishing theorems for torsion-free modules and divisible modules. In these results Q is a nilpotent group, R is a principal ideal domain and M is an RQ-module with finite R-ranks.

10.3.8 (Robinson 1980) *If M is R-torsion-free and $M_Q = 0$, then $H_n(Q, M) = 0 = H^n(Q, M)$ for all $n \geq 0$.*

10.3.9 (Robinson 1980) *If M is R-divisible and $M^Q = 0$, then $H_n(Q, M) = 0 = H^n(Q, M)$ for all $n \geq 0$.*

Proof of 10.3.8 As usual we deal only with cohomology. From 10.3.6 we see that M^Q is R-bounded and hence $M^Q = 0$. By 10.3.6 there exists $r_n \neq 0$ in R such that $r_n \cdot H^n(Q, M) = 0$. Taking the cohomology sequence for the exact sequence $M \xrightarrow{r_n} M \to M/r_n M$, we obtain the exact sequence $0 \to H^n(Q, M) \to H^n(Q, M/r_n M)$. But $M/r_n M$ is a bounded R-module with finite R-ranks, so it is the direct sum of finitely many cyclic R-modules. Therefore $M/r_n M$ is R-noetherian. Since $(M/r_n M)_Q = 0$, we may deduce from 10.3.1 that $H^n(Q, M/r_n M) = 0$ and hence $H^n(Q, M) = 0$. ∎

Proof of 10.3.9 From 10.3.7 we see that M_Q is bounded, whence $M_Q = 0$ by divisibility of M. Thus $r_n \cdot H^n(Q, M) = 0$ for some $r_n \neq 0$ in R by 10.3.6. From the exact sequence $M[r_n] \rightarrowtail M r_n \twoheadrightarrow M$, where $M[r_n] = \{a \in M \mid r_n \cdot a = 0\}$, we obtain $H^n(Q, M[r_n]) \to H^n(Q, M) \to 0$, which is exact. Here $M[r_n]$ is R-artinian, being a direct sum of cyclic torsion R-modules. Hence $H^n(Q, M[r_n]) = 0$ by 10.3.2, and thus $H^n(Q, M) = 0$. ∎

Finally, we record a vanishing theorem which will turn out to be a useful tool in the study of complete soluble groups—see 10.4 for this theory.

10.3.10 (Robinson 1980) *Let Q be a nilpotent group, R a principal ideal domain and M an RQ-module with finite total R-rank. If $H^0(Q, M) = 0 = H^1(Q, M)$, then $H_n(Q, M) = 0 = H^n(Q, M)$ for all $n \geq 0$.*

Proof Consider the case of cohomology. Let T denote the R-torsion submodule of M; then T is an artinian R-module and $T^Q = 0$, so we may apply 10.3.2 to obtain $H^n(Q, T) = 0$ for $n \geq 0$. Taking the cohomology sequence in dimensions 0 and 1 for $T \rightarrowtail M \twoheadrightarrow M/T$, we obtain the exact sequence

$$0 = M^Q \to (M/T)^Q \to 0 \to H^1(Q, M) = 0 \to H^1(Q, M/T) \to H^2(Q, T) = 0.$$

It follows at once that $(M/T)^Q = 0 = H^1(Q, M/T)$. Therefore, it is enough to prove the theorem for the module M/T and we may assume that M is R-torsion-free.

From 10.3.6 we see that M_Q is bounded. Now if $M_Q = 0$, the result follows at once from 10.3.8. Assuming $M_Q \neq 0$, we may be sure, since M_Q is torsion, that M_Q has non-trivial p-component for some irreducible element p of R. Then $pM_Q \neq M_Q$, which shows that $M \neq pM + [M, Q]$, that is, $(M/pM)_Q \neq 0$.

On the other hand, from the exact sequence $M \xrightarrow{p} M \twoheadrightarrow M/pM$ and the cohomology sequence, we derive the exact sequence

$$0 = M^Q \to (M/pM)^Q \to H^1(Q, M) = 0,$$

which gives $(M/pM)^Q = 0$ and hence the contradiction $(M/pM)_Q = 0$. ∎

Several generalized versions of the bounded (co)homology theorems of this section are known. For example, in 10.3.8 and 10.3.9 the ring R can be a Dedekind domain, while it is enough if the group G is locally nilpotent: for these see Robinson (1987*b*). In addition a version of 10.3.6 for locally supersoluble groups is known (Robinson 1987*a*).

10.4 Applications to infinite soluble groups

In dimensions 1 and 2 the theorems proved in 10.3 yield (near) splitting and conjugacy theorems for infinite soluble groups. In this section these theorems are applied to obtain structural information about such groups.

Nilpotent supplements

Recall that a *supplement* for a subgroup H in a group G is a subgroup X such that $G = XH$. If in addition $X \cap H = 1$, then X is *complement* of H in G. Our first result establishes the existence of nilpotent supplements in metanilpotent groups with $\min -n$, the minimal condition for normal subgroups.

10.4.1 *Let G be a group satisfying $\min -n$ and suppose that N is a nilpotent normal subgroup such that G/N is nilpotent. Then*

(i) *N has a nilpotent supplement which is a Černikov group;*
(ii) *if in addition N has no non-trivial G-central factors, then N has a nilpotent Černikov complement.*

Proof

(i) We show first that N may be assumed abelian. For suppose this case has been taken care of and put $N_i = \gamma_i(N)$. We prove that there is a sequence of subgroups $G = X_0, X_1, \ldots$ such that $X_i = X_{i+1}N_{i+1}$ and X_i/N_{i+1} is a nilpotent Černikov group. Suppose $X_0, \ldots, X_n$ have already been constructed, with $X_n \geq N_{n+1}$, and consider the group X_n/N_{n+2}. Note that $G = X_nN$, so that N_{n+1}/N_{n+2} has $\min -X_n$ since it has $\min -G$. Therefore X_n/N_{n+2} satisfies $\min -n$. Of course N_{n+1}/N_{n+2} is abelian, so by the abelian case there is a nilpotent subgroup X_{n+1}/N_{n+2} such that $X_n = X_{n+1}N_{n+1}$. Now

$$X_{n+1}/X_{n+1} \cap N_{n+1} \simeq X_{n+1}N_{n+1}/N_{n+1} = X_n/N_{n+1},$$

which is nilpotent and Černikov. Also $(X_{n+1} \cap N_{n+1})/N_{n+2}$ satisfies $\min -X_{n+1}$ since $G = X_{n+1}N$ and thus N_n/N_{n+1} has $\min -X_{n+1}$. It follows that X_{n+1}/N_{n+2} is a nilpotent Černikov group and our construction has been effected. If c denotes the nilpotent class of N, then $N_{c+1} = 1$ and $X = X_c$ is a nilpotent Černikov subgroup such that $G = XN$.

From now on N is assumed to be abelian. Let H denote the G-hypercentre of N, that is, H is the union of the chain $\{N_\alpha\}$, where $N_{\alpha+1}/N_\alpha = (N/N_\alpha) \cap Z(G/N_\alpha)$. By a theorem of Baer (1949)—see also Robinson (1972b)—the hypercentre of a group satisfying min $-n$ is a Černikov group; thus H is a Černikov group.

Now consider $\bar{G} = G/H$ and regard $\bar{N} = N/H$ as an artinian $\mathbb{Z}(G/N)$-module. Since $\bar{N}^{G/N} = 0$ and G/N is nilpotent, we may apply 10.3.2 to show that $H^2(G/N, \bar{N}) = 0$, so that $\bar{G}$ splits over $\bar{N}$ and there is a subgroup Y such that $G = YN$ and $Y \cap N = H$. Notice that Y/H is nilpotent here.

Let F be the finite residual of the Černikov group H. Thus H/F is finite and F is a radicable abelian group with min. Hence there exists $r \geq 0$ such that $[F, {}_rY] = [F, {}_{r+1}Y] = E$, say. Notice that Y/E is nilpotent because Y/H is nilpotent and H is G-hypercentral. Since $E = [E, Y]$, we may apply 10.3.6 and 10.1.16 to show that there is a subgroup X for which $Y = XE$ and $X \cap E$ is finite. Hence $G = YN = XN$. Since $X \cap N \leq Y \cap N = H$, we see that $X \cap E$ lies in a finite term of the upper central series of X and, since $X/(X \cap E)$ is nilpotent, it follows that X is nilpotent. Also $X/X \cap N$ is Černikov, as is $X \cap N$, whence X is Černikov.

(ii) Now assume that N has no non-trivial G-central factors. As in (i), we see that there is a nilpotent Černikov subgroup X such that $G = XN$, and an easy induction on j shows that

$$[X \cap Z_{i+1}(N), {}_jG] \leq (\gamma_{j+1}(X) \cap Z_{i+1}(N))Z_i(N)$$

for all i. If d is the nilpotent class of X, it follows that

$$[X \cap Z_{i+1}(N), {}_dG] \leq Z_i(N)$$

for all i. Thus $(X \cap Z_{i+1}(N))Z_i(N)/Z_i(N)$, being G-hypercentral, must be trivial, so that $X \cap Z_{i+1}(N) = X \cap Z_i(N)$. Since X is nilpotent, it follows that $X \cap N = 1$ and the proof is complete. ∎

Since by 1.5.4 the Fitting subgroup of a metanilpotent group with min $-n$ is nilpotent, we obtain:

Corollary 10.4.2 *The Fitting subgroup of a metanilpotent group with* min $-n$ *has a nilpotent Černikov supplement.*

Next we prove a nilpotent supplementation theorem which is applicable to soluble FATR-groups.

10.4.3 (Zaicev 1977; Lennox and Robinson 1980) *Let G be a soluble group with finite abelian total rank. If N is normal subgroup such that G/N is virtually nilpotent, there is a nilpotent subgroup X, such that $|G : XN|$ is finite.*

Proof Clearly it may be assumed that G/N is nilpotent. By 5.2.1 there is a characteristic series $1 = N_0 \triangleleft N_1 \triangleleft \cdots \triangleleft N_k = N$ in which each factor is torsion-free abelian or abelian with min. The proof is by induction on $k > 0$. Write $A = N_1$: then by the induction hypothesis there is a nilpotent section Y/A such that $|G : YN|$ is finite.

Suppose first that A is torsion-free. Define a sequence of Y-admissible subgroups A_i by $A_0 = 0$ and $A_{i+1}/A_i = (A/A_i) \cap Z(Y/A_i)$. Then A/A_i is torsion-free. Since A has finite rank, there is an i such that $A_i = A_{i+1} = \cdots = B$ say, and A/B is torsion-free. Since $(A/B)^Y = 0$, we may deduce that Y/B nearly splits over A/B by 10.3.6 and 10.1.16. So there is a section X/B such that $|Y : XA|$ is finite and $X \cap A = B$. Since Y/A is nilpotent, X is nilpotent and $|G : XN|$ is finite.

Now assume A satisfies min; then there is an $r \geq 0$ such that $[A, {}_rY] = [A, {}_{r+1}Y] = C$, say. Now Y/C is nilpotent and $C = [C, Y]$, so by 10.3.6 and 10.1.16 there is a subgroup X_0 such that $|Y : X_0C|$ and $|X_0 \cap C|$ are finite. Here X_0 is finite-by-nilpotent, so it is nilpotent-by-finite. Let $X \lhd X_0$, where X is nilpotent and $|X_0 : X|$ is finite. Then $|G : XN|$ is finite.   ∎

Corollary 10.4.4 *Let G be a soluble group with finite abelian total rank. Then there is a nilpotent subgroup X, such that $|G : X \operatorname{Fit}(G)|$ is finite.*

This is a consequence of 10.4.3 since $\operatorname{Fit}(G)$ is nilpotent and $G/\operatorname{Fit}(G)$ is virtually abelian by 5.2.2. That there are limitations to the validity of such near supplementation theorems, is shown by the following examples from Lennox and Robinson (1980).

(a) *There is a supersoluble group of Hirsch length 3 which cannot be expressed as a product of pairwise permutable nilpotent subgroups.*

This is in contrast to 10.4.3 since supersoluble groups are nilpotent-by-abelian.

(b) *There is a torsion-free polycyclic group with no subgroup of finite index which is a split extension of nilpotent groups.*

Thus nilpotent complements are not to be expected in these situations.

Nearly maximal subgroups

A subgroup of a group G is said to be *nearly maximal* if it has infinite index, but every strictly larger subgroup has finite index in G. Thus nearly maximal means 'maximal of infinite index'. Such subgroups were introduced by Riles (1969). If G is a finitely generated infinite group, Zorn's Lemma shows that every subgroup of infinite index is contained in a nearly maximal subgroup of G.

Recall that a finitely generated soluble group all of whose maximal subgroups are normal is nilpotent—see 4.4.5. Our next objective is to prove a corresponding theorem for nearly maximal subgroups.

10.4.5 (Lennox and Robinson 1982) *The following conditions are equivalent for a finitely generated virtually soluble group G:*

  (i) *each nearly maximal subgroup has finitely many conjugates in G;*
 (ii) *each nearly maximal subgroup is normal in G;*
(iii) *G is finite-by-nilpotent.*

A simple auxiliary lemma is required in the proof.

10.4.6 *Let G be a finitely generated group in which every nearly maximal subgroup has finitely many conjugates. Then every subgroup with finite index in G has the same property.*

Proof Let H be a subgroup with finite index in G and suppose that M is a nearly maximal subgroup of H. Certainly M is contained in some nearly maximal subgroup L of G. Now $H \cap L = M$ since otherwise $|H : H \cap L|$, and hence $|G : H \cap L|$, would be finite. By hypothesis $|G : N_G(L)|$ is finite, whence so is $|G : N_H(L)|$. But $N_H(L) \leq N_H(M)$ since $M = H \cap L$, and therefore $|H : N_H(M)|$ is finite, as required.   ∎

In the next result a special case of the main theorem will be established.

10.4.7 *Let G be a finitely generated, virtually nilpotent group in which every nearly maximal subgroup has finitely many conjugates. Then G is finite-by-nilpotent.*

Proof Plainly G satisfies max, so we may assume the result to be true for every proper quotient of G, but not for G itself. Let A denote the Fitting subgroup of G: thus G/A is finite. If $A' \neq 1$, then G/A', being a proper quotient of G, is finite-by-nilpotent: now, since G is finitely generated, the tensor product argument of 1.2.17 may be applied to show that G is finite-by-nilpotent. Therefore, A is abelian and indeed it must be free abelian. We complete the proof by induction on $|G : A|$.

Suppose first that G/A is not cyclic. If $x \in G$, then $H = \langle x, A \rangle \neq G$ and by 10.4.6 the subgroup H inherits the conditions on G. Therefore, H is finite-by-nilpotent by the induction hypothesis. Since A is torsion-free, it follows that $[A, {}_r H] = 1$ for some $r > 0$. Hence each element of G acts unipotently on A and so, by a standard result from linear algebra, G acts unipotently on A, that is, A is finitely hypercentral in G. It now follows by a theorem of Baer (1952)—see also (Robinson 1972b)—that G is finite-by-nilpotent.

The argument of the last paragraph demonstrates that G/A must be cyclic: write $G = \langle x, A \rangle$ and $X = \langle x \rangle$. We will argue by induction on the rank of A that G is actually nilpotent, and for this purpose it may be assumed that $C_A(X) = 1$ and hence $X \cap A = 1$. If A were $\mathbb{Z}X$-rationally reducible, there would exist a proper non-trivial $\mathbb{Z}G$-submodule B such that A/B is torsion-free and rationally irreducible. By induction on the rank of A it follows that G/B is nilpotent and hence $[A, X] \leq B$. But this implies that $C_A(X) \neq 1$ since otherwise $A \simeq [A, X]$. Thus A must be $\mathbb{Z}X$-rationally irreducible, from which we conclude that X is nearly maximal in G. It follows from the hypothesis of the theorem that $|G : N_G(X)|$ is finite. However $N_G(X) = XC_A(X) = X$, which gives the contradiction $A = 1$.   ∎

Proof of 10.4.5 We show first that (iii) implies (ii). Let G be finite-by-nilpotent and suppose M is nearly maximal in G. Since the torsion subgroup T of G is finite, $|G : MT|$ is infinite and therefore $T \leq M$. This allows us to assume that G is torsion-free and nilpotent. Therefore $M \neq N_G(M)$, which shows that $|G : N_G(M)|$ is finite. Denote the isolator of M in G by I. Obviously $|I : M|$ is finite, whence $I = M$ and M is isolated. By 2.3.7 the subgroup $N_G(M)$ is also isolated. However $|G : N_G(M)|$ is finite and it follows that $N_G(M) = G$ and $M \lhd G$. Clearly (ii) implies (i), so what remains to be shown is that (i) implies (iii), which is in fact the main step in the proof.

Assume that (i) is valid. By hypothesis there is a soluble normal subgroup S with finite index in G. We may assume $S \neq 1$ and use induction on the derived length of S. Then the induction hypothesis implies that G is abelian-by-finite-by-nilpotent, and of course finitely generated. By 4.2.2 the group G satisfies $\max -n$. Therefore we may suppose that every proper quotient of G, but not G itself, is finite-by-nilpotent.

Let A denote the Fitting subgroup of G, which is nilpotent because of the property $\max -n$. If $A' \neq 1$, then G/A' is finite-by-nilpotent and, as has already been remarked, this implies that G is finite-by-nilpotent by the argument of 1.2.17. Thus A must be abelian. Next let T be the torsion subgroup of A. Then T has finite exponent since G has $\max -n$, and thus A^e is torsion-free for some $e > 0$. It follows that $T \cap A^e = 1$ and G embeds in $(G/T) \times (G/A^e)$. This indicates that either T or A^e must be trivial. Thus either A is torsion-free or it has finite exponent. In the latter event A will clearly have to be a p-group for some prime p. In addition $A^p = 1$: for otherwise A/A^p would be finite, which implies that A is finite. In short, A is either torsion-free or an elementary abelian p-group.

Consider first the case where A is torsion-free. Since G/A is finitely generated and finite-by-nilpotent, and A is finitely generated as a $\mathbb{Z}(G/A)$-module, 4.3.3 may be applied to yield a free abelian subgroup B such that A/B is a π-group for some finite set of primes π. Choosing a prime q not in π, we have $A^q \cap B = B^q$. Now G/A^q is finitely generated and finite-by-nilpotent, so A/A^q is finite. From this we deduce that B/B^q is finite, so that B has finite rank: consequently A has finite rank. Before pursuing this situation any further, we examine the other possibility, A is an elementary abelian p-group.

Observe that $Q = G/A$ cannot be finite since otherwise G would be finite. Since Q is finite-by-nilpotent, some $Q/Z_i(Q)$ is finite. This means that $Z(Q)$ contains an element of infinite order, say $\bar{z} = zA$: for indeed the alternative is that $Z(Q)$, and hence $Z_i(Q)$, is finite (see 1.2.20).

Write $J = \mathbb{Z}_p \langle \bar{z} \rangle$, which is a principal ideal domain, and regard A as a J-module via conjugation by z. Since $\bar{z} \in Z(Q)$, just as in the proof of 4.3.4, we may regard the subgroup A as a JQ-module. We claim that A is J-torsion-free. Indeed suppose this is false and consider the J-torsion submodule S of A: notice that $S \lhd G$ since $\bar{z} \in Z(Q)$. If $a \in S$, then $a^{f(z)} = 1$, where $0 \neq f \in J[t]$, which is easily seen to imply that $a^{\langle z \rangle}$ is finite: consequently $[a, z^{k(z)}] = 1$ for some $k(z) > 0$. Since $\bar{z} \in Z(Q)$, it follows that $[a^G, z^{k(z)}] = 1$. Now S is finitely generated as a JQ-module because A is $\mathbb{Z}G$-noetherian. It follows that $[S, z^k] = 1$ for some $k > 0$. Next G/S is finite-by-nilpotent. Also A/S is J-torsion-free, so it cannot contain non-trivial finite J-submodules. It follows that $\langle z, A \rangle /S$ is nilpotent and hence $\langle z^k, A \rangle$ is nilpotent. Since also $\langle z^k, A \rangle \lhd G$, we deduce that $z^k \in \mathrm{Fit}(G) = A$, a contradiction which shows that S is trivial.

Now that A is known to be J-torsion-free, we can prove that A has finite J-rank, just as in the case where A is $\mathbb{Z}$-torsion-free. From now on in the proof we will cover the torsion-free case and the p-case simultaneously, using J to denote either $\mathbb{Z}$ or $\mathbb{Z}_p \langle \bar{z} \rangle$.

Next G/A is finite-by-nilpotent and so virtually nilpotent. Let $M \lhd G$ where M/A is nilpotent and G/M finite. Notice that M inherits the hypothesis on G by 10.4.6; also A has finite rank as a $J(M/A)$-module. Furthermore, if the result holds for M, it will hold for G by 10.4.7. Therefore we may replace G by M and assume G/A to be nilpotent. Now choose G to be a counterexample with A of least J-rank.

If $C_A(G) \neq 1$, then $G/Z(G)$ is finite-by-nilpotent and G will be virtually nilpotent, which is impossible by 10.4.7. Therefore $C_A(G) = 1$. This puts us in a position to apply 10.3.6 and conclude that $H^2(G/A, A)$ is a bounded J-module. But J/rJ is finite if $0 \neq r \in J$. Thus we may apply 10.1.16 for the principal ideal domain R and consequently G nearly splits over A, so there is a subgroup X such that $|G : XA|$ is finite and $X \cap A = 1$ (since A is J-torsion-free). By 10.4.6 we may assume that $G = XA$.

Suppose next that A is JX-rationally reducible and let $1 \neq B < A$ where $B \lhd G$ and A/B is J-torsion-free and JX-rationally irreducible. Then G/B is finite-by-nilpotent, from which it follows that A/B is G-central. By 10.3.6 we deduce that $C_A(X) \neq 1$, that is, $C_A(G) \neq 1$, which was seen to be untenable.

We conclude that A must be JX-rationally irreducible and in consequence X is nearly maximal in $G = XA$. Hence $|G : N_G(X)|$ is finite, which leads to $[A, X] = 1$, a contradiction which completes the proof. ∎

Similar characterizations of virtually nilpotent groups and supersoluble groups in terms of their nearly maximal subgroups are mentioned for comparison with 10.4.5 (see Lennox and Robinson 1982).

10.4.8 *Let G be a finitely generated virtually soluble group. Then G is virtually nilpotent if and only if every nearly maximal subgroup has a subgroup of finite index with finitely many conjugates in G.*

10.4.9 *A finitely generated soluble group G is supersoluble if and only if each nearly maximal subgroup has a subgroup of index at most 2 with finitely many conjugates in G.*

These theorems can be proved by similar methods.

Complete abelian-by-nilpotent groups

Cohomology has already been employed to show that certain infinite soluble groups cannot be complete (see 10.1.19 and 10.1.20). By using the final vanishing theorem of 10.3, we are able to obtain deeper results.

Let us consider an abelian-by-nilpotent group G. Then there is a maximal abelian normal subgroup A such that G/A is nilpotent. Clearly $C_G(A) = A$, so that $Q = G/A$ may be identified with a subgroup of $\operatorname{Aut}(A)$. If $\gamma \in N_{\operatorname{Aut}(G)}(A)$, then γ induces an automorphism α in A. For $a \in A$ and $g \in G$ we have

$$(a^g)^\alpha = (a^g)^\gamma = (a^\gamma)^{g^\gamma} = (a^\alpha)^{g^\gamma}.$$

On replacing a by $a^{\alpha^{-1}}$, we find that $a^{g^\gamma} = a^{\alpha^{-1} g \alpha}$. Since $A = C_G(A)$, it follows that $g^\alpha \equiv (\alpha^{-1} g \alpha) \operatorname{mod}(A)$ and hence α normalizes $Q \leq \operatorname{Aut}(G)$. Consequently the assignment $\gamma \mapsto \alpha$ is a homomorphism

$$\theta : N_{\operatorname{Aut} G}(A) \to N_{\operatorname{Aut} A}(Q).$$

Next, if $\gamma \in \operatorname{Ker}(\theta)$, then γ acts trivially on A and hence on Q. Therefore $\operatorname{Ker}(\theta) \simeq \operatorname{Der}(Q, A)$ by 10.1.12, which means that we have an exact sequence

$$0 \to \operatorname{Der}(Q, A) \to N_{\operatorname{Aut}(G)}(A) \xrightarrow{\theta} N_{\operatorname{Aut}(A)}(Q).$$

On restricting to inner automorphisms, we obtain by 10.1.12 again

$$0 \to \operatorname{Inn}(Q, A) \to \operatorname{Inn}(G) \to Q \to 1.$$

On passing to quotients, we arrive at the exact sequence

$$0 \to H^1(Q, A) \to N_{\operatorname{Out}(G)}(A) \to N_{\operatorname{Aut}(A)}(Q)/Q. \quad (*).$$

Now suppose that G is a complete group. Thus $Z(G) = 1 = \operatorname{Out}(G)$, and so by the exact sequence $(*)$ above $H^0(Q, A) = 0 = H^1(Q, A)$. At this point we will specialize to soluble FATR-groups and establish:

10.4.10 *Let G be an abelian-by-nilpotent group with finite abelian total rank. Then G is complete if and only if $G = Q \ltimes A$, where:*

(i) *$A = C_G(A)$ and Q is a self-normalizing nilpotent subgroup of $\operatorname{Aut}(A)$;*
(ii) *$A = [A, Q]$ and $C_A(Q) = 1$.*

Proof Let A be a maximal abelian normal subgroup of G with $Q = G/A$ a nilpotent group: thus $A = C_G(A)$. If G is complete, the foregoing discussion shows that $H^0(Q, A) = 0 = H^1(Q, A)$. Now we can apply 10.3.10 and deduce that $H_0(Q, A) = 0$, that is, $A = [A, Q]$, and also $H^2(Q, A) = 0$, so that G splits over A: write $G = Q \ltimes A$. Here Q is nilpotent and, since $C_G(A) = A$, we may identify Q with a subgroup of $\operatorname{Aut}(A)$. Now each element α of $N_{\operatorname{Aut}(A)}(Q)$ extends to an automorphism γ of G, where $(x, a)^\gamma = (\alpha^{-1} x \alpha, a^\alpha)$, $x \in Q$, $a \in A$. By hypothesis γ must be inner, so it follows that $\alpha \in Q$ and therefore $Q = N_{\operatorname{Aut}(A)}(Q)$.

Conversely, assume that conditions (i) and (ii) hold for $G = Q \ltimes A$. Then $Z(G) \leq C_G(A) = A$ and $C_A(Q) = 1$, so that $Z(G) = 1$. Also, if c is the nilpotent class of Q, then $A = \gamma_{c+1}(G)$ by (ii). Therefore, A is characteristic in G and the exact sequence $(*)$ above becomes

$$0 \to H^1(Q, A) \to \mathrm{Out}(G) \to N_{\mathrm{Aut}(A)}(Q)/Q.$$

Let T be the torsion subgroup of A; thus T has min. Since $T^Q = 1$, we obtain $H^1(Q, T) = 0$ for all $n \geq 0$ by 10.3.2. Also $(A/T)_Q = 0$, so $H^1(Q, A/T) = 0$ by 10.3.8 and it follows that $H^1(Q, A) = 0$. Finally, $Q = N_{\mathrm{Aut}(A)}(Q)$ and hence $\mathrm{Out}(G) = 1$. ∎

Corollary 10.4.11 *A non-trivial torsion-free abelian-by-nilpotent group of finite abelian total rank cannot be complete.*

Proof For, if G is such a group, then $G = Q \ltimes A$, where A is abelian and Q is nilpotent. The automorphism of A in which $a \mapsto a^{-1}$ clearly normalizes Q, so it must belong to Q. Since G is torsion-free, it follows that $A = 1$ and $G = 1$. ∎

On the other hand, it is known that there are complete torsion-free polycyclic groups with Hirsch length 7 (Robinson 1980).

Observe that 10.4.10 provides a scheme for classifying complete abelian-by-nilpotent FATR-groups. Let A be an abelian group with finite total rank. Choose a self-normalizing nilpotent subgroup Q of $\mathrm{Aut}(A)$ with finite total rank such that $A = [A, Q]$ and $C_A(Q) = 1$. Then form the semidirect product $G = Q \ltimes A$: this is a complete abelian-by-nilpotent FATR-group and, furthermore, all such groups arise in this way.

By such methods all finite complete metabelian groups were classified by Gagen and Robinson (1979), and this result was later extended to finite abelian-by-nilpotent groups by Gagen (1980).

Constructing complete polycyclic groups

As a further illustration of the utility of these techniques, we show how to construct a complete metabelian polycyclic group, starting from an algebraic number field F. Let U be the group of algebraic units of F and denote by A the additive subgroup of F generated by U. Then we have an embedding $\tau : U \to \mathrm{Aut}(A)$ given by $(a)u^\tau = au, (a \in A, u \in U)$. Now form the semidirect product $G = U \underset{\tau}{\ltimes} A$. Regarding this group we prove:

10.4.12 *Assume that F has trivial Galois group over $\mathbb{Q}$ and that $A = [A, U^\tau]$. Then $G = U \ltimes A$ is a complete metabelian polycyclic group.*

Proof First note that A is a finitely generated free abelian group and U is finitely generated and abelian. Thus G is metabelian and polycyclic. To show that G is complete, we need only verify the conditions of 10.4.10. Note that $C_A(U^\tau) = 0$ since $-1 \in U$, while $A = [A, U^\tau]$ is given. So all that remains to be proved is that $U^\tau = N_{\mathrm{Aut}(A)}(U^\tau)$.

Let $\alpha \in N_{\mathrm{Aut}(A)}(U^\tau)$ and $u \in U$. Denote by f the irreducible polynomial of u, which is also the minimum polynomial of u^τ. Now $(u^\tau)^\alpha = v^\tau$, where $v \in U$, and v is also a root of f. Hence v is a root of f, so there is a field automorphism sending u to v. Since $\mathrm{Gal}(F)$ is trivial, it follows that $v = u$ and $\alpha \in C_{\mathrm{Aut}(A)}(U^\tau)$.

For any $a \in A$ we have $(a)u^\tau \alpha = (a)\alpha u^\tau$, that is, $(au)\alpha = (a\alpha)u$. Setting $a = 1$, we deduce that $(u)\alpha = wu$, where $w = (1)\alpha \in A$. Since $(u)\alpha^{-1} = w^{-1}u$, we also have $w^{-1} \in A$, so that $w \in U$.   ∎

For example, we could take F to be the field $Q(a)$ where a is the real root of $t^3 + t - 1$. Clearly $\mathrm{Gal}(F) = 1$, while $a - 1 = -a^3$, so that $A = [A, U^\tau]$.

The theory just described can be extended to complete metanilpotent groups with the aid of an exact sequence of automorphism groups due to Wells (1971). For this and further results on complete infinite soluble groups the reader is referred to Robinson (1982), where additional references are given.

Soluble products of polycyclic groups

We conclude our applications by establishing the following theorem of Lennox and Roseblade (1980).

10.4.13 *Let G be a soluble group which is the product of two polycyclic subgroups H and K. Then G is polycyclic.*

Proof Let d denote the derived length of G: if $d \leq 1$, then G is abelian and the result is obvious. So let $d > 1$ and put $A = G^{d-1}$. Then by induction on d, we have G/A polycyclic, so that G is abelian-by-polycyclic.

At this point we make a reduction which is standard in the theory of factorized groups. Let

$$X = HA \cap KA,$$

a subgroup usually referred to as *the factorizer* of A. Since $G = HK$, we easily see that

$$X = (H \cap KA)(HA \cap K) = (H \cap KA)A = (HA \cap K)A.$$

This allows us to replace G by X. In other words we may asume we have a triple factorization

$$G = HK = HA = KA.$$

Next G is finitely generated and abelian-by-polycyclic, so it satisfies $\max{-n}$ by 4.2.2. Assuming that G is not polycyclic, we may suppose that all proper quotients of G are polycyclic, that is, G is a just non-polycyclic group. Let $F = \mathrm{Fit}(G)$. Then F is abelian and self-centralizing, by 7.4.2. Observe that $H \cap F \triangleleft HF = G$ and similarly $K \cap F \triangleleft G$. Thus $(H \cap F)(K \cap F)$ is a polycyclic normal subgroup of G and by the just non-polycyclic property we must have

$$H \cap F = 1 = K \cap F.$$

Next let $N = \mathrm{Fit}(H)$ and consider the N-module F. Suppose that $[F, {}_r N] = 1$ for some $r > 0$. Then NF is nilpotent and $NF \lhd HF = G$. Hence $N \leq H \cap F = 1$, so that $H = 1$, which is clearly impossible. It now follows from 7.3.2 that there is a maximal H-submodule B of F such that $[F, H] \not\leq B$. Notice that $B \lhd HF = G$.

Now write $\bar{G} = G/B$, $\bar{F} = F/B$ etc. Then $\bar{F}$ is a simple $\bar{H}$-module, so it is finite by 7.1.1. Also $\bar{F} = [\bar{F}, \bar{N}]$. By 10.3.1 we obtain $H^1(\bar{N}, \bar{F}) = 0 = H^0(\bar{N}, \bar{F})$. The LHS spectral sequence now shows that $H^1(\bar{H}, \bar{F}) = 0$.

This conclusion means that complements of $\bar{F}$ in the semidirect product $\bar{G} = \bar{H}\,\bar{F} = \bar{K}\,\bar{F}$ are conjugate by 10.1.9. Therefore there exist $\bar{h} \in H$ and $\bar{k} \in K$ such that $\bar{H}^{\bar{h}\,\bar{k}} = \bar{K}$. But this implies that $\bar{H} = \bar{K} = \bar{G}$ and $\bar{F} = 1$, a final contradiction. ∎

There is a very large body of work on factorized soluble groups. For example, an important result of Wilson (1988a) states that *a soluble group which is the product of two minimax groups is also minimax*. For a full account of the theory of factorized soluble groups we refer the reader to Amberg, Franciosi and de Giovanni (1992).

10.5 Kropholler's theorem on soluble minimax groups

One of the most impressive demonstrations of the power of cohomological techniques when applied to infinite soluble groups is the following theorem of Kropholler (1984).

10.5.1 *A finitely generated soluble group is minimax if and only if it has no sections isomorphic with $\mathbb{Z}_p \, \mathrm{wr} \, \mathbb{Z}$ for any prime p.*

The proof of this theorem is difficult and makes great play with cohomology groups in the case where the coefficient module is expressed as a direct limit. Indeed it is the failure of cohomology to commute with direct limits which causes the complications in the proof.

Kropholler's theorem has several applications of which we mention two at this point.

Corollary 10.5.2 *If every 2-generator subgroup of a finitely generated soluble group G is minimax, then G is minimax.*

This follows at once since a section of G of type $\mathbb{Z}_p \, \mathrm{wr} \, \mathbb{Z}$ would be a quotient of a 2-generator subgroup, which cannot then be minimax. Another application, which demonstrates the central position occupied by finitely generated soluble minimax groups, is:

Corollary 10.5.3 (Robinson 1975) *A finitely generated soluble group with finite abelian ranks is minimax.*

Here the point to note is that the base group of the wreath product $\mathbb{Z}_p \, \mathrm{wr} \, \mathbb{Z}$ has infinite rank.

In addition to cohomological techniques, the proof of Kropholler's Theorem makes use of the following interesting property of finitely generated soluble minimax groups.

10.5.4 (Kropholler 1984) *A finitely generated virtually soluble minimax group is expressible as a product of finitely many cyclic groups.*

The proof of 10.5.4 hinges on a technical lemma.

10.5.5 *Let A be an abelian minimax group and assume that $A = \langle S \rangle$, where the subset S has the property that $\langle s \rangle \subseteq S$ whenever $s \in S$. Then there is a positive integer k such that $A = SS \cdots S$ with k factors.*

Proof Suppose first of all that A is a torsion group. Since A satisfies min, we may write $A = A_1 \times A_2 \times \cdots \times A_\ell$, where A_i is the p_i-primary component of A. We claim that if X is any finite subset of S, then $\langle X \rangle$ can be generated by m or fewer elements of S, where $m = \max\{r_{p_i}(A) \mid i = 1, 2, \ldots, k\}$. Here one can assume that A is a p-group. Then clearly A/pA can be generated by $\{x + pA \mid x \in X_0\}$, where X_0 is an m-element subset of X. But this implies that X_0 generates A, as required. Finally, $\langle X \rangle \subseteq SS \cdots S$, with m factors, and thus $A = SS \cdots S$.

Now consider the general case. Since $A = \langle S \rangle$, there is a linearly independent subset of r_0 elements of infinite order in S, where $r_0 = r_0(A)$. These elements generate a subgroup B which is contained in the product $SS \cdots S$ with r_0 factors. Finally, A/B is a torsion group, so the result follows via the first paragraph. ∎

Proof of 10.5.4 As a first step, consider the case where G is virtually metabelian. Thus G has an abelian normal subgroup A with G/A virtually abelian. Clearly we may write $G = TA$, where T is a product of finitely many cyclic subgroups. Since A is minimax, it has a finitely generated subgroup B such that A/B is radicable. Now put

$$S = \bigcup_{g \in G} B^g.$$

For any g in G, we have $g = ta$ with $t \in T$, $a \in A$, and thus $B^g = B^t \subseteq TBT$. It follows that $S \subseteq TBT$. Since $B \subseteq S$, the group $A/\langle S \rangle$ is radicable. But $\langle S \rangle \triangleleft G$ and G satisfies max $-n$ by 4.2.2, so that $A = \langle S \rangle$. If $s \in S$, then plainly $\langle s \rangle \subseteq S$ and therefore 10.5.5 may be applied to yield $A = SS \cdots S$, with finitely many factors S. Since $S \subseteq TBT$ and $G = TA$, the result follows.

We are now ready to handle the general case. By 5.2.2 the group G has a normal nilpotent subgroup N such that G/N is virtually abelian. Let c denote the class of N: then we may assume that $c > 1$, by the first part of the proof, and proceed by induction on c. Thus $G = T\gamma_c(N)$ where T is a product of finitely many cyclic subgroups. Hence $N = T_1\gamma_c(N)$, where $T_1 = N \cap T$. Next $\gamma_c(N)$ is generated by the subset $S = \{[x, y] \mid x \in \gamma_{c-1}(N), y \in N\}$. Furthermore, if $s \in S$, then $s^n \in S$ for all n since $[x, y]^n = [x^n, y]$. Therefore 10.5.5 may be applied to give $\gamma_c(N) = SS \cdots S$, with ℓ factors say. In addition, if $x \in \gamma_{c-1}(N)$ and $y \in N$, we may write $[x, y] = [t, t']$ with $t, t' \in T_1$ since $N = T_1 \gamma_c(N)$. Consequently $S \subseteq T^{-1} T^{-1} T T$. Since $G = TSS \cdots S$, with ℓ factors S, the result follows. ∎

Before becoming enmeshed in the details of the proof of Kropholler's Theorem, we need to explain the relevance of the cyclic product theorem (10.5.4) and show how we are led to consider the cohomology of modules which are direct limits.

Let G be a finitely generated soluble group with no sections of type $\mathbb{Z}_p \, wr \, \mathbb{Z}$ for any p. It is natural to try to prove that G is minimax by induction on the derived length. Thus we may assume that G has an abelian normal subgroup M which is torsion or torsion-free such that $Q = G/M$ is a finitely generated soluble minimax group. Let $a \in M$ and $x \in G$. If M is a torsion group, then $a^{\langle x \rangle}$ has finite exponent. In fact it must be finite: for otherwise $\langle a \rangle$ would contain an element a' of some prime order p such that $a'^{\langle x \rangle}$ is infinite, which is easily seen to imply that $\langle x, a' \rangle \simeq \mathbb{Z}_p \, wr \, \mathbb{Z}$. On the other hand, if M is torsion-free, $a^{\langle x \rangle}$ must have finite rank, since otherwise $\langle x, a \rangle \simeq \mathbb{Z} \, wr \, \mathbb{Z}$, which has $\mathbb{Z}_p \, wr \, \mathbb{Z}$ as an image for any prime p.

Next 10.5.4 can be applied to show that $G = \langle x_1 \rangle \langle x_2 \rangle \cdots \langle x_k \rangle M$ for some $x_i \in G$; it therefore follows from the conclusions of the previous paragraph that

$$a^G = \left(\cdots \left(a^{\langle x_1 \rangle} \right)^{\langle x_2 \rangle} \cdots \right)^{\langle x_k \rangle}$$

is either finite or of finite rank. What this shows is that either M is a locally finite $\mathbb{Z}Q$-module or else $M \otimes_{\mathbb{Z}} \mathbb{Q}$ is locally a finite dimensional $\mathbb{Q}Q$-module. Thus *either M is a direct limit of finite $\mathbb{Z}Q$-submodules or else $M \otimes_{\mathbb{Z}} \mathbb{Q}$ is a direct limit of finite dimensional $\mathbb{Q}Q$-modules.*

Now suppose that G splits over M, with $G = X \ltimes M$; then G is generated by elements $x_1 a_1, \ldots, x_k a_k$, where $x_i \in X, a_i \in M$. Thus $M = \langle a_1, \ldots, a_k \rangle^G$, that is, M is a finitely generated $\mathbb{Z}Q$-module. By the previous paragraph we may infer that M, and hence G, has finite total rank: the result is now a consequence of 5.2.8.

Unfortunately the splitting assumed in the last paragraph is too much to hope for. However these considerations do suggest that we study the cohomology of a finitely generated soluble minimax group with coefficients in a module which is a direct limit of submodules that are finite or of finite $\mathbb{Q}$-dimension. This turns out to be a delicate matter, since in general the cohomology functor does not commute with direct limits.

Let Q be any group and let $\mathbf{F}$ be a covariant functor from the category of $\mathbb{Q}Q$-modules to the category of abelian groups. If $(M_i)_{i \in I}$ is a direct system of $\mathbb{Z}Q$-modules, there is corresponding direct system $(\mathbf{F}(M_i))_{i \in I}$, and a homomorphism

$$\varinjlim \mathbf{F}(M_i) \to \mathbf{F}(\varinjlim M_i).$$

If this map is an isomorphism for all direct systems $(M_i)_{i \in I}$, then $\mathbf{F}$ is said to *commute with direct limits*. While the functor $H^n(Q, -)$ does not in general commute with direct limits, it does so in certain special cases, as we now note.

10.5.6

(i) *If Q is a finitely generated group, then $\varinjlim H^1(Q, M_i) \to H^1(Q, \varinjlim M_i)$ is injective.*

(ii) *If Q is a polycyclic group, the functors $H^n(Q, -)$ commute with direct limits for all n.*

For a proof of this result we refer the reader to Bieri (1981). The first cohomological tool is needed to establish 10.5.1 will handle the torsion-free case. It is:

10.5.7 *Let Q be a soluble minimax group and let $(M_i)_{i \in I}$ be a direct system of finite dimensional $\mathbb{Q}Q$-modules. Then the natural map*

$$\varinjlim H^n(Q, M_i) \to H^n(Q, \varinjlim M_i)$$

is an isomorphism for all $n \geq 0$.

In order to prove this, an auxiliary result must be established.

10.5.8 *If Q is a soluble minimax group and M is $\mathbb{Q}Q$-module with finite $\mathbb{Q}$-dimension, then $H^n(Q, M)$ has finite $\mathbb{Q}$-dimension for $n \geq 0$.*

Proof We argue by induction on $m(Q)$, the minimality of Q, (i.e. the number of infinite factors in a series with cyclic or quasicyclic factors: see Section 5.1). Then the LHS-spectral sequence for cohomology allows us to reduce to the case where Q is cyclic or quasicyclic. If Q is cyclic, the standard formulas for $H^n(Q, M)$ in Section 10.1 show at once that this $\mathbb{Q}$-space has finite dimension. Suppose Q is quasicyclic. Recalling that torsion subgroups of $GL_n(\mathbb{Q})$ are finite by the theorem of Schur, we see that Q must act trivially on M. Therefore, $H^n(Q, M) \simeq \mathrm{Hom}(Q, M)$ or $\mathrm{Ext}(Q, M)$ by 10.2.6. Both of these are zero since M is torsion-free and divisible. ∎

Proof of 10.5.7 If Q is cyclic, the result is true by virtue of 10.5.6 (ii) or by the formulas for $H^n(Q, M)$. Suppose Q is quasicyclic. As in the previous proof, we see that M_i and M are trivial Q-modules and that $H^n(Q, M_i) = 0 = H^n(Q, \varinjlim M_i) = 0$. So the result is true in this case too.

In general we argue by induction on $m(Q) \geq 1$. Let $N \triangleleft Q$, where G/N is infinite cyclic or quasicyclic, so that $m(N) = m(Q) - 1$. By 10.5.8 each $H^q(N, M_i)$ has finite $\mathbb{Q}$-dimension and it follows via the induction hypothesis that the natural maps

$$\varinjlim H^q(N, M_i) \to H^q(N, \varinjlim M_i)$$

and

$$\varinjlim H^p(Q/N, H^q(N, M_i)) \to H^p(Q/N, \varinjlim H^q(N, M_i))$$

are isomorphisms. As a consequence the natural map

$$\alpha : \varinjlim H^p(Q/N, H^q(N, M_i)) \to H^p(Q/N, H^q(N, \varinjlim M_i))$$

is an isomorphism for all p and q.

Finally, since the functor $\varinjlim$ is exact, there are two direct limit forms of the LHS-spectral sequence,

$$\varinjlim H^p(Q/N, H^q(N, M_i)) \underset{p+q=n}{\Rightarrow} \varinjlim H^n(Q, M_i)$$

and

$$H^p(Q/N, H^q(N, \varinjlim M_i)) \underset{p+q=n}{\Rightarrow} H^n(Q, \varinjlim M_i).$$

It follows that α induces the natural map

$$\varinjlim H^n(Q, M_i) \to H^n(Q, \varinjlim M_i),$$

which is therefore an isomorphism. ∎

The torsion case

A more subtle cohomological tool is needed to dispose of the case where the abelian normal subgroup M is a torsion group. We have to deal with the situation of a soluble minimax group Q and a locally finite $\mathbb{Z}Q$-module M.

Suppose that A is an abelian normal subgroup of Q and let an ascending sequence of submodules of M be defined by repeatedly taking A-fixed points; thus

$$M_0 = 0, \quad M_{i+1}/M_i = (M/M_i)^A, \quad i = 0, 1, \dots.$$

Put $U = \bigcup_{i=0,1,\dots} M_i$. We are interested in the natural homomorphisms

$$\varinjlim H^n(Q, M_i) \to H^n(Q, U) \quad \text{and} \quad H^n(Q, U) \to H^n(Q, M),$$

and their composite

$$\nu_n : \varinjlim H^n(Q, M_i) \to H^n(Q, M).$$

Our immediate aim is to establish the following facts about the maps ν_1 and ν_2.

10.5.9 *The map ν_1 is an isomorphism, while ν_2 is an isomorphism provided that Q is finitely generated.*

Proof Our first observation is that the map

$$H^n(Q, U) \to H^n(Q, M)$$

is an isomorphism for all n. To prove this, write $\bar{M} = M/U$ and apply the cohomology functor to the exact sequence $U \rightarrowtail M \twoheadrightarrow \bar{M}$ to produce the exact sequence

$$\cdots \to H^{m-1}(Q, \bar{M}) \to H^m(Q, U) \to H^m(Q, M) \to H^{m+1}(Q, \bar{M}) \to \cdots$$

From this we see that it suffices to prove that $H^n(Q, \bar{M}) = 0$ for $n \geq 0$. Furthermore, it is clear from the LHS-spectral sequence for $A \rightarrowtail Q \twoheadrightarrow Q/A$ that it suffices to establish $H^n(A, \bar{M}) = 0$ for $n \geq 0$.

Let A_0 be a finitely generated subgroup of A such that A/A_0 is radicable and torsion: this exists because A is a minimax group. By the LHS-spectral sequence for the exact sequence $A_0 \rightarrowtail A \twoheadrightarrow A/A_0$ we need only prove that $H^n(A_0, \bar{M}) = 0$ for $n \geq 0$. Next, since A/A_0 is radicable, while $\bar{M}$ is a locally finite $\mathbb{Z}Q$-module, $\bar{M}^{A_0}$ must be a trivial $\mathbb{Z}A$-module, that is, $\bar{M}^{A_0} = \bar{M}^A$. Let $x + U \in \bar{M}^A$, then $(x)\mathbb{Z}Q$ is finite and clearly A acts nilpotently on it. Hence $x \in M_i$ for some i and therefore $\bar{M}^{A_0} = \bar{M}^A = 0$. Suppose now that $\bar{F}_j$ is any finite $\mathbb{Z}Q$-submodule of $\bar{M}$. Then $\bar{F}_j^{A_0} = 0$ and so by 10.3.2 we have $H^n(A_0, \bar{F}_j) = 0$. By 10.5.6(ii) applied to $\bar{M} = \varinjlim \bar{F}_j$, it follows that

$$H^n(A_0, \bar{M}) \simeq \varinjlim H^n(A_0, \bar{F}_j) = 0,$$

as required. As a consequence of what has just been proved, we may replace M by U and assume that $M = \varinjlim M_i$.

After this reduction we are ready to confront the task of proving that ν_n is an isomorphism for $n = 1$, and for $n = 2$ if Q is finitely generated. Apply the cohomology functor $H^*(Q, -)$ to the exact sequence $M_i \rightarrowtail M \twoheadrightarrow M/M_i$ and then take $\varinjlim$ to get the exact sequence

$$0 \to \varinjlim M_i^Q \to M^Q \to \varinjlim (M/M_i)^Q \to \varinjlim H^1(Q, M_i) \overset{\nu_1}{\to} H^1(Q, M) \to$$

$$\varinjlim H^1(Q, M/M_i) \to \varinjlim H^2(Q, M_i) \overset{\nu_2}{\to} H^2(Q, M) \to \varinjlim H^2(Q, M/M_i) \to \cdots.$$

Notice that $(M/M_i)^Q \subseteq M_{i+1}/M_i$, so that

$$\varinjlim (M/M_i)^Q \subseteq \varinjlim (M_{i+1}/M_i) = 0,$$

since all the maps in the last direct limit are zero. So what we really want to establish is

$$\varinjlim H^1(Q, M/M_i) = 0 \quad (*)$$

and, if Q is finitely generated,

$$\varinjlim H^2(Q, M/M_i) = 0 \quad (**).$$

Apply the LHS-spectral sequence for the extension $A \rightarrowtail Q \twoheadrightarrow Q/A$ in conjunction with the functor $\varinjlim$. The terms which are relevant to $H^1(Q, M/M_i)$ are

$$E_2^{10} = \varinjlim H^1(Q/A, (M/M_i)^A)$$

and

$$E_2^{01} = \varinjlim (H^1(A, M/M_i))^Q.$$

Since $(M/M_i)^A = M_{i+1}/M_i$ and all maps in the direct system (M_{i+1}/M_i) are zero, we have $E_2^{10} = 0$. To analyse E_2^{01}, choose a finitely generated subgroup A_0 with A/A_0 radicable and apply the spectral sequence, in the direct limit form, to $A_0 \rightarrowtail A \twoheadrightarrow A/A_0$. The terms relevant to $H^1(A, M/M_i)$ are

$$\varinjlim(H^1(A/A_0, (M/M_i)^{A_0})) \quad \text{and} \quad \varinjlim(H^1(A_0, M/M_i))^A.$$

The first of these vanishes since $(M/M_i)^{A_0} = (M/M_i)^A$ and in (M_{i+1}/M_i) all the maps are 0. The second direct limit embeds in $\varinjlim H^1(A_0, M/M_i)$, which is isomorphic with $H^1(A_0, \varinjlim(M/M_i)) = 0$ by 10.5.6(ii), since A_0 is finitely generated. Thus $(*)$ has been proved.

Next we must establish $(**)$ under the assumption that Q is finitely generated. By the direct limit form of the LHS-spectral sequence for $A \rightarrowtail Q \twoheadrightarrow Q/A$, it is sufficient to prove that

$$\varinjlim H^2(A, M/M_i)^Q = 0 \quad (***).$$

For clearly $\varinjlim H^2(Q/A, (M/M_i)^A) = 0$, while by 10.5.6 (i)

$$\varinjlim H^1(Q/A, H^1(A, M/M_i)) \simeq H^1(Q/A, \varinjlim H^1(A, M/M_i)) = 0$$

by $(*)$, thus proving $(***)$.

With A_0 as before, apply the direct limit form of the spectral sequence for $A_0 \rightarrowtail A \twoheadrightarrow A/A_0$. The terms which are relevant to $H^2(A, M/M_i)$ are

$$\varinjlim H^2(A/A_0, (M/M_i)^{A_0}) = \varinjlim H^2(A/A_0, M_{i+1}/M_i),$$

which is 0,

$$\varinjlim H^1(A/A_0, H^1(A_0, M/M_i))$$

and

$$\varinjlim H^2(A_0, M/M_i)^A \simeq H^2(A_0, \varinjlim(M/M_i))^A = 0$$

by 10.5.6 (ii). So only the second term need be examined.

Next M/M_i is the direct limit of its finite submodules, as is $H^1(A_0, M/M_i)$. Since A/A_0 is radicable, A acts trivially on $H^1(A_0, M/M_i)$. It follows via the spectral sequence that

$$\varinjlim H^2(A, M/M_i) \simeq \varinjlim \text{Hom}(A/A_0, H^1(A_0, M/M_i)).$$

From $A_0 \rightarrowtail A \twoheadrightarrow A/A_0$ we obtain the exact sequence

$$0 \to \text{Hom}(A/A_0, H^1(A_0, M/M_i)) \to \text{Hom}(A, H^1(A_0, M/M_i))$$
$$\to \text{Hom}(A_0, H^1(A_0, M/M_i)).$$

On applying $\varinjlim$, the term on the right vanishes since A_0 is finitely generated and $\varinjlim H^1(A_0, M/M_i) = 0$. Therefore

$$\varinjlim \text{Hom}(A/A_0, H^1(A_0, M/M_i)) \simeq \varinjlim \text{Hom}(A, H^1(A_0, M/M_i)).$$

Now consider the restriction map res : $H^1(A, M/M_i) \to H^1(A_0, M/M_i)$. This occurs in the standard exact sequence

$$0 \qquad\qquad\qquad\qquad 0$$
$$\searrow \qquad\qquad \nearrow$$
$$\text{Im(res)}$$
$$\nearrow \qquad\qquad \searrow$$
$$0 \to H^1(A/A_0, M_{i+1}/M_i) \to H^1(A, M/M_i) \overset{\text{res}}{\to} H^1(A_0, M/M_i) \to H^2(A/A_0, M_{i+1}/M_i).$$

This sequence can be expressed as the composite of two shorter exact sequences obtained by inserting the term Im(res) as shown in the diagram. Apply $\text{Hom}(A, -)$ to these sequences and then take $\varinjlim$. The conclusion is that restriction induces an isomorphism

$$\varinjlim \text{Hom}(A, H^1(A, M/M_i)) \simeq \varinjlim \text{Hom}(A, H^1(A_0, M/M_i)),$$

essentially because of the trivial nature of the direct system (M_{i+1}/M_i). Putting together the relevant maps, we find that

$$\varinjlim H^2(A, M/M_i) \simeq \varinjlim \text{Hom}(A, H^1(A, M/M_i)).$$

It is now necessary to verify that the above isomorphism is one of $\mathbb{Z}Q$-modules. This involves a careful examination of the spectral sequences involved (see Kropholler 1984). Since A is a finitely generated $\mathbb{Z}Q$-module, we have

$$\varinjlim H^2(A, M/M_i)^Q \simeq \varinjlim(\text{Hom}(A, H^1(A, M/M_i))^Q$$
$$= \varinjlim(\text{Hom}_{\mathbb{Z}Q}(A, H^1(A, M/M_i))$$
$$\subseteq \text{Hom}_{\mathbb{Z}Q}(A, \varinjlim H^1(A, M/M_i))$$
$$= 0,$$

which establishes $(***)$ and completes the proof of 10.5.9.   ∎

We are finally in a position to prove Kropholler's theorem.

[Proof of 10.5.1] Let G be a finitely generated soluble group with no sections of type $\mathbb{Z}_p \, wr \, \mathbb{Z}$. By induction on the derived length of G we may assume that G has an abelian normal subgroup M such that $Q = G/M$ is minimax. Evidently, it suffices to deal separately with the two cases: M torsion and M torsion-free.

Case 1: M is torsion-free. Write $\bar{M}$ for $M \otimes_{\mathbb{Z}} \mathbb{Q}$. The push-out of the natural injection $M \to \bar{M}$ gives rise to a commutative diagram

$$\begin{array}{ccccc} M & \rightarrowtail & G & \twoheadrightarrow & Q \\ \downarrow & & \downarrow & & \| \\ \bar{M} & \rightarrowtail & \bar{G} & \twoheadrightarrow & Q \end{array}$$

We have seen that $\bar{M}$ is locally a $\mathbb{Q}Q$-module with finite $\mathbb{Q}$-dimension. View $\bar{M}$ as the limit of a direct system $(\bar{M}_i)_{i \in I}$ of finite $\mathbb{Q}$-dimensional $\mathbb{Q}Q$-submodules. By 10.5.7 the natural map

$$\varinjlim H^2(Q, \bar{M}_i) \to H^2(Q, \bar{M})$$

is an isomorphism. It follows from this that the extension $\bar{M} \rightarrowtail \bar{G} \twoheadrightarrow Q$ arises from an extension of $\bar{M}_i$ by Q for some i. Therefore, $\bar{G}/\bar{M}_i$ splits over $\bar{M}/\bar{M}_i$. Since G is finitely generated and may be regarded as a subgroup of $\bar{G}$, it follows that $(G \cap \bar{M})\bar{M}_i/\bar{M}_i$ is contained in some finitely generated $\mathbb{Z}Q$-submodule of $\bar{M}/\bar{M}_i$, which must have finite $\mathbb{Q}$-dimension. We deduce that $(G \cap \bar{M})\bar{M}_i/\bar{M}_i$ has finite torsion-free rank, as must $G \cap \bar{M}$. Thus G is a finitely generated soluble FATR-group and it follows from 5.2.8 that G is minimax.

Case 2: M is torsion. Suppose that G is a counterexample. Then G/M is infinite, since otherwise G would be finite. Assume the pair (G, M) has been chosen so that G is a counterexample with the minimality $m(G/M)$ *least*. Notice that G/M contains an infinite abelian normal subgroup $A = B/M$, for example, the centre of its Fitting subgroup. We may further assume that A has been chosen with *largest* minimality.

Define $M_0 = 1$ and $M_{i+1}/M_i = (M/M_i)^A$. Then by 10.5.9 the natural map

$$\nu_2 : \varinjlim H^2(G/M, M_i) \to H^2(G/M, M)$$

is an isomorphism, from which we deduce that G/M_i splits over M/M_i for some i. Since G is finitely generated, it follows that M/M_i is a finitely generated $\mathbb{Z}(G/M)$-module, and, since M is locally finite as a $\mathbb{Z}(G/M)$-module, M/M_i is therefore finite. Put $B_0 = C_B(M/M_i)$.

Now B/B_0 is finite and B_0 is nilpotent since $[B_0', B_0] \leq [M, B_0] \leq M_i$ and $[M_i, {}_iB_0] = 1$. Suppose that B_0/B_0' is minimax: then 1.2.12 shows that B_0 is minimax, and hence so is G. Therefore, B_0/B_0' is not minimax and we may pass to G/B_0', that is, we may assume B_0 is abelian. Let T/M be the torsion subgroup of B/M; if this were infinite, so would T be and then $m(G/T) < m(G/M)$, in contradiction to the minimality of $m(G/M)$. Hence T/M is finite and we may replace B by a suitable power of B_0 so that B/M is torsion-free and M is the torsion subgroup of B. Note that B is a $\mathbb{Z}(G/M)$-module.

Since B/M has finite rank and M is torsion, there is a finitely generated $\mathbb{Z}(G/M)$-submodule W such that B/W is torsion. We claim that $W \cap M$ cannot be minimax. For if it were, W would be minimax and hence G/W could not be minimax. But $m(G/B) < m(G/M)$, so that the pair $(G/W, B/W)$ would violate the minimality of $m(G/M)$. Therefore $W \cap M$ is not minimax. Now $W/W \cap M$ is torsion-free, so we may choose $W_0/W \cap M$ to be $\mathbb{Q}$-rationally irreducible and finitely generated as a $\mathbb{Z}(G/M)$-module, with $W_0 \leq W$. Thus $T = W \cap M$ is the torsion subgroup of W_0. Now form the semidirect product

$$G_0 = (G/B) \ltimes W_0 :$$

this is a finitely generated soluble group satisfying the hypothesis on G. In addition, since $W_0/T \simeq W_0 M/M$, we have

$$m(G_0/T) = m(W_0/T) + m(G/B) \leq m(B/M) + m(G/B) = m(G/M).$$

But G_0 is not minimax, so by minimality of $m(G/M)$ we must have $m(G_0/T) = m(G/M)$ and it follows that $m(W_0/T) = m(B/M)$.

Next let $C = C_{G_0}(W_0/T)$ and assume first that C/W_0 is infinite. Since G/W_0 is minimax, 5.2.5 shows that C/W_0 has an infinite characteristic abelian subgroup C_0/W_0. We observe also that $C_0/T = (C_0 \cap (G/B)) \times W_0)/T$ is abelian.

Now consider the pair (C_0, T). For this we have

$$m(C_0/T) = m(C_0/W_0) + m(W_0/T) > m(W_0/T) = m(B/M),$$

which contradicts the maximality of $m(B/M)$ for the pair (G, M).

It therefore follows that C/W_0 is finite. But G_0/C is an irreducible soluble $\mathbb{Q}$-linear group, and as such it is abelian-by-finite by 3.1.8. Hence G_0 is finitely generated, metabelian-by-finite and so it has max $-n$ by 4.2.2. Since T is a locally finite $\mathbb{Z}G$-module, it follows that T is finite, our final contradiction. ∎

We mention that a subsequent paper of Kropholler (1986a) contains the following result, which is more powerful than 10.5.9.

10.5.10 *Let Q be a soluble group of type $(FP)_n$ which has finite abelian subgroup rank and let M be a locally finite $\mathbb{Z}G$-module. If $(M_i)_{i \in I}$ is the direct system of all submodules of finite total rank, then for all $n \geq 0$ the natural map*

$$\varinjlim H^n(Q, M_i) \to H^n(Q, M)$$

is surjective.

(Here a group G is said to be of type $(FP)_n$ if there is a projective $\mathbb{Z}G$-resolution $\mathbf{P} \twoheadrightarrow \mathbb{Z}$ such that P_i is a finitely generated $\mathbb{Z}G$-module for $i \leq n$.)

In conclusion, here is another application of Kropholler's Theorem.

10.5.11 *A finitely generated soluble group has the property that every finitely generated subgroup is a product of finitely many cyclic groups if and only if it is minimax.*

Proof By 10.5.1 it is enough to show that $W = \mathbb{Z}_p \, wr \, \mathbb{Z}$ is not a product of finitely many cyclic groups. Suppose that in fact W is a product of k cyclic subgroups. Let q be any prime and note that W has as an image the group $\bar{W} = \mathbb{Z}_p \, wr \, \mathbb{Z}_q$. Each element of $\bar{W}$ has order dividing pq, so it follows that $|\bar{W}| \leq (pq)^k$ and hence $p^q q \leq p^k q^k$. But this is clearly untenable for large q. ∎

One can also prove a version of 10.5.11 for finitely generated linear groups by using the Tits Alternative (see 3.1.2) since a free group of rank 2 is clearly not the product of finitely many cyclic groups.

Boundedly generated groups

A group is called *boundedly generated*, or BG, if it is the product of finitely many cyclic subgroups. We have seen that this property plays a crucial role in the proof of Kropholler's Theorem. The property BG has been studied recently in connection with subgroup growth problems. A group G is said to have *polynomial index growth*, or to be PIG, if there is a constant c such that

$$|G : G^n| \leq n^c$$

for all $n > 0$. It is clear that a group which is BG is PIG. A related property is that of *polynomial subgroup growth*, PSG. A group G is said to have PSG if there is a constant d such that

$$s_n(G) \leq n^d$$

for all $n > 0$, where $s_n(G)$ is the number of subgroups of G with index at most n. It is known that the implications below are strict for finitely generated residually finite groups:

$$\text{PSG} \Longrightarrow \text{BG} \Longrightarrow \text{PIG},$$

(Balog, Pyber and Mann 2000). It was shown by Lubotzky, Mann and Segal (1993) that *a finitely generated residually finite group has PSG if and only if it is a virtually soluble minimax group*. Now it is simple to prove that a finitely generated reduced soluble minimax group has PIG. Therefore the properties PSG, BG and PIG coincide for finitely generated residually finite soluble groups. For a detailed discussion of these properties the reader is referred to Dixon *et al.* (1999), Pyber (2002), and to Segal (1986).

It is an open question (due to J. C. Lennox) whether a soluble group that is BG need be minimax. Segal (1986) has obtained a partial result in this direction: *a soluble virtually residually nilpotent group which is a product of finitely many cyclic subgroups is a minimax group*.

11

FINITELY PRESENTED SOLUBLE GROUPS

For the group theorist with an interest in structure there are few properties more frustrating than finite presentability. By Higman's Embedding Theorem nothing can be said about the finitely generated subgroups of finitely presented groups apart from the existence of a recursive presentation. For finitely presentable soluble groups the situation is, if anything, worse since no version of Higman's Theorem is known for this class. Thus the nature of the subgroups of these groups remains mysterious. A further sign of the complexity of the class of finitely presented soluble groups is the fact that all the standard decision problems have a negative solution, as was observed in 9.1.

However, despite these pessimistic observations, it can be said that the work of several researchers, including G. Baumslag, R. Bieri, J. R. G. Groves and R. Strebel over the last 30 years has enormously increased our understanding of finitely presented soluble groups, especially in the metabelian case. The introduction of the Bieri–Strebel geometric invariant in 1978 was a key event in the development of the theory. Today the subject remains an attractive one, with many challenging open questions.

11.1 Some finitely presented and infinitely presented soluble groups

Our aim here is to notice some types of finitely generated soluble group which are finitely presented and also to record some significant examples of finitely generated infinitely presented soluble groups.

We begin with some elementary remarks. It was shown by P. Hall that the class of finitely presented groups is extension closed (P. Hall 1954). Since cyclic groups and finite groups are certainly finitely presented, we deduce at once:

1. *Every virtually polycyclic group is finitely presented.*

Examples of finitely presented soluble groups that are not polycyclic are not difficult to find.

2. *Let $G = \langle x, a \mid a^x = a^m \rangle$ where m is a non-zero integer. Then G is metabelian and it is polycyclic if and only if $m = \pm 1$.*

Let $m \neq \pm 1$ and write π for the set of prime divisors of m. Then it is easily seen that $a^G \simeq \mathbb{Q}_\pi$, the additive group of π-adic rationals, and that

$$G = \langle x \rangle \ltimes \mathbb{Q}_\pi,$$

where x has infinite order and induces in $\mathbb{Q}_\pi$ the automorphism $b \mapsto mb$. Notice that G is a minimax group, while it is polycyclic if and only if π is empty.

A notable feature of the group G is that it is an ascending HNN-extension with base group $\langle a \rangle$ and stable letter x. Indeed

$$\langle a \rangle < \langle a \rangle^{x^{-1}} < \langle a \rangle^{x^{-2}} < \cdots$$

and $a^G = \bigcup_{i=1,2,\ldots} \langle a \rangle^i$. We will see later that many finitely presented soluble groups arise from such HNN-extensions.

It can be a difficult task to prove that a given finitely generated soluble group is infinitely presented (i.e. not finitely presented). For this reason the next result is useful.

11.1.1 (P. Hall 1954) *Let G be a finitely generated group with a normal subgroup N such that G/N is finitely presented. Then N is finitely generated as a G-operator group.*

To prove this, one forms a surjective homomorphism $F \to G$, where F is a free group of finite rank, and takes the preimage M of N. By a result of B. H. Neumann (see Robinson 1996: 2.2.3) $G/N \simeq F/M$ can be finitely presented by any finite set of generators, and so 11.1.1 follows. This result will now be applied to prove:

3. *The finitely generated metabelian groups $\mathbb{Z} \operatorname{wr} \mathbb{Z}$ and $\mathbb{Z}_m \operatorname{wr} \mathbb{Z}$, $(m > 1)$, are infinitely presented.*

For the proof we refer to P. Hall's construction of finitely generated centre-by-metabelian groups in Chapter 4. By 4.1.3 there is a group $H = \langle \alpha \rangle \ltimes X$, where $X = \langle x_i \mid i \in \mathbb{Z} \rangle$ is nilpotent of class 2 and $x_i^\alpha = x_{i+1}$. Also $c_r = [x_i, x_{i+r}]$ is independent of i and $Z(H) = H' = \langle c_r \mid r = 1, 2, \ldots \rangle$. Clearly $H/Z(H) \simeq \mathbb{Z} \operatorname{wr} \mathbb{Z}$. Since $Z(H)$ is free abelian of infinite rank, $\mathbb{Z} \operatorname{wr} \mathbb{Z}$ cannot, by 11.1.1, be finitely presented.

If we adjoin to H the additional relations $x_i^m = 1 = c_r^m$ for $i \in \mathbb{Z}$, $r \in \mathbb{N}$, we obtain a group $\bar{H}$, which is an image of H. Now it is the case that $\bar{H}/Z(\bar{H}) \simeq \mathbb{Z}_m \operatorname{wr} \mathbb{Z}$ and $Z(\bar{H})$ is not finitely generated. Therefore $\mathbb{Z}_m \operatorname{wr} \mathbb{Z}$ is infinitely presented.

In fact wreath products are of very little use in constructing finitely presented groups, as the next result indicates.

11.1.2 (Baumslag 1960) *The standard wreath product $H \operatorname{wr} K$ is finitely presented if and only if either $H = 1$ and K is finitely presented or K is finite and H is finitely presented.*

Proof In the first place the conditions are clearly sufficient. Suppose that $H \operatorname{wr} K$ is finitely presented and that $H \neq 1$ and K is infinite. Set $F = H * K$, the free product. Then $H^F = H^K$ is the free product $\operatorname{Fr}_{k \in K} H^k$. Define

$$N = \langle [H^k, H^{k'}] \mid k \neq k' \in K \rangle^F.$$

Then clearly $F/N \simeq H \, wr \, K$ and consequently N is the normal closure in F of subgroups

$$[H^{k_i}, H^{k'_i}], \quad i = 1, 2, \dots, r,$$

for finitely many elements $k_i \neq k'_i$ in K. Since K is infinite, we may choose $\bar{k}$ in K such that $\bar{k} \notin \{k_i, k'_i \mid i = 1, 2, \dots, r\}$.

Consider the natural projection $\theta : H^F \to H * H^{\bar{k}}$ in which $H^k \to 1$ if $k \neq 1$ or $\bar{k}$. Then $[H^{k_i}, H^{k'_i}] \leq \mathrm{Ker}(\theta)$ for all i, whence $N \leq \mathrm{Ker}(\theta)$. However $[H, H^{\bar{k}}] \nleq \mathrm{Ker}(\theta)$, a contradiction. ∎

The following is a well-known example of a finitely generated metabelian group which is not finitely presented. Recall that $\mathbb{Q}_\pi$ denotes the ring of all π-adic rationals, $\{m/n \mid m, n \in \mathbb{Z}, n = \text{a } \pi - \text{number}\}$.

4. *Let $G = \langle t \rangle \ltimes A$, where $A = \mathbb{Q}_{\{2,3\}}$ and t induces the automorphism $a \mapsto \frac{3}{2}a$ in A. Then G is not finitely presented.*

The easiest way to prove this result is to apply 11.4.1 below: the details are given in 11.4.3. A noteworthy feature of this group is that its Schur multiplier is zero. Now it follows from Hopf's formula (see 10.1.2) that the multiplier of a finitely presented group is always finitely generated. In fact the last example shows that the converse statement is false, even for finitely generated metabelian groups.

5. Another source of finitely generated, infinite presented soluble groups is the free groups in a soluble variety. For example, there is the well-known result of Šmel'kin (1965):

11.1.3 *If F is a free soluble group of derived length $d > 1$, then F is infinitely presented.*

A wide generalization of Šmel'kin's Theorem has been given by Bieri and Strebel (1978). *Let $\mathbf{V}$ be a variety of soluble groups. Then every finitely generated free $\mathbf{V}$-group is infinitely presented unless all $\mathbf{V}$-groups are nilpotent-by-finite.*

6. *Abels' finitely presented soluble minimax group.*

A key example of a finitely presented soluble group was published by H. Abels in 1979. This group, which defeats a wide array of tempting conjectures, admits a very simple description.

Let p be any prime and define G to be the subgroup of $GL_4(\mathbb{Q}_p)$ consisting of all matrices of the form

$$\begin{bmatrix} 1 & * & * & * \\ 0 & p^i & * & * \\ 0 & 0 & p^j & * \\ 0 & 0 & 0 & 1 \end{bmatrix}.$$

Here i and j are any integers and $*$ denotes an arbitrary element of $\mathbb{Q}_p$. Since G is a subgroup of the triangular group $T_4(\mathbb{Q}_p)$, it is nilpotent-by-abelian (by 3.1.5). The main features of the group are listed in the next result.

11.1.4 (Abels 1979)

(i) G is finitely presented and nilpotent of class 3-by-abelian.
(ii) G is a torsion-free minimax group and it is residually finite.
(iii) $Z(G) \simeq \mathbb{Q}_p$, so G does not satisfy $\max -n$.
(iv) G has a non-hopfian[1] finitely presented quotient.

Proof Clearly G is torsion-free and minimax, and it is residually finite by 5.3.2. It is easy to see that $Z(G)$ consists of upper unitriangular matrices with arbitrary $(1,4)$ entries and other off-diagonal entries 0. If $1 \neq z \in Z(G)$, then $G/\langle z \rangle$ is finitely presented if G is, and $Z(G/\langle z \rangle) = Z(G)/\langle z \rangle$, which is of type p^∞. Clearly $G/\langle z \rangle$ is non-hopfian and also it is not residually finite.

All that remains to be done is to show that G is finitely presented. To this end we introduce the matrices

$$D_2 = \begin{bmatrix} 1 & 0 & 0 & 0 \\ 0 & p^{-1} & 0 & 0 \\ 0 & 0 & 1 & 0 \\ 0 & 0 & 0 & 1 \end{bmatrix} \quad \text{and} \quad D_3 = \begin{bmatrix} 1 & 0 & 0 & 0 \\ 0 & 1 & 0 & 0 \\ 0 & 0 & p^{-1} & 0 \\ 0 & 0 & 0 & 1 \end{bmatrix} :$$

also let E_{ij} be the 4×4 matrix with 1 as its (i,j) entry which has 1's on the diagonal and 0's elsewhere. One quickly verifies the relations

$$E_{12}^{D_2} = E_{12}^p, \quad E_{23}^{D_2} = E_{23}^p, \quad E_{34}^{D_2} = E_{34},$$

$$E_{12}^{D_3} = E_{12}, \quad E_{23}^{D_3} = E_{23}^p, \quad E_{34}^{D_3} = E_{34}^p,$$

and also that $[D_i, D_j] = 1$. In addition there are the standard relations in the E_{ij},

$$[E_{ij}, E_{jk}] = E_{ik} \quad \text{and} \quad [E_{rs}, E_{uv}] = 1$$

if $r \neq v$, $s \neq u$. From these relations it is clear that

$$G = \langle D_2, D_3, E_{12}, E_{23}, E_{34} \rangle.$$

Next we form the group $\bar{G}$ with generators d_2, d_3, e_{ij}, where $(i, j) = (1, 2)$, $(2, 3)$ or $(3, 4)$, subject to the above relations in d_2, d_3 and the e_{ij}, instead of D_2, D_3 and the E_{ij}. By Von Dyck's Theorem there is a surjective homomorphism $\theta : \bar{G} \to G$ such that $d_i^\theta = D_i$ and $e_{ij}^\theta = E_{ij}$.

Now it is easy to see that θ maps the subgroup $\bar{U} = \langle e_{12}, e_{23}, e_{34} \rangle$ isomorphic-ally to $U_4(\mathbb{Z})$ and that $V = \bar{U}^{\bar{G}} \simeq U_4(\mathbb{Q}_p)$. Also $\bar{G} = \langle d_2, d_3 \rangle \ltimes V$ is a torsion-free soluble minimax group. Since $h(\bar{G}) \leq 8 = h(G)$, it follows that $\mathrm{Ker}(\theta) = 1$ and $G \simeq \bar{G}$. ∎

7. So far all our examples of finitely presented soluble groups have been minimax groups. The best known example which is not minimax is due to Baumslag (1972) and Remeslennikov (1973a).

---

[1] A group is called *hopfian* if it is not isomorphic with a proper quotient group.

11.1.5 *Let G be the group with the presentation*

$$\langle a, s, t \mid a^t = aa^s, [a, a^s] = 1 = [s, t] \rangle.$$

Then G is a finitely presented metabelian group and G' is free abelian with countably infinite rank.

Proof We begin with the group $H = \mathbb{Z} \, wr \, \mathbb{Z}$, which has the infinite presentation

$$H = \langle a, s \mid [a, a^{s^i}] = 1, \, i = 1, 2, \ldots \rangle.$$

An injective endomorphism σ of H is defined by the rules

$$a^\sigma = aa^s \quad \text{and} \quad s^\sigma = s.$$

Thus we may form the HNN-extension

$$G = \langle t, H \mid h^\sigma = h^t, \, h \in H \rangle,$$

and plainly G has the presentation

$$\langle a, s, t \mid a^t = aa^s, [a, a^{s^i}] = 1, [s, t] = 1, i = 1, 2, \ldots \rangle.$$

Clearly G is finitely generated and metabelian, and in fact $G' = \langle a^{s^i} \mid i \in \mathbb{Z} \rangle$ is free abelian of infinite rank.

Actually the relations $[a, a^{s^i}] = 1$ in this presentation are redundant for $i > 1$. Indeed, assume that the equations $[a, a^{s^j}] = 1$, $j = 1, 2, \ldots, i$, follow from $[a, a^s] = 1$. Then we have

$$1 = [a, a^{s^i}]^t = [aa^s, (aa^s)^{s^i}] = [aa^s, a^{s^i} a^{s^{i+1}}] = [a, a^{s^{i+1}}].$$

Therefore G has the finite presentation

$$\langle a, s, t \mid a^t = aa^s, [a, a^s] = 1 = [s, t] \rangle.$$

∎

A noteworthy feature of the preceding proof is the embedding of the finitely generated, infinitely presented group $\mathbb{Z} \, wr \, \mathbb{Z}$ in the finitely presented group G. This foreshadows the theorem of Baumslag and Remeslennikov that any finitely generated metabelian group embeds in a finitely presented metabelian group, which is 11.3.1 below.

8. Up to this point all our examples of finitely presented soluble groups have been nilpotent-by-abelian-by-finite. The first example which is not of this type was given by Robinson and Strebel (1982):

11.1.6 *Let ℓ be an integer such that $|\ell| > 1$ and let G be the group with the presentation*

$$\langle a, s, t \mid a^t = aa^s, a^{[s,t]} = a^\ell, [a, a^s] = [s, t, s] = [s, t, t] = 1 \rangle.$$

Then G is abelian-by-(nilpotent of class 2), but it is not nilpotent-by-abelian-by-finite. Thus G is not linear over any field.

More general examples of this sort has been constructed by Strebel in an unpublished work (Strebel 1981*a*).

9. *One-relator groups.*

A *1-relator group* is a group with a presentation having a single defining relator. Such groups, which were first studied by Magnus in the 1930s, exhibit much better behaviour than finitely presented groups in general: for example, the word problem is known to be soluble (Magnus 1932).

In fact 1-relator groups are quite close to being free. One sign of this is a 'Tits Alternative' established by Moldavanskiĭ (1969) and Karrass and Solitar (1971): *a subgroup of a 1-relator group either contains a free subgroup of rank 2 or is soluble.* Now intuitively one would expect there to be few soluble 1-relator groups. After all it must be difficult for the solubility of a group to be a consequence of a single defining relator. Thus most 1-relator groups should have free subgroups of rank 2.

The examples of soluble 1-relator groups that come to mind are cyclic groups and the groups of example 2 above, $G(m) = \langle x, a \,|\, a^x = a^m \rangle$, where m is a non-zero integer: note that these include the free abelian group of rank 2. In fact there are no others, as we will briefly indicate. First note that a soluble 1-relator group must be finitely generated: for otherwise it would have a free group of infinite rank as a free factor.

It turns out that even the soluble subgroups of 1-relator groups are quite restricted. Indeed it follows from work of Moldavanskiĭ (1969) and B. B. Newman (1968, 1973) that *a soluble subgroup of a 1-relator group is cyclic, a rational group of type $\mathbb{Q}_\pi$ with π a finite set of primes, or one of the groups $G(m)$.* Therefore *the only soluble 1-relator groups are cyclic groups and the groups $G(m)$.* For a thorough treatment of 1-relator groups, the reader is referred to Lyndon and Schupp (1977).

Finally, we mention that soluble groups with deficiency 1 have been studied by Wilson 1996*a*; recall here that a finitely presented group has *deficiency equal to 1* if the maximum difference between the number of generators and the number of relators in a presentation is 1. It was shown that such groups are necessarily of the type $G(m)$.

11.2   Constructible soluble groups

The topic of this section is a class of finitely presented soluble groups which arises by the natural procedure of forming successive finite extensions and ascending *HNN*-extensions.

More precisely, a group G is said to be *constructible* if there is a finite sequence of groups

$$1 = G_0, G_1, \ldots, G_n = G$$

in which each G_i is a finite extension of the fundamental group of a finite graph of groups whose edge and vertex groups occur among $G_0, G_1, \ldots, G_{i-1}$. This amounts to requiring that G arise from the trivial group by forming success-ive finite extensions, generalized free products $H \underset{L}{*} K$, and HNN-extensions $\langle t, B \mid s^t = s^\sigma, s \in S \rangle$, where H, K, L, the base group B and the associated subgroups S, T of the HNN-extension have been previously constructed: here of course $\sigma : S \to T$ is an isomorphism. Note that the number of such operations must be finite.

As examples of constructible groups we immediately notice free groups and polycyclic-by-finite groups. While constructible groups can have a complex structure, they are always finitely presented.

11.2.1 *Every constructible group is finitely presented.*

Proof Assume that the constructible group G arises from a sequence of groups $1 = G_0, G_1, \ldots, G_n = G$, by means of finite extensions, generalized free products and HNN-extensions. By induction on n we may suppose that G_i is finitely presented for $i < n$. If G is a finite extension of G_{n-1}, then G is certainly finitely presented since the class of finitely presented groups is exten-sion closed. If $G = H \underset{L}{*} K$, where H, K, L are finitely presented, it is easy to write down a finite presentation of G from presentations of H, K, L. Like-wise, if $G = \langle t, B \mid s^t = s^\sigma, s \in S \rangle$ is an HNN-extension and B, S, T are finitely presented, we can write down a finite presentation of G.   ■

In a similar way it may be shown that every constructible group G has the stronger property $(FP)_\infty$, that is, there is a projective $\mathbb{Z}G$-resolution $\mathbf{P} \twoheadrightarrow \mathbb{Z}$ in which each module P_i is $\mathbb{Z}G$-finitely generated.

Since our brief is with soluble group theory, our interest lies with constructible soluble groups, which were first studied by Baumslag and Bieri (1976). Now it is quite hard for a constructible group to be soluble since non-cyclic free subgroups can easily occur during the construction. Thus we should expect to find structural restrictions on a soluble constructible group.

Consider a generalized free product $F = H \underset{L}{*} K$. Using standard results about subgroups of F, (see Lyndon and Schupp 1977), it may be shown that F has non-cyclic free subgroups unless $|H : L| \leq 2$ and $|K : L| \leq 2$: under these cir-cumstances $L \vartriangleleft F$ and F/L, being generated by two elements of order at most 2, is virtually abelian. Thus if L is virtually soluble, then so will F be.

In the case of an HNN-extension $E = \langle t, B \mid s^t = s^\sigma, s \in S \rangle$, where B is the base group and S, T are the associated subgroups, non-cyclic free subgroups will occur unless $B = S$ or T: (for HNN-extensions see Lyndon and Schupp 1977 and Serre 1980). If, say, $B = S$, then σ is an injective endomorphism of S and there is a chain of subgroups

$$S \leq S^{t^{-1}} \leq S^{t^{-2}} \leq \cdots .$$

Clearly $S^E = \bigcup_{i=1,2,\ldots} S^{t^{-i}}$ and of course $E = \langle t \rangle \ltimes S^E$. Such groups are called *ascending HNN-extensions*. Notice that, if S is virtually soluble, then so is E since $S \simeq S^{t^{-i}}$.

On the basis of these observations, we may formulate a *Tits Alternative* for constructible groups—cf. Tits's Theorem on linear groups 3.1.2.

11.2.2 *For any constructible group G, either G is virtually soluble or else it has a free subgroup of rank 2.*

Proof Let G arise from a sequence of groups $1 = G_0, G_1, \ldots, G_n = G$ and assume there are no non-cyclic free subgroups of G. The same is true of G_i for $i < n$, so by induction each of these is virtually soluble. The preceding discussion shows that G is also virtually soluble. ∎

A further application of these ideas leads to a simplified definition of constructible soluble groups.

11.2.3 (Baumslag and Bieri 1976) *The constructible soluble groups form the smallest class of groups which is closed with respect to forming extensions by finite soluble groups and ascending HNN-extensions with base group in the class.*

Proof Let G be a soluble constructible group. It is enough to show that G arises from the trivial group by finitely many finite extensions and ascending HNN-extensions. By the discussion above, any HNN-extension occurring in the construction of G is of ascending type. Thus we may reduce to the situation where $G = H \underset{L}{*} K$. Then, as we saw, $|H : L| \leq 2$ and $|K : L| \leq 2$. Here L is soluble and normal in G and the assertion is true for L. Also either G/L is a finite soluble group or else it is infinite dihedral. In the latter case G is a finite extension of a split extension of L by an infinite cyclic group, and so the result follows. ∎

Structure of constructible soluble groups

That constructible soluble groups form a very restricted class of finitely presented soluble groups is made clear by the following fundamental theorem.

11.2.4 (Baumslag and Bieri 1976) *Let G be a constructible soluble group. Then G is a finitely presented, reduced soluble minimax group. Thus G is residually finite and virtually torsion-free.*

Proof First we prove that G has finite abelian total rank (FATR) and that finite subgroups have bounded order. By 11.2.3 we may assume that G is an ascending HNN-extension $\langle t, S \mid s^t = s^\sigma, s \in S \rangle$, where σ is an injective endomorphism of S, and, by the induction hypothesis, the group S is a reduced soluble minimax group with all its finite subgroups of order at most d. Since S^G is the union of the chain of subgroups $S^{t^{-i}}$, it follows that S^G has FATR and finite subgroups of order at most d. Hence the same holds for G.

Since G is finitely generated, we may now deduce from 5.2.8 that G is a minimax group. Furthermore there can be no Prüfer subgroups of G since its finite subgroups have bounded order. Thus G is reduced and by 5.3.2 it is residually finite. Finally, G is virtually torsion-free by 5.2.5. ∎

The class of constructible soluble groups has some useful closure properties, including closure under homomorphic images.

11.2.5 (Baumslag and Bieri 1976) *The class of constructible soluble groups is closed with respect to forming*

(i) *subgroups of finite index,*
(ii) *extensions,*
(iii) *homomorphic images.*

Proof

(i) Let G be a constructible soluble group with H a subgroup of finite index. By 11.2.3 we may assume G to be an ascending HNN-extension $\langle t, S \mid s^t = s^\sigma, s \in S \rangle$, where the assertion is true for S. Since G/H_G is finite, there is a $k > 0$ such that $t^k \in H_G$. Thus H_G contains the subgroup $H_0 = \langle t^k, H_G \cap S \rangle$. Since $(H_G \cap S)^{\sigma^k} = (H_G \cap S)^{t^k} \leq H_G \cap S$, we recognize that H_0 is constructible: for S is constructible and $|S : H_G \cap S|$ is finite and the result is assumed to be true for S. But $|H : H_0|$ is finite, so H is constructible.

(ii) Let $N \triangleleft G$ and assume that N and G/N are constructible soluble groups. Thus G is soluble. By induction on the number of groups in a sequence leading to G/N (in the sense of 11.2.3), we may suppose G/N to be an ascending HNN-extension, say $\langle tN, V/N \rangle$, where V/N is constructible in fewer steps than G/N. By induction V is constructible. Also $V \geq V^t N = V^t$, so that $G = \langle t, V \rangle$ is an ascending HNN-extension and G is constructible.

(iii) Suppose that G is obtained in n steps from the trivial group by means of finite extensions and ascending HNN-extensions and let $N \triangleleft G$. We argue that G/N is constructible by induction on $n > 0$. Suppose first that G has a normal subgroup L of finite index which is constructible in $n - 1$ steps. Then by induction LN/N is constructible, whence so is G/N.

We may therefore assume that G is an ascending HNN-extension $\langle t, S \mid s^t = s^\sigma, s \in S \rangle$, where σ is an injective endomorphism of S and the latter is constructible in $n - 1$ steps. Conjugation by t in $\bar{S} = S/S \cap N$ induces an injective endomorphism, $\bar{\sigma}$ say. Thus we may form an ascending HNN-extension

$$\bar{G} = \left\langle \bar{t}, \bar{S} \mid (s(S \cap N))^{\bar{t}} = s^t(S \cap N), s \in S \right\rangle.$$

Now $\bar{S}$ is constructible by induction on n and consequently $\bar{G}$ is constructible.

Next assume for the moment that $N \leq S^G$. A homomorphism $\varphi : \bar{G} \to G/N$ is defined by the rules

$$\bar{t}^\varphi = tN \quad \text{and} \quad (s(S \cap N))^\varphi = sN, \quad s \in S.$$

This is a homomorphism because, if $s \in S$, we have

$$((s(S \cap N))^{\bar{t}})^\phi = (s^t(S \cap N))^\varphi = s^t N,$$

which equals

$$(sN)^t = ((s(S \cap N))^\varphi)^{\bar{t}^\varphi}.$$

Clearly φ is surjective and $\mathrm{Ker}(\varphi) \leq \bar{S}$. If $(s(S \cap N))^{\bar{t}^i} \in \mathrm{Ker}(\varphi)$, then $s^{t^i} N = N$ and $s \in S \cap N$. Thus φ is an isomorphism and G/N is constructible in this case. Since the argument certainly applies to the normal subgroup $S^G \cap N$, it follows that $G/(S^G \cap N)$ is constructible, which shows that we may assume $S^G \cap N = 1$.

Therefore we may suppose N to be infinite cyclic, so that $G/C_G(N)$ is finite. It is enough to show that $C_G(N)/N$ is constructible since by (i) $C_G(N)$ is constructible. So we may also assume that $N \leq Z(G)$.

Next $G/(S^G N)$ is finite, G/S^G is cyclic, and $N \not\leq S^G$. Hence $t^k \in S^G N$ for some $k > 0$, say $t^k = a\, s^{t^{-i}}$, where $a \in N$ and $s \in S$. Since $a \in Z(G)$, the element t^k normalizes $S^{t^{-i}}$, from which it follows that t^k normalizes S. This implies that σ is an *automorphism* of S and so $S \lhd G$. Hence G/N is a finite extension of SN/N, whence G/N is constructible. ∎

Corollary 11.2.6 *A constructible soluble group G satisfies* $\max -n$.

For, if $N \lhd G$, then G/N is constructible and hence is finitely presented. By 11.1.1 the normal subgroup N is finitely generated as a G-operator group, which shows that G satisfies $\max -n$.

Although constructible soluble groups are finitely presented, it is far from being the case that every finitely presented soluble group is constructible. Indeed the Baumslag–Remeslennikov group (see 11.1.5) has infinite rank, so it cannot be constructible. Also Abel's group G (see 11.1.4) has non-finitely generated centre, so it does not satisfy $\max -n$ and by 11.2.6 it cannot be constructible, even though it is minimax. In fact neither is G/G'' constructible (see 11.5.17 below), even though this is a finitely presented, metabelian minimax group (by 11.5.14).

Subgroups of constructible soluble groups

One property that is missing from 11.2.5 is subgroup closure. Now it is plain from 11.2.4 that a subgroup of a constructible soluble group is necessarily a reduced soluble minimax group. In fact there is nothing more that can be said of such subgroups, as is shown by the following interesting theorem of Kilsch (1978). It generalizes an earlier result of Baumslag and Bieri (1976), in which the group was assumed to be finitely generated.

11.2.7 *Every reduced soluble minimax group G can be embedded in a constructible soluble group.*

Proof We begin by applying a theorem of Wehrfritz (1974) to embed G in $GL_n(\mathbb{Q}_\pi)$ for some n and a finite set of primes π—see 5.1.8 in this connection. Let G act in the natural way on a free $\mathbb{Q}_\pi$-module V of rank n and form a $\mathbb{Q}_\pi G$-composition series in V, say

$$0 = V_0 \leq V_1 \leq \cdots \leq V_m = V.$$

Thus V_i/V_{i-1} is a $\mathbb{Q}_\pi G$-module on which G acts *rationally irreducibly*.

By 3.1.6 and the argument of 5.2.3, there is a normal subgroup T_i with finite index in G such that $T_i/C_G(V_i/V_{i-1})$ is finitely generated and abelian. Writing T for $T_1 \cap T_2 \cap \cdots \cap T_m$, we have G/T finite. Now G embeds in the wreath product $T\,wr\,(G/T)$, from which it follows quickly that it suffices to embed T in a constructible soluble group. On the basis of these remarks we may replace G by T, that is, we may suppose that each $G/C_G(V_i/V_{i-1})$ is finitely generated and abelian. Also, by refining the series of V_i's we may continue to assume the G-rational irreducibility of V_i/V_{i-1}.

Choosing a suitable $\mathbb{Q}_\pi$-basis of V, we represent each $g \in G$ by a block matrix over $\mathbb{Q}_\pi$,

$$\begin{bmatrix} A_1 & * & * & \cdots & * \\ 0 & A_2 & * & \cdots & * \\ \cdot & \cdot & \cdot & \cdot & \cdot \\ 0 & 0 & \cdot & \cdot & A_m \end{bmatrix},$$

where A_i represents the action of G on V_i/V_{i-1}. Since $G/C_G(V_i/V_{i-1})$ is abelian, we may identify each V_i/V_{i-1} with an additive subgroup of some algebraic number field F, in such a way that elements of G act on V_i/V_{i-1} via multiplication by non-zero elements of F.

It follows that we may take G to be a subgroup of the triangular group $T_n(R)$, where R is the integral closure of $\mathbb{Q}_\pi$ in F. Notice that the additive group R^+ is a free $\mathbb{Q}_\pi$-module of finite rank. Thus 11.2.7 will follow at once from the following.

11.2.8 *Let R be an integral domain such that R^+ is a torsion-free π-minimax group. Then $T_n(R)$ is a constructible soluble group.*

Proof In the first place $T_1(R)$ is isomorphic with $U(R)$, the group of units in the ring R. By a result of Baer (1968)—see also Robinson (1972b)—this group is finitely generated, so it is certainly constructible. Assume that $n > 1$ and proceed by induction on n.

By mapping a matrix onto the submatrix obtained by deleting row 1 and column 1, we obtain a surjective homomorphism $T_n(R) \to T_{n-1}(R)$ whose kernel K consists of all matrices of the form

$$\begin{bmatrix} a & b_1 & b_2 & \cdots & b_n \\ 0 & 1 & 0 & \cdots & 0 \\ \cdot & \cdot & \cdot & \cdots & \cdot \\ 0 & 0 & 0 & \cdots & 1 \end{bmatrix} \quad (*).$$

Therefore $T_n(R)/K$ is constructible by induction on n and by 11.2.5 it is enough to show that K is constructible.

Set $L = K \cap U_n(R)$, so that L consists of all unitriangular matrices of the form $(*)$. Thus L is a direct sum of $n - 1$ copies of R^+. Since R^+ is π-minimax, so is L. Hence there is a finitely generated subgroup L_0 such that L/L_0 is a π-group. Writing $\pi = \{p_1, p_2, \ldots, p_r\}$, we define P_i to be the $n \times n$ diagonal matrix with diagonal entries $p_i^{-1}, 1, 1, \ldots, 1$; thus $P_i \in K$. For any X in L we clearly have $P_i^{-1} X P_i = X^{p_i}$ and thus conjugation by P_i induces an injective endomorphism of L.

Our next move is to introduce a sequence of ascending HNN-extensions $L_1, L_2, \ldots, L_r$. These are defined recursively by

$$L_i = \langle t_i, L_{i-1} \mid a^{t_i} = a^{\sigma_i}, a \in L_{i-1} \rangle$$

where σ_i is the injective endomorphism of L_{i-1} given by $X^{\sigma_i} = X^{p_i}$ for $X \in L_0$ and $t_j^{\sigma_i} = t_j$ for $j = 1, 2, \ldots, r-1$. Since L_0 is constructible, so is L_r. Notice also that L_r is soluble.

Next, the assignments $a \mapsto a$, and $t_i \mapsto X_i$, $(a \in L_0, i = 1, 2, \ldots, r)$, determine a surjective homomorphism from L_r to the group $M = \langle L_0, P_1, \ldots, P_r \rangle$. Since L_r is a constructible soluble group, M is constructible by 11.2.5. In addition we have $K' \leq L$, and also $L \leq M$. For, if $\ell \in L$, then $\ell^m \in L_0 \leq M$ for some positive π-number m. Since conjugation in L by P_i sends an element to its p_ith power and $\ell^m = \ell^{p_1^{s_1} \cdots p_r^{s_r}}$ for certain integers s_j, we have $\ell \in M$ since $P_j \in M$.

It follows that $M \lhd K$. Now K/L is isomorphic with a subgroup of $U(R)$, so it is finitely generated and thus K/M is too. Therefore K is constructible and the proof is complete. ∎

By a similar argument Kilsch (1978) proved the following result. *Let G be a reduced metabelian minimax group with an abelian normal subgroup A such that G/A is finitely generated abelian. Then G embeds in a constructible metabelian group.* In particular this implies a result of Boler (1976b): *every finitely generated metabelian minimax group embeds in a finitely presented metabelian minimax group.*

On the other hand, not every finitely generated soluble minimax group can be embedded in a finitely presented soluble group. This is because there are uncountably many non-isomorphic finitely generated soluble minimax groups—see 5.3.15 and the subsequent remarks. These cannot all have recursive presentations, whereas a finitely generated subgroup of a finitely presented group must have a recursive presentation. Finally, note that there are soluble minimax groups that cannot be embedded in even a finitely generated soluble minimax group: for there are soluble minimax groups which are not nilpotent-by-polycyclic—see the example following 5.2.3.

11.3 Embedding in finitely presented metabelian groups

Although the finitely generated metabelian groups that are finitely presented might seem to be very special, their finitely generated subgroups can be complicated and far from finitely presented. For example, the Baumslag–Remeslennikov group of 11.1.5 contains $\mathbb{Z} \, wr \, \mathbb{Z}$ as a subgroup. In fact it is a remarkable result of G. Baumslag (1973b) and Remeslennikov (1973a) that any finitely generated metabelian group can be realized as a subgroup of some finitely presented metabelian group. Our main object is to prove this theorem.

11.3.1 *Every finitely generated metabelian group can be embedded in a finitely presented metabelian group.*

This result should be compared with Kilsch's theorem (11.2.7) and also, of course, with the Higman Embedding Theorem.

An important reduction in the proof can be made with the aid of the well-known *Magnus embedding* (Magnus 1939).

11.3.2 *Let F be the free group on a set $\{x_i \mid i \in I\}$ and let R be a normal subgroup of F. Let there be given an isomorphism from F/R to a group H in which $x_i R \mapsto h_i$, $(i \in I)$. If A is the free abelian group on a set $\{a_i \mid i \in I\}$, then the assignment $x_i R' \mapsto h_i a_i$ determines an embedding of F/R' into the wreath product $A \operatorname{wr} H$.*

The Magnus Embedding will now be used to establish:

11.3.3 *Let G be a finitely generated metabelian group. Then G can be embedded in a split extension $G_{ab} \ltimes A$, where A is a finitely generated $\mathbb{Z}G_{ab}$-module.*

Proof Let $G = \langle g_1, \ldots, g_n \rangle$ and denote by F the free group on a set $\{x_1, \ldots, x_n\}$. The assignment $x_i \mapsto g_i$ determines a surjective homomorphism $\theta : F \to G$. If K denotes the kernel of θ, we have $F'' \leq K$ since G is metabelian. Let R be the preimage of G' under θ; thus $R = F'K$ and $R' \leq K$.

Let A_0 be the free abelian group on a set $\{a_1, \ldots, a_n\}$. Then by 11.3.2, applied with $H = G_{ab}$, the assignment $x_i R' \mapsto (g_i G')a_i$ determines an embedding

$$\psi : F/R' \to W = A \operatorname{wr} G_{ab}.$$

Notice here that W is a finitely generated metabelian group. Put $N = (K/R')^{\psi}$. By definition of ψ we see that N is contained in B, the base group of W. Also $N \triangleleft \operatorname{Im}(\psi)$, and the latter contains all the elements $(g_i G')a_i$. It follows that N is normalized by each $g_i G'$. Since W is generated by B, which is abelian, and the $g_i G'$, we conclude that $N \triangleleft W$.

Finally, $G \simeq (F/R')/(K/R')$, so that ψ induces an embedding of G in W/N. Also $W/N = G_{ab} \ltimes (B/N)$ and $A = B/N$ is a finitely generated $\mathbb{Z}G_{ab}$-module. ∎

With the aid of 11.3.3 we immediately reduce the proof of 11.3.1 to the case of a finitely generated metabelian group of the form

$$G = H \ltimes A,$$

where H and A are abelian. The problem is to embed G in a finitely presented metabelian group, and here we may assume H to be infinite, since otherwise G is polycyclic and therefore finitely presented.

The idea behind the proof is to embed G in a sequence of ascending HNN-extensions, culminating in a finitely presented metabelian group. In constructing these HNN-extensions we shall find the following result useful.

11.3.4 *Let H be a finitely generated abelian group and A a finitely generated $\mathbb{Z}H$-module. If h is an element of H, there is a polynomial*

$p = 1 + c_1 x + \cdots + c_{r-1} x^{r-1} + x^r \in \mathbb{Z}[x]$ *such that the assignment* $a \mapsto ap(h)$ *determines an injective* $\mathbb{Z}H$-*endomorphism of* A.

Proof For the purposes of this proof let us call a polynomial of the specified form *special*. Let A_0 be the set of all a in A such that $a\,p(h) = 0$ for some special polynomial p. Then A_0 is a $\mathbb{Z}H$-submodule of A because the product of two special polynomials is clearly special.

Next A_0 is finitely generated as a $\mathbb{Z}H$-module, say by $b_1, \ldots, b_s$, since A satisfies $\max - \mathbb{Z}H$ by 4.2.3. Then $b_i p_i(h) = 0$ for some special polynomial p_i. Now define

$$p = x p_1 \cdots p_s + 1,$$

which is plainly a special polynomial. Suppose that $ap(h) = 0$ for some a in A. Then $a \in A_0$, so that $a = \sum_{i=1}^{s} b_i f_i$, where $f_i \in \mathbb{Z}H$. Now $b_i p(h) = b_i h p_1(h) \cdots p_s(h) + b_i = b_i$. Therefore

$$0 = ap(h) = \sum_{i=1}^{s} b_i f_i p(h) = \sum_{i=1}^{s} b_i f_i = a,$$

and the proof is complete. ∎

Proof of 11.3.1 We assume that $G = H \ltimes A$ where H and A are abelian and G is finitely generated. Recall that H may be assumed infinite and write

$$H = \langle h_1 \rangle \times \langle h_2 \rangle \times \cdots \times \langle h_r \rangle \times \cdots \times \langle h_n \rangle,$$

where $h_1, h_2, \ldots, h_r$ have infinite order and $h_{r+1}, \ldots, h_n$ have finite orders $q_{r+1}, \ldots, q_n$. By 11.3.4 for $i = 1, 2, \ldots, r$ there exist polynomials

$$p_i = 1 + c_{i1} x + \cdots + c_{id_i - 1} x^{d_i - 1} + x^{d_i}$$

in $\mathbb{Z}[x]$ such that $a \mapsto a\,p_i(h_i)$ is an injective $\mathbb{Z}H$-endomorphism of A, say τ_i. These endomorphisms will now be used to construct a sequence of HNN-extensions.

Let $G_0 = G = H \ltimes A$. Since H is abelian, τ_1 can be extended to an injective endomorphism of G_0, also denoted by τ_1, such that τ_1 acts as the identity on H. Hence there is an ascending HNN-extension

$$G_1 = \left\langle t_1, G_0 \mid g_0^{t_1} = g_0^{\tau_1}, \ g_0 \in G_0 \right\rangle.$$

The next step is to extend τ_1 from G_0 to an injective endomorphism τ_2 of G_1, by requiring it to act like the identity on the abelian subgroup $\langle t_1, H \rangle$. Now form the ascending HNN-extension

$$G_2 = \left\langle t_2, G_1 \mid g_1^{t_2} = g_1^{\tau_2}, \ g_1 \in G_1 \right\rangle.$$

Continuing in this manner r times, we obtain a group $\bar{G} = G_r$ such that

$$\bar{G} = Q \ltimes \bar{A},$$

where $Q = H \times \langle t_1 \rangle \times \cdots \times \langle t_r \rangle$ and $\bar{A} = A^{\langle t_1, \dots, t_r \rangle}$. Let $a_1, \dots, a_m$ be finitely many $\mathbb{Z}H$-module generators for A. Then $\bar{G}$ is clearly generated by the elements

$$h_1, \dots, h_n, \qquad t_1, \dots, t_r, \qquad a_1, \dots, a_m.$$

Furthermore, G embeds in $\bar{G}$ and $\bar{G}$ is metabelian. Our remaining task is to find a finite presentation of $\bar{G}$.

We begin by writing down relations which we see to hold in $\bar{G}$. As before $H = \langle h_1, \dots, h_n \rangle$ and A is the $\mathbb{Z}H$-module generated by $a_1, \dots, a_m$: first of all

$$[h_i, h_j] = [t_i, t_j] = [h_i, t_j] = [a_i, a_j] = 1$$

for all relevant i and j: also

$$h_i^{q_i} = 1, \quad (i = r+1, \dots, n),$$

and

$$a_i^{t_j} = a_i^{p_j(h_j)}, \quad (i = 1, 2, \dots, m, j = 1, 2, \dots, r).$$

Of course the module A is being written multiplicatively here. Now the $\mathbb{Z}H$-module A has a finite $\mathbb{Z}H$-presentation in the a_j since $\mathbb{Z}H$ is noetherian, let us say with relations

$$a_1^{r_{i1}} a_2^{r_{i2}} \cdots a_m^{r_{im}} = 1, \quad i = 1, 2, \dots, k,$$

where $r_{ij} \in \mathbb{Z}H$.

Finally, we need relations which will ensure that the normal closure of $\langle a_1, \dots, a_m \rangle$ in $\bar{G}$ is abelian. It will in fact be sufficient to adjoin the following *finite* set of relations:

$$[a_i^v, a_j^w] = 1,$$

where v and w are elements of the form $h_1^{u_1} \cdots h_n^{u_n}$, with $0 \le u_i \le d_i$ for $1 \le i \le r$ and $0 \le u_i < q_i$ for $r + 1 \le i \le n$. Recall here that $d_i = \deg(p_i)$ and q_i is the order of h_i for $r + 1 \le i \le n$.

The next step is to form the group G^* generated by elements $h_1, \dots, h_n$, $t_1, \dots, t_r, a_1, \dots, a_m$ with the relations listed above as defining relations. Certainly G^* is finitely presented and there is a surjective homomorphism $G^* \to \bar{G}$, since all the defining relations of G^* hold in $\bar{G}$.

It will be sufficient to prove that G^* is metabelian. For then G^* will satisfy $\max -n$, so that $\bar{G}$, as a quotient of G^* by the normal closure of a finite subset, will be finitely presented. Let $A^* = \langle a_1, a_2, \dots, a_m \rangle^{G^*}$: clearly it suffices to prove A^* to be abelian. Let $A_0^* = \langle a_1, a_2, \dots, a_m \rangle^H$. Then A^* is the union of a chain of subgroups of the form

$$(A_0^*)^{t_1^{-\ell_1} \cdots t_r^{-\ell_r}} \simeq A_0^*, \quad \ell_i \ge 0.$$

Therefore, what really needs to be established is that A_0^* is abelian. For this purpose a simple lemma is helpful.

11.3.5 *Let a, b, h, t be elements of a group such that $[a^{h^i}, b^{h^j}] = 1$ if $0 \le i, j \le d$, (for some fixed $d > 0$), and $[h, t] = 1$. Assume in addition that $a^t = a^p$ and*

$b^t = b^p$ *where* $p = 1 + c_1 h + \cdots + c_{d-1} h^{d-1} + h^d \in \mathbb{Z} \langle h \rangle$. *Then* $[a^{h^i}, b^{h^j}] = 1$ *for all integers i and j.*

Proof We argue by induction on n that $[a^{h^i}, b^{h^j}] = 1$ if $0 \leq i, j \leq n + d$. Of course this is true if $n = 0$. Assume that $n > 0$ and the statement is valid for all integers $m < n$. Since $n < n + d$, we have $[a, b^{h^n}] = 1$ by induction. Therefore

$$1 = [a, b^{h^n}]^t = [a^t, (b^t)^{h^n}],$$

which equals

$$[a^{1 + c_1 h + \cdots + c_{d-1} h^{d-1} + h^d}, \; b^{h^n + c_1 h^{n+1} + \cdots + c_{d-1} h^{n+d-1} + h^{n+d}}] = [a, b^{h^{n+d}}],$$

by the induction hypothesis. In the same way it can be shown that $[a^{h^{n+d}}, b] = 1$. Consequently the statement is true for n ∎

Proof of 11.3.1 (concluded) We resume the previous notation of the proof. As was observed above, it is sufficient to show that A_0^* is abelian, that is, that

$$[a_u^{h_1^{\ell_1} \cdots h_n^{\ell_n}}, a_v^{h_1^{m_1} \cdots h_n^{m_n}}] = 1, \quad 1 \leq u, v \leq m,$$

for all integers ℓ_i, m_i, where $0 \leq \ell_i, m_i < q_i$ if $i = r + 1, \ldots, n$. This is of course true if $0 \leq \ell_i, m_i \leq d$ for all $i = 1, 2, \ldots, r$, since these are defining relations.

Assume now that $0 \leq \ell_i, m_i \leq d_i$ for $i = 1, 2, \ldots, r - 1$, but ℓ_r and m_r are arbitrary. Put

$$\bar{a}_u = a^{h_1^{\ell_1} \cdots h_{r-1}^{\ell_{r-1}} h_{r+1}^{\ell_{r+1}} \cdots h_n^{\ell_n}} \quad \text{and} \quad \bar{a}_v = a_v^{h_1^{m_1} \cdots h_{r-1}^{m_{r-1}} h_{r+1}^{m_{r+1}} \cdots h_n^{m_n}}.$$

Then

$$\left[\bar{a}_u^{h_r^i}, \bar{a}_v^{h_r^j} \right] = 1 \quad \text{for } 0 \leq i, j \leq d_r.$$

Therefore 11.3.4 may be applied to show that

$$\left[\bar{a}_u^{h_r^i}, \bar{a}_v^{h_r^j} \right] = 1 \text{ for all integers } i \text{ and } j.$$

The preceding argument shows that the commutator relation is valid when ℓ_r and m_r are arbitrary integers. In the same we may prove that the relation is valid when ℓ_i and m_i are arbitrary for $i = 1, 2, \ldots, r - 1$. Hence A_0^* is abelian and the proof of Baumslag's Theorem is complete. ∎

It is tempting to try to extend 11.3.1 to other varieties of soluble groups, but this is not always possible. Indeed Groves (1978a) has shown that a finitely presented centre-by-metabelian group is abelian-by-polycyclic and hence by 7.2.1 satisfies $\max -n$: for this result see 11.5.10 below. Now there exist finitely generated centre-by-metabelian groups which have infinitely generated centres, and these do not satisfy $\max -n$: for example, Hall's construction in 4.3 yields such groups. These groups cannot be embedded in finitely presented centre-by-metabelian groups.

The converse of Grove's Theorem is true by a result of Strebel (1983).

11.3.6 *A finitely generated centre-by-metabelian group which is abelian-by-polycyclic can be embedded in a finitely presented centre-by-metabelian group.*

The general problem of embedding finitely generated soluble groups in finitely presented soluble groups is very difficult. For this to be possible the group must at least have a finitely generated recursive presentation: but no other necessary conditions are known.

Finally we mention a deep theorem of Brookes, Roseblade and Wilson (1997).

11.3.7 *A finitely presented abelian-by-polycyclic group is virtually metanilpotent.*

Now there are many finitely generated abelian-by-polycyclic groups which are not virtually metanilpotent, for example, $\mathbb{Z} \, wr \, G$ where G is polycyclic but not virtually nilpotent. Such groups cannot be embedded in a finitely presented abelian-by-polycyclic group, although they might conceivably be embeddable in finitely presented soluble groups.

Another interesting restrictive property due to Groves and Wilson (1994) asserts that *a finitely presented metanilpotent 2-generator group is nilpotent-by-(nilpotent of class at most 2)-by-finite.*

11.4   Structural properties of finitely presented soluble groups

As has been observed, it is rarely possible to deduce structural properties from a finite presentation of a group. Thus the following fundamental result, although elementary, is quite remarkable.

11.4.1 (Bieri and Strebel 1978) *Suppose that G is a finitely presented soluble group with a normal subgroup N such that G/N is infinite cyclic. Then G is an ascending HNN-extension with a finitely generated base group which is contained in N*

First of all it should be observed that any finitely presented infinite soluble group G has a subgroup G_0 of finite index, which has an infinite cyclic quotient: moreover G_0 is finitely presented. For example, we could take G_0 to be the smallest term of the derived series of G which has finite index. Thus 11.4.1 is of wide applicability in the study of finitely presented soluble groups.

It will be a simple matter to deduce 11.4.1 from the next result.

11.4.2 *Let G be a finitely presented group with a normal subgroup N such that $G/N = \langle gN \rangle$ is infinite cyclic. Then there exist finitely generated subgroups B, S, T of N, with $S \leq B$ and $T \leq B$, such that conjugation by g induces an isomorphism $\tau : S \to T$. Furthermore, if $\bar{G}$ is the corresponding HNN-extension $\langle t, B \mid s^t = s^\tau, s \in S \rangle$, then the assignment $t \mapsto g$ induces an isomorphism $\psi : \bar{G} \to G$.*

Thus the theorem identifies G as an *HNN*-extension with finitely generated base group and associated subgroups.

Proof of 11.4.2 Since G is finitely generated and $G = \langle g \rangle \ltimes N$, there exist elements $b_1, \ldots, b_n$ of N such that $G = \langle g, b_1, \ldots, b_n \rangle$. Let F be the free group on a set $\{x, a_1, \ldots, a_n\}$ and let

$$\pi : F \to G$$

be the surjective homomorphism such that $x^\pi = g$ and $a_i^\pi = b_i$, $(i = 1, 2, \ldots, n)$. If $R = \mathrm{Ker}(\pi)$, then $R \leq \langle a_1, \ldots, a_n \rangle^F$, since gN has infinite order.

Since G is finitely presented, R is the normal closure in F of some finite subset $\{r_1, \ldots, r_k\}$. Replacing r_i by a conjugate if necessary, we may assume that

$$r_j \in \langle a_i^{x^\ell} \, | \, 0 \leq \ell \leq m, 1 \leq i \leq n \rangle$$

for $j = 1, 2, \ldots, k$, where m is some positive integer. Next define

$$S = \langle b_i^{g^\ell} \, | \, 0 \leq \ell \leq m - 1, 1 \leq i \leq n \rangle$$

and put $T = S^g$: in addition write

$$B = \langle S, b_i^{g^m} \, | \, 1 \leq i \leq n \rangle.$$

Here B, S, and T are finitely generated subgroups of N and $\langle S, T \rangle \leq B$.

Conjugation in S by g yields an isomorphism $\tau : S \to T$ and this enables us to form the HNN-extension

$$\bar{G} = \langle t, B \, | \, s^t = s^\tau, \, s \in S \rangle.$$

The inclusion map $B \hookrightarrow G$ and $t \mapsto g$ determine a homomorphism

$$\psi : \bar{G} \to G,$$

which is surjective because g and the b_i generate G.

Next the assignments $x \mapsto t$ and $a_i \mapsto b_i$ yield a surjective homomorphism

$$\chi : F \to \bar{G}.$$

We claim that $b_i^{t^\ell} = b_i^{g^\ell}$ where $0 \leq \ell \leq m$. Now this is obviously true if $\ell = 0$: assuming that $\ell > 0$ and the assertion holds for $\ell - 1$, we have $b_i^{t^{\ell-1}} = b_i^{g^{\ell-1}}$. Therefore

$$b_i^{t^\ell} = \left(b_i^{g^{\ell-1}} \right)^t = \left(b_i^{g^{\ell-1}} \right)^\tau = \left(b_i^{g^{\ell-1}} \right)^g = b_i^{g^\ell},$$

since $b_i^{g^{\ell-1}} \in S$. From this result we deduce that for $j = 1, \cdots, n$

$$\left(r_j \left(a_i^{x^\ell} \right) \right)^\chi = r_j \left(b_i^{t^\ell} \right) = r_j \left(b_i^{g^\ell} \right) = 1$$

in the group $\bar{G}$. Hence $R^\chi = 1$ and $R \leq \mathrm{Ker}(\chi)$.

It follows directly from the definitions of the mappings that $\pi = \chi\psi$, which implies that $\mathrm{Ker}(\chi) \leq \mathrm{Ker}(\pi) = R$. Therefore $\mathrm{Ker}(\chi) = R$ and we have $G \simeq F/R \simeq \bar{G}$, as required. ∎

Proof of 11.4.1 By 11.4.2 the group G is an HNN-extension with finitely generated base group B and associated subgroups S and T, all contained in B. Now

G is soluble, so it cannot have a non-cyclic free subgroup. As was observed in 11.2, this means that $B = S$ or T, and G is an ascending HNN-extension. ∎

Here is a simple application of this powerful theorem.

11.4.3 *Let r be a non-zero rational number and let π denote the (possibly empty) set of primes involved in r. Define $G = \langle t \rangle \ltimes \mathbb{Q}_\pi$, where $\mathbb{Q}_\pi$ is the additive group of π-adic rationals, $\langle t \rangle$ is infinite cyclic and t induces $a \mapsto ra$ in $\mathbb{Q}_\pi$. Then the finitely generated metabelian group G is finitely presented if and only if r or r^{-1} is an integer.*

Proof It is easy to see that $G = \langle t, 1 \rangle$. Suppose that G is finitely presented. Apply 11.4.1 with $N = \mathbb{Q}_\pi$ to conclude that multiplication by r or r^{-1} is an injective endomorphism of some finitely generated, non-trivial subgroup of $\mathbb{Q}_\pi$, and hence of $\mathbb{Z}$. But this can only be true if r or $r^{-1} \in \mathbb{Z}$. The converse is clear. ∎

We remark that the multiplier of G can be shown to be 0 if $r \neq 1$. For example, suppose that $r = \frac{3}{2}$, so that $\pi = \{2, 3\}$: in this case the group G is not finitely presented (see Section 11.1, example 4 in this connection).

Application to coherent soluble groups

A group is termed *coherent* if every finitely generated subgroup is finitely presented. Obvious examples of coherent groups are free groups, locally finite groups and virtually polycyclic groups. On the other hand, the Baumslag–Remeslennikov group (example 7 in Section 11.1) is a finitely presented metabelian group which has $\mathbb{Z}\, wr\, \mathbb{Z}$ as a subgroup, so it is not coherent.

We will apply 11.4.1 to establish the following very satisfying characterization of finitely generated coherent soluble groups.

11.4.4 (Groves 1978*b*; Bieri and Strebel 1979*b*) *Let G be a finitely generated soluble group. Then the following statements about G are equivalent:*

(i) *G is coherent;*
(ii) *every finitely generated subgroup of G is constructible;*
(iii) *G is either polycyclic or an ascending HNN-extension with polycyclic base group.*

Proof In the first place (ii) implies (i) since constructible groups are finitely presented. In addition (iii) implies (ii). To see this, suppose that G is an ascending HNN-extension $\langle t, B \rangle$, where B is polycyclic, and let H be a finitely generated subgroup of G. Writing $\bar{B}$ for $B^{\langle t \rangle}$, we see that either $H \leq \bar{B}$, and thus H is polycyclic, or $H/(H \cap \bar{B})$ is infinite cyclic. In the latter case 11.4.1 shows that H is an ascending HNN-extension of a finitely generated, and hence polycyclic, subgroup of $\bar{B}$, so that H is finitely presented. Therefore G is coherent.

Thus it remains only to show that (i) implies (iii). From now on assume that G is coherent. There are four steps in the proof, the first being:

(a) *G is a minimax group.* Certainly we can assume that G is infinite, so it has a normal subgroup H of finite index which has an infinite cyclic quotient H/N_1. By 11.4.1 the group H is an ascending HNN-extension with finitely generated base group S_1 contained in N_1. Since $H' \le N_1$, we have $h(S_1 H'/H') < h(H_{ab})$. Suppose that $S_1 H'/H'$ is infinite; then there exists a subgroup N_2 such that $H' \cap S_1 \le N_2 < S_1$ and S_1/N_2 is infinite cyclic. Since G is coherent, S_1 is finitely presented and 11.4.1 may be applied to show that S_1 is an ascending HNN-extension with finitely generated base group $S_2 \le N_2$. Thus

$$h(S_2 H'/H') < h(S_1 H'/H') < h(H_{ab}).$$

This argument may be applied repeatedly. After at most $h(H_{ab})$ applications, an integer r is obtained such that $S_r H'/H'$ is finite. Now $S_r \cap H'$ is finitely generated and hence is coherent, so by induction on the derived length of G the group $S_r \cap H'$ is minimax. Therefore S_r is minimax.

By construction $S_{r-1} = \langle t_{r-1}, S_r \rangle$ is an ascending HNN-extension with base group S_r, and $S_r^{\langle t_{r-1} \rangle}$ is the union of an ascending chain of conjugates of S_r. Consequently $S_r^{\langle t_{r-1} \rangle}$ is a soluble FATR-group, as will be S_{r-1}. Working back up to H, we may argue that H, and hence G, is a soluble FATR-group. But G is finitely generated, so it is minimax by 5.2.8, and thus (a) is proven.

We may now apply 5.2.2 to show that G has a nilpotent normal subgroup N such that $Q = G/N$ is abelian-by-finite. We may assume Q to be infinite since otherwise G is polycyclic. Let

$$V = N_{ab} \otimes_{\mathbb{Z}} \mathbb{Q},$$

which is a finite dimensional $\mathbb{Q}Q$-module.

Each q in Q induces a linear operator q' in V. Let f_q denote the characteristic polynomial of q'. Our next objective is to establish:

(b) *For each q in Q either f_q or $f_{q^{-1}}$ has integer coefficients.* Choose elements $b_1, b_2, \ldots, b_n$ of N such that $\{b_1 N' \otimes 1, b_2 N' \otimes 1, \ldots, b_n N' \otimes 1\}$ is a $\mathbb{Q}$-basis of V. Let $q \in Q$ and write $q = gN$. If q has finite order m, then f_q is a product of cyclotomic polynomials with orders dividing m. Thus f_q certainly has integer coefficients.

Now suppose that q has infinite order. The group $L = \langle g, b_1, \ldots, b_n \rangle$ is finitely presented and L/M is infinite cyclic, where $M = \langle b_1, \ldots, b_n \rangle^{\langle g \rangle}$. By 11.4.1 we may assume that L is an ascending HNN-extension, with stable letter g or g^{-1}, whose base group is a finitely generated subgroup U of M. Thus M is the union of an ascending chain of conjugates, namely $U^{g^{-i}}$ or U^{g^i}, $i = 1, 2, \ldots$.

It follows from the last paragraph that either g or g^{-1} induces an automorphism in $M_{ab} \otimes_{\mathbb{Z}} \mathbb{Q}$ whose characteristic polynomial has integral coefficients. Now the natural map $M_{ab} \to N_{ab}$ has torsion kernel and cokernel. Thus, on applying the functor $- \otimes_{\mathbb{Z}} \mathbb{Q}$, we obtain a $\mathbb{Z}\langle g \rangle$-isomorphism

$$M_{ab} \otimes_{\mathbb{Z}} \mathbb{Q} \to N_{ab} \otimes_{\mathbb{Z}} \mathbb{Q} = V.$$

Hence (b) follows.

In the next part of the proof let $\mathbb{Q}^\dagger$ denote the multiplicative group of positive rationals: recall that q' is the linear operator induced by q in V. Let $\theta : Q \to \mathbb{Q}^\dagger$ be the homomorphism defined by

$$q^\theta = |\det(q')|.$$

(c) $\mathrm{Im}(\theta)$ *is a cyclic subgroup of* $\mathbb{Q}^\dagger$. If $q \in Q$, then f_q or $f_{q^{-1}}$ has integer coefficients by (b): therefore q^θ or $(q^\theta)^{-1}$ is an integer. Assuming that $H = \mathrm{Im}(\theta) \neq 1$, we may write

$$H = \left\{ 1, \ell_1, \ell_2, \dots, \frac{1}{\ell_1}, \frac{1}{\ell_2}, \dots \right\}$$

where $1 < \ell_1 < \ell_2 < \cdots$ and ℓ_i is an integer. Assume we have shown that ℓ_i is a power of ℓ_1 for all $i < j$. Since $1 < \ell_j/(\ell_{j-1}) < \ell_j$ and $\ell_j/(\ell_{j-1}) \in H$, we conclude that $\ell_j/(\ell_{j-1})$ must equal one of $\ell_1, \dots, \ell_{j-1}$. Hence ℓ_j is a power of ℓ_1. It now follows by induction on j that $H = \langle \ell_1 \rangle$.

(d) *If* $\mathrm{Im}(\theta) = 1$, *then* G *is polycyclic.* Choose elements $q_1, \dots, q_r$ of Q which freely generate a free abelian subgroup F of finite index in Q. By (b) we may suppose that each $f_i = f_{q_i}$ has integer coefficients. Let $\bar{N}$ denote the quotient of N_{ab} by its torsion subgroup.

Consider a cyclic $\mathbb{Z}F$-submodule $\bar{N}_0$ of $\bar{N}$. Now $f_i(q_i) \in \mathbb{Z}F$ annihilates $\bar{N}$ since $f_i(q_i') = 0$. Consequently $\bar{N}_0 \overset{\mathbb{Z}F}{\simeq} \mathbb{Z}F/I$, where I is an ideal containing $f_i(q_i)$ for $i = 1, 2, \dots, r$. Put $A_i = \mathbb{Z}\langle q_i \rangle /(f_i(q_i))$; then the canonical isomorphism

$$\mathbb{Z}\langle q_1 \rangle \otimes_{\mathbb{Z}} \cdots \otimes_{\mathbb{Z}} \mathbb{Z}\langle q_r \rangle \to \mathbb{Z}F$$

induces a surjective homomorphism

$$A_1 \otimes_{\mathbb{Z}} \cdots \otimes_{\mathbb{Z}} A_r \to \frac{\mathbb{Z}F}{I} \overset{\sim}{\to} \bar{N}_0.$$

However each A_i is a finitely generated abelian group because the highest and lowest coefficients of f_i are ± 1: here we use the fact that $1 = q_i^\theta = |\det(q_i)|$. It follows that $\bar{N}_0$ is a finitely generated abelian group.

Next N_{ab} is a noetherian $\mathbb{Z}Q$-module by 4.2.3. Consequently $\bar{N}$ is a finitely generated Q-module, from which it follows that $\bar{N}$ is a finitely generated abelian group, in view of the conclusion of the previous paragraph. Also the torsion subgroup of N_{ab} has finite exponent and, in addition, it is a minimax group, so it is finite. We deduce that N_{ab} is a finitely generated abelian group and thus G is polycyclic, since N is nilpotent.

(e) *The final step.* By (c) and (d) we may assume that $\mathrm{Im}(\theta)$ is infinite cyclic. Put $K/N = \mathrm{Ker}(\theta)$; then G/K is infinite cyclic, whence G is an ascending HNN-extension $\langle g, S \mid s^g = s^\sigma, s \in S \rangle$, where σ is an injective endomorphism of S, a finitely generated subgroup of K. Then $G = \langle g \rangle S^{\langle g \rangle}$ and $S^{\langle g \rangle} \leq K$: hence

$$N = N \cap \left(\langle g \rangle S^{\langle g \rangle} \right) = N \cap S^{\langle g \rangle}$$

since $\langle g \rangle \cap K = 1$. Write $L = N \cap S$ and $L_i = L^{g^{-i}}$. Then

$$N = \bigcup_{i=1,2,\ldots} N \cap S^{g^{-i}} = \bigcup_{1,2,\ldots} L_i$$

and $L_1 \leq L_2 \leq \cdots$ is an ascending chain of isomorphic minimax groups. Hence $|L_{i+1} : L_i|$ is finite for all i by 5.1.5.

Consider the natural map $L_{ab} \to N_{ab}$ and let U and W be its kernel and image respectively. Since $N = \bigcup_{i=1,2,\ldots} L_i$, we see that N/W is a torsion group. Also $U = (L \cap N')/L'$, and in fact U is a torsion group. For, if $x \in L \cap N'$, then $x \in L_i'$ for some i, and since each $|L_{j+1}' : L_j'|$ is finite, $x^k \in L'$ for some $k > 0$. It follows on applying the functor $- \otimes_{\mathbb{Z}} \mathbb{Q}$ to $L_{ab} \to N_{ab}$ that the induced mapping

$$L_{ab} \otimes_{\mathbb{Z}} \mathbb{Q} \to N_{ab} \otimes_{\mathbb{Z}} \mathbb{Q}$$

is a $\mathbb{Z}(S/L)$-isomorphism.

Now consider the finitely presented group S and its normal subgroup $L = S \cap N$. Since $SN/N \leq K/N = \mathrm{Ker}(\theta)$, we are in the situation of (d). Thus it follows that S is a polycyclic group and $G = \langle g, S \rangle$ is the required ascending HNN-extension. ∎

In the light of 11.4.4, it is natural to ask which polycyclic groups can occur as the base group B of the HNN-extension in part (iii) of the statement, when the group G is not polycyclic. Recall that group is called *cohopfian* if it cannot be isomorphic with a proper subgroup. Thus in part (iii) of 11.4.4, if G is not polycyclic, the base group B must be non-cohopfian. The question of interest is therefore: *which infinite polycyclic groups are non-cohopfian?*

Obviously a finitely generated infinite abelian group is non-cohopfian; in fact it is not hard to prove that *every finitely generated, infinite nilpotent group of class 2 is non-cohopfian* (see G. C. Smith 1985). On the other hand, there are examples of cohopfian, finitely generated, infinite nilpotent groups of class 3, (Belegradek 2002).

Finitely presented nilpotent-by-cyclic groups

One class of groups for which it is possible to give a clear criterion for finite presentability is the class of finitely generated nilpotent-by-cyclic groups.

11.4.5 (Bieri and Strebel 1978; Strebel 1984) *A finitely generated nilpotent-by-cyclic group G is finitely presented if and only if it is polycyclic or else an ascending HNN-extension whose base group is finitely generated and nilpotent.*

Proof Assume that G is finitely presented and let $N \triangleleft G$, where N is nilpotent and G/N is cyclic. If G/N is finite, then G is polycyclic. Otherwise G is an ascending HNN-extension $\langle t, S \rangle$ whose base group S is a finitely generated subgroup of N. The converse statement is obviously true. ∎

Corollary 11.4.6 *The following conditions on a finitely generated nilpotent-by-cyclic group G are equivalent:*

(i)   *G is finitely presented;*
(ii) *G is coherent;*
(iii) *finitely generated subgroups of G are constructible.*

Proof Let G satisfy (i). Then either G is polycyclic or it is an ascending HNN-extension with finitely generated nilpotent base group, by 11.4.5, and hence G is coherent by 11.4.4. Also (ii) implies (iii) by the same result. Finally it is clear that (iii) implies (i).   ∎

11.5   The Bieri–Strebel invariant

A crucial event in the development of infinite soluble group theory was the introduction of a certain geometric invariant of a finitely presented soluble group by Bieri and Strebel (1980). This invariant has played a prominent role in the problem of recognizing those finitely generated soluble groups which are finitely presented. In particular it was the key to the solution of Baumslag's problem, to determine which finitely generated metabelian groups have a finite presentation (Baumslag 1974).

Valuation spheres

Let Q be a finitely generated abelian group, written multiplicatively. A homomorphism of groups

$$\nu : Q \to \mathbb{R}$$

is called a *valuation* on Q. If Q has Hirsch number, that is torsion-free rank, n, then

$$\mathrm{Hom}(Q, \mathbb{R}) \simeq \mathbb{R}^n,$$

so that the valuation group $\mathrm{Hom}(Q, \mathbb{R})$ may be regarded as a topological vector space.

A valuation ν on Q can be extended to a function $\nu : \mathbb{Z}Q \to \mathbb{R} \cup \{\infty\}$ by defining 0^ν to be ∞ and

$$\left(\sum_{q \in Q} \ell_q q \right)^\nu = \min\{q^\nu \mid q \in Q, \ell_q \neq 0\},$$

where $\ell_q \in \mathbb{Z}$. It is easy to verify that for any r, s in $\mathbb{Z}Q$,

$$(r + s)^\nu \geq \min\{r^\nu, s^\nu\} \quad \text{and} \quad (rs)^\nu \geq r^\nu + s^\nu.$$

Two valuations ν and ν' on Q are said to be *equivalent* if $\nu' = a\nu$ for some positive real number a. This is clearly an equivalence relation on $\mathrm{Hom}(Q, \mathbb{R})$: we denote the equivalence class of ν by $[\nu]$.

Here we are concerned with the set of all equivalence classes of non-zero valuations

$$S(Q) = \{[\nu] \mid 0 \neq \nu \in \mathrm{Hom}(Q, \mathbb{R})\}.$$

Notice that $S(Q)$ is empty precisely when $n = 0$, that is, Q is finite. Evidently $S(Q)$ may be identified with the unit sphere S^{n-1} in $\mathbb{R}^n$. Thus $S(Q)$ is a compact manifold, which is connected if $n > 1$. We will refer to $S(Q)$ as the *valuation sphere of Q*.

For any valuation ν on Q define

$$Q_\nu = \{q \in Q \mid q^\nu \geq 0\}.$$

This is evidently a submonoid of Q and from it we may form $\mathbb{Z}Q_\nu$, the monoid ring, which is a subring of $\mathbb{Z}Q$.

We are now ready to define the *Bieri–Strebel invariant* of a finitely generated $\mathbb{Z}Q$-module A, denoted by $\sum_A$: this consists of all $[\nu]$ in $S(Q)$ such that A is finitely generated as a $\mathbb{Z}Q_\nu$-module. Bieri and Strebel showed that, for any exact sequence $0 \rightarrow A \rightarrow G \rightarrow Q \rightarrow 1$, the condition for the group G to be finitely presented can be expressed in terms of Σ_A: roughly speaking this should be a 'large' subset of $S(Q)$.

As the reader may already suspect, it is rarely possible to compute Σ_A directly from the definition. But fortunately there is an alternative description of Σ_A in terms of the *centralizer ring* of A in $\mathbb{Z}A$, that is,

$$C(A) = \{r \in \mathbb{Z}Q \mid ar = a, \forall a \in A\}.$$

Thus $C(A)$ is just $1 + \text{Ann}_{\mathbb{Z}Q}(A)$.

The connection with the invariant Σ_A is shown by the following result.

11.5.1 (Bieri and Strebel 1980) *Let Q be a finitely generated abelian group and let A be a finitely generated $\mathbb{Z}Q$-module. Suppose that ν is a non-zero valuation on Q. Then A is a finitely generated $\mathbb{Z}Q_\nu$-module if and only if $C(A)$ contains an element r such that $r^\nu > 0$.*

Proof Suppose there is an r in $C(A)$ such that $r^\nu > 0$. If $0 \neq s \in \mathbb{Z}Q$, then $(sr^m)^\nu \geq s^\nu + mr^\nu$, which is non-negative for large enough m. Then $sr^m \in \mathbb{Z}Q_\nu$ and it follows that $as = asr^m \in a(\mathbb{Z}Q_\nu)$ for all a in A: therefore $a(\mathbb{Z}Q) = a(\mathbb{Z}Q_\nu)$. Consequently each $\mathbb{Z}Q$-generating subset of A is also a $\mathbb{Z}Q_\nu$-generating set and thus $\nu \in \Sigma_A$.

Conversely, suppose that A is generated as a $\mathbb{Z}Q_\nu$-module by a finite subset $\{a_1, \ldots, a_m\}$. Since $\nu \neq 0$, we may choose q in Q such that $q^\nu > 0$. Now $a_i q^{-1} = \sum_{j=1}^m a_j r_{ij}$, where $r_{ij} \in \mathbb{Z}Q_\nu$, and hence

$$\sum_{j=1}^m a_j(\delta_{ij} - qr_{ij}) = 0$$

for $i = 1, 2, \ldots, m$. Thus A is annihilated by $\det(\delta_{ij} - qr_{ij})$ and the latter has the form $1 - qu$, where $u \in \mathbb{Z}Q_\nu$. Put $r = qu$ and observe that $r \in C(A)$, while $r^\nu \geq q^\nu + u^\nu > 0$ since $q^\nu > 0$ and $u^\nu \geq 0$. $\blacksquare$

As a consequence of 11.5.1, the definition of Σ_A may be reformulated in the following very useful form.

11.5.2 *Let Q be a finitely generated abelian group and let A be a finitely generated $\mathbb{Z}Q$-module. Then*

$$\Sigma_A = \bigcup_{r \in C(A)} \{[\nu] \mid 0 \neq \nu \in \mathrm{Hom}(Q, \mathbb{R}), r^\nu > 0\}.$$

Furthermore Σ_A is an open subset of the valuation sphere $S(Q)$.

The expression for Σ_A here is an immediate consequence of 11.5.1. If $r \in C(A)$, the set $\{[\nu] \mid r^\nu > 0\}$ is open since it is the intersection of finitely many open sets $\{[\nu] \mid q^\nu > 0\}$, where $q \in Q$ occurs in the expression for r. Hence Σ_A is open.

Two examples

It is instructive to look at two examples which give some insight into the link between the Bieri–Strebel invariant and finite presentability.

Let $Q = \langle x \rangle$ be infinite cyclic; then clearly

$$S(Q) = \{[-1], [1]\}.$$

We will consider two different Q-modules. First let $A = \mathbb{Q}_2$, where the module action is given by $a \cdot x = 2a$. Let $0 \neq \nu \in \mathrm{Hom}(Q, A)$: here we may suppose that ν, that is, x^ν, equals -1 or $+1$. If $\nu = -1$, then $r = 2x^{-1} \in C(A)$ and $r^\nu = (x^{-1})^\nu = 1 > 0$. Thus $[-1] \in \Sigma_A$.

Now suppose that $\nu = 1$. Let $r \in C(A)$ and assume $r^\nu > 0$; then we can write $r = 1 + (x - 2)f(x)$, where $f(x) = \sum_n a_n x^n \in \mathbb{Z}[x]$. Observe that r can only involve positive powers of x since $r^\nu > 0$. Equating the coefficients of non-positive powers of x to 0, we obtain equations

$$1 + a_{-1} - 2a_0 = 0 \quad \text{and} \quad a_{-j-1} - 2a_{-j} = 0, \quad j > 0.$$

Hence $a_{-j} = 0$ if $j > 0$, and $1 - 2a_0 = 0$, which is impossible. Therefore $[1] \notin \Sigma_A$ and we conclude that

$$\Sigma_A = \{[-1]\} \quad \text{and} \quad S(Q) = \Sigma_A \cup (-\Sigma_A),$$

where $-\Sigma_A = \{-[\nu] \mid [\nu] \in \Sigma_A\}$. Notice that the group $Q \ltimes A$ is finitely presented.

In our second example let the module be $B = \mathbb{Q}_{\{2,3\}}$ with the action given by $a \cdot x = \frac{3}{2}a$. A similar calculation reveals that $\Sigma_B = \emptyset$. Thus $S(Q) \neq \Sigma_B \cup (-\Sigma_B)$. Notice that in this case the group $G = Q \ltimes B$ is not finitely presented, by 11.4.3.

So the lesson here would appear to be that, if we want $Q \ltimes A$ to be finitely presented, Σ_A should not be too small. This will be made precise in the subsequent discussion.

Next we consider the relation between the valuation spheres of two finitely generated abelian groups Q and $\bar{Q}$ when a homomorphism between them is given, say

$$\varphi : \bar{Q} \to Q.$$

If $\nu \in \mathrm{Hom}(Q, \mathbb{R})$, then $\varphi\nu \in \mathrm{Hom}(\bar{Q}, \mathbb{R})$ and the assignment $\nu \mapsto \varphi\nu$ yields a continuous homomorphism

$$\varphi^* : S(Q) \to S(\bar{Q}).$$

Also, if A is a $\mathbb{Z}Q$-module, we may make A into a $\mathbb{Z}\bar{Q}$-module, say $\bar{A}$, by means of φ. With the notation just described, we will establish a useful reduction property.

11.5.3 *If* $\mathrm{Ker}(\varphi)$ *and* $\mathrm{Coker}(\varphi)$ *are both finite, then* φ^* *is a homeomorphism from* $S(Q)$ *to* $S(\bar{Q})$. *Furthermore* $(\Sigma_A)\varphi^* = \Sigma_{\bar{A}}$.

Proof In the first place φ^* is bijective since $\mathbb{R}$ is torsion-free. Thus φ^* is a homeomorphism. Next let $\nu \in \mathrm{Hom}(Q, \mathbb{R})$. We will show that A is finitely generated as a $\mathbb{Z}Q_\nu$-module if and only if it is finitely generated as a $\mathbb{Z}(Q_\nu \cap \bar{Q}^\varphi)$-module: of course sufficiency is obvious here, so assume A is a finitely generated $\mathbb{Z}Q_\nu$-module and $\nu \neq 0$. Noting that $Q/\bar{Q}^\varphi$ is finite, we choose a transversal $\{q_1, \ldots, q_k\}$ to $\bar{Q}^\varphi$ in Q. Now $(\bar{Q}^\varphi)^\nu \neq 0$, since otherwise $Q^\nu = 0$ and $\nu = 0$. Hence there exists $q \in \bar{Q}^\varphi$ such that $q^\nu \neq 0$. Replacing q by a suitable power, we may suppose that $q^\nu \leq -q_i^\nu$ for $i = 1, 2, \ldots, k$, so that $(q_i q)^\nu = q_i^\nu + q^\nu \leq 0$. By replacing q_i by $q_i q$, we may assume that $q_i^\nu \leq 0$ for $i = 1, 2, \ldots, k$.

With this choice of transversal we have

$$Q_\nu = Q_\nu \cap \bigcup_{i=1}^k q_i \bar{Q}^\varphi$$

$$= \bigcup_{i=1}^k Q_\nu \cap (q_i \bar{Q}^\varphi)$$

$$= \bigcup_{i=1}^k q_i \left(q_i^{-1} Q_\nu \cap \bar{Q}^\varphi \right)$$

$$\subseteq \bigcup_{i=1}^k q_i \left(Q_\nu \cap \bar{Q}^\varphi \right)$$

since $(q_i^{-1})^\nu = -q_i^\nu \geq 0$. It follows at once that A can be finitely generated as a $\mathbb{Z}\left(Q_\nu \cap \bar{Q}^\varphi\right)$-module.

Consequently, A is finitely generated over $\mathbb{Z}Q_\nu$ if and only if it is finitely generated over the preimage of $\mathbb{Z}Q_\nu$ under φ, that is, over $\mathbb{Z}Q_{\varphi\nu}$. Therefore

$$(\Sigma_A)\varphi^* = \Sigma_{\bar{A}} \cap (S(Q))\varphi^* = \Sigma_{\bar{A}} \cap S(\bar{Q}) = \Sigma_{\bar{A}}.$$

∎

The usefulness of 11.5.3 derives from the fact that it sometimes allows us to reduce to the case where Q is torsion-free.

Discrete valuations

If Q is a finitely generated abelian group, a valuation $\nu : Q \to \mathbb{R}$ is called *discrete* if $\mathrm{Im}(\nu) = \mathbb{Z}$. Let $S_{\mathrm{dis}}(Q)$ denote the set of all discrete valuations on Q. The next result is the first indication that finite presentability of a soluble group can force the associated Bieri–Strebel invariant to be large.

11.5.4 (Strebel 1984) *Let G be a finitely presented soluble group and let $N \lhd G$. Assume $Q = G/N$ is abelian and write $A = N_{ab}$. If ν is a discrete valuation on Q, then either $[\nu]$ or $[-\nu]$ belongs to Σ_A*

Here $[-\nu] = -[\nu]$ is the antipodal point of $[\nu]$ on the valuation sphere $S(Q)$. Thus what 11.5.4 actually asserts is that

$$S_{\mathrm{dis}}(Q) \subseteq \Sigma_A \cup (-\Sigma_A).$$

Proof of 11.5.4 Write $K/N = \mathrm{Ker}(\nu)$. Then G/K is infinite cyclic since ν is discrete. Thus by 11.4.1 the group G is an ascending HNN-extension, say

$$G = \langle t, S \mid s^t = s^\sigma, s \in S \rangle,$$

where S is a finitely generated subgroup of K and σ is an injective endomorphism of S. Also $K = \bigcup_{j=0,1,\dots} S^{t^{-j}}$. Since $\mathrm{Im}(\nu) = \mathbb{Z}$, we have $(tN)^\nu = \epsilon = \pm 1$. Our aim is to prove that $[-\epsilon\nu] \in \Sigma_A$, which will show that $[\nu] \in \Sigma_A \cup (-\Sigma_A)$.

First observe that SN'/N' is a finitely generated metabelian group, whence $(S \cap N)N'/N'$ is a finitely generated module over $SN'/(S \cap N)N'$ and so over SN/N. This implies that A is a finitely generated module over $U = Rg \langle SN/N, t^{-1}N \rangle$: for

$$A = (N/N') \cap \left(\left(\bigcup_{j=0,1,\dots} S^{t^{-j}} \right) N/N' \right) = \bigcup_{j=0,1,\dots} (S \cap N)^{t^{-j}} N/N'.$$

Now $SN/N \leq K/N \leq Q_{(-\epsilon\nu)}$, while $(t^{-1}N)^{-\epsilon\nu} = -\epsilon(t^{-1}N)^\nu = (-\epsilon)^2 = 1$. Hence $t^{-1}N \in Q_{(-\epsilon\nu)}$ and consequently $U \subseteq \mathbb{Z}Q_{(-\epsilon\nu)}$. Therefore A is a finitely generated $\mathbb{Z}Q_{(-\epsilon\nu)}$-module and $[-\epsilon\nu] \in \Sigma_A$. ∎

Tame modules

If Q is a finitely generated abelian group and A is a finitely generated $\mathbb{Z}Q$-module, then A is said to be *tame* if

$$S(Q) = \Sigma_A \cup (-\Sigma_A).$$

Thus for each non-zero valuation ν, either $[\nu]$ or its antipodal point $-[\nu]$ must belong to Σ_A.

The following result, which is a wide generalization of 11.5.4, shows that tame modules are inextricably involved in finitely presented soluble groups.

11.5.5 (Bieri and Strebel 1980) *Let G be a finitely presented soluble group, let $N \lhd G$ with $Q = G/N$ abelian and put $A = N_{ab}$. Then N_{ab} is a tame $\mathbb{Z}Q$-module.*

Since the proof of this theorem is both lengthy and complex, we can present only the barest outline of it here.

Choose a free abelian subgroup with finite index in Q, say $\bar{Q}$, and write $\bar{Q} = \bar{G}/N$. Apply 11.5.3 with φ the inclusion map $\bar{Q} \hookrightarrow Q$ to conclude that $\varphi^* : S(Q) \to S(\bar{Q})$ is a homeomorphism: furthermore $(\Sigma_A)\varphi^* = \Sigma_{\bar{A}}$ where $\bar{A}$ is

A when it is regarded as a $\mathbb{Z}\bar{Q}$-module. Since $([-\nu])\varphi^* = -([\nu])^{\varphi^*}$, we also have $(-\Sigma_A)\varphi^* = -\Sigma_{\bar{A}}$. Consequently A is a tame $\mathbb{Z}Q$-module if and only if $\bar{A}$ is tame as a $\mathbb{Z}\bar{Q}$-module. Since $\bar{G}$ is finitely presented, we reduce to the case where Q is free abelian.

Let $0 \neq \nu \in \mathrm{Hom}(Q, \mathbb{R})$: it must be shown that A is finitely generated as a $\mathbb{Z}Q_\nu$ or a $\mathbb{Z}Q_{(-\nu)}$-module: of course A is certainly a finitely generated $\mathbb{Z}Q$-module. The idea of the proof is to use the finite presentation of G to obtain an infinite presentation of N by means of the Reidemeister–Schreier procedure. This can be done in such a way that N is seen to be a generalized free product $H \underset{S}{*} K$. Since G is soluble, we must have $|H : S| \leq 2$ and $|K : S| \leq 2$. This permits us to identify N with H or K. A careful analysis then reveals that A is finitely generated over $\mathbb{Z}Q_\nu$ or $\mathbb{Z}Q_{(-\nu)}$. For details of the proof see Bieri and Strebel (1980).

The discussion so far, and especially 11.5.5, suggests that if G is a finitely presented soluble group with $N \lhd G$ and $Q = G/N$ abelian, then $\Sigma_{N_{ab}}$ is a large subset of $S(Q)$. One can ask to what extent the converse is true. The extreme case would be when $\Sigma_A = S(Q)$, a situation which will now be examined in detail.

11.5.6 *Let Q be a finitely generated abelian group and A a finitely generated $\mathbb{Z}Q$-module. Then A is finitely generated as an abelian group if and only if $\Sigma_A = S(Q)$.*

In proving this theorem we shall make use of an elementary geometric lemma. If α is a positive real number, let B_α denote the open ball of radius α in $\mathbb{R}^n$, that is, the set of x in $\mathbb{R}^n$ such that $\|x\| < \alpha$: here the standard inner product on $\mathbb{R}^n$ is used.

11.5.7 *Let $\mathcal{F}$ be a finite collection of finite subsets of $\mathbb{R}^n$ and assume that for each $x \neq 0$ in $\mathbb{R}^n$ there exists F_x in $\mathcal{F}$ such that $\langle x, y \rangle > 0$ for all y in F_x. Then there exist a positive real number α_0 and a function $\varepsilon : \{\alpha \in \mathbb{R} \mid \alpha > \alpha_0\} \to \mathbb{R}^+$ such that the following is true: if $\alpha > \alpha_0$ and $x \in B_{\alpha + \varepsilon(\alpha)} \backslash B_\alpha$, then $x + L \subseteq B_\alpha$ for some $L \in \mathcal{F}$.*

Deferring for the present the proof of this lemma, we proceed to prove 11.5.6.

Proof of 11.5.6 Let n denote the Hirsch number of Q and let T be its torsion subgroup: choose a basis $\{x_1 T, \ldots, x_n T\}$ for Q/T. Denote by θ the homomorphism from Q to $\mathbb{R}^n$ such that $\mathrm{Ker}(\theta) = T$ and $x_i^\theta = [0, \ldots, 0, 1, \ldots, 0]$, with 1 in the ith position, for $i = 1, \ldots, n$. Thus $Q^\theta = \mathbb{Z}^n$. For each $\alpha \in \mathbb{R}^+$ we define X_α to be the preimage of B_α under θ. So $\{X_\alpha \mid \alpha \in \mathbb{R}^+\}$ is an ascending chain of finite sets whose union is equal to Q.

Assume that $\Sigma_A = S(Q)$. By 11.5.2 there is a finite subset C_0 of $C(A)$ such that

$$S(Q) = \bigcup_{r \in C_0} \{[\nu] \mid r^\nu > 0\}.$$

If $r \in \mathbb{Z}Q$, we may write in a unique way

$$r \equiv \sum_{\ell_1,\ldots,\ell_n \in \mathbb{Z}} m_{\ell_1,\ldots,\ell_n} x_1^{\ell_1} \cdots x_n^{\ell_n} (\operatorname{mod} T)$$

with $m_{\ell_1 \ldots \ell_n} \in \mathbb{Z}$. Define $\sigma(r)$ to be the finite set

$$\{(\ell_1,\ldots,\ell_n) \mid m_{\ell_1 \ldots \ell_n} \neq 0\} \subseteq \mathbb{Z}^n.$$

Next we apply 11.5.7 to the collection of sets

$$\mathcal{F} = \{\sigma(r) \mid r \in C_0\} :$$

this is possible since, if $0 \neq x \in \mathbb{R}^n$, there exists a $y \in \mathbb{Z}^n$ for which the inner product $\langle x, y \rangle$ is positive and the entries of y yield an element r in $\mathbb{Z}Q$ such that $y \in \sigma(r)$.

It follows that there exist $\alpha_0 \in \mathbb{R}^+$ and a function

$$\varepsilon : \{\alpha \in \mathbb{R} \mid \alpha > \alpha_0\} \to \mathbb{R}^+$$

such that for any $q \in X_{\alpha+\varepsilon(\alpha)} \backslash X_\alpha$ we have $q^\theta + F \subseteq B_\alpha$ for some $F = \sigma(r) \in \mathcal{F}, r \in C_0$. Suppose that $(\ell_1,\ldots,\ell_n) \in F$; then

$$(qx_1^{\ell_1} \cdots x_n^{\ell_n})^\theta = q^\theta + (\ell_1,\ldots,\ell_n) \in q^\theta + F \subseteq B_\alpha.$$

Hence $qr \in \mathbb{Z}X_\alpha$.

For any a in A we have $aq = aqr \in a(\mathbb{Z}X_\alpha)$, since $r \in C_0$, and it follows that

$$a\left(\mathbb{Z}X_{\alpha+\varepsilon(\alpha)}\right) = a(\mathbb{Z}X_\alpha)$$

for all $\alpha > \alpha_0$. Now let k be a positive integer such that $k > \alpha_0^2$. Notice that if $\sqrt{k} < p \leq \sqrt{k+1}$, then $X_p = X_{\sqrt{k+1}}$. Hence $a\left(\mathbb{Z}X_{\sqrt{k+1}}\right) \subseteq a\left(\mathbb{Z}X_{\sqrt{k}}\right)$ and therefore

$$a\left(\mathbb{Z}X_{\sqrt{k}}\right) = a\left(\mathbb{Z}X_{\sqrt{k+1}}\right)$$

for all $k > \alpha_0^2$. It follows that $(a)(\mathbb{Z}Q) = a\left(\mathbb{Z}X_{\sqrt{k_0}}\right)$, where k_0 is the smallest integer exceeding α_0^2. Since A is a finitely generated $\mathbb{Z}Q$-module, we may conclude that it is finitely generated as an abelian group, since $X_{\sqrt{k_0}}$ is finite.

Conversely, if A is finitely generated as an abelian group, it is obviously a finitely generated $\mathbb{Z}Q_\nu$-module for all valuations ν; hence $\nu \in \Sigma_A$ and $\Sigma_A = S(Q)$. ∎

As an application we give a criterion for a finitely generated nilpotent-by-abelian-by-finite group to be virtually polycyclic.

11.5.8 *Let G be a finitely generated group with normal subgroups $N \leq M$. Assume that N is nilpotent, M/N is abelian and G/M is finite. Then G is virtually polycyclic if and only if $\Sigma_{N_{ab}} = S(M/N)$.*

Proof Clearly G will be virtually polycyclic if and only if N is, and, since N is nilpotent, this occurs precisely when N_{ab} is finitely generated by 1.2.16. The result is now a direct consequence of 11.5.6. ∎

For the sake of completeness a proof will now be given of the geometric lemma on which 11.5.6 depends.

Proof of 11.5.7 We define a function $f : S^{n-1} \to \mathbb{R}$ by the rule

$$f(u) = \max_L \min_y \{\langle u, y \rangle \mid y \in L \in \mathcal{F}\},$$

where $\langle \ \rangle$ is the standard inner product on $\mathbb{R}^n$: here $f(u)$ is finite since $\mathcal{F}$ is a finite collection of finite sets. Evidently f is continuous and $f(u) > 0$ for all $u \in S^{n-1}$. Since S^{n-1} is compact, it follows that

$$c = \min\{f(u) \mid u \in S^{n-1}\} > 0.$$

A second constant d is determined by the equation

$$d = \max_L \max_y \{\|y\| \mid y \in L \in \mathcal{F}\},$$

that is, d is the radius of the smallest ball which contains $\bigcup_{L \in \mathcal{F}} L$.

Now define a real number $\alpha_0 > 0$ and a function $\varepsilon : \{\alpha \in \mathbb{R} \mid \alpha > \alpha_0\} \to \mathbb{R}$ by the rules

$$\alpha_0 = \frac{d^2}{2c} \quad \text{and} \quad \varepsilon(\alpha) = c - \frac{d^2}{2\alpha}.$$

It will be shown that with this α_0 and ε the assertion of 11.5.7 is true. Observe that ε is an increasing function of α and that, if $\alpha > \alpha_0$,

$$\varepsilon(\alpha) > c - \frac{d^2}{2\alpha_0} = 0.$$

Thus ε has positive values.

Let $x \in \mathbb{R}^n$ satisfy $\|x\| \geq \alpha_0$. By definition of c there exists $L_x \in \mathcal{F}$ such that

$$\min \left\{ \left\langle \frac{-x}{\|x\|}, y \right\rangle \mid y \in L_x \right\} \geq c,$$

since otherwise $f(-x/\|x\|) < c$. This is equivalent to

$$\max \left\{ \left\langle \frac{x}{\|x\|}, y \right\rangle \mid y \in L_x \right\} \leq -c.$$

Therefore for any y in L_x we have

$$\|x + y\|^2 = \|x\|^2 + 2 \left\langle \frac{x}{\|x\|}, y \right\rangle \|x\| + \|y\|^2 \leq \|x\|^2 - 2c\|x\| + d^2.$$

Since $\|x\| \geq \alpha_0$, we deduce that

$$\|x + y\|^2 \leq \|x\|^2 - 2c\alpha_0 + d^2 = \|x\|^2.$$

Consequently

$$\|x+y\| - \|x\| = \frac{\|x+y\|^2 - \|x\|^2}{\|x+y\| + \|x\|} \leq \frac{-2c\|x\| + d^2}{\|x+y\| + \|x\|} = \frac{-\varepsilon(\|x\|)(2\|x\|)}{\|x+y\| + \|x\|} \leq -\varepsilon(\|x\|).$$

Now suppose that
$$-\alpha_0 < \alpha \le \|x\| < \alpha + \varepsilon(\alpha),$$
so that $x \in B_{\alpha+\varepsilon(\alpha)} \backslash B_\alpha$. Then for all $y \in L_x$, we have from the inequality above
$$\|x + y\| \le \|x\| - \varepsilon(\|x\|) < \alpha + \varepsilon(\alpha) - \varepsilon(\|x\|) \le \alpha,$$
since $\|x\| \ge \alpha_0$ and ε is an increasing function. It now follows that $x + y \in B_\alpha$ and so $x + L_x \subseteq B_\alpha$. ∎

For comparison with 11.5.6, we remark that Strebel (1984: 300) has established the following result.

11.5.9 *Let Q be a finitely generated abelian group and A a finitely generated $\mathbb{Z}Q$-module. If A is a minimax group, then $S(Q) \backslash \Sigma_A$ is finite.*

For the Bieri–Strebel invariants for finitely generated soluble minimax groups see Åberg (1986) and Meinert (1996, 1998).

Finitely presented centre-by-metabelian groups

Recall from 4.1.3 that the centre of a finitely generated centre-by-metabelian group can be any non-trivial countable abelian group. That the situation is entirely different for finitely presented centre-by-metabelian groups is made clear by the following important result.

11.5.10 (Groves 1978a) *If G is a finitely presented centre-by-metabelian group, then $G'/Z(G')$ is finitely generated. Thus G is abelian-by-polycyclic and therefore it satisfies* $\max -n$ *and is residually finite.*

Proof Put $Q = G_{ab}$ and $A = (G')_{ab}$: thus A is a finitely generated $\mathbb{Z}Q$-module (written multiplicatively). Also write B for $G'/Z(G')$; then B is a $\mathbb{Z}Q$-image of A since $G'' \le Z(G')$.

If x, y belong to G', the assignment $(xZ(G'), yZ(G')) \mapsto [x, y]$ determines a well-defined function
$$\alpha : B \times B \to G''.$$
Clearly α is a non-degenerate, skew symmetric, bilinear form on B. Also, if $x, y \in G'$ and $g \in G$, we have $[x^g, y] = [x, y^{g^{-1}}]$ because $G'' \le Z(G)$.

Define a function $\tau : \mathbb{Z}Q \to \mathbb{Z}Q$ by the rule
$$\left(\sum_q \ell_q q \right)^\tau = \sum_q \ell_q q^{-1}, \quad (\ell_q \in \mathbb{Z});$$
thus τ is an isomorphism of additive groups. The commutator relation obtained above now implies that
$$\alpha((b_1^r, b_2)) = \alpha((b_1, b_2^{r^\tau})),$$
where $b_i \in B$ and $r \in \mathbb{Z}Q$. It follows that τ maps the annihilator $\mathrm{Ann}_{\mathbb{Z}Q}(B)$ onto itself: indeed, if $r \in \mathrm{Ann}_{\mathbb{Z}Q}(B)$, then
$$0 = \alpha((b_1^r, b_2)) = \alpha((b_1, b_2^{r^\tau}))$$

for all $b_i \in B$, whence $b_2^{r^\tau} = 0$ by non-degeneracy of α, and thus $r^\tau \in \mathrm{Ann}_{\mathbb{Z}Q}(B)$. Since $C(B) = 1 + \mathrm{Ann}_{\mathbb{Z}Q}(B)$, it follows that $(C(B))^\tau = C(B)$.

Let $\nu \in \mathrm{Hom}(Q, \mathbb{R})$. Then $q^{-\nu} = (q^{-1})^\nu = q^{\tau\nu}$ for $q \in Q$, and thus $-\nu = \tau\nu$. Using 11.5.2, we have

$$-\Sigma_B = \{[-\nu] \mid [\nu] \in \Sigma_B\} = \{[\tau\nu] \mid [\nu] \in \Sigma_B\} = \bigcup_{r \in C(B)} \{[\tau\nu] \mid [\nu] \in S(Q), r^\nu > 0\}.$$

On writing $r = r_1^\tau$ with $r_1 \in C(B)$, we deduce that

$$-\Sigma_B = \bigcup_{r_1 \in C(B)} \{[\nu] \mid [\nu] \in S(Q), r_1^\nu > 0\} = \Sigma_B.$$

Now $\Sigma_A \subseteq \Sigma_B$, because B is a quotient of A. Therefore $-\Sigma_A \subseteq -\Sigma_B$ and hence

$$\Sigma_B = \Sigma_B \cup (-\Sigma_B) \supseteq \Sigma_A \cup (-\Sigma_A).$$

Since G is finitely presented, we have $S(Q) = \Sigma_A \cup (-\Sigma_A)$ by 11.5.5, and it follows at once that $S(Q) = \Sigma_B$. Therefore B is a finitely generated abelian group by 11.5.6.

Finally, we conclude that $G/Z(G')$ is polycyclic, so that G is abelian-by-polycyclic. That G satisfies $\max -n$ and is residually finite are consequences of 4.2.2 and 7.2.1. ∎

Characterizing finitely presented metabelian groups

In 1980 Bieri and Strebel succeeded in characterizing those finitely generated metabelian groups which are finitely presented, thus solving Baumslag's problem stated above. The definitive result is:

11.5.11 (Bieri and Strebel 1980) *Let Q be a finitely generated metabelian group and let A be a tame $\mathbb{Z}Q$-module. Then every extension of A by Q inducing the given module structure is finitely presented.*

On combining this with 11.5.5, one obtains a criterion for a finitely generated metabelian group to have a finite presentation.

11.5.12 *If G is a finitely generated metabelian group, then G is finitely presented if and only if $S(G_{ab}) = \Sigma_{G'} \cup (-\Sigma_{G'})$.*

From 11.5.11 and 11.5.12 we recognize that whether an extension G of A by Q is finitely presented does not depend on the cohomology class of the extension. Furthermore, G is finitely presented if and only if the split extension $Q \ltimes A$ is finitely presented. At this point one can really appreciate the power of the Bieri–Strebel invariant: indeed without this it is difficult to imagine how finitely presented metabelian groups could be characterized.

In the proof of 11.5.11 one first reduces to the case where Q is free abelian, using 11.5.3. Let $\{x_1, \ldots, x_n\}$ be a basis of Q and let $a_1, \ldots, a_m$ be $\mathbb{Z}Q$-module generators of A. One can easily write down a presentation of G in the generators $x_1, \ldots, x_n, a_1, \ldots, a_m$, with relations as follows:

$$[x_i, x_j] = c_{ij}, \quad 1 \leq i < j \leq n, \qquad [a_k, a_\ell^w] = 1, \quad 1 \leq k < \ell \leq m,$$

$$x_i^{-1} a_k x_i = b_{ik}, \quad x_i a_k x_i^{-1} = b'_{ik}, \qquad 1 \leq i \leq n, \quad 1 \leq k \leq m,$$

and also $r_1 = r_2 = \cdots = r_\ell = 1$. Here $c_{ij}, b_{ik}, b'_{ik}, r_q$ are $\mathbb{Z}Q$-words in the a_k and $r_1, \ldots, r_\ell$ are defining relations for the $\mathbb{Z}Q$-module A: also w is an element of Q, and so this is an infinite presentation. The thrust of the proof is to show that the tameness of the $\mathbb{Z}Q$-module A implies that some finite subset of these relations suffices to present G. The argument is complex: for a detailed account the reader is referred to Bieri and Strebel (1980).

We now present some applications of 11.5.11, the first being a generalization of 11.5.12.

11.5.13 (Strebel 1984) *Let G be a finitely generated centre-by-metabelian group. Then G is finitely presented if and only if $(G')_{ab}$ is a tame $\mathbb{Z}G_{ab}$-module.*

Proof If G'_{ab} is tame, G/G'' is finitely presented by 11.5.12, and therefore G'' is finitely generated as a G-operator group. But $G'' \leq Z(G)$, so G'' is finitely generated and G is certainly finitely presented. The converse follows from 11.5.5. ∎

The next result is a simple observation, but it has great significance.

11.5.14 (Bieri and Strebel 1980) *If G is any finitely presented soluble group, then G/G'' is finitely presented.*

Proof By 11.5.5 $(G')_{ab}$ is a tame $\mathbb{Z}G_{ab}$-module. Therefore G/G'' is finitely presented, by 11.5.11. ∎

A further application of 11.5.11 is a structural restriction which applies to any finitely presented soluble group.

11.5.15 (Bieri and Strebel 1980) *If G is a finitely presented soluble group, then $G/[G'', G']$ satisfies $\max -n$ and is residually finite.*

Proof By 11.5.14 the group G/G'' is finitely presented and thus G'' is finitely generated as a G-operator group. Hence $G''/[G'', G']$ is a finitely generated $\mathbb{Z}G_{ab}$-module and so it satisfies $\max -G$ by 4.2.3. Since G/G'' has $\max -n$, so does $G/[G'', G']$.

To establish residual finiteness, we first note that $G''/[G'', G']$ is residually finite as a $\mathbb{Z}G_{ab}$-module since

$$G_{ab} \ltimes (G''/[G'', G'])$$

is a finitely generated metabelian group: here 4.3.1 has been applied. Because of this observation, we recognize that it is sufficient to prove G/L residually finite where $[G'', G'] \leq L \leq G''$ and G''/L is finite.

Put $C = C_G(G''/L)$; thus G/C is finite. Since G/G'' is finitely presented, G/L is finitely presented, as must be C/L. In addition C/L is centre-by-metabelian. We may therefore pass to the group G/L and suppose G'' to be finite and that $C = C_G(G'')$ is finitely presented and centre-by-metabelian.

Applying 11.5.10, we conclude that $C'/Z(C')$ is finitely generated. In addition C' is nilpotent of class at most 2 and C'' is finite, from which it follows that $C'/Z(C')$ has finite exponent. Thus $C'/Z(C')$ is finite and, since C is finitely generated, it follows quickly that $C/Z(C')$ is virtually abelian. We now have the situation that C is finitely generated and virtually metabelian, as is G. Therefore G is residually finite. ∎

We remark that the last part of the argument shows a little more, namely that every quotient of $G/[G'', G']$ is residually finite.

Tame modules and tensor powers

Suppose that G is a finitely generated nilpotent-by-abelian group and let $N \vartriangleleft G$ where N is nilpotent and $Q = G/N$ is abelian. Then N_{ab} is a finitely generated $\mathbb{Z}Q$-module. Now the other lower central factors of N are $\mathbb{Z}Q$-images of tensor powers of N_{ab} by 1.2.11. However such factors may not be finitely generated $\mathbb{Z}Q$-modules because the property of being finitely generated as a $\mathbb{Z}Q$-module does not pass to a tensor power (where the action is diagonal). Thus G need not satisfy $\max - n$. On the other hand, the situation is better for tame modules, as was observed by Bieri and Groves (1982).

Let Q be a finitely generated abelian group and let A be a finitely generated $\mathbb{Z}Q$-module. Then A is called *m-tame*, where $m > 1$ is an integer, if every m-subset of $S(Q)\backslash\Sigma_A$ is contained in an open hemisphere of $S(Q)$. It is clear that $(m+1)$-tame implies m-tame and also that A is 2-tame precisely when $S(Q)\backslash\Sigma_A$ does not contain a pair of antipodal points, that is, $S(Q) = \Sigma_A \cup (-\Sigma_A)$. Thus 2-tame is the same as tame and m-tame is a strong version of tameness. It is not difficult to prove the following result.

11.5.16 (Bieri and Groves 1982) *If A is m-tame as a $\mathbb{Z}Q$-module, the mth tensor power $\otimes A^m$ is a finitely generated $\mathbb{Z}Q$-module.*

From this we may easily deduce the following fact. *Let G be a finitely generated group with $N \vartriangleleft G$ and assume that N is nilpotent of class c. Suppose that $Q = G/N$ is abelian. If N_{ab} is c-tame as $\mathbb{Z}Q$-module, then G satisfies $\max - n$.* When $c = 2$, this provides another proof of 11.5.15.

There have been many articles investigating tameness. The interested reader may consult Bieri and Groves (1982), Kropholler and Stammbach (1990), and Kochloukova (1996, 1999a, b, 2000).

Finally, we mention a condition on the Bieri–Strebel invariant which implies constructibility.

11.5.17 (Bieri and Strebel 1982) *Let G be a finitely generated group and let $N \leq M$ be normal subgroups of G. Assume that N is nilpotent, M/N is abelian, and G/M is finite. Then the following are equivalent:*

(i) *G is constructible;*
(ii) *G/N' is constructible;*
(iii) *$S(M/N)\backslash\Sigma_{N_{ab}}$ is contained in an open hemisphere of $S(M/N)$.*

As the foregoing discussion has made clear, the Bieri–Strebel invariant has had a major impact on the theory of finitely presented soluble groups and has provided techniques for attacking problems that had previously seemed impenetrable. On the other hand, the invariant $\Sigma_{G'_{ab}}$ occurs quite high up in a finitely presented soluble group G and so there are limits to the amount of information about G that it can carry. This is reflected in our almost complete lack of knowledge of the structure of $[G'', G']$.

Generalized Bieri–Strebel invariants

A generalization of the Bieri-Strebel invariant was given by Bieri, Neumann, and Strebel (1987). The effect of this is to extend the theory significantly beyond metabelian groups. We shall briefly describe the main ideas involved.

Let Q be a finitely generated group and let n denote the Hirsch number of Q_{ab}. Let A be a finitely generated Q-operator group such that elements of Q' act as inner automorphisms on A. Then the generalized Bieri–Strebel invariant Σ_A is defined to consist of all $[\nu] \in S(Q)$ such that A is finitely generated over a finitely generated submonoid of Q_ν: here $S(Q)$ and Q_ν are defined in the same way as in the abelian case. It can be shown that Σ_A is an open subset of $S(Q)$. When A and Q are abelian, this reduces to the ordinary Bieri–Strebel invariant defined above.

Next let H be a subgroup of Q and define

$$S(Q, H) = \{[\nu] \in S(Q) \,|\, H^\nu = 0\}.$$

We record two fundamental results.

11.5.18 *Let Q be a finitely generated group, H a finitely generated subgroup of Q and A a finitely generated Q-operator group with elements of Q' acting as inner automorphisms on A. Then A is finitely generated as an H-operator group if and only if $S(Q, H) \subseteq \Sigma_A$.*

When A and Q are abelian and $H = Q$, this is 11.5.6. Another noteworthy result, which generalizes 11.5.5, is:

11.5.19 *Let G be a finitely presented group with no non-abelian free subgroups. Then $S(G) = \Sigma_{G'} \cup (-\Sigma_{G'})$.*

For further developments of the theory see Kochloukova (2002).

12

SUBNORMALITY AND SOLUBILITY

The topic of this chapter is the interplay between solubility and properties of the partially ordered set of subnormal subgroups of a group. Since soluble groups tend to have many subnormal subgroups, one would expect that properties of the subnormal subgroups would have a noticeable impact on the structure of a soluble group. This is already seen in the simple observation that the maximal condition for subnormal subgroups forces a soluble group to be polycyclic: many similar results were established in 6.3. There are of course numerous conditions other than chain conditions that could be imposed on the partially ordered set of subnormals.

In Section 12.1 we consider the effect of requiring intersections of subnormal subgroups to be subnormal, that is, greatest lower bounds must exist in the partially ordered set of subnormal subgroups. The finitely generated soluble groups with this subnormal intersection property (SIP) are characterized in 12.1.2 as the finitely generated finite-by-nilpotent groups (Robinson 1965).

One can also take the opposite point of view and ask about the consequences of a group having many subnormal subgroups. The natural question is about the groups with every subgroup subnormal. In a remarkable sequence of papers in the 1980s, W. Möhres proved that a group in which every subgroup is subnormal is soluble. In Sections 12.2–12.5 we give an account of the very difficult proof of Möhres's Theorem, although without presenting every detail. However, recent methods due to C. Casolo are utilized to prove in full that torsion-free groups with every subgroup subnormal are nilpotent, a result due to Casolo and H. Smith.

12.1 Soluble groups and the subnormal intersection property

It is a simple observation that the intersection of any *finite* set of subnormal subgroups of a group is itself subnormal. However the intersection of an *infinite* set of subnormals may fail to be subnormal, even in a polycyclic group. The simplest example of this phenomenon is provided by the infinite dihedral group,

$$G = \langle x, a \mid x^2 = 1, a^x = a^{-1} \rangle.$$

For, if $H_i = \langle x, a^{2^i} \rangle$, it is straightforward to show that H_i is subnormal in G. However, $\bigcap_{i=1,2,\ldots} H_i = \langle x \rangle$ is self-normalizing in G and so cannot be subnormal.

Let us say that the group G has the SIP if $\bigcap_{\lambda \in \Lambda} H_\lambda$ is subnormal in G whenever each H_λ is subnormal in G.

A related property is that of having *bounded subnormal defects:* a group G has this property if there is an integer $d \geq 0$ such that for any subnormal subgroup H of G there is a subnormal series from H to G of length at most d, that is, the subnormal defect of H in G is at most d: in symbols

$$H \vartriangleleft {}^d G.$$

At this point we recall the definition of the *chain of successive normal closures* of a subgroup H of a group G: the terms of this descending chain are defined recursively by

$$H^{G,0} = G \quad \text{and} \quad H^{G,i+1} = H^{H^{G,i}}.$$

It is a simple matter to prove that

$$H^{G,i} = H\,[G,{}_iH]$$

and that H is subnormal in G precisely when some $H^{G,i}$ equals H. The length of the chain $\{H^{G,i} \mid i = 0, 1, \dots \}$ is then equal to the subnormal defect of H in G.

We note the elementary lemma:

12.1.1 *Let G be any group.*

(i) *G has the SIP if and only if for each subgroup H of G there is an $i = i(H) \geq 0$, such that $H^{G,i} = H^{G,i+1} = \cdots$.*

(ii) *G has bounded subnormal defects if and only if there is an integer i such that $H^{G,i} = H^{G,i+1}$ for all subgroups H of G.*

The easy proof may be found in Lennox and Stonehewer (1987: 209), which is also a good general reference for subnormality. It is an immediate consequence that *every group with bounded subnormal defects has the SIP*. Now the converse is false even for metabelian groups: indeed the group $\mathbb{Z}_p \, wr \, \mathbb{Z}_{p^\infty}$ is a counter-example for every prime p (see the previous reference). Here we will prove that for finitely generated soluble groups the situation is entirely different.

12.1.2 (Robinson 1965) *Let G be a finitely generated soluble group. Then the following statements about G are equivalent:*

(i) *G has the SIP;*

(ii) *G has bounded subnormal indices;*

(iii) *G is finite-by-nilpotent.*

We begin the proof by establishing a special, but crucial, case of the theorem.

12.1.3 *Let $G = \langle x, A \rangle$ be a group with the SIP, where $A \vartriangleleft G$ and A is free abelian with finite rank. Then G is nilpotent.*

Proof We argue by induction on the rank of A, which can be assumed positive. Suppose first that A is G-rationally reducible, so that there exists a $B \vartriangleleft G$ with $1 < B < A$ and A/B torsion-free. By the induction hypothesis G/B is nilpotent, whence $\langle x, B \rangle$ is subnormal in G and so inherits the SIP. Therefore $\langle x, B \rangle$ is nilpotent, by induction again, and it follows that G is nilpotent.

Next assume A is rationally irreducible with respect to G. Then $C_A(x) \triangleleft G$ and $A/C_A(x)$ is torsion-free. Now $C_A(x) \neq 1$ would imply that $C_A(x) = A$ and G is abelian. Therefore we may assume that $C_A(x) = 1$ and hence that $\langle x \rangle \cap A = 1$.

Choose $t > 0$ so that $A \neq [A, x^t]$: for example, we could choose x^t to centralize the finite group A/A^2. Write $y = x^t$ and let ξ be the G-endomorphism of A which maps a to $[a, y]$. Then $\langle y \rangle^{G,r} = \langle y \rangle^{G,r+1}$ for some r by the SIP and 12.1.1. Since $\langle y \rangle \cap A = 1$, it follows that $A^{\xi^r} = A^{\xi^{r+1}}$. Now if $\mathrm{Ker}(\xi) = 1$, we could deduce that $A = A^{\xi} = [A, y]$, a contradiction. Hence $\mathrm{Ker}(\xi) \neq 1$ and by rational irreducibility we obtain $\mathrm{Ker}(\xi) = A$ and $[A, y] = 1$. Thus $y \in Z(G)$ and so we may pass to the group $G/\langle y \rangle$, that is, we can assume that x has finite order, say m.

Suppose that $m = p^i$ where p is a prime. Then G/A^p is a finite p-group, so that $[A, x] < A$, and, since $C_A(x) = 1$, we have $[A, {}_{i+1}x] < [A, {}_i x]$ for all $i \geq 1$. On the other hand, $[A, {}_r x] = [A, {}_{r+1}x]$ for some r by the SIP. By this contradiction m is not a prime power.

Write $m = uv$ where $1 < u, v < m$ and u, v are relatively prime. By induction on $|G : A|$ we see that $\langle x^u, A \rangle$ and $\langle x^v, A \rangle$ are nilpotent and hence, by rational irreducibility of A, abelian. Therefore G is abelian, which completes the proof. ∎

Proof of 12.1.2 Recall that G is a finitely generated soluble group. Suppose that G has a finite normal subgroup N, where G/N is nilpotent of class c. If H is subnormal in G, then clearly $H \triangleleft {}^d G$ where $d = |N| + c$. Hence G has bounded subnormal defects and (iii) implies (ii); also (ii) implies (i) by 12.1.1. Thus we are left to prove that (i) implies (iii).

Assume that the group G satisfies the SIP, but it is not finite-by-nilpotent, and suppose further that of such groups, G has minimal derived length $d > 1$. Let $B = G^{(d-1)}$: then G/B is finite-by-nilpotent and G satisfies $\max -n$ by 4.2.2. Hence there is a normal subgroup M which is maximal subject to G/M not being finite-by-nilpotent. Thus by replacing G by G/M, we may assume that all proper quotients of G, but not G itself, are finite-by-nilpotent. Also we have seen that G satisfies $\max -n$, so the Fitting subgroup A of G is nilpotent.

Suppose that $A' \neq 1$. Then G/A' is finite-by-nilpotent, so that the G-module A/A' is finite-by-polytrivial: it is also finitely generated as an abelian group, so G is polycyclic. It follows via the tensor product argument of 1.2.11 that each term of the lower central series of A is finite-by-polytrivial as a G-module. As a consequence, there is a normal series of G in which every infinite factor is G-central, say $1 = G_0 \triangleleft G_1 \triangleleft \cdots \triangleleft G_k = G$, where $k > 1$. Then G/G_1 is finite-by-nilpotent: let N/G_1 be finite and normal with G/N nilpotent. Since G_1 cannot be finite, it must be contained in $Z(G)$. Therefore, $N/Z(N)$ is finite and N' is finite. But proper quotients of G are finite-by-nilpotent, so N must be torsion-free and abelian. Next $N^\ell \leq Z(G)$ for some $\ell > 0$. Hence $1 = [N^\ell, G] = [N, G]^\ell$, from which we deduce that $[N, G] = 1$ and G is nilpotent. By this contradiction A must be abelian.

Now suppose G/A is finite. If $xA \in \text{Fit}(G/A)$, then $\langle x, A \rangle$ is subnormal in G and therefore it has the SIP. Also A is finitely generated and clearly it must be torsion-free. Applying 12.1.3, we conclude that $\langle x, A \rangle$ is nilpotent and hence that $x \in \text{Fit}(G) = A$, that is, $G = A$. By this contradiction G/A must be infinite.

Since G/A is finite-by-nilpotent, some (finite) term of the upper central series has finite index by a theorem of P. Hall (1956), (see also Robinson 1972b). Thus there must exist an element of infinite order in $Z(G/A)$, say xA. Then $\langle x, A \rangle$ has the SIP.

Let $t > 0$ and let ξ_t denote the G-endomorphism of A in which $a \mapsto [a, x^t]$. Since $\langle x \rangle \cap A = 1$, the SIP implies that $A^{\xi_t^r} = A^{\xi_t^{r+1}} = \text{etc.} = J_t$ say, for some $r > 0$. Also $\text{Ker}(\xi_t^i) \lhd G$ and G has max $-n$; therefore $\text{Ker}(\xi_t^s) = \text{Ker}(\xi_t^{s+1}) = \text{etc.} = K_t$ say, for large enough s. We may now invoke Fitting's Lemma to show that $A = J_t \times K_t$.

Now let t be chosen so that K_t is maximal. If $K_t = A$, then $\langle x^t, A \rangle$ is nilpotent and normal in G: but this implies that $x^t \in A$, which is impossible. Therefore $K_t < A$. Now the group G/K_t is residually finite by 4.3.1. Consequently, there exists $A_1 \lhd G$ such that $K_t \leq A_1 < A$ and A/A_1 is finite. Now for some $m > 0$ the element x^{tm} centralizes A/A_1, so $J_{tm} \leq A_1$. Also $K_t = K_{tm}$ by maximality of K_t. Therefore

$$A = J_{tm} \times K_{tm} \leq A_1 K_t = A_1 < A,$$

a final contradiction which completes the proof of 12.1.2.   ∎

We remark that 12.1.2 cannot be extended as it stands to finitely generated virtually soluble groups. Indeed the standard wreath product $\mathbb{Z}\,wr\,A_5$ has all subnormal defects at most 2, but it is not finite-by-nilpotent.

There has been considerable work on other types of soluble group with the SIP. McDougall (1970b) proved that *a soluble minimax group with the SIP is an extension of a radical abelian group with min by a virtually torsion-free nilpotent group*. Subsequently McCaughan (1974a) showed that *a soluble minimax group with the SIP has bounded subnormal defects*. However a complete characterization of these groups is still lacking. McCaughan also proved that for soluble groups with finite Prüfer rank the SIP does not imply that a bound for subnormal defects exists.

Soluble groups with bounded subnormal defects

Let $\mathbf{B}_d$ denote the class of groups in which every subnormal subgroup has defect at most d. The best known of these classes is $\mathbf{B}_1$, which is the class of groups in which normality is a transitive relation, or *T-groups*. Infinite soluble *T*-groups have been studied in detail by Robinson (1964). For example, there is the well-known result: *soluble T-groups are metabelian*. Also *a finitely generated soluble T-group is either finite or abelian*. (A complete classification of finite soluble *T*-groups had been given earlier by Gaschütz 1957.)

On the other hand, there are torsion-free soluble $\mathbf{B}_2$-groups of arbitrary derived length (Robinson 1967a). The situation is better for finite soluble groups since Casolo (1985) has shown that *a finite soluble group in the class* $\mathbf{B}_2$ *has derived length at most 5,* a bound which is best possible. For more on finite soluble $\mathbf{B}_d$-groups we refer the reader to Lennox and Stonehewer (1987).

Finally, we mention that the class of groups in which every subnormal subgroup is permutable has been the subject of investigation. (Recall that a subgroup H of a group G is said to be *permutable* if $HK = KH$ for all subgroups K of G.) This class of groups represents a considerable generalization of the class of T-groups. Infinite soluble groups with all subnormal subgroups permutable were classified by Menegazzo (1968, 1969) in a similar way to soluble T-groups.

12.2 Groups with every subgroup subnormal

For the remainder of this chapter we will be concerned with the following theorem of W. Möhres (1990), which must be regarded as one of the triumphs of infinite soluble group theory.

12.2.1 *A group in which every subgroup is subnormal is soluble.*

Before considering how the theorem might be proved, we shall give some account of its provenance. In the first place, it is a well-known consequence of Sylow's Theorem that a finite group in which every subgroup is subnormal is nilpotent. More generally Baer (1955a) established:

12.2.2 *A group which is generated by its subnormal abelian subgroups is locally nilpotent.*

This already provides valuable information about the groups with every subgroup subnormal: since such groups are locally nilpotent, the elements of finite order form a subgroup.

A critical development in the theory was the following theorem of Roseblade (1965), see also Robinson (1972b: vol. 2).

12.2.3 *Let G be a group in which every subgroup is subnormal with defect at most d. Then G is nilpotent with class bounded by some function of d.*

Any hope that groups in which all subgroups are subnormal but with unbounded defects would be nilpotent was dashed by Heineken and Mohamed (1968), who gave an example of a metabelian p-group in which every proper subgroup is subnormal and nilpotent, but which has trivial centre.

Subsequently many other counterexamples were given. For example, H. Smith (1983) found a non-nilpotent hypercentral group with every subgroup subnormal: this group has elements of infinite order. Then Menegazzo (1995) gave examples of soluble p-groups of arbitrary derived length which have all their subgroups subnormal. Other examples may be found in Meldrum (1973), Hartley (1974), and H. Smith (2002a).

Trivially a group with every subgroup subnormal has the SIP, but it will not have bounded subnormal defects unless it is nilpotent, by 12.2.3. Thus the counterexamples just mentioned provide more instances of groups with the SIP which have unbounded subnormal defects.

The conjecture that nilpotence is a consequence of all subgroups being subnormal having failed, there remained the question of solubility. Möhres's achievement was to answer this very difficult question in the positive.

Reductions

We begin by pointing out that it suffices to prove Möhres's theorem in two special cases.

12.2.4 *In order to prove 12.2.1 it suffices to deal with the case of a p-group and the case of a torsion-free group.*

Proof Assume that these cases have been dealt with. Let G be a group with every subgroup subnormal and let T denote its torsion subgroup. By hypothesis G/T and each primary component G_p are soluble.

We claim that for all but a finite number of primes p the groups G_p have subnormal defects not exceeding some number d. Indeed, if this were not so, there would exist an infinite sequence of primes $p_1, p_2, \ldots$ such that G_{p_i} has a subgroup H_i with subnormal defect d_i, where the d_i are unbounded. Set $H = \langle H_1, H_2, \ldots \rangle$: then $H \lhd^d G$ for some d, which implies that $H_i \lhd^d G_{p_i}$ for all i, a contradiction.

By the previous paragraph all but a finite number of the G_p's have bounded subnormal defects and so are nilpotent of bounded class by 12.2.3. It now follows that G is soluble. ∎

Our objective for much of the remainder of the chapter will be to establish a stronger result in the torsion-free case.

12.2.5 (Casolo 2001; H. Smith 2001*a*) *A torsion-free group in which every subgroup is subnormal is nilpotent.*

In proving this result we follow the methods of Casolo. To complete the proof of Möhres's theorem 12.2.1, it is also necessary to show that a *p*-group in which every subgroup is subnormal is soluble. This will not be done here since the proof is very long and complex: for the details the reader is referred to the original articles (Möhres 1989*a*, 1989*b*, 1990). The proof of 12.2.5 will be accomplished in 12.3 and 12.4. In the current section a number of special cases of Möhres's Theorem will handled.

The derived series

An important move in proving 12.2.5 is to establish:

12.2.6 (Casolo 1986) *The derived series of a group with all its subgroups subnormal terminates in finitely many steps.*

In the proof we will need two preliminary results.

12.2.7 *Let* $\mathbf{X}$ *be a class of groups and let* G *be an* $\mathbf{X}$*-group with all its subgroups subnormal. Then there exist a positive integer* r*, an* $\mathbf{X}$*-subgroup* K *and a finitely generated subgroup* H *of* K *such that* $L \triangleleft^r K$ *holds whenever* $H \leq L \leq K$ *and* $L \in \mathbf{X}$*.*

Proof Assume the result is false. We define two sequences of subgroups H_i, K_i, $i = 1, 2, \ldots$, such that H_i is finitely generated, $K_i \in \mathbf{X}$, $K_{i+1} \leq K_i$, and $H_j \leq K_i$ for all i, j.

To start the construction put $K_1 = G$ and $H_1 = 1$, and suppose that $H_1, \ldots, H_{i-1}, K_1, \ldots, K_{i-1}$ have already been suitably defined, where $i > 1$. Our task is to show how to define H_i and K_i. Now $K_{i-1} \in \mathbf{X}$ contains the finitely generated subgroup $J = \langle H_1, \ldots, H_{i-1} \rangle$. Since the result is assumed to be false, there must exist an $\mathbf{X}$-subgroup K_i of K_{i-1}, containing J, such that K_i has subnormal defect i or greater in K_{i-1}. Consequently, there exists a finite subset S of K_i such that $[K_{i-1,i-1}S] \not\leq K_i$. Now set $H_i = \langle S \rangle \leq K_i$. Then K_i contains $\langle H_1, \ldots, H_i \rangle$, so that the construction is complete.

Finally, put $H^* = \langle H_i \mid i = 1, 2, \ldots \rangle$; thus $H^* \leq K_i$ for all i. Now $H^* \triangleleft^d G$ for some d by hypothesis. Therefore

$$[K_{i-1,d}S] \leq [G,_d H^*] \leq H^* \leq K_i,$$

which shows that $d > i - 1$ for all $i > 1$, a contradiction. ∎

The second preliminary result needed is:

12.2.8 *Let* H *be a subgroup of a group* G *such that every subgroup containing* H *is subnormal in* G *with defect at most* n*. Then* $G^{(g(n))} \leq H$ *for some function* g*.*

Proof If $n \leq 1$, then $H \triangleleft G$ and G/H is a Dedekind group, that is, all its subgroups are normal. The structure of such groups is well known and in particular they are metabelian, (see Robinson 1996: 5.3): thus we may take $g(n)$ to be 2. Now let $n > 1$ and argue by induction on n.

If $H \leq L \leq H^G$, then $L^G = H^G$ and L has subnormal defect at most $n - 1$ in H^G. By the induction hypothesis $(H^G)^{(g(n-1))} \leq H$. Every subgroup of G/H^G is subnormal with defect at most n, so 12.2.3 may be applied to show that G/H^G is nilpotent of class at most $f(n) > 0$, where f is the function mentioned in 12.2.3. Therefore G/H^G is soluble with derived length at most $s = \log_2(f(n)) + 1$, and it follows that

$$G^{(s+g(n-1))} \leq H.$$

To complete the proof define $g(n)$ to be $s + g(n-1)$. ∎

Proof of 12.2.6 Let G be a residually soluble group all of whose subgroups are subnormal. Clearly it suffices to prove that G is soluble.

Let $\mathbf{X}$ be the class of all insoluble groups (including the trivial group) and assume that G is insoluble, so that $G \in \mathbf{X}$. According to 12.2.7 there exist a positive integer r, an insoluble subgroup K and a finitely generated subgroup H

of K such that $L \lhd^r K$ holds whenever $H \le L \le K$ and L is insoluble. For such a subgroup L, every subgroup of K containing L is also insoluble and so has subnormal defect at most r in K. Applying 12.2.8, we conclude that $K^{(t)} \le L$ for some $t = t(n)$ and any such L. If V is any insoluble subgroup of K, then $R = K^{(t)} \le \langle V, H \rangle$. Hence $R \le K^{(n)} H$ for all n since $K^{(n)}$ is insoluble. Now G is locally nilpotent, so H is nilpotent and hence soluble, say with derived length d. Writing $e = d + t + 1$, we have $R \le K^{(e)} H$, whence

$$R^{(d)} \le K^{(e)} H^{(d)} = K^{(e)} = (K^{(t)})^{(d+1)} = R^{(d+1)} = (R^{(d)})'.$$

Hence $R^{(d)}$ is perfect, so by the residual solubility of G we have $1 = R^{(d)} = K^{(e-1)}$ and K is soluble, a contradiction. ∎

We conclude by mentioning two consequences of 12.2.6.

12.2.9 *A hypercentral group all of whose subgroups are subnormal is soluble.*

This follows from the known fact that hypercentral groups are residually soluble. Another easy deduction is:

12.2.10 *A residually finite group all of whose subgroups are subnormal is soluble.*

Here the point is that every finite quotient of the group is nilpotent, so the group is residually nilpotent and hence residually soluble. Now apply 12.2.6.

12.3 Torsion-free groups with all subgroups subnormal — solubility

In this section we accomplish the first major step in proving the nilpotence of torsion-free groups with every subgroup subnormal by showing that such groups are soluble.

12.3.1 *A torsion-free group with every subgroup subnormal is soluble.*

In Section 12.4 we will complete the proof of 12.2.5 by establishing nilpotence. The proof of 12.3.1 is a good illustration of the power of isolator theory, as developed in Chapter 2. We begin with a number of preliminary lemmas.

12.3.2 *Let G be a countable locally nilpotent group, H a finitely generated subgroup and S a finite subset of G such that $H \cap S = \emptyset$. Then there is a subgroup K containing H such that $I_G(K) = G$ and $K \cap S = \emptyset$.*

Proof Let $G = \{x_i \mid i = 1, 2, \dots\}$. We show how to find a sequence of positive integers $m_1, m_2, \dots$ such that $\langle H, x_1^{m_1}, \dots, x_n^{m_n} \rangle$ has empty intersection with S for $n = 1, 2, \dots$. Assume that $m_1, \dots, m_{n-1}$ have been suitably defined and set $L = \langle H, x_1^{m_1}, \dots, x_{n-1}^{m_{n-1}} \rangle$. Then $M = \langle L, x_n \rangle$ is finitely generated and hence nilpotent. By 1.3.10 the subgroup L is the intersection of all the subgroups of finite index in M that contain it. Since S is finite and $L \cap S = \emptyset$, it follows that there is a subgroup V of finite index in M such that $L \le V$ and $V \cap S = \emptyset$. Now $x_n^{m_n} \in V$

for some $m_n > 0$ because $|M : V|$ is finite. Hence $\langle H, x_1^{m_1}, \ldots, x_n^{m_n} \rangle \cap S = \emptyset$ and the construction has been effected.

Finally define $K = \langle H, x_i^{m_i} \mid i = 1, 2, \ldots \rangle$. Then by construction $I_G(K) = G$, while

$$K \cap S = \bigcup_{n=1,2,\ldots} (\langle H, x_1^{m_1}, x_2^{m_2}, \ldots, x_n^{m_n} \rangle \cap S) = \emptyset,$$

as required.   ■

We now come to the key lemma, which is an extension of 12.2.7.

12.3.3 *Let G be a non-nilpotent torsion-free group with all of its subgroups subnormal. Then there exist a positive integer r, a non-nilpotent subgroup K and a finitely generated subgroup H of K such that $U \lhd^r K$ whenever $H \leq U \leq K$. Furthermore, if G is countable, K may be chosen so that $I_G(K) = G$.*

Proof Since G must have a countable non-nilpotent subgroup, there is no loss in assuming G to be countable. Suppose the result is false for G.

Assume we have defined a chain $H_1 \leq H_2 \leq \cdots \leq H_{i-1}$ of finitely generated subgroups and elements $x_1, x_2, \ldots, x_{i-1}$ such that $x_j \notin H_{i-1}$ for $j = 1, 2, \ldots, i - 1$. Applying 12.3.2, we may assert the existence of a subgroup K_i such that $I_G(K_i) = G$, $H_{i-1} \leq K_i$, and $x_j \notin K_i$ for $1 \leq j < i$. Since G is a non-nilpotent, torsion-free locally nilpotent group, we may apply 2.3.9(iii) to show that K_i cannot be nilpotent.

Since we are supposing the result to be false, there is a subgroup H_i of K_i such that $H_{i-1} \leq H_i$ and the subnormal defect of H_i in K_i is greater than i. Here we may take H_i to be finitely generated, because H_{i-1} is finitely generated. Now choose x_i from $[K_i, {}_i H_i] \backslash H_i$; then $x_j \notin H_i$ for $1 \leq j \leq i$ since $H_i \leq K_i$. This completes the construction of the H_i and x_i.

Now put $H = \bigcup_{i=1,2,\ldots} H_i$ and suppose that $H \lhd^d G$. Then by construction

$$x_d \in [K_d, {}_d H_d] \leq [G, {}_d H] \leq H,$$

so that $x_d \in H_i$ for sufficiently large i, which is a contradiction.   ■

Next comes an elementary result about hyperabelian groups, that is, groups which have an ascending series of normal subgroups with abelian factors. These groups turn out to be the key to proving solubility.

12.3.4 *Let G be an insoluble hyperabelian group. Then G contains an insoluble normal subgroup which is the union of an ascending chain of soluble normal subgroups of G.*

Proof Let $\{G_\alpha \mid \alpha \leq \beta\}$ be an ascending normal series in G with abelian factors. Since G is insoluble, there is a first ordinal γ such that G_γ is insoluble, and obviously γ must be a limit ordinal. For $n = 1, 2, \ldots$ we define

$$\gamma_n = \min\{\alpha \mid G_\alpha^{(n)} \neq 1\}.$$

Let $x \in G_\gamma$, so that $x \in G_\alpha$ for some $\alpha < \gamma$. Now, by definition of γ, the group G_α is soluble, with derived length n say. Then $\alpha < \gamma_n$ and so $x \in G_{\gamma_n}$, which shows that $G_\gamma \leq \bigcup_{n=1,2,\ldots} G_{\gamma_n}$.

Next suppose that $\gamma \leq \gamma_n$. Then

$$G_\gamma^{(n)} = \bigcup_{\alpha < \gamma} G_\alpha^{(n)} \leq \bigcup_{\alpha < \gamma_n} G_\alpha^{(n)} = 1,$$

which contradicts the definition of γ. Hence $\gamma_n < \gamma$ and

$$G_\gamma = \bigcup_{n=1,2,\ldots} G_{\gamma_n}.$$

Here G_γ is insoluble and the $G_{\gamma_n}, n = 1, 2, \ldots$, form the required ascending normal chain in G. ∎

We are now in a position to establish 12.3.1 for hyperabelian groups.

12.3.5 *A torsion-free hyperabelian group G with all its subgroups subnormal is soluble.*

Proof Clearly we can assume G to be countable. Suppose G is not soluble; then 12.3.4 shows that G contains an insoluble normal subgroup L which is the union of an ascending chain $\{L_n\}$ of soluble normal subgroups.

By 12.3.3 there exist a positive integer r, a (non-nilpotent) subgroup K of L and a finitely generated subgroup H of K such that $I_L(K) = L$ and $U \triangleleft^r K$ for all subgroups U satisfying $H \leq U \leq K$. Since H is finitely generated, $H \leq K \cap L_n$ for some n. Each subgroup of $K/(K \cap L_n)$ has subnormal defect at most r, so this group is nilpotent by 12.2.3. Thus K is soluble since $K \cap L_n$ is soluble. We may now apply 2.3.9(iv), together with $I_L(K) = L$, to obtain the contradiction that L is soluble. ∎

One further preliminary result is required, which demonstrates the essential role played by hyperabelian groups.

12.3.6 *Let G be a group in which every subgroup is subnormal. If G is insoluble, then it contains an insoluble hyperabelian subgroup.*

Proof Assume that all hyperabelian subgroups of G are soluble. We will show that G contains a non-trivial abelian normal subgroup.

First of all consider the situation

$$1 \neq U \triangleleft V \triangleleft W \leq G,$$

where U is abelian. We argue that W has a non-trivial abelian normal subgroup. Let F be the Fitting subgroup of V, so that $U \leq F \triangleleft W$. Furthermore F is hyperabelian, being generated by nilpotent normal subgroups of V. Thus F is soluble by the hypothesis, say of derived length n. Then $F^{(n-1)}$ is a non-trivial abelian normal subgroup of W.

Now let A be a non-trivial abelian subgroup of G. Then A is subnormal, so there is a series $A = A_0 \triangleleft A_1 \triangleleft \cdots \triangleleft A_{r-1} \triangleleft A_r = G$, where we may assume $r > 1$. By induction on r we may suppose that A_{r-1} contains a non-trivial abelian normal subgroup B. Now apply the result of the previous paragraph to the subgroups $B \triangleleft A_{r-1} \triangleleft G$ to reach the desired conclusion.

By repeatedly applying the fact just established, we see that G is hyperabelian. ∎

The truth of 12.3.1 is now an immediate consequence of 12.3.5 and 12.3.6.

12.4 Torsion-free groups with all subgroups subnormal — nilpotence

In this section we complete the proof of 12.2.5, by showing that a torsion-free group with all its subgroups subnormal is nilpotent.

Let f be the function in Roseblade's theorem (12.2.3): thus if $H \triangleleft^n G$ for all $H \leq G$, then G is nilpotent with class at most $f(n)$. It will be convenient to define for each prime p a related function g_p, of positive integral variables k, n, by

$$g_p(k,n) = (f(n) + 1)p^{[\log_p(k(f(n)+1))]+1}.$$

The key result for the proof is:

12.4.1 *Let $G = \langle x \rangle A$ be a nilpotent group, where A is a normal elementary abelian p-subgroup of G. Assume there is a finite subgroup F of A such that $F^G = A$ and let $|F| = p^k$. Suppose further that $H \triangleleft^n G$ whenever $F \leq H \leq G$. Then*

$$\left[A, {}_{g_p(k,n)-1}\, x\right] = 1.$$

Proof Set $r = f(n)$, $s = g_p(k,n)$, and $m = \left[\log_p(k(r+1))\right] + 1$: then $p^m > k(r+1)$ and $s = (r+1)p^m$. Certainly $[A, {}_t x] = 1$ for some t; we need to prove that $[A, {}_{s-1} x] = 1$ and by induction we may assume $[A, {}_s x] = 1$. Suppose that $[A, {}_{s-1} x] \neq 1$; then the subgroups $[A, {}_i x]$, $i = 0, 1, \ldots, s$, must all be distinct, since otherwise $[A, {}_{s-1} x] = [A, {}_s x] = 1$. From this it follows that

$$\left| [A, {}_{rp^m} x] \right| \geq p^{s - rp^m} = p^{(r+1)p^m - rp^m} = p^{p^m} > p^{k(r+1)} = |F|^{r+1}.$$

Since A is elementary abelian p, we have $[A, x^p] = [A, {}_p x]$. Therefore

$$\left[A, {}_{r+1} x^{p^m}\right] = [A, {}_{(r+1)p^m} x] = [A, {}_s x] = 1,$$

so that $\left[F, {}_{r+1} x^{p^m}\right] = 1$. By an easy commutator calculation it is seen that $F^{\langle x^{p^m} \rangle}$ is generated by all $\left[F, {}_i x^{p^m}\right]$ for $i = 0, 1, \ldots, r$. Since $\left| [F, {}_i x^{p^m}] \right| \leq |F|$, it follows that

$$\left| F^{\langle x^{p^m} \rangle} \right| \leq |F|^{r+1}.$$

Next put $H = \left\langle x^{p^m}, A \right\rangle$; then $F^H = F^{\langle x^{p^m} \rangle}$ since $F \leq A$ and A is abelian. By hypothesis each subgroup of H/F^H is subnormal with defect at most m, so H/F^H is nilpotent of class at most $r = f(n)$. Therefore

$$\left[A, _r x^{p^m} \right] \leq F^H = F^{\langle x^{p^m} \rangle}.$$

Since $\left[A, _r x^{p^m} \right] = [A, _{rp^m} x]$, we obtain the inequality

$$\left|[A, _{rp^m} x]\right| \leq \left| F^{\langle x^{p^m} \rangle} \right| \leq |F|^{r+1},$$

in contradiction to the inequality of the first paragraph. Hence $[A, _{s-1} x] = 1$. ∎

One further auxiliary result is needed for our proof of 12.2.5.

12.4.2 *Let G be a torsion-free locally nilpotent group and assume that A is an abelian normal subgroup of G such that G/A is abelian. Suppose further that there exists a finitely generated subgroup F of A and a non-negative integer n such that $F \leq H \leq G$ implies that $H \lhd^n G$. Then G is nilpotent with class bounded by a function of n and the rank of F.*

Proof Let r denote the rank of F. Choose any element x of G and put $X = F^{\langle x \rangle} \leq A$. Since G is locally nilpotent, $\langle x, F \rangle$ is finitely generated nilpotent and X is free abelian with finite rank, say s.

Set $Y = X^2$, so that $Y \lhd \langle x, F \rangle$ and $\bar{X} = X/Y$ is elementary abelian of order 2^s. Also write $\bar{F} = FY/Y$ and note that $|\bar{F}| \leq 2^r$: furthermore every subgroup of $\langle x, F \rangle /Y$, which contains $\bar{F}$ is subnormal with defect at most n. Observe also that

$$\bar{X} = X/Y = F^{\langle x \rangle} Y/Y = (\bar{F})^{\langle x \rangle}.$$

Apply 12.4.1 to the group $\langle xY, \bar{F} \rangle$ with $p = 2$. The conclusion is that $\left[\bar{X}, _t x \right] = 1$ where $t = g_2(r, n) - 1$. Let h be the smallest integer such that $2^h \geq t$. Then

$$\left[\bar{X}, x^{2^h} \right] = \left[\bar{X}, _{2^h} x \right] = 1,$$

since $\bar{X}$ is an elementary abelian 2-group. Consequently $\bar{F}$ has at most 2^h conjugates in $\langle x, F \rangle /Y$. Since $\bar{X}$ is generated by conjugates of $\bar{F}$ in $\langle x, F \rangle /Y$, it follows that

$$2^s = |\bar{X}| \leq |\bar{F}|^{2^h} \leq 2^{r2^h}.$$

Hence $s \leq r2^h$: at this point it is necessary to observe that h depends on r and n, and not on x.

Now we know that $X = F^{\langle x \rangle}$ is free abelian of rank at most $u = r2^h$. Since $\langle x, F \rangle$ is nilpotent, we may deduce that $[\langle x, F \rangle, _u x] = 1$ for all $x \in G$. Now $\langle x, F \rangle \lhd^n G$ by hypothesis. Therefore for any $x, g \in G$ we have

$$[g, _{n+u} x] = [[g, _n x], _u x] \in [\langle x, F \rangle, _u x] = 1.$$

It follows that G is an $(n + u)$—Engel group. Since G is also torsion-free and metabelian, we may deduce from a result of Gupta and Newman (1966)—see also (Robinson 1972b: vol. 2, theorem 7.36)—that G is nilpotent of class at most $n + u$, an integer that depends only on r and n. ∎

Finally, we are able to prove 12.2.5.

Proof of 12.2.5 Let G be a torsion-free group with every subgroup subnormal; then G is soluble by 12.3.1, say with derived length $d > 1$. We argue by induction on d. Suppose first that G is metabelian but non-nilpotent. By 12.3.3 there exist a non-nilpotent subgroup K, a finitely generated subgroup F of K and a positive integer n such that all subgroups of K containing F are subnormal of defect at most n. Clearly we may assume that $K = G$ here.

Put $H = FG'$ and $I = I_G(H')$; then $H \triangleleft G$ and hence $I \triangleleft G$. In addition G/I is torsion-free. If $x \in F$, then $\langle x, G' \rangle$ is nilpotent, because G' is abelian and $\langle x \rangle$ is subnormal in G. Since F is finitely generated, it follows that H is nilpotent and therefore, by 2.3.15, the group G/I cannot be nilpotent. This shows that we may pass to G/I, that is, we may assume that H is abelian. We are now in a position to apply 12.4.2 and deduce that G is nilpotent, a contradiction.

In the general case write $N = G'$. By induction on the derived length, N is nilpotent while $G/I_G(N')$ is nilpotent by the metabelian case. Therefore, G is nilpotent by a further application of 2.3.15. ∎

12.5 Torsion groups with all subgroups subnormal—recent developments

In this section we make some brief remarks on the proof of Möhres's theorem in the torsion case and we survey some recent results on the structure of groups with each subgroup subnormal.

First recall that, by 12.2.4, in order to establish solubility of a group G with every subgroup subnormal it is necessary to deal with two cases, G torsion-free and G a p-group. The first of these tasks has been accomplished, so we may assume G to be a p-group. Furthermore, by 12.3.6 one can also assume that G is hyperabelian. However, despite these reductions, the task of establishing solubility is still a formidable one and the proof is significantly more complicated than in the torsion-free case. We will merely outline some of the steps in the proof: the first of these is:

12.5.1 (Möhres 1989b,c) *A soluble group of finite exponent which has every subgroup subnormal is nilpotent.*

As had been noted previously by Heineken and Mohamed (1972), this theorem is essentially a problem about metabelian p-groups. It turns out that extensions of elementary abelian p-groups by elementary abelian p-groups play a crucial role, and Möhres (1989a) first investigated how in such a group the abelian quotient acts on the elementary abelian normal subgroup. The outcome of this investigation is applied to complete the proof of 12.5.1 in Möhres (1989b,c). The proof is a long sequence of delicately balanced arguments.

The next step in the proof is to extend 12.5.1 to:

12.5.2 (Möhres 1989*c*) *Let G be an extension of a nilpotent torsion group by a soluble group of finite exponent. If every subgroup of G is subnormal, then G is nilpotent.*

This time the proof reduces to dealing with metabelian p-groups which are abelian-by-elementary abelian. It uses 12.2.6 and Engel properties of metabelian p-groups.

Further developments

Since the appearance of Möhres's work there has been considerable activity in investigating the structure of groups with all subgroups subnormal. We will survey the main accomplishments without giving proofs.

Casolo (2002) settled a long standing open question by proving:

12.5.3 *A group in which every subgroup is subnormal is a Fitting group, that is, each element lies in a nilpotent normal subgroup.*

An even stronger result of Casolo (2003) has appeared recently, stating that even the normal closure of a nilpotent subgroup is nilpotent.

Casolo's arguments also yield an alternative proof of a result of H. Smith (2001*b*): *a hypercentral group of length ω with all subgroups subnormal is nilpotent.* Smith (1983) had previously given an example of a non-nilpotent, hypercentral metabelian group of length $\omega + 1$ with all subgroups subnormal: interestingly Smith's group has Prüfer rank 2.

Smith's example just mentioned contains elements of infinite order and in fact for hypercentral torsion groups nilpotence is valid.

12.5.4 (Möhres 1991) *A hypercentral torsion group with every subgroup subnormal is nilpotent.*

A further result of H. Smith (2000) may be mentioned at this point: *a nilpotent-by-(finite exponent) group with every subgroup subnormal is nilpotent.*

The structure of an arbitrary group in which every subgroup is subnormal is elucidated in an article of Casolo (2002) as follows.

12.5.5 *A torsion group with every subgroup subnormal has a nilpotent normal subgroup whose quotient is a radicable abelian group of finite rank.*

From this and 12.2.5 one can deduce a general structure theorem.

12.5.6 *A group G in which every subgroup is subnormal is an extension of a nilpotent torsion group by a nilpotent group.*

Proof By 12.2.5 and 12.5.5 we may factor out by a nilpotent normal subgroup: this allows us to assume that G has a radicable abelian torsion normal subgroup R of finite rank such that G/R is nilpotent. We need to show that G is nilpotent. If $g \in G$, then, since $\langle g \rangle$ is subnormal in G, we have $[R, {}_r g] = 1$, where r is the rank of R. From this it is easy to show that $R \leq Z_r(G)$, using a well-known theorem of Burnside on unipotent linear groups. Hence G is nilpotent.   ∎

In conclusion we mention that various generalizations of groups with all subgroups subnormal have been studied. Casolo and Mainardis (2003) have considered groups with every subgroup f-subnormal. Here *f-subnormal* means that there is a finite chain of subgroups leading from the subgroup to the group such that, for each pair of consecutive terms in the chain, either the index is finite or there is normality. It is shown that *groups with all their subgroups f-subnormal are finite-by-soluble*. Finally, groups in which all their non-nilpotent subgroups are subnormal are considered by H. Smith (2002c).

BIBLIOGRAPHY

Abels, H. (1979). An example of a finitely presented solvable group. In *Homological Group Theory* (Proceedings of Symposium, Durham, 1977), pp. 205–11, London Mathematical Society Lecture Note Series, 36. Cambridge University Press, Cambridge and New York.

Abels, H. and Brown, K. S. (1987). Finiteness properties of solvable S-arithmetic groups: an example. Proceedings of the Northwestern Conference on Cohomology of Groups (Evanston, Ill., 1985). *J. Pure Appl. Algebra*, **44**(1–3), 77–83.

Åberg, H. (1986). Bieri–Strebel valuations (of finite rank). *Proc. London Math. Soc.*, (3) **52**(2), 269–304.

Adyan, S. I. (1957). Unsolvability of some algorithmic problems in the theory of groups. (Russian). *Trudy Moskov. Mat. Obšč.*, **6**, 231–98.

Akhavan-Malayeri, M. and Rhemtulla, A. (1998). Commutator length of abelian-by-nilpotent groups. *Glasgow Math. J.*, **40**(1), 117–21.

Amberg, B., Franciosi, S., and de Giovanni, F. (1992). *Products of Groups.* Oxford University Press, New York.

Amberg, B. and Robinson, D. J. S. (1984). Soluble groups which are products of nilpotent minimax groups. *Arch. Math. (Basel)*, **42**(5), 385–90.

Andreadakis, S. and Gupta, C. K. (1991). Automorphism groups of free metabelian nilpotent groups. *Algebra and Logic*, **29**(6), 480–83.

Artemovich, O. D. (1991). Indecomposable metabelian groups. *Ukrainian Math. J.*, **42**(9), 1114–16.

Artemovich, O. D. (1998). Solvable groups with the minimality condition for non-hypercentral subgroups. (Ukrainian). *Dopov. Nats. Akad. Nauk Ukr. Mat. Prirodozn. Tekh. Nauki*, **11**, 7–9.

Artemovich, O. D. (1999). On solvable periodic groups with the maximality condition for subgroups that are not almost hypercentral. (Ukrainian). *Visn. Kiev. Univ. Ser. F Mat. Nauki*, **4**, 9–11.

Artemovich, O. D. (2000a). Solvable groups with many BFC-subgroups. *Publ. Mat.*, **44**(2), 491–501.

Artemovich, O. D. (2000b). Solvable groups with many conditions on nilpotent-by-Chernikov subgroups. *Mat. Stud.*, **13**(1), 23–32.

Atiyah, M. F. and Macdonald, I. G. (1969). *Introduction to Commutative Algebra.* Addison-Wesley Publishing Co., Reading, MA.

Auslander, L. (1960). Discrete solvable matrix groups. *Proc. Amer. Math. Soc.*, **11**, 687–8.

Auslander, L. (1961). A characterization of discrete solvable matrix groups. *Bull. Amer. Math. Soc.*, **67**, 235–6.

Auslander, L. (1967). On a problem of Philip Hall. *Ann. of Math.*, (2) **86**, 112–16.

Auslander, L. (1969). The automorphism group of a polycyclic group. *Ann. of Math.*, (2) **89**, 314–22.

Bachmuth, S. (1965). Automorphisms of free metabelian groups. *Trans. Amer. Math. Soc.*, **118**, 93-104.

Bachmuth, S. (1986). *Automorphisms of Solvable Groups I* (Proceedings of Groups—St. Andrews, 1985), pp. 1–14, London Mathematical Society Lecture Note Series, 121. Cambridge.

Bachmuth, S., Baumslag, G., Dyer, J., and Mochizuki, H. Y. (1987). Automorphism groups of two generator metabelian groups. *J. London Math. Soc.*, (2) **36**(3), 393–406.

Bachmuth, S. and Mochizuki, H. Y. (1967). Automorphisms of a class of metabelian groups. II. *Trans. Amer. Math. Soc.*, **127**, 294–301.

Bachmuth, S. and Mochizuki, H. Y. (1975). Automorphisms of solvable groups. *Bull. Amer. Math. Soc.*, **81**, 420–2.

Bachmuth, S. and Mochizuki, H. Y. (1978). IA-automorphisms of the free metabelian group of rank 3. *J. Algebra*, **55**(1), 106–15.

Bachmuth, S. and Mochizuki, H. Y. (1982). GL_n *and the Automorphism Groups of Free Metabelian Groups and Polynomial Rings* (Proceedings of Groups—St. Andrews, 1981), pp. 160–8, London Mathematical Society Lecture Note Series, 71. Cambridge.

Baer, R. (1937). Abelian groups without elements of finite order. *Duke Math J.*, **3**, 68–122.

Baer, R. (1949). Groups with descending chain condition for normal subgroups. *Duke Math. J.*, **16**, 1–22.

Baer, R. (1952). Endlichkeitskriterien für Kommutatorgruppen. *Math. Ann*, **124**, 161–77.

Baer, R. (1955*a*). Nilgruppen. (German). *Math. Z.*, **62**, 402–37.

Baer, R. (1955*b*). Supersoluble groups. *Proc. Amer. Math. Soc.*, **6**, 16–32.

Baer, R. (1955*c*). Auflösbare Gruppen mit Maximalbedingung. (German). *Math. Ann.*, **129**, 139–73.

Baer, R. (1955*d*). Finite extensions of Abelian groups with minimum condition. *Trans. Amer. Math. Soc.*, **79**, 521–40.

Baer, R. (1959). Überauflösbare Gruppen. (German). *Abh. Math. Sem. Univ. Hamburg*, **23**, 11–28.

Baer, R. (1964*a*). Groups with minimum condition. *Acta Arithmetica*, **9**, 117–32.

Baer, R. (1964*b*). Irreducible groups of automorphisms of abelian groups. *Pacific J. Math.*, **14**, 385–406.

Baer, R. (1965). Die Sternbedingung: Eine Erweiterung der Engelbedingung. (German). *Math. Ann.*, **162**, 54–73.

Baer, R. (1966*a*). Local and global hypercentrality and supersolubility. I, II. *Nederl. Akad. Wetensch. Proc. Ser. A*, **69**; *Indag. Math.*, **28**, 93–110, 111–26.

Baer, R. (1966*b*). Noethersche Gruppen. II. (German). *Math. Ann.*, **165**, 163–80.

Baer, R. (1967*a*). Normalisatorreiche Gruppen. (German). *Rend. Sem. Mat. Univ. Padova*, **38**, 358–450.

Baer, R. (1967*b*). Soluble artinian groups. *Canad. J. Math.*, **19**, 904–23.

Baer, R. (1967*c*). Auflösbare, artinsche, noethersche Gruppen. (German). *Math. Ann.*, **168**, 325–63.

Baer, R. (1967*d*). Noetherian soluble groups. In *Proceedings of International Conference Theory of Groups* (Canberra, 1965), pp. 17–32. Gordon and Breach, New York.

Baer, R. (1968). Polyminimaxgruppen. (German). *Math. Ann.*, **175**, 1–43.

Baer, R. (1969). Lokal endlich-auflösbare Gruppen mit endlichen Sylowuntergruppen. (German). *J. Reine Angew. Math.*, **239/240**, 109–44.

Baer, R. (1970). Fast-zyklische Gruppen. (German). *Arch. Math. (Basel)*, **21**, 225–39.

Baer, R. (1973). Einbettungseigenschaften von Normalteilern: der Schluß vom Endlichen aufs Unendliche. (German). In *Proceedings of the Second International Conference on the Theory of Groups* (Australian National University, Canberra, 1973), pp. 13–62, Lecture Notes in Mathematics, Vol. 372. Springer, Berlin, 1974.

Baer, R. and Heineken, H. (1972). Radical groups of finite abelian subgroup rank. *llinois J. Math.*, **16**, 533–80.

Balog, A., Pyber, L., and Mann, A. (2000). Polynomial index growth groups. *Int. J. Algebra Comput.*, **10**(6), 773–82.

Baudisch, A. (1988*a*). On stable solvable groups of bounded exponent. In *Proceedings of the 6th Easter Conference on Model Theory* (Wendisch Rietz, 1988), pp. 7–27, Seminarberichte 98. Humboldt University, Berlin.

Baudisch, A. (1988*b*). On the model theory of free metabelian groups of bounded exponent. In *Logic Colloquium '88* (Padova 1988), pp. 1–10, Studies in Logic Foundations of Mathematics, 127, North-Holland, Amsterdam, 1989.

Baudisch, A. and Wilson, J. S. (1992). Stable actions of torsion groups and stable soluble groups. *J. Algebra*, **153**(2), 453–7.

Baumslag, G. (1959). Wreath products and *p*-groups. *Proc. Cambridge Philos. Soc.*, **55**, 224–31.

Baumslag, G. (1960). Wreath products and finitely presented groups. *Math. Z.*, **7**, 22–8.

Baumslag, G. (1961). A remark on hyperabelian groups. *Arch. Math.*, **12**, 321–3.

Baumslag, G. (1967). Finitely presented groups. In *Proceedings of International Conference on Theory of Groups* (Canberra, 1965), pp. 37–50. Gordon and Breach, New York.

Baumslag, G. (1971). Lecture notes on nilpotent groups. In *Regional Conference Series in Mathematics*, No. 2 American Mathematical Society, Providence, RI.

Baumslag, G. (1972). A finitely presented metabelian group with a free abelian derived group of infinite rank. *Proc. Amer. Math. Soc.*, **35**, 61–2.

Baumslag, G. (1973*a*). A finitely presented solvable group that is not residually finite. *Math. Z.*, **133**, 125–7.

Baumslag, G. (1973*b*). Subgroups of finitely presented metabelian groups. Collection of articles dedicated to the memory of Hanna Neumann I. *J. Austral. Math. Soc.*, **16**, 98–110.

Baumslag, G. (1974). Finitely presented metabelian groups. In *Proceedings of the Second International Conference on the Theory of Groups* (Australian National University, Canberra, 1973), pp. 65–74. Lecture Notes in Mathematics, Vol. 372. Springer, Berlin.

Baumslag, G. (1984). *Algorithmically Insoluble Problems About Finitely Presented Solvable Groups, Lie and Associative Algebras* (Proceedings of Groups— Korea and Kyoungju, 1983, pp. 1–14. Lecture Notes in Math., 1098. Springer, Berlin.

Baumslag, G. (1990). Some reflections on finitely generated metabelian groups. In *Combinatorial Group Theory* (College Park, MD, 1988), pp. 1–9. Contemporary Mathematics, 109. American Mathematical Society, Providence, RI.

Baumslag, G. (1999). Finitely generated residually torsion-free nilpotent groups. I. *J. Austral. Math. Soc. Ser. A*, **67**(3), 289–317.

Baumslag, G. and Bieri, R. (1976). Constructable solvable groups. *Math. Z.*, **151**(3), 249–57.

Baumslag, G., Cannonito, F. B., and Miller, C. F., III. (1981*a*). Computable algebra and group embeddings. *J. Algebra*, **69**(1), 186–212.

Baumslag, G., Cannonito, F. B., and Miller, C. F., III. (1981*b*). Some recognizable properties of solvable groups. *Math. Z.*, **178**(3), 289–95.

Baumslag, G., Cannonito, F. B., and Robinson, D. J. S. (1994). The algorithmic theory of finitely generated metabelian groups. *Trans. Amer. Math. Soc.*, **344**(2), 629–48.

Baumslag, G., Cannonito, F. B., Robinson, D. J. S., and Segal, D. (1991). The algorithmic theory of polycyclic-by- finite groups. *J. Algebra*, **142**(1), 118–49.

Baumslag, G. and Dyer, E. (1982). The integral homology of finitely generated metabelian groups. I. *Amer. J. Math.*, **104**(1), 173–82.

Baumslag, G., Dyer, E., and Groves, J. R. J. (1987). The integral homology of finitely generated metabelian groups. II. *Amer. J. Math.*, **109**(1), 133–55.

Baumslag, G., Dyer, E., and Miller, C. F. III. (1983). On the integral homology of finitely presented groups. *Topology*, **22**(1), 27–46.

Baumslag, G., Gildenhuys, D., and Strebel, R. (1985*a*). Algorithmically insoluble problems about finitely presented solvable groups, Lie and associative algebras II. *J. Algebra*, **97**(1), 278–85.

Baumslag, G., Gildenhuys, D., and Strebel, R. (1986*b*). Algorithmically insoluble problems about finitely presented solvable groups, Lie and associative algebras. I. *J. Pure Appl. Algebra*, **39**(1–2), 53–94.

Baumslag, G., Gildenhuys, D., and Strebel, R. (1988). Algorithmically insoluble problems about finitely presented solvable groups, Lie and associative algebras. III. *J. Pure Appl. Algebra*, **54**(1), 1–35.

Baumslag, G. and Mahler, K. (1965). Equations in free metabelian groups. *Michigan Math. J.*, **12**, 417–20.

Bavard, C. and Meigniez, G. (1992). Commutateurs dans les groupes métabéliens. (French. English summary) (Commutators in metabelian groups). *Indag. Math. (N.S.)*, **3**(2), 129–35.

Beidleman, J. C. and Robinson, D. J. S. (1991). On the structure of the normal subgroups of a group: nilpotency. *Forum Math.*, **3**(6), 581–93.

Beidleman, J. C. and Robinson, D. J. S. (1992). On the structure of the normal subgroups of a group: supersolubility. *Rend. Sem. Mat. Univ. Padova*, **87**, 139–49.

Beidleman, J. C. and Robinson, D. J. S. (1998). The permutizer condition in infinite soluble groups. *J. Algebra*, **210**(1), 311–19.

Beidleman, J. C. and Smith, H. (1992). On supersolubility in some groups with finitely generated Fitting radical. *Proc. Amer. Math. Soc.*, **114**(2), 319–24.

Beidleman, J. C. and Smith, H. (1993*a*). On Frattini-like subgroups. *Glasgow Math. J.*, **35**(1), 95–8.

Beidleman, J. C. and Smith, H. (1993*b*). Corrigendum: 'On Frattini-like subgroups'. *Glasgow Math. J.*, **35**(3), 409.

Beidleman, J. C. and Smith, H. (1993*c*). On nonsupersoluble and nonpolycyclic normal subgroups. *Rend. Sem. Mat. Univ. Padova*, **89**, 47–56.

Belegradek, I. (2002). On Co-Hopfian nilpotent groups. *Bull London Math. Soc.*, **35**(2003), 805–11.

Bergman, George M. (1971). The logarithmic limit-set of an algebraic variety. *Trans. Amer. Math. Soc.*, **157**, 459–69.

Bergman, George M. (1989). HSP $\neq$ SHPS for metabelian groups, and related results. *Algebra Universalis*, **26**(3), 267–83.

Beuerle, J. R. and Kappe, L.-C. (2000). Infinite metacyclic groups and their non-abelian tensor squares. *Proc. Edinburgh Math. Soc.*, (2) **43**(3), 651–62.

Bieri, R. (1972). Über die cohomologische Dimension der auflösbaren Gruppen. (German). *Math. Z.*, **128**, 235–43.

Bieri, R. (1979). Finitely presented soluble groups. In *Séminaire d'Algèbre Paul Dubreil 31ème année* (Paris, 1977–1978), pp. 1–8, Lecture Notes in Mathematics, 40. Springer, Berlin.

Bieri, R. (1981). *Homological Dimension of Discrete Groups*, 2nd edn. Queen Mary College Mathematical Notes. Queen Mary College, Department of Pure Mathematics, London.

Bieri, R. (1985). Tensor powers of modules over finitely generated abelian groups. *J. Algebra*, **97**(1), 68–78.

Bieri, R. (1993). The geometric invariants of a group. A survey with emphasis on the homotopical approach. In *Geometric Group Theory*, Vol. 1 (Sussex, 1991), pp. 24–36, London Mathematical Society Lecture Note Series, 181. Cambridge University Press, Cambridge.

Bieri, R. and Groves, J. R. J. (1982). Metabelian groups of type $(FP)_\infty$ are virtually of type (FP). *Proc. London Math. Soc.*, (3) **45**(2), 365–84.

Bieri, R., Neumann, W. D., and Strebel, R. (1987). A geometric invariant of discrete groups. *Invent. Math.*, **90**(3), 451–77.

Bieri, R. and Renz, B. (1986). Invariants géométriques supérieurs d'un groupe discret. (French). *C. R. Acad. Sci. Paris Sr. I Math.*, **303**(10), 435–7.

Bieri, R. and Strebel, R. (1978). Almost finitely presented soluble groups. *Comment. Math. Helv.*, **53**(2), 258–78.

Bieri, R. and Strebel, R. (1979a). Metabelian quotients of finitely presented soluble groups are finitely presented. In *Homological Group Theory* (Proceedings of Symposium, Durham, 1977), pp. 231–4, London Mathematical Society Lecture Note Series, 36. Cambridge University Press, Cambridge and New York.

Bieri, R. and Strebel, R. (1979b). Soluble groups with coherent group rings. In *Homological Group Theory* (Proceedings of Symposium, Durham, 1977), pp. 235–40, London Mathematical Society Lecture Note Series, 36. Cambridge University Press, Cambridge and New York.

Bieri, R. and Strebel, R. (1980). Valuations and finitely presented metabelian groups. *Proc. London Math. Soc.*, (3) **41**(3), 439–64.

Bieri, R. and Strebel, R. (1981). A geometric invariant for modules over an abelian group. *J. Reine Angew. Math.*, **322**, 170–89.

Bieri, R. and Strebel, R. (1982). A geometric invariant for nilpotent-by-abelian-by-finite groups. *J. Pure Appl. Algebra*, **25**(1), 1–20.

Boler, J. (1976a). Conjugacy in abelian-by-cyclic groups. *Proc. Amer. Math. Soc.*, **55**(1), 17–21.

Boler, J. (1976b). Subgroups of finitely presented metabelian groups of finite rank. *J. Austral. Math. Soc. Ser. A*, **22**(4), 501–8.

Boone, W. W. (1955). Certain simple unsolvable problems of group theory. *Indig. Math.*, **16**, 231–7, 492–7; **17**, 252–6; **19**, **22–27**, 227–32.

Borevič, Z. I. and Šafarevič, I. R. (1966). *Number Theory*. Academic Press, New York.

Botto Mura, R. and Rhemtulla, A. H. (1974). Solvable groups in which every maximal partial order is isolated. *Pacific J. Math.*, **51**, 509–14.

Botto Mura, R. and Rhemtulla, A. H. (1975). Solvable R^*-groups. *Math. Z.*, **142**, 293–8.

Botto Mura, R. and Rhemtulla, A. H. (1977). *Orderable Groups*. Lecture Notes in Pure and Applied Mathematics, Vol. 27. Marcel Dekker, Inc., New York and Basel.

Bovdi, A. (1992). On group algebras with solvable unit groups. In *Proceedings of the International Conference on Algebra*, Part 1 (Novosibirsk, 1989), pp. 81–90, Contemporary Mathematics, 131, Part 1. American Mathematics Society, Providence, RI.

Bovdi, A. and Khripta, I. (1985). Solvable normal subgroups of the multiplicative group of a group ring of a periodic group. (Russian). *Tartu Riikl. Ül. Toimetised*, **700**, 3–10.

Bovdi, A. and Khripta, I. (1986). Group algebras with a polycyclic multiplicative group. (Russian). *Ukrain. Mat. Zh.*, **38**(3), 373–5, 407.

Bowers, J. F. (1960). On composition series of polycyclic groups. *J. London Math. Soc.*, **35**, 433–44.

Bowers, J. F. (1974). Completely infinite groups. *Proc. London Math. Soc.*, (3) **28**, 595–613.

Bowers, J. F. (1986*a*). The residual finiteness of completely infinite polycyclic groups. *Arch. Math. (Basel)*, **46**(2), 108–13.

Bowers, J. F. (1986*b*). Normal series of soluble groups of finite rank. *Arch. Math. (Basel)*, **46**(4), 289–98.

Bowers, J. F. and Stonehewer, S. E. (1973). A theorem of Mal'cev on periodic subgroups of soluble groups. *Bull. London Math. Soc.*, **5**, 323–4.

Brandl, R. (1983). Infinite soluble groups with Engel cycles; a finiteness condition. *Math. Z.*, **182**(2), 259–64.

Brandl, R. (1987). Infinite soluble groups with the Bell property: a finiteness condition. *Monatsh. Math.*, **104**(3), 191–7.

Brandl, R. (1993). Commutator sequences in soluble groups; an algorithmic approach. *Bull. Soc. Math. Belg. Sér. B*, **45**(2), 137–50.

Brandl, R. A. and Brookes, C. J. B. (1989). Engel-like elements in infinite soluble groups. *Proc. Edinburgh Math. Soc.*, (2) **32**(3), 337–43.

Brandl, R., Franciosi, S., and de Giovanni, F. (1990). On the Wielandt subgroup of infinite soluble groups. *Glasgow Math. J.*, **32**(2), 121–5.

Brookes, C. J. B. (1983). Groups with every subgroup subnormal. *Bull. London Math. Soc.*, **15**(3), 235–8.

Brookes, C. J. B. (1985). Abelian subgroups and Engel elements of soluble groups. *J. London Math. Soc.*, (2) **32**(3), 467–76.

Brookes, C. J. B. (1986). Engel elements of soluble groups. *Bull. London Math. Soc.*, **18**(1), 7–10.

Brookes, C. J. B. (1988). Modules over polycyclic groups. *Proc. London Math. Soc.*, (3) **57**(1), 88–108.

Brookes, C. J. B., Roseblade, J. E., and Wilson, J. S. (1997). Exterior powers of modules for group rings of polycyclic groups. *J. London Math. Soc.*, (2) **56**(2), 231–44.

Brown, K. A. (1979). The derived subgroup of a free metabelian group. *Arch. Math. (Basel)*, **32**(6), 526–9.

Brown, K. A. (1981*a*). Primitive group rings of soluble groups. *Arch. Math. (Basel)*, **36**(5), 404–13.

Brown, K. A. (1981*b*). Modules over polycyclic groups have many irreducible images. *Glasgow Math. J.*, **22**(2), 141–50.

Brown, K. A. (1981*c*). The structure of modules over polycyclic groups. *Math. Proc. Cambridge Philos. Soc.*, **89**(2), 257–83.

Brown, K. A. (1993). Finitely presented groups and the finite generation of exterior Powers. In *Combinatorial and Geometric Group Theory* (Edinburgh, 1993), pp. 16–28, London Mathematical Society Lecture Note Series, 204. Cambridge University Press, Cambridge, 1995.

Brown, K. A. and Wehrfritz, B. A. F. (1984). Division rings associated with polycyclic groups. *J. London Math. Soc.*, (2) **30**(3), 465–7.

Bruno, B. and Phillips, R. E. (1991). On multipliers of Heineken–Mohamed type groups. *Rend. Sem. Mat. Univ. Padova*, **85**, 133–46.

Bryant, R. M. (1977). The verbal topology of a group. *J. Algebra*, **48**(2), 340–6.

Bryant, R. M. and Gupta, C. K. (1984). Characteristic subgroups of free centre-by-metabelian groups. *J. London Math. Soc.*, (2) **29**(3), 435–40.

Bryant, R. M. and Gupta, C. K. (1989). Automorphism groups of free nilpotent groups. *Arch. Math. (Basel)*, **52**(4), 313–20.

Bryant, R. M. and Gupta, C. K. (1993). Automorphisms of free nilpotent-by-abelian groups. *Math. Proc. Cambridge Philos. Soc.*, **114**(1), 143–7.

Buckley, J. T., Lennox, J. C., Neumann, B. H., Smith, H., and Wiegold, J. (1995). Groups with all subgroups normal-by-finite. *J. Austral. Math. Soc. Ser. A*, **59**(3), 384–98.

Buckley, J. and Wiegold, J. (1978). On the number of outer automorphisms of an infinite nilpotent *p*-group. *Arch. Math. (Basel)*, **31**(4), 321–8.

Buckley, J. and Wiegold, J. (1981). On the number of outer automorphisms of an infinite nilpotent *p*-group. II. *Arch. Math. (Basel)*, **36**(1), 1–5.

Buckley, J. and Wiegold, J. (1986). Nilpotent extensions of abelian *p*-groups. *Canad. J. Math.*, **38**(5), 1025–52.

Burns, R. G., Okoh, F., Smith, H., and Wiegold, J. (1984). On the number of normal subgroups of an uncountable soluble group. *Arch. Math. (Basel)*, **42**(4), 289–95.

Campbell, C. M. and Robertson, E. F. (1980). On 2-generator 2-relation soluble groups. *Proc. Edinburgh Math. Soc.*, (2) **23**(3), 269–73.

Cannonito, F. B. (1980). Two decidable Markov properties over a class of solvable groups. *Algebra i Logika*, **19**(6), 646–58, 745.

Cannonito, F. B. and Gupta, N. D. (1983). On centre-by-free solvable groups. *Arch. Math. (Basel)*, **41**(6), 493–7.

Cannonito, F. B. and Robinson, D. J. S. (1984). The word problem for finitely generated soluble groups of finite rank. *Bull. London Math. Soc.*, **16**(1), 43–6.

Čarin, V. S. (1949). A remark on the minimal condition for subgroups. (Russian). *Dokl. Akad. Nauk SSSR (N.S.)*, **66**, 575–6.

Čarin, V. S. (1954). On groups of automorphisms of nilpotent groups. (Russian). *Ukrain. Mat. Ž.*, **6**, 295–304.

Čarin, V. S. (1956). On the theory of nilpotent groups. (Russian). *Učen. Zap. Ural. Gos. Univ.*, (vyp.) **19**, 21–5.

Čarin, V. S. (1957). On locally solvable groups of finite rank. (Russian). *Mat. Sb. (N.S.)*, **41**(83), 37–48.

Čarin, V. S. (1960). Solvable groups of type A_4. (Russian). *Mat. Sb. (N.S.)*, **52**(94), 895–914.

Čarin, V. S. (1961). Solvable groups of type A_3. (Russian). *Mat. Sb. (N.S.)*, **54**(96), 489–99.

Casolo, C. (1985). On groups with all subgroups subnormal. *Bull. London Math. Soc.*, **17**(4), 397.

Casolo, C. (1986). Groups in which all subgroups are subnormal. *Rend. Accad. Naz. Sci. XL Mem. Mat.*, (5) **10**(1), 247–9.

Casolo, C. (1987). Groups with subnormal subgroups of bounded defect. *Rend. Sem. Mat. Univ. Padova*, **77**, 177–87.

Casolo, C. (2001). Torsion-free groups in which every subgroup is subnormal. *Rend. Circ. Mat. Palermo*, (2) **50**(2), 321–4.

Casolo, C. (2002). On the structure of groups with all subgroups subnormal. *J. Group Theory*, **5**(3), 293–300.

Casolo, C. (2003). Nilpotent subgroups of groups with all subgroups subnormal. *Bull. London Math. Soc.*, **35**(1), 15–22.

Casolo, C. and Mainardis, M. (2002). Groups in which every subgroup is f-subnormal. *J. Group Theory*, **4**(3), 341–65.

Černikov, S. N. (1940). Infinite locally soluble Groups. (Russian). *Mat. Sb.*, **7**, 35–64.

Černikov, S. N. (1946). Complete groups with an ascending central series. (Russian). *Mat. Sb.*, **18**, 397–422.

Černikov, S. N. (1948). On the theory of complete groups. (Russian). *Mat. Sb.*, **22**, 319–48, 455–6. *Amer. Math. Soc. Translations*, (1) **56**, 3–49, 1951.

Černikov, S. N. (1950). On the centralizer of a complete Abelian normal divisor in an infinite periodic group. (Russian) *Dokl. Akad. Nauk SSSR (N.S.)*, **72**, 243–6.

Chamberlain, R. F. (1988). Solvable groups with certain finite homomorphic images cyclic. *Arch. Math. (Basel)*, **51**(1), 1–12.

Chamberlain, R. F. and Kappe, L.-C. (1987). Nilpotent groups with every finite homomorphic image cyclic. *Arch. Math. (Basel)*, **49**(1), 1–11.

Cherlin, G. L. and Felgner, U. (1991). Homogeneous solvable groups. *J. London Math. Soc.*, (2) **44**(1), 102–20.

Cid, C. F. (2002). Torsion-free metabelian groups with commutator quotient $C_{p^n} \times C_{p^m}$. *J. Algebra*, **248**(1), 15–36.

Cossey, J. (1991). The Wielandt subgroup of a polycyclic group. *Glasgow Math. J.*, **33**(2), 231–4.

Crawley-Boevey, W. W., Kropholler, P. H., and Linnell, P. A. (1988). Torsion-free soluble groups, completions, and the zero divisor conjecture. *J. Pure Appl. Algebra*, **54**(2–3), 181–96.

Curtis, C. W. and Reiner, I. (1966). *Representation Theory of Groups and Associative Algebras*, 2nd edn. Interscience, New York.

Curzio, M., Lennox, J. C., Rhemtulla, A. H., and Wiegold, J. (1990). Groups with many permutable subgroups. *J. Austral. Math. Soc. Ser. A*, **48**(3), 397–401.

Dark, R. S. and Rhemtulla, A. H. (1970). On R_0-closed classes and finitely generated groups. *Canad. J. Math.*, **22**, 176–84.

Dekimpe, K. (1999). *Polycyclic-by-finite Groups: From Affine to Polynomial Structures* (Proceedings of Groups—St. Andrews, 1997) Bath I, pp. 219–36. London Mathematics Society Lecture Note Series, 260. Cambridge University Press, Cambridge.

Dekimpe, K. (2000). Polynomial structures on polycyclic groups: recent developments. In *Crystallographic Groups and Their Generalizations* (Kortrijk, 1999), pp. 99–120, Contemporary Mathematics, 262. Amer. Math. Soc., Providence, RI.

Dekimpe, K. and Igodt, P. (1997*a*). Polycyclic-by-finite groups admit a bounded-degree polynomial structure. *Invent. Math.*, **129**(1), 121–40.

Dekimpe, K. and Igodt, P. (1997*b*). Polynomial structures on polycyclic groups. *Trans. Amer. Math. Soc.*, **349**(9), 3597–610.

Dekimpe, K., Igodt, P., and Lee, K. B. (1996). Polynomial structures for nilpotent groups. *Trans. Amer. Math. Soc.*, **348**(1), 77–97.

Delizia, C. (1994). Finitely generated soluble groups with a condition on infinite subsets. *Istit. Lombardo Accad. Sci. Lett. Rend. A*, **128**(2), 201–8.

Delizia, C. (1995). On groups with a nilpotence condition on infinite subsets. *Algebra Colloq.*, **2**(2), 97–104.

Delizia, C. (1996). On certain residually finite groups. *Comm. Algebra*, **24**(11), 3531–5.

Delizia, C. (1999). A nilpotency condition for finitely generated soluble groups. *Atti Accad. Naz. Lincei Cl. Sci. Fis. Mat. Natur. Rend. Lincei* (9) *Mat. Appl.*, **9**(4), 237–9.

Delizia, C. and Nicotera, C. (2001). On residually finite groups with an Engel condition on infinite subsets. *Houston J. Math.*, **27**(4), 757–61.

Delizia, C., Rhemtulla, A., and Smith, H. (2000). Locally graded groups with a nilpotency condition on infinite subsets. *J. Austral. Math. Soc. Ser. A*, **69**(3), 415–20.

Detomi, E. (2001). On the nilpotent length of some residually finite groups, *J. Group Theory*, **4**(2), 193–7.

Dixon, J. D. (1967). The Fitting subgroup of a linear solvable group. *J. Austral. Math. Soc.*, **7**, 417–24.

Dixon, J. D. (1968). The solvable length of a solvable linear group. *Math. Z.*, **107**, 151–8.

Dixon, M. R. (1994). *Sylow Theory, Formations and Fitting Classes in Locally Finite Groups.* World Scientific Publishing, Singapore.

Dixon, J. D., du Sautoy, M. P. F., Mann, A., and Segal, D. (1999). *Analytic Pro-p Groups*, 2nd edn. Cambridge Studies in Advanced Mathematics, 61. Cambridge University Press, Cambridge.

Doerk, K. and Hawkes, T. O. (1992). *Finite Soluble Groups.* De Gruyter Expositions in Mathematics, 4. Walter de Gruyter & Co., Berlin.

Donkin, S. (1981). Polycyclic groups, Lie algebras and algebraic groups. *J. Reine Angew. Math.*, **326**, 104–23.

Donkin, S. (1982). Locally finite representations of polycyclic-by-finite groups. *Proc. London Math. Soc.*, (3) **44**(2), 333–48.

Duguid, A. M. and McLain, D. H. (1956). FC-nilpotent and FC-soluble groups. *Proc. Cambridge Philos. Soc.*, **52**, 391–8.

Eckmann, B. (1946). Der Cohomologie-Ring einer beliebigen Gruppe. (German). *Comment. Math. Helv.*, **18**, 232–82.

Eick, B. (2001). On the Fitting subgroup of a polycyclic-by-finite group and its applications. *J. Algebra*, **242**(1), 176–87.

Eilenberg, S. and MacLane, S. (1947). Cohomology theory in abstract groups I, II. *Ann. of Math.*, (2) **48**, 51–78, 326–41.

Eisenbud, D. (1995). *Commutative Algebra. With a View Toward Algebraic Geometry.* Graduate Texts in Mathematics, 150. Springer-Verlag, New York, xvi+785 pp.

Endimioni, G. A. (1994). Groups covered by finitely many nilpotent subgroups. *Bull. Austral. Math. Soc.*, **50**(3), 459–64.

Endimioni, G. A. (1997). A characterization of nilpotent-by-finite groups in the class of finitely generated soluble groups. *Comm. Algebra*, **25**(4), 1159–68.

Endimioni, G. A. (1998). On the nilpotent length of polycyclic groups. *J. Algebra*, **203**(1), 125–33.

Enochs, K. and Nesin, A. (1990). On 2-step solvable groups of finite Morley rank. *Proc. Amer. Math. Soc.*, **110**(2), 479–89.

Evans, M. J. (1994). Presentations of the free metabelian group of rank 2. *Canad. Math. Bull.*, **37**(4), 468–72.

Evans, M. J. (1998). Presentations of free abelian-by-(nilpotent of class 2) groups. *Bull. London Math. Soc.*, **30**(2), 136–44.

Evans, M. J. (1999). Primitive elements in the free metabelian group of rank 3. *J. Algebra*, **220**(2), 475–91.

Farkas, D. R. (1982). Endomorphisms of polycyclic groups. *Math. Z.*, **181**(4), 567–74.

Farkas, D. R. and Snider, R. L. (1979). Induced representations of polycyclic groups. *Proc. London Math. Soc.*, (3) **39**(2), 193–207.

Feit, W. and Thompson, J. G. (1963). Solvability of groups of odd order. *Pacific J. Math.*, **13**, 775–1029.

Fitting, H. (1938). Beiträge zur Theorie der Gruppen endlicher Ordnung. *Jahresber. Deutsch. Math. Verein.*, **48**, 77–141.

Formanek, E. W. (1976). Conjugate separability in polycyclic groups. *J. Algebra*, **42**(1), 1–10.

Franciosi, S. and de Giovanni, F. (1986). Soluble groups with many Černikov quotients. *Atti Accad. Naz. Lincei Rend. Cl. Sci. Fis. Mat. Natur.*, (8) **79**(1–4), 1985, (19–24).

Franciosi, S. and de Giovanni, F. (1989). Soluble groups with many nilpotent quotients. *Proc. Roy. Irish Acad. Sect. A*, **89**(1), 43–52.

Franciosi, S. and de Giovanni, F. (1996). Frattini properties of groups with polycyclic-by-finite conjugacy classes. *Boll. Un. Mat. Ital. A*, (7) **10**(3), 653–9.

Franciosi, S. and de Giovanni, F. (1998). On groups with many nearly maximal subgroups. *Atti Accad. Naz. Lincei Cl. Sci. Fis. Mat. Natur. Rend. Lincei* (9) *Mat. Appl.*, **9**(1), 19–23.

Franciosi, S., de Giovanni, F., Heineken, H., and Newell, M. L. (1991). On the Fitting length of a soluble product of nilpotent groups. *Arch. Math. (Basel)*, **57**(4), 313–18.

Franciosi, S., de Giovanni, F., and Newell, M. L. (2000). Groups with polycyclic non-normal subgroups. *Algebra Colloq.*, **7**(1), 33–42.

Franciosi, S., de Giovanni, F., and Sysak, Y. P. (1999). Groups with many polycyclic-by-nilpotent subgroups. *Ricerche Mat.*, **48**(2), 361–78.

Frattini, G. (1885). Intorno alle generazione dei gruppi di operazioni I, II. (Italian). *Rend. Accad. Naz. Lincei*, (4) **1**, 281–5, 455–7.

Frick, M. and Newman, M. F. (1972). Soluble linear groups. *Bull. Austral. Math. Soc.*, **6**, 31–44.

Fuchs, L. (1960). *Abelian Groups*. Pergamon Press. Oxford.

Fuchs, L. (1970). *Infinite Abelian Groups*, Vols. 1, 2. Pure and Applied Mathematics, Vol. 36. Academic Press, New York and London, 1970–3.

Gagen, T. M. (1980). Some finite solvable groups with no outer automorphisms. *J. Algebra*, **65**(1), 84–94.

Gagen, T. M. and Robinson, D. J. S. (1979). Finite metabelian groups with no outer automorphisms. *Arch. Math. (Basel)*, **32**(5), 417–23.

Gaglione, A. M. and Waldinger, H. (1990). The commutator collection process. In *Combinatorial Group Theory* (College Park, MD, 1988), pp. 43–58, Contemporary Mathematics, 109. Amer. Math. Soc., Providence, RI.

Garrison, D. J. and Kappe, L.-C. (1996). Metabelian groups with all cyclic subgroups subnormal of bounded defect. In *Infinite groups 1994* (Ravello), pp. 73–85. de Gruyter, Berlin.

Gaschütz, W. (1953). Über die Φ-Untergruppe endlicher Gruppen. (German). *Math. Z.*, **58**, 160–70.

Gaschütz, W. (1957). Gruppen, in denen das Normalteilersein transitiv ist. (German). *J. Reine Angew. Math.*, **198**, 87–92.

Gaschütz, W. (1965). Kohomologische Trivialitäten und äussere Automorphismen von p-Gruppen. (German). *Math. Z.*, **88**, 432–3.

Gaschütz, W. (1966). Nichtabelsche p-Gruppen besitzen äussere p-Automorphismen. (German). *J. Algebra*, **4**, 1–2.

Gildenhuys, D. (1979). Classification of soluble groups of cohomological dimension two. *Math. Z.*, **166**(1), 21–5.

Gildenhuys, D. and Strebel, R. (1981). On the cohomological dimension of soluble groups. *Canad. Math. Bull.*, **24**(4), 385–92.

Gildenhuys, D. and Strebel, R. (1982). On the cohomology of soluble groups. II. *J. Pure Appl. Algebra*, **26**(3), 293–323.

de Giovanni, F. (1991). Soluble groups with many min-by-max quotients. *Boll. Un. Mat. Ital. B*, (7) **5**(2), 449–62.

de Giovanni, F, Paek, D. H., Robinson, D. J. S., and Russo, A. (2004). The maximal and minimal conditions for normal subgroups of infinite order or index. *Commun in Algebra*. To appear.

Gluškov, V. M. (1952). On the central series of infinite groups. (Russian). *Mat. Sbornik N.S.*, **31**(73), 491–6.

Groves, D. P. (2001a). Some properties of free groups of some soluble varieties of groups. *J. London Math. Soc.*, (2) **63**(3), 592–606.

Groves, D. P. (2001b). Free groups of outer commutator varieties of groups. *J. London Math. Soc.*, (2) **64**(2), 423–35.

Groves, J. R. J. (1971*a*). On varieties of soluble groups. *Bull. Austral. Math. Soc.*, **5**, 95–109.

Groves, J. R. J. (1971*b*). Varieties of soluble groups and a dichotomy of P. Hall. *Bull. Austral. Math. Soc.*, **5**, 391–410.

Groves, J. R. J. (1972). An extension of a dichotomy of P. Hall to some varieties of soluble groups. *Arch. Math. (Basel)*, **23**, 573–80.

Groves, J. R. J. (1978*a*). Finitely presented centre-by-metabelian groups. *J. London Math. Soc.*, (2) **18**(1), 65–9.

Groves, J. R. J. (1978*b*). Soluble groups in which every finitely generated subgroup is finitely presented. *J. Austral. Math. Soc. Ser. A*, **26**(1), 115–25.

Groves, J. R. J. (1978*c*). Soluble groups with every proper quotient polycyclic. *Illinois J. Math.*, **22**(1), 90–5.

Groves, J. R. J. (1982). Metabelian groups with finitely generated integral homology. *Quart. J. Math. Oxford Ser.*, (2) **33**(132), 405–20.

Groves, J. R. J. (1983). A conjecture of Lennox and Wiegold concerning supersoluble groups. *J. Austral. Math. Soc. Ser. A*, **35**(2), 218–20.

Groves, J. R. J. (1985). Some examples of finiteness conditions in centre-by-metabelian groups. *J. Austral. Math. Soc. Ser. A*, **38**(2), 171–4.

Groves, J. R. J. (1991). Some finitely presented nilpotent-by-abelian groups. *J. Algebra*, **144**(1), 127–66.

Groves, J. R. J. (1993). HNN-extensions of finitely presented soluble groups. *J. Algebra*, **162**(1), 12–27.

Groves, J. R. J. and Smith, G. C. (1993). Soluble groups with a finite rewriting system. *Proc. Edinburgh Math. Soc.*, (2) **36**(2), 283–8.

Groves, J. R. J. and Wilson, J. S. (1994). Finitely presented metanilpotent groups. *J. London Math. Soc.*, (2) **50**(1), 87–104.

Gruenberg, K. W. (1953). Two theorems on Engel groups. *Proc. Cambridge Philos. Soc.*, **49**, 377–80.

Gruenberg, K. W. (1957). Residual properties of infinite soluble groups. *Proc. London Math. Soc.*, (3) **7**, 29–62.

Gruenberg, K. W. (1959). The Engel elements of a soluble group. Illinois *J. Math.*, **3**, 151–68.

Gruenberg, K. W. (1960). Resolutions by relations. *J. London Math. Soc.*, **35**, 481–94.

Gruenberg, K. W. (1961). The upper central series in soluble groups. Illinois *J. Math.*, **5**, 436–66.

Gruenberg, K. W. (1967). *Some Cohomological Topics in Group Theory*. Queen Mary College Mathematics Notes Queen Mary College, London.

Gruenberg, K. W. (1970). *Cohomological Topics in Group Theory*. Lecture Notes in Mathematics, Vol. 143. Springer-Verlag, Berlin and New York.

Gruenberg, K. W. and Roseblade, J. E. (1984). Group theory. In *Essays for Philip Hall.* (ed. K. W. Gruenberg and J. E. Roseblade). Academic Press, Inc. (Harcourt Brace Jovanovich, Publishers), London.

Grunewald, F. J., Pickel, P. F., and Segal, D. (1979). Finiteness theorems for polycyclic groups. *Bull. Amer. Math. Soc. (N.S.)*, **1**(30), 575–78.

Grunewald, F. J., Pickel, P. F., and Segal, D. (1980). Polycyclic groups with isomorphic finite quotients. *Ann. of Math.*, (2) **111**(1), 155–95.

Grunewald, F. and Segal, D. (1975). Residual nilpotence in polycyclic groups. *Math. Z.*, **142**, 229–41.

Grunewald, F. and Segal, D. (1978*a*). Conjugacy in polycyclic groups. *Comm. Algebra*, **6**(8), 775–98.

Grunewald, F. and Segal, D. (1978*b*). On polycyclic groups with isomorphic finite quotients. *Math. Proc. Cambridge Philos. Soc.*, **84**(2), 235–46.

Grunewald, F. and Segal, D. (1978*c*). A note on arithmetic groups. *Bull. London Math. Soc.*, **10**(3), 297–302.

Grunewald, F. and Segal, D. (1979*a*). Remarks on injective specializations. *J. Algebra*, **61**(2), 538–47.

Grunewald, F. and Segal, D. (1979*b*). The solubility of certain decision problems in arithmetic and algebra. *Bull. Amer. Math. Soc. (N.S.)*, **1**(6), 915–18.

Grunewald, F. and Segal, D. (1980*a*). Some general algorithms. I. Arithmetic groups. *Ann. of Math.*, (2) **112**(3), 531–83.

Grunewald, F. and Segal, D. (1980*b*). Some general algorithms. II. Nilpotent groups. *Ann. of Math.*, (2) **112**(3), 585–617.

Grunewald, F. and Segal, D. (1984). *Reflections on the Classification of Torsion-free Nilpotent Groups. Group Theory*, pp. 121–58. Academic Press, London.

Grunewald, F. and Segal, D. (1994). On affine crystallographic groups. *J. Differential Geom.*, **40**(3), 563–94.

Grunewald, F. J., Segal, D., and Smith, G. C. (1988). Subgroups of finite index in nilpotent groups. *Invent. Math.*, **93**(1), 185–223.

Grunewald, F., Segal, D., and Sterling, L. S. (1982). Nilpotent groups of Hirsch length six. *Math. Z.*, **179**(2), 219–35.

Gupta, C. K. (1969*a*). On certain soluble groups. *Proc. Cambridge Philos. Soc.*, **66**, 1–4.

Gupta, C. K. (1969*b*). A faithful matrix representation for certain centre-by-metabelian groups. *J. Austral. Math. Soc.*, **10**, 451–64.

Gupta, C. K. (1990). Conjugacy in certain solvable groups. In *Algebraic Structures and Number Theory* (Hong Kong, 1988), pp. 124–31. World Scientific Publishing, Teaneck, NJ.

Gupta, C. K. and Gupta, N. D. (1973). On the linearity of free nilpotent-by-abelian groups. *J. Algebra*, **24**, 293–302.

Gupta, C. K., Gupta, N. D., and Levin, F. (1989). On conjugacy *p*-separability of free centre-by-metabelian groups. *J. Austral. Math. Soc. Ser. A*, **47**(2), 334–42.

Gupta, C. K., Gupta, N. D., and Rhemtulla, A. H. (1971). Dichotomies in certain finitely generated soluble groups. *J. London Math. Soc.*, (2) **3**, 517–25.

Gupta, N. D. (1972). Third-Engel 2-groups are soluble. *Canad. Math. Bull.*, **15**, 523–4.

Gupta, N. D. (1982). On the dimension subgroups of metabelian groups. *J. Pure Appl. Algebra*, **24**(1), 1–6.

Gupta, N. D. (1987). Sjøgren's theorem for dimension subgroups—the metabelian case. In *Combinatorial Group Theory and Topology* (Alta, Utah, 1984), pp. 197–211, Annals of Mathematics Studies, 111. Princeton University Press, Princeton, NJ.

Gupta, N. D., Hales, A. W., and Passi, I. B. S. (1984). Dimension subgroups of metabelian groups. *J. Reine Angew. Math.*, **346**, 194–8.

Gupta, N. D., Hurley, T C., and Levin, F. (1985). On the lower central factors of free centre-by-metabelian groups. *J. Austral. Math. Soc. Ser. A*, **38**(1), 65–75.

Gupta, N. D. and Levin, F. (1976). Separating laws for free centre-by-metabelian nilpotent groups. *Comm. Algebra*, **4**(3), 249–70.

Gupta, N. D. and Levin, F. (1980). On soluble Engel groups and Lie algebras. *Arch. Math. (Basel)*, **34**(4), 289–95.

Gupta, N. D. and Newman, M. F. (1966). On metabelian groups. *J. Austral. Math. Soc.*, **6**, 362–8.

Gupta, N. D. and Newman, M. F. (1968). Engel congruences in groups of prime-power exponent. *Canad. J. Math.*, **20**, 1321–3.

Gupta, N. D., Newman, M. F., and Tobin, S. J. (1968). On metabelian groups of prime-power exponent. *Proc. Roy. Soc. Ser. A*, **302**, 237–42.

Gupta, N. D. and Sidki, S. (1999). On torsion-free metabelian groups with commutator quotients of prime exponent. *Int. J. Algebra Comput.*, **9**(5), 493–520.

Gupta, N. D. and Tahara, K. (1985). Dimension and lower central subgroups of metabelian *p*-groups. *Nagoya Math. J.*, **100**, 127–33.

Hall, M. (1959). *The Theory of Groups*. MacMillan, New York.

Hall, P. (1940). The classification of prime-power groups. *J. Reine Angew. Math.*, **182**, 130–41.

Hall, P. (1954). Finiteness conditions for soluble groups. *Proc. London Math. Soc.*, (3) **4**, 419–36.

Hall, P. (1956). Finite-by-nilpotent groups. *Proc. Cambridge Philos. Soc.*, **52**, 611–16.

Hall, P. (1958). Some sufficient conditions for a group to be nilpotent. *Illinois J. Math.*, **2**, 787–801.

Hall, P. (1959). On the finiteness of certain soluble groups. *Proc. London Math. Soc.*, (3) **9**, 595–622.

Hall, P. (1961). The Frattini subgroups of finitely generated groups. *Proc. London Math. Soc.*, (3) **11**, 327–52.

Hall, P. (1964). A note on $\overline{SI}$-groups. *J. London Math. Soc.*, **39**, 338–44.

Hall, P. (1969). *The Edmonton Notes on Nilpotent Groups*. Queen Mary College Mathematics Notes. Mathematics Department, Queen Mary College, London.

Hall, P. (1988). *The Collected Works of Philip Hall*. (Compiled and with a preface by K. W. Gruenberg and J. E. Roseblade. With an obituary by Roseblade). Oxford Science Publications. Oxford University Press, New York.

Hall, P. and Hartley, B. (1966). The stability group of a series of subgroups. *Proc. London Math. Soc.*, (3) **16**, 1–39.

Hartley, B. (1974). The normalizer condition and minitransitive permutation groups. (Russian) *Algebra i Logika*, **13**(5), 589–602, 606.

Hartley, B. (1977). A dual approach to Černikov modules. *Math. Proc. Cambridge Philos. Soc.*, **82**(2), 215–39.

Hartley, B., Lennox, J. C., and Rhemtulla, A. H. (1982). Cyclically separated groups. *Bull. Austral. Math. Soc.*, **26**(3), 355–84.

Hartley, B. and McDougall, D. (1971). Injective modules and soluble groups satisfying the minimal condition for normal subgroups. *Bull. Austral. Math. Soc.*, **4**, 113–35.

Hartley, B. and Robinson, D. J. S. (1980). On finite complete groups. *Arch. Math. (Basel)*, **35**(1–2), 67–74.

Hasse, H. (1980). *Number Theory*. Springer, Berlin.

Heilbronn, H. A. (1967). Zeta-functions and L-functions. In 1967 *Algebraic Number Theory* (Proceedings of Instructional Conference, Brighton, 1965), pp. 204–30. Thompson, Washington, DC.

Heineken, H. (1971). Normalizer condition and nilpotent normal subgroups. *J. London Math. Soc.* (2), **4**, 458–60.

Heineken, H. (1989). On E-groups in the sense of Peng. *Glasgow Math. J.*, **31**(2), 231–42.

Heineken, H. and Lennox, J. C. (1983). A note on products of abelian groups. *Arch. Math. (Basel)*, **41**(6), 498–501.

Heineken, H. and Mohamed, I. J. (1968). A group with trivial centre satisfying the normalizer condition. *J. Algebra*, **10**, 368–76.

Heineken, H. and Mohamed, I. J. (1972). Groups with normalizer condition. *Math. Ann.*, **198**, 179–87.

Heineken, H. and Mohamed, I. J. (1974). Non-nilpotent groups with normalizer condition. In *Proceedings of the Second International Conference on the Theory of Groups*. (Australian National University, Canberra, 1973), pp. 357–60, Lecture Notes in Mathematics, Vol. 372. Springer-Berlin.

Heislbetz, H. P. and Mutzbauer, O. (1990). Invariante Typen in torsionsfreien, auflösbaren Gruppen endlichen Ranges. (German) *Arch. Math. (Basel)*, **55**(1), 10–24.

Hermann, G. (1926). Die Frage der endlich vielen Schritte in der Theorie der Polynomideale. (German). *Math. Ann.*, **96**, 736–88.

Higman, G. (1955). A remark on finitely generated nilpotent groups. *Proc. Amer. Math. Soc.*, **6**, 284–5.

Higman, G. (1961). Subgroups of finitely presented groups. *Proc. Roy. Soc. Ser. A*, **262**, 455–75.

Hilton, P. J., Mislin, G., and Roitberg, J. (1975). *Localization of Nilpotent Groups and Spaces*. North-Holland Mathematics Studies, No. 15. Notas de Matemática, No. 55. (Notes on Mathematics, No. 55) North-Holland Publishing Co., Amsterdam-Oxford; American Elsevier Publishing Co., Inc., New York, x+156 pp.

Hilton, P. J. and Stammbach, U. (1997). *A Course in Homological Algebra*. 2nd edn. Graduate Texts in Mathematics, 4. Springer-Verlag, New York.

Hirsch, K. A. (1938a). On infinite soluble groups. I. *Proc. London Math. Soc.*, (2) **44**, 53–60.

Hirsch, K. A. (1938b). On infinite soluble groups. II. *Proc. London Math. Soc.*, (2) **44**, 336–44.

Hirsch, K. A. (1946). On infinite soluble groups. III. *Proc. London Math. Soc.*, (2) **49**, 184–94.

Hirsch, K. A. (1950). Sur les groupes résolubles à condition maximale. (French). Algèbre et Théorie des Nombres. In Colloques Internationaux du Centre National de la Recherche Scientifique, No. 24, pp. 209–10. Centre National de la Recherche Scientifique, Paris.

Hirsch, K. A. (1952). On infinite soluble groups. IV. *J. London Math. Soc.*, **27**, 81–5.

Hirsch, K. A. (1954). On infinite soluble groups. V. *J. London Math. Soc.*, **29**, 250–51.

Holt, D. F. (1979). An interpretation of the cohomology groups $H^n(G, M)$. *J. Algebra*, **60**(2), 307–20.

Huebschmann, J. (1980). Crossed n-fold extensions of groups and cohomology. *Comment. Math. Helv.*, **55**(2), 302–13.

Hursey, R. J. (1971). On ordered polycyclic groups. *Proc. Amer. Math. Soc.*, **28**, 391–4.

Hursey, R. J. (1989). On nilpotent and polycyclic groups. *Bull. Austral. Math. Soc.*, **40**(1), 119–22.

Janusz, G. J. (1996). *Algebraic Number Fields*, 2nd. edn. Graduate Studies in Mathematics, 7. American Mathematical Society, Providence, RI.

Jategaonkar, A. V. (1974). Integral group rings of polycyclic-by-finite groups. *J. Pure Appl. Algebra*, **4**, 337–43.

Jeanes, S. C. and Wilson, J. S. (1978). On finitely generated groups with many profinite-closed subgroups. *Arch. Math.* (*Basel*), **31**(2), 120–2.

Kalužnin, L. A. (1953). Über gewisse Beziehungen zwischen einer Gruppe und ihren Automorphismen. (German). Bericht über die Mathematiker-Tagung in Berlin, Januar, 1953, pp. 164–72. Deutscher Verlag der Wissenschaften, Berlin.

Kappe, L.-C. (1981). Right and left Engel elements in metabelian groups. *Comm. Algebra*, **9**(12), 1295–306.

Kappe, L.-C. (1983). Engel margins in metabelian groups. *Comm. Algebra*, **11**(17), 1965–87.

Kappe, L.-C. and Kappe, W. P. (1974). Metabelian Levi-formations. *Arch. Math.* (*Basel*), **25**, 454–62.

Kappe, L.-C. and Kappe, W. P. (1993). Engel elements in center-by-metabelian groups. In *Group Theory* (Granville, OH, 1992), pp. 198–205, World Scientific Publishing, River Edge, NJ.

Kappe, L.-C. and Morse, R. F. (1990). Levi-properties in metabelian groups. In *Combinatorial Group Theory* (College Park, MD, 1988), pp. 59–72, Contemporary Mathematics, 109. Amer. Math. Soc., Providence, RI.

Karbe, M. J. (1987). Groups satisfying the weak chain conditions for normal subgroups. *Rocky Mountain J. Math.*, **17**(1), 41–7.

Karbe, M. J. and Kurdačenko, L. A. (1988). Just infinite modules over locally soluble groups. *Arch. Math.* (*Basel*), **51**(5), 401–11.

Kargapolov, M. I. (1962). On solvable groups of finite rank. (Russian). *Algebra i Logika*, **1**(5), 37–44.

Kargapolov, M. I. (1974). Some questions in the theory of soluble groups. In *Proceedings of the Second International Conference on the Theory of Groups* (Australian National University, Canberra, 1973), pp. 389–94, Lecture Notes in Mathematics, Vol. 372. Springer, Berlin.

Kargapolov, M. I. and Remeslennikov, V. N. (1966). The conjugacy problem for free solvable groups. (Russian). *Algebra i Logika Sem.*, **5**(6), 15–25.

Karrass, A. and Solitar, D. (1971). Subgroups of HNN-groups and groups with one defining relation. *Canad. J. Math.*, **23**, 627–43.

Kegel, O. H. (1966). Über den Normalisator von subnormalen und erreichbaren Untergruppen, *Math. Ann.* (1966), 248–58.

Kharlampovich, O. (1981). A finitely presented soluble group with insoluble word problem. (Russian). *Izv. Akad. Nauk. Ser. Math.*, **45**, 852–73.

Kharlampovich, O. (1989). Finitely presented solvable groups and Lie algebras with an unsolvable word problem. *Math. Notes*, **46**(3–4), 731–8.

Kharlampovich, O. (1990). The word problem for solvable Lie algebras and groups. *Math. USSR-Sb.*, **67**(2), 489–525.

Kharlampovich, O. (1992*a*). The word problem for solvable groups and Lie algebras. In *Algorithms and Classification in Combinatorial Group Theory* (Berkeley, CA, 1989), pp. 61–7, Mathematical Sciences Research Institute Publication, 23. Springer, New York.

Kharlampovich, O. (1992*b*). The word problem for solvable groups and Lie algebras, a boundary between solvability and unsolvability. In *Proceedings of the International Conference on Algebra*, Part 2 (Novosibirsk, 1989), 53–7, Contemp. Math., 131, Part 2. Amer. Math. Soc., Providence, RI.

Khaĭkin Zapiraĭn, A. and Khukhro, E. I. (2000). A connection between nilpotent groups and Lie rings. *Siberian Math. J.*, **41**(5), 994–1004.

Khukhro, E. I. (1981). Nilpotency and solvability in varieties of groups with operators. (Russian). *Sibirsk. Mat. Zh.*, **22**(5), 209–11, 224.

Khukhro, E. I. (1982). Nilpotent subdirect products. (Russian) *Sibirsk. Mat. Zh.*, **23**(6), 178–80, 207.

Khukhro, E. I. (1987). Nilpotent periodic groups with an almost regular automorphism of prime order. (Russian) *Algebra i Logika*, **26**(4), 502–17, 526.

Khukhro, E. I. (1989). Nilpotent *p*-groups and their automorphisms. (Russian). In *Problems in Algebra*, No. 4 (Russian). (Gomel 1986), pp. 51–60, "Universitetskoe", Minsk.

Khukhro, E. I. (1992). Nilpotency in varieties of groups with operators. *Math. Notes*, **50**(1–2), 869–71.

Kilsch, D. (1978). On minimax groups which are embeddable in constructible groups. *J. London Math. Soc.*, (2) **18**(3), 472–4.

Kilsch, D. (1981). Profinitely closed subgroups of soluble groups of finite rank. *J. Algebra*, **70**(1), 162–72.

Kilsch, D. (1983). Conjugacy separability and separable orbits. *J. Pure Appl. Algebra*, **30**(2), 167–79.

Kilsch, D. (1984). Nilpotent-by-finite groups with the same finite images. *Arch. Math. (Basel)*, **43**(4), 303–11.

Kilsch, D. (1986). Fitting subgroups of profinite completions of soluble minimax groups. *J. Algebra*, **101**(1), 120–6.

Kim, P. S. and Rhemtulla, A. H. (1989). Permutable word products in groups. *Bull. Austral. Math. Soc.*, **40**(2), 243–54.

Kim, P. S., Rhemtulla, A. H., and Smith, H. (1991). A characterization of infinite metabelian groups. *Houston J. Math.*, **17**(3), 429–37.

Kim, Y. K. and Rhemtulla, A. H. (1995). Weak maximality condition and polycyclic groups. *Proc. Amer. Math. Soc.*, **123**(3), 711–14.

Kirkinskiĭ, A. S. and Remeslennikov, V. N. (1975). The isomorphism problem for solvable groups. (Russian). *Mat. Zametki*, **18**(3), 437–43.

Kleene, S. C. (1936). General recursive functions of natural numbers. *Math. Ann.*, **112**, 727–42.

Kochloukova, D. H. (1996). The FP_m-conjecture for a class of metabelian groups. *J. Algebra*, **184**(3), 1175–204.

Kochloukova, D. H. (1999*a*). Finite generation of exterior and symmetric powers. *Math. Proc. Cambridge Philos. Soc.*, **125**(1), 21–9.

Kochloukova, D. H. (1999*b*). A new characterisation of m-tame groups over finitely generated abelian groups. *J. London Math. Soc.*, (2) **60**(3), 802–16.

Kochloukova, D. H. (2000). Some finitely presented nilpotent-by-abelian groups. *Comm. Algebra*, **28**(2), 949–57.

Kochloukova, D. H. (2001*a*). More about the geometric invariants $\Sigma^m(G)$ and $\Sigma^m(G, \mathbf{Z})$ for groups with normal locally polycyclic-by-finite subgroups. *Math. Proc. Cambridge Philos. Soc.*, **130**(2), 295–306.

Kochloukova, D. H. (2001*b*). Geometric invariants and modules of type FP_∞ over constructible nilpotent-by-abelian groups. *J. Pure Appl. Algebra*, **159**(2–3), 187–202.

Kochloukova, D. H. (2002). Subgroups of constructible nilpotent-by-abelian groups and a generalization of a result of Bieri, Neumann and Strebel. *J. Group Theory*, **5**(2), 219–31.

Kopytov, V. M. (1968). Matrix groups. (Russian). *Algebra i Logika*, **7**(3), 51–9.

Kopytov, V. M. (1971). Solvability of the occurrence problem in finitely generated solvable matrix groups over a number field. (Russian). *Algebra i Logika*, **10**, 169–82.

Kovács, L. G. and Newman, M. F. (1996). Torsionfree varieties of metabelian groups. In *Infinite Groups* 1994 (Ravello), pp. 125–8. de Gruyter, Berlin.

Kropholler, P. H. (1984). On finitely generated soluble groups with no large wreath product sections. *Proc. London Math. Soc.*, (3) **49**(1), 155–69.

Kropholler, P. H. (1985). A note on the cohomology of metabelian groups. *Math. Proc. Cambridge Philos. Soc.*, **98**(3), 437–45.

Kropholler, P. H. (1986*a*). The cohomology of soluble groups of finite rank. *Proc. London Math. Soc.*, (3) **53**(3), 453–73.

Kropholler, P. H. (1986*b*). Cohomological dimension of soluble groups. *J. Pure Appl. Algebra*, **43**(3), 281–7.

Kropholler, P. H. (1993). Soluble groups of type $(FP)_\infty$ have finite torsion-free rank. *Bull. London Math. Soc.*, **25**(6), 558–66.

Kropholler, P. H. and Stammbach, U. (1990). Some remarks on tensor, symmetric and exterior powers of modules over an abelian group. *Comm. Algebra*, **18**(11), 3765–73.

Kurdachenko, L. A. (1979). Groups that satisfy weak minimality and maximality conditions for normal subgroups (Russian). *Sibirsk. Mat. Zh.*, **20**(5), 1068–76, 1167.

Kurdachenko, L. A. (1981). Groups satisfying weak minimality and maximality conditions for subnormal subgroups. (Russian). *Mat. Zametki*, **29**(1), 19–30, 154.

Kurdachenko, L. A. (1993). Artinian modules over groups of finite rank, and groups with some finiteness conditions. (Russian). In *Infinite Groups and Related Algebraic Structures*, pp. 144–59. Akad. Nauk Ukrainy, Inst. Mat., Kiev.

Kurdachenko, L. A. (1995). Artinian modules over groups of finite rank and the weak minimal condition for normal subgroups. *Ricerche Mat.*, **44**(2), 303–35.

Kurdachenko, L. A., Otal, J., and Subbotin, I. Ya. (2002a).On some criteria of nilpotency. *Comm. Algebra*, **30**(8), 3755–76.

Kurdachenko, L. A., Otal, J., and Subbotin, I. Ya. (2002b). *Groups with Prescribed Quotient Groups and Associated Module Theory*. Series in Algebra, 8. World Scientific Publishing Co., Inc., River Edge, NJ.

Kurdachenko, L. A. and Smith, H. (1996). Groups with the maximal condition on nonsubnormal subgroups. *Boll. Un. Mat. Ital. B*, (7) **10**(2), 441–60.

Kurdachenko, L. A. and Smith, H. (1998a). The nilpotency of some groups with all subgroups subnormal. *Publ. Mat.*, **42**(2), 411–21.

Kurdachenko, L. A. and Smith, H. (1998b). Groups with the weak maximal condition for non-subnormal subgroups. *Ricerche Mat.*, **47**(1), 29–49.

Kurdachenko, L. A. and Tushev, A. V. (1985). Two-step solvable groups with the weak minimality condition for normal subgroups. (Russian). *Ukrain. Mat. Zh.*, **37**(3), 300–6, 403.

Kuroš, A. G. and Černikov, S. N. (1947). Solvable and nilpotent groups. (Russian) *Uspehi Matem. Nauk* (*N.S.*), **2**, **3**(19), 18–59; (1953). Solvable and nilpotent groups. *Amer. Math. Soc. Translations*, **80**.

Kuzennyĭ, N. F. and Semko, N. N. (1983). The structure of solvable nonnilpotent meta-Hamiltonian groups. (Russian). *Mat. Zametki*, **34**(2), 179–88.

Kuzennyĭ, N. F. and Semko, N. N. (1985). The structure of solvable meta-Hamiltonian groups. (Russian). *Dokl. Akad. Nauk Ukrain. SSR Ser. A*, **2**(6–8), 87.

Kuzennyĭ, N. F. and Semko, N. N. (1987). The structure of periodic metabelian and meta-Hamiltonian groups with a non-elementary commutator group. (Russian). *Ukrain. Mat. Zh.*, **39**(2), 180–5, 271.

Kuzennyĭ, N. F. and Semko, N. N. (1988). The structure of periodic metabelian meta-Hamiltonian groups with an elementary commutator subgroup of rank two. *Ukrainian Math. J.*, **40**(6), 627–33.

Kuzennyĭ, N. F. and Semko, N. N. (1989). The structure of periodic nonabelian meta-Hamiltonian groups with an elementary commutator subgroup of rank three. *Ukrainian Math. J.*, **41**(2), 153–8.

Kuzennyĭ, N. F. and Semko, N. N. (1990). Meta-Hamiltonian groups with elementary commutator subgroup of rank two. *Ukrainian Math. J.*, **42**(2), 149–54.

Laue, R. (1979). Stability groups and central series. *J. Reine Angew. Math.*, **306**, 42–8.

Learner, A. (1962). The embedding of a class of polycyclic groups. *Proc. London Math. Soc.*, (3) **12**, 496–510.

Learner, A. (1964). Residual properties of polycyclic groups. *Illinois J. Math.*, **8**, 536–42.

Lennox, J. C. (1971). On a centrality property of finitely generated torsion-free soluble groups. *J. Algebra*, **18**, 541–8.

Lennox, J. C. (1973a). Bigenetic properties of finitely generated hyper-(abelian-by-finite) groups. *J. Austral. Math. Soc.*, **16**, 309–15.

Lennox, J. C. (1973b). Polycyclic Frattini factors in certain finitely generated groups. *Arch. Math. (Basel)*, **24**, 571–8.

Lennox, J. C. (1976a). Some normality properties of finitely generated nilpotent groups. *Bull. Calcutta Math. Soc.*, **68**(2), 91–6.

Lennox, J. C. (1976b). On the solubility of a product of permutable subgroups. *J. Austral. Math. Soc. Ser. A*, **22**(2), 252–5.

Lennox, J. C. (1976c). Finitely generated metabelian groups are not subnormality separable. *Math. Z.*, **149**(2), 201–2.

Lennox, J. C. (1977). On groups in which every subgroup is almost subnormal. *J. London Math. Soc.*, (2) **15**(2), 221–31.

Lennox, J. C. (1978). Lower central depth in finitely generated soluble-by-finite groups. *Glasgow Math. J.*, **19**(2), 153–4.

Lennox, J. C. (1983). On quasinormal subgroups of certain finitely generated groups. *Proc. Edinburgh Math. Soc.*, (2) **26**(1), 25–8.

Lennox, J. C. (1984). A fixed point theorem for modules over finitely generated nilpotent groups. *Bull. London Math. Soc.*, **16**(3), 289–91.

Lennox, J. C. (1996). On soluble groups in which centralizers are finitely generated. *Rend. Sem. Mat. Univ. Padova*, **96**, 131–5.

Lennox, J. C., Mohammadi Hassanabadi, A., Stewart, A. G. R., and Wiegold, J. (1990). Nilpotent extensibility and centralizers in infinite 2-groups. Proceedings of the Second International Group Theory Conference (Bressanone, 1989). *Rend. Circ. Mat. Palermo*, (2) (Suppl.) **23**, 209–19.

Lennox, J. C., Neumann, P. M., and Wiegold, J. (1990). Nilpotent subgroups and the hypercentre of infinite soluble groups. *Arch. Math. (Basel)*, **54**(5), 417–21.

Lennox, J. C. and Robinson, D. J. S. (1980). Soluble products of nilpotent groups. *Rend. Sem. Mat. Univ. Padova*, **62**, 261–80.

Lennox, J. C. and Robinson, D. J. S. (1982). Nearly maximal subgroups of finitely generated soluble subgroups. *Arch. Math. (Basel)*, **38**(4), 289–95.

Lennox, J. C. and Roseblade, J. E. (1970). Centrality in finitely generated soluble groups. *J. Algebra*, **16**, 399–435.

Lennox, J. C. and Roseblade, J. E. (1980). Soluble products of polycyclic groups. *Math. Z.*, **170**(2), 153–4.

Lennox, J. C. and Stonehewer, S. E. (1987). *Subnormal Subgroups of Groups*. Oxford University Press, New York.

Lennox, J. C. and Wiegold, J. (1974a). Some remarks on coherent soluble groups. *Bull. Austral. Math. Soc.*, **10**, 277–9.

Lennox, J. C. and Wiegold, J. (1974b). Converse of a theorem of Mal'cev on nilpotent groups. *Math. Z.*, **139**, 85–6.

Lennox, J. C. and Wiegold, J. (1976). Lower Frattini series in finitely generated soluble groups. *Proc. Amer. Math. Soc.*, **57**(1), 43–4.

Lennox, J. C. and Wilson, J. S. (1979). On products of subgroups in polycyclic groups. *Arch. Math. (Basel)*, **33**(4), 305–9.

Levič, E. M. (1969a). The representation of solvable groups by matrices over a certain field of characteristic zero. (Russian). *Dokl. Akad. Nauk SSSR*, **188**, 520–21.

Levič, E. M. (1969b). The representation of solvable groups by matrices over a certain field of characteristic zero. (Russian). *Proceedings of the Riga Seminar on Algebra*, pp. 74–97. Latv. Gos. Univ., Riga.

Levič, E. M. (1970). The representation of two-step-solvable groups by matrices over a field of characteristic zero. (Russian). *Mat. Sb. (N.S.)*, **81**(123), 352–7.

Lichtman, A. I. (1989). On nilpotent and soluble subgroups of linear groups over fields of fractions of enveloping algebras and of group rings I. In *Representation Theory, Group Rings, and Coding Theory*, pp. 247–81, Contemporary Mathematics, 93. Amer. Math. Soc., Providence, RI.

Lichtman, A. I. (1993). The soluble subgroups and the Tits alternative in linear groups over rings of fractions of polycylic group rings. I. *J. Pure Appl. Algebra*, **86**(3), 231–87.

Lichtman, A. I. (1995). Automorphism groups of free soluble groups. *J. Algebra*, **174**(1), 132–49.

Liu, H. G. (1992). Solvable SD_2-groups of infinite order. (Chinese). *Acta Math. Sinica*, **35**(3), 339–49.

Liu, H. G. (1994a). Infinite polycyclic groups such that all the proper normal subgroups of their finite quotients are 2-generated. (Chinese). *Chinese Ann. Math. Ser. A*, **15**(3), 255–61.

Liu, H. G. (1994b). Solvable SD_2 groups of infinite order. II. The finite residue case. (Chinese). *Acta Math. Sinica*, **37**(6), 721–7.

Liu, H. G. (1995). A class of infinite polycyclic groups. (Chinese). *Chinese Ann. Math. Ser. A*, **16**(3), 320–7.

Liu, H. G. (1996). Finitely generated solvable groups whose subnormal subgroups have defect at most 2. (Chinese). *Chinese Ann. Math. Ser. A*, **17**(1), 21–4.

Liu, H. G. (1997). Infinite solvable SD_3-groups. (Chinese). *J. Math. (Wuhan)*, **17**(2), 155–8.

Liu, H. G. (1998*a*). Solvable groups whose finitely generated subgroups are 3-generated. (Chinese). *Chinese Ann. Math. Ser. A*, **19**(1), 21–6.

Liu, H. G. (1998*b*). Fitting subgroups of polycyclic groups. (Chinese). *Acta Math. Sinica*, **41**(2), 299–302.

Liu, H. G. (1999). Two applications of Baer's theorem on supersolvability. (Chinese). *Chinese Ann. Math. Ser. A*, **20**(4), 403–6.

Liu, H. G. (2000*a*). Residually finite properties of soluble groups of finite rank. (Chinese). *Acta Math. Sinica*, **43**(1), 163–6.

Liu, H. G. (2000*b*). Solvable groups of finite rank. (Chinese). *Adv. Math. (China)*, **29**(1), 55–60.

Liu, H. G. (2002). The strongly residually-finite properties of some infinite soluble groups. (Chinese). *Chinese Ann. Math. Ser. A*, **23**(3), 321–4.

Longobardi, P., MacHenry, T., Maj, M., and Wiegold, J. (1996). On absolutely-nilpotent of class k groups. *Atti Accad. Naz. Lincei Cl. Sci. Fis. Mat. Natur. Rend. Lincei*, (9) *Mat. Appl.*, **6**(4), 201–9.

Longobardi, P. and Maj, M. (1986). Finitely generated subgroups of some solvable groups. (Italian). *Ann. Mat. Pura Appl.*, (4) **143**, 373–83.

Longobardi, P. and Maj, M. (1993). Finitely generated soluble groups with an Engel condition on infinite subsets. *Rend. Sem. Mat. Univ. Padova*, **89**, 97–102.

Longobardi, P., Maj, M., Mann, A., and Rhemtulla, A. (1996). Groups with many nilpotent subgroups. *Rend. Sem. Mat. Univ. Padova*, **95**, 143–52.

Longobardi, P., Maj, M., and Rhemtulla, A. H. (1995). Groups with no free subsemigroups. *Trans. Amer. Math. Soc.*, **347**(4), 1419–27.

Longobardi, P., Maj, M., and Smith, H. (2001). Torsion-free groups isomorphic to all of their non-nilpotent subgroups. *J. Aust. Math. Soc.*, **71**(3), 339–47.

Lubotzky, A., Mann, A., and Segal, D. (1993). Finitely generated groups of polynomial subgroup growth. *Israel J. Math.*, **82**(1–3), 363–71.

Lyndon, R. C. (1952). Two notes on nilpotent groups. *Proc. Amer. Math. Soc.*, **3**, 579–83.

Lyndon, R. C. and Schupp, P. E. (1977). *Combinatorial Group Theory*. Ergebnisse der Mathematik und ihrer Grenzgebiete, Band 89. Springer-Verlag, Berlin and New York.

MacLane, S. (1949). Cohomology theory in abstract groups III. Operator homomorphisms of kernels. *Ann. of Math.*, (2) **50**, 736–61.

MacLane, S. (1963). *Homology*. Die Grundlehren der mathematischen Wissenschaften, Bd. 114. Academic Press, Inc., Publishers, New York; Springer-Verlag, Berlin, Göttingen, and Heidelberg.

Magid A. (1982). Analytic Representations of polycyclic groups, *J. Algebra*, **74**, 149–58.

Magnus, W. (1930). Über diskontinuierliche Gruppen mit einer definierenden Relation (Der Freiheitsatz). (German). *J. Reine Angew. Math*, **163**, 141–65.

Magnus, W. (1932). Das Identitätsproblem für Gruppen mit einer definierenden Relation. (German). *Math. Ann.*, **106**, 295–307.

Magnus, W. (1939). On a theorem of Marshall Hall. *Ann. of Math.*, (2) **40**, 764–8.

Magnus, W., Karrass, A., and Solitar, D. (1976). *Combinatorial Group Theory: Presentations of Groups in Terms of Generators and Relations*, 2nd edn. Interscience Publishers (John Wiley & Sons, Inc.), New York, London, and Sydney.

Mal'cev, A. I. (1940). On faithful matrix representations of infinite groups. (Russian). *Rec. Math. (Mat. Sbornik) N.S.*, **8**(50), 405–22. *Amer. Math. Soc. Translations*, (2) **45**, 1–18, 1965.

Mal'cev, A. I. (1949). Nilpotent torsion-free groups. (Russian). *Izv. Akad. Nauk. SSSR. Ser. Mat.*, **13**, 201–12.

Mal'cev, A. I. (1951). On some classes of infinite soluble groups. (Russian). *Mat. Sbornik N.S.*, **28**(70), 567–588; (1956). On some classes of infinite soluble groups. *Amer. Math. Soc. Translations*, (2) **2**, 1–21, 1956.

Mal'cev, A. I. (1955). Two remarks on nilpotent groups. (Russian). *Mat. Sb. (N.S.)*, **37**(79), 567–72.

Mal'cev, A. I. (1958). Homomorphisms onto finite groups. (Russian). *Ivanov. Gos. Ped. Inst. Ucen. Zap*, **18**, 49–60.

Mal'cev, A. I. (1960). On free soluble groups. *Soviet Math. Dokl.*, **1**, 65–8.

Mal'cev, A. I. (1970). Algorithms and Recursive Functions. (Translated from the first Russian edition). Wolters-Noordhoff Publishing, Groningen.

Mann, A. and Segal, D. (1990). Uniform finiteness conditions in residually finite groups. *Proc. London Math. Soc.*, (3) **61**(3), 529–45.

Matijasevič, Ju. V. (1970). The Diophantineness of enumerable sets. *Soviet Math. Dokl.*, **11**, 354–8.

McCaughan, D. J. (1973). Subnormal structure in some classes of infinite groups. *Bull. Austral. Math. Soc.*, **8**, 137–50.

McCaughan, D. J. (1974a). Subnormality in soluble minimax groups. Collection of articles dedicated to the memory of Hanna Neumann, V. *J. Austral. Math. Soc.*, **17**, 113–28.

McCaughan, D. J. (1974b). On subnormality in soluble minimax groups. In *Proceedings of the Second International Conference on the Theory of Groups*

(Australian National University, Canberra, 1973), pp. 443–5, Lecture Notes in Mathematics, Vol. 372. Springer, Berlin, 1974.

McCaughan, D. J. and McDougall, D. (1972). The subnormal structure of metanilpotent groups. *Bull. Austral. Math. Soc.*, **6**, 287–306.

McCaughan, D. J. and McDougall, D. (1977). Criteria for subnormality. *Arch. Math. (Basel)*, **29**(5), 451–4.

McCutcheon, J. J. (1969). On certain polycyclic groups. *Bull. London Math. Soc.*, **1**, 179–86.

McCutcheon, J. J. and Newstead, P. E. (1972). Some groups of derived length six. *J. London Math. Soc.*, (2) **5**, 32–8.

McDougall, D. (1970*a*). Soluble groups with the minimum condition for normal subgroups. *Math. Z.*, **118**, 157–67.

McDougall, D. (1970*b*). Soluble minimax groups with the subnormal intersection property. *Math. Z.*, **114**, 241–4.

McDougall, D. (1972). The subnormal structure of some classes of soluble groups. *J. Austral. Math. Soc.*, **13**, 365–77.

McLain, D. H. (1956). Remarks on the upper central series of a group. *Proc. Glasgow Math. Assoc.*, **3**, 38–44.

Meier, D. and Rhemtulla, A. H. (1984). On torsion-free groups of finite rank. *Canad. J. Math.*, **36**(6), 1067–80.

Meier, D. and Rhemtulla, A. H. (1985). Rank restricting properties of finitely generated soluble groups. *Arch. Math. (Basel)*, **44**(3), 216–24.

Meinert, H. (1995). The higher geometric invariants of modules over Noetherian group rings. In *Combinatorial and Geometric Group Theory* (Edinburgh, 1993), pp. 247–54, London Mathematical Society Lecture Note Series, 204. Cambridge University Press, Cambridge.

Meinert, H. (1996). The homological invariants for metabelian groups of finite Prüfer rank: a proof of the Σ^m-conjecture. *Proc. London Math. Soc.*, (3) **72**(2), 385–424.

Meinert, H. (1998). On the geometric invariants of soluble groups of finite Prüfer rank. In *Geometry and Cohomology in Group Theory* (Durham, 1994), pp. 249–62, London Mathematical Society Lecture Note Series, 252. Cambridge University Press, Cambridge.

Meldrum, J. D. P. (1971). On central series of a soluble group. *J. London Math. Soc.*, (2) **3**, 633–9.

Meldrum, J. D. P. (1973). On the Heineken–Mohamed groups. *J. Algebra*, **27**, 437–44.

Mendelson, E. (1997). *Introduction to Mathematical Logic*, 4th edn. Chapman & Hall, London.

Menegazzo, F. (1968). Gruppi nei quali la relazione di quasi-normalità è transitiva. (Italian). *Rend. Sem. Mat. Univ. Padova*, **40**, 347–61.

Menegazzo, F. (1969). Gruppi nei quali la relazione di quasi-normalità è transitiva. II. (Italian). *Rend. Sem. Mat. Univ. Padova*, **42**, 389–99.

Menegazzo, F. (1970). Sui gruppi relativamente complementati. (Italian). *Rend. Sem. Mat. Univ. Padova*, **43**, 209–14.

Menegazzo, F. (1995). Groups of Heineken-Mohamed. *J. Algebra*, **171**(3), 807–25.

Menegazzo, F. and Puglisi, O. (2000). Outer automorphisms of supersoluble groups. *Glasgow Math. J.*, **42**(1), 115–20.

Menegazzo, F. and Stonehewer, S. E. (1985). On the automorphism group of a nilpotent p-group. *J. London Math. Soc.*, (2) **31**(2), 272–6.

Merzljakov, J. I. (1964). Locally solvable groups of finite rank. (Russian). *Algebra i Logika*, **3**(2), 5–16.

Merzljakov, J. I. (1968*a*). Groups which can be almost approximated by finite p-groups. (Russian). *Algebra i Logika*, **7**(1), 105–11.

Merzljakov, J. I. (1968*b*). Matrix representation of automorphisms, extensions and solvable groups. (Russian). *Algebra i Logika*, **7**(3), 63–104.

Merzljakov, J. I. (1969*a*). Matrix representation of groups of outer automorphisms of Černikov groups. (Russian). *Algebra i Logika*, **8**, 478–82.

Merzljakov, J. I. (1969*b*). Locally solvable groups of finite rank. II. (Russian). *Algebra i Logika*, **8**, 686–90.

Merzljakov, J. I. (1970). Integer representation of the holomorphs of polycyclic groups. (Russian). *Algebra i Logika*, **9**, 539–58.

Merzljakov, J. I. (1984). On the theory of locally polycyclic groups. *J. London Math. Soc.*, (2) **30**(1), 67–72.

Merzljakov, J. I. (1988). Exact matrix representation of the holomorphs of the Abels groups. (Russian). *Ukrainian Math. J.*, **40**(3), 295–8.

Meskin, S. (1974). A finitely generated residually finite group with an unsolvable word problem. *Proc. Amer. Math. Soc.*, **43**, 8–10.

Mihaĭlova, K. A. (1966). The occurrence problem for direct products of groups. (Russian). *Mat. Sb. (N.S.)*, **70**(112), 241–51.

Miller, C. F., III. (1971). *On Group-Theoretic Decision Problems and Their Classification.* Annals of Mathematics Studies, No. 68. Princeton University Press, Princeton, NJ; University of Tokyo Press, Tokyo.

Miller, J. I. (1979). Center-by-metabelian groups of prime exponent. *Trans. Amer. Math. Soc.*, **249**(1), 217–24.

Möhres, W. (1987). *Torsion-Free Nilpotent Groups with Bounded Ranks of the Abelian Subgroups. Group Theory* (Bressanone, 1986), 115–17, Lecture Notes in Mathematics, 1281. Springer, Berlin.

Möhres, W. (1988). Gruppen deren Untergruppen alle subnormal sind. (German). PhD dissertation, University of Würzburg.

Möhres, W. (1989*a*). Torsionsfreie Gruppen, deren Untergruppen alle subnormal sind. (German). *Math. Ann.*, **284**(2), 245–9.

Möhres, W. (1989*b*). Torsionsgruppen, deren Untergruppen alle subnormal sind. (German). *Geom. Dedicata*, **31**(2), 237–44.

Möhres, W. (1989*c*). Auflösbare Gruppen mit endlichem Exponenten, deren Untergruppen alle subnormal sind. I, II. (German). *Rend. Sem. Mat. Univ. Padova*, **81**, 255–87.

Möhres, W. (1990). Auflösbarkeit von Gruppen, deren Untergruppen alle subnormal sind. (German). *Arch. Math. (Basel)*, **54**(3), 232–5.

Möhres, W. (1991). Hyperzentrale Torsionsgruppen, deren Untergruppen alle subnormal sind. (German). *Illinois J. Math.*, **35**(1), 147–57.

Moldavanskiĭ, D. I. (1969). A certain theorem of Magnus. (Russian). *Ivanov. Gos. Ped. Inst. Učen. Zap.*, **44**(mat), 26–8.

Nazzal, S. H. and Rhemtulla, A. H. (1991). Centrality in abelian by polycyclic groups. *Arch. Math. (Basel)*, **56**(4), 333–42.

Neumann, B. H. (1954). Groups covered by permutable subsets. *J. London Math. Soc.*, **29**, 236–48.

Neumann, B. H. (1993). A large nilpotent group without large abelian subgroups. *Bull. London Math. Soc.*, **25**(4), 305–8.

Neumann, B. H. and Neumann, H. (1959). Embedding theorems for groups, *J. London Math. Soc.*, **34**, 465–79.

Neumann, H. (1967). *Varieties of Groups*. Springer-Verlag, New York.

Neumann, P. M. (1980). Endomorphisms on infinite soluble groups. *Bull. London Math. Soc.*, **12**(1), 13–16.

Neumann, P. M. (1989). *Pathology in the Representation Theory of Infinite Soluble Groups* (Proceedings of Groups—Pusan, Korea 1988), pp. 124–39, Lecture Notes in Mathematics, 1398. Springer, Berlin.

Newell, M. L. (1970*a*). On soluble min-by-max groups. *Math. Ann.*, **186**, 282–96.

Newell, M. L. (1970*b*). Finiteness conditions in generalized soluble groups. *J. London Math. Soc.*, (2) **2**, 593–6.

Newell, M. L. (1970*c*). A subgroup characterisation of soluble min-by-max groups. *Arch. Math. (Basel)*, **21**, 128–31.

Newell, M. L. (1970*d*). On normal coverings of groups. *Arch. Math. (Basel)*, **21**, 337–43.

Newell, M. L. (1972). The eliminant of a group. *Proc. London Math. Soc.*, (3) **24**, 432–48.

Newell, M. L. (1973*a*). Subgroups with almost-complements. *J. London Math. Soc.*, (2), **6**, 761–8.

Newell, M. L. (1973*b*). Homomorphs and formats in polycyclic groups. *J. London Math. Soc.*, (2) **7**, 317–27.

Newell, M. L. (1974). The nilpotent-by-finite projectors of polycyclic groups. *J. London Math. Soc.*, (2), **7**, 540–46.

Newell, M. L. (1975*a*). Nilpotent projectors in $\mathfrak{S}_1$-groups. *Proc. Roy. Irish Acad. Sect. A*, **75**(11), 107–14.

Newell, M. L. (1975*b*). Some splitting theorems for infinite supersoluble groups. *Math. Z.*, **144**(3), 265–75.

Newell, M. L. (1975*c*). Supplements in abelian-by-nilpotent groups. *J. London Math. Soc.*, (2) **11**(1), 74–80.

Newell, M. L. (1976). Ideals with hypercentral action on modules. *J. Algebra*, **42**(2), 600–3.

Newman, B. B. (1968). Some results on one-relator groups. *Bull. Amer. Math. Soc.*, **74**, 568–71.

Newman, B. B. (1973). The soluble subgroups of a one-relator group with torsion. Collection of articles dedicated to the memory of Hanna Neumann, III. *J. Austral. Math. Soc.*, **16**, 278–85.

Newman, M. F. (1960*a*). On a class of metabelian groups. *Proc. London Math. Soc.*, (3) **10**, 354–64.

Newman, M. F. (1960*b*). On a class of nilpotent groups. *Proc. London Math. Soc.*, (3) **10**, 365–75.

Newman, M. F. (1966). Another non-Hopf group. *J. London Math. Soc.*, **41**, 292.

Newman, M. F. (1972). The soluble length of soluble linear groups. *Math. Z.*, **126**, 59–70.

Newman, M. F. (1984). *Metabelian Groups of Prime-power Exponent* (Proceedings of Groups— Korea, Kyoungju 1983), pp. 87–98, Lecture Notes in Mathematics, 1098. Springer, Berlin.

Newman, M. F. and Wiegold, J. (1964). Groups with many nilpotent subgroups. *Arch. Math.*, **15**, 241–50.

Noskov, G. A. (1974). Almost approximability of finitely generated torsion-free *AP*-groups by finite *p*-groups. (Russian). *Algebra i Logika*, **13**(6), 676–84, 720.

Noskov, G. A. (1982). On conjugacy in metabelian groups. (Russian). *Mat. Zametki*, **31**(4), 495–507, 653.

Noskov, G. A. (1997). The Bieri–Strebel invariant and homological finiteness conditions for metabelian groups. *Algebra and Logic*, **36**(2), 117–32.

Novikov, P. S. (1955). On the algorithmic unsolvability of the word problem in group theory. *Trudy Mat. Inst. Steklov*, **44**. *Amer. Math. Soc. Translations*, (7) 9 (1958), 1–122.

Ol'šanskiĭ, A. Yu. (1991). Geometry of defining relations in groups. Kluwer Academic Publishers. Dordrecht.

Onishchuk, V. A. and Zaĭcev, D. I. (1991). Locally nilpotent groups with a centralizer that satisfies a finiteness condition. *Ukrainian Math. J.*, **43**(7–8), 1018–21.

Paek, D. H. (2001). Chain conditions for subnormal subgroups of infinite order or index. *Commun. Algebra*, **29**, 3069–81.

Paek, D. H. (2002). Chain conditions for subgroups of infinite order or index. *J. Algebra*, **249**, 291–305.

Passman, D. S. (1984). Group rings of polycyclic groups. In *Group Theory*. Academic Press, London.

Passman, D. S. (1985). The Algebraic Structure of Group Rings. (Reprint of the 1977 original). R.E. Krieger Publishing Co., Inc., Melbourne, FL.

Peng, T. A. (1969). On groups with nilpotent derived groups. *Arch. Math. (Basel)*, **20**, 251–3.

Peng, T. A. (1975). A criterion for subnormality. *Arch. Math. (Basel)*, **26**, 225–30.

Peng, T. A. (1976). A note on subnormality. *Bull. Austral. Math. Soc.*, **15**(1), 59–64.

Plotkin, B. I. (1955a). On the theory of solvable groups with finiteness conditions. (Russian). *Dokl. Akad. Nauk SSSR (N.S.)*, **100**, 417–20.

Plotkin, B. I. (1955b). Radical groups. (Russian). *Mat. Sb. (N.S.)*, **37**(79), 507–26.

Plotkin, B. I. (1956). On groups with finiteness conditions for abelian subgroups. (Russian). *Dokl. Akad. Nauk SSSR (N.S.)*, **107**, 648–51.

Polovickiĭ, Ja. D. (1962). Locally extremal and layer-extremal groups. (Russian). *Mat. Sb. (N.S.)*, **58**(100), 685–94.

Pyber, L. (2002). Bounded generation and subgroup growth. *Bull. London Math. Soc.*, **34**(1), 55–60.

Rabin, M. O. (1958). Recursive unsolvability of group theoretic problems. *Ann. of Math.*, (2) **67**, 172–94.

Rabin, M. O. (1960). Computable algebra, general theory and theory of computable fields. *Trans. Amer. Math. Soc.*, **95**, 341–60.

Remeslennikov, V. N. (1968). Finite approximability of metabelian groups. (Russian). *Algebra i Logika*, **7**(4), 106–13.

Remeslennikov, V. N. (1969a). Conjugacy in polycyclic groups. (Russian). *Algebra i Logika*, **8**, 712–25.

Remeslennikov, V. N. (1969b). Representation of finitely generated metabelian groups by matrices. (Russian). *Algebra i Logika*, **8**, 72–5.

Remeslennikov, V. N. (1972). An example of a finitely presented solvable group without the maximality condition for normal subgroups. (Russian). *Mat. Zametki*, **12**, 287–93.

Remeslennikov, V. N. (1973a). On finitely presented groups. In *Proceedings of the Fourth All-Union Symposium on the Theory of Groups* (Russian), pp 164–9. *Novosibirsk*.

Remeslennikov, V. N. (1973*b*). An example of a group, finitely presented in the variety $\mathfrak{A}^5$, with an undecidable word problem. (Russian). *Algebra i Logika*, **12**, 577–602, 618.

Remeslennikov, V. N. (1974). A finitely presented group whose center is not finitely generated. (Russian). *Algebra i Logika*, **13**(4), 450–9, 488.

Remeslennikov, V. N. (1975). Studies on infinite solvable and finitely approximable groups. (Russian). *Mat. Zametki*, **17**(5), 819–24.

Remeslennikov, V. N. (1979). An algorithmic problem for nilpotent groups and rings. (Russian). *Sibirsk. Mat. Zh.*, **20**(5), 1077–81, 1167.

Remeslennikov, V. N. and Romanovskiĭ, N. S. (1980). Algorithmic problems for solvable groups. In *Word Problems, II* (Conference on Decision Problems in Algebra, Oxford, 1976), pp. 337–46, Studies in Logic Foundations of Mathematics, 95. North-Holland, Amsterdam-New York.

Rhemtulla, A. H. (1967). A minimality property of polycyclic groups. *J. London Math. Soc.*, **42**, 456–62.

Rhemtulla, A. H. (1969). Commutators of certain finitely generated soluble groups. *Canad. J. Math.*, **21**, 1160–64.

Rhemtulla, A. H. (1972). Right-ordered groups. *Canad. J. Math.*, **24**, 891–5.

Rhemtulla, A. H. (1981). Polycyclic right-ordered groups. In *Proceedings of Conference on Algebra*, (Southern Illinois University, Carbondale, Ill., 1980), pp. 230–34, Lecture Notes in Math., 848, Springer, Berlin.

Rhemtulla, A. H. (1985). Characteristic properties of soluble groups of finite rank. *J. Korean Math. Soc.*, **22**(2), 135–42.

Rhemtulla, A. H. (1989). *Groups with Many Elliptic Subgroups*. (Proceedings of Groups—Pusan, Korea, 1988), pp. 156–62, Lecture Notes in Mathematics, 1398. Springer, Berlin.

Rhemtulla, A. H. and Sidki, S. (1989). Factorizable infinite solvable groups. *J. Algebra*, **122**(2), 397–409.

Rhemtulla, A. H. and Smith, H. (1984). A finite index property of certain solvable groups. *Canad. Math. Bull.*, **27**(4), 485–9.

Rhemtulla, A. H. and Smith, H. (1993). *On Infinite Solvable Groups. Infinite Groups and Group Rings* (Tuscaloosa, AL, 1992), pp. 111–21, Series in Algebra, 1, World Scientific Publishing, River Edge, NJ.

Rhemtulla, A. H. and Wehrfritz, B. A. F. (1984). Isolators in soluble groups of finite rank. Rocky Mountain *J. Math.*, **14**(2), 415–21.

Rhemtulla, A. H. and Weiss, A. (1989). Groups with permutable subgroup products. In *Group Theory* (Singapore, 1987), pp. 485–95. de Gruyter, Berlin.

Rhemtulla, A. H., Weiss, A., and Yousif, M. (1984). Solvable groups with π-isolators. *Proc. Amer. Math. Soc.*, **90**(2), 173–7.

Rhemtulla, A. H. and Wilson, J. S. (1987). On elliptically embedded subgroups of soluble groups. *Canad. J. Math.*, **39**(4), 956–68.

Rhemtulla, A. H. and Wilson, J. S. (1988). Elliptically embedded subgroups of polycyclic groups. *Proc. Amer. Math. Soc.*, **102**(2), 230–34.

Ribes, L., Segal, D., and Zalesskii, P. A. (1999). Conjugacy separability and free products of groups with cyclic amalgamation. *J. London Math. Soc.*, (2) **57**(3), 609–28.

Riles, J. B. (1969). The near Frattini subgroups of infinite groups. *J. Algebra*, **12**, 155–71.

Robinson, D. J. S. (1964). Groups in which normality is a transitive relation. *Proc. Cambridge Philos. Soc.*, **60**, 21–38.

Robinson, D. J. S. (1965). On finitely generated soluble groups. *Proc. London Math. Soc.*, (3) **15**, 508–16.

Robinson, D. J. S. (1967*a*). Wreath products and indices of subnormality. *Proc. London Math. Soc.*, (3) **17**, 13–40, 257–70.

Robinson, D. J. S. (1967*b*). *Infinite Soluble and Nilpotent Groups*. Queen Mary College Mathematics Notes, Queen Mary College, London.

Robinson, D. J. S. (1967*c*). On soluble minimax groups. *Math. Z.*, **101**, 13–40.

Robinson, D. J. S. (1968*a*). Residual properties of some classes of infinite soluble groups. *Proc. London Math. Soc.*, (3) **18**, 495–520.

Robinson, D. J. S. (1968*b*). A property of the lower central series of a group. *Math. Z.*, **107**, 225–31.

Robinson, D. J. S. (1968*c*). Finiteness conditions on subnormal and ascendant abelian subgroups. *J. Algebra*, **10**, 333–59.

Robinson, D. J. S. (1969). A note on groups of finite rank. *Compositio Math.*, **21**, 240–46.

Robinson, D. J. S. (1970). A theorem on finitely generated hyperabelian groups. *Invent. Math.*, **10**, 38–43.

Robinson, D. J. S. (1972*a*). Intersections of primary powers of a group. *Math. Z.*, **124**, 119–32.

Robinson, D. J. S. (1972*b*). *Finiteness Conditions and Generalized Soluble Groups*, Parts 1 and 2. Ergebnisse der Mathematik und ihrer Grenzgebiete, Band 62. Springer-Verlag, New York and Berlin.

Robinson, D. J. S. (1974). Hypercentral ideals, noetherian modules and a theorem of Stroud. *J. Algebra*, **32**, 234–9.

Robinson, D. J. S. (1975). On the cohomology of soluble groups of finite rank. *J. Pure Appl. Algebra*, **6**(2), 155–64.

Robinson, D. J. S. (1976*a*). The vanishing of certain homology and cohomology groups. *J. Pure Appl. Algebra*, **7**(2), 145–67.

Robinson, D. J. S. (1976*b*). Splitting theorems for infinite groups. In *Symposia Mathematics*, Vol. XVII (Convegno sui Gruppi Infiniti, INDAM, Rome, 1973), pp. 441–70. Academic Press, London.

Robinson, D. J. S. (1976c). A new treatment of soluble groups with finiteness conditions on their abelian subgroups. *Bull. London Math. Soc.*, **8**(2), 113–29.

Robinson, D. J. S. (1979). Homology of group extensions with divisible abelian kernel. *J. Pure Appl. Algebra*, **14**(2), 145–65.

Robinson, D. J. S. (1980). Infinite soluble groups with no outer automorphisms. *Rend. Sem. Mat. Univ. Padova*, **62**, 281–94.

Robinson, D. J. S. (1982). *Applications of Cohomology to the Theory of Groups* (Proceedings of Groups—St. Andrews, 1981), pp. 46–80, London Mathematical Society Lecture Note Series, 71. Cambridge University Press, Cambridge and New York.

Robinson, D. J. S. (1983). Infinite factorized groups. *Rend. Sem. Mat. Fis. Milano*, **53**, 347–55.

Robinson, D. J. S. (1984a). Finiteness, solubility and nilpotence. In *Group Theory*, pp. 159–206, Academic Press, London.

Robinson, D. J. S. (1984b). *Decision Problems for Infinite Soluble Groups* (Proceedings of Groups—Kyoungju, Korea, 1983), pp. 111–17, Lecture Notes in Mathematics, 1098, Springer, Berlin.

Robinson, D. J. S. (1986a). Decision Problems for soluble groups of finite rank. *Illinois J. Math.*, **30**(2), 197–213.

Robinson, D. J. S. (1986b). Soluble products of nilpotent groups. *J. Algebra*, **98**(1), 183–96.

Robinson, D. J. S. (1987a). Homology and cohomology of locally supersoluble groups. *Math. Proc. Cambridge Philos. Soc.*, **102**(2), 233–50.

Robinson, D. J. S. (1987b). Cohomology of locally nilpotent groups. *J. Pure Appl. Algebra*, **48**(3), 281–300.

Robinson, D. J. S. (1987c). Vanishing theorems for cohomology of locally nilpotent groups. In *Group Theory* (Bressanone, 1986), pp. 120–29, Lecture Notes in Mathematics, 1281. Springer, Berlin.

Robinson, D. J. S. (1989a). Cohomology in infinite group theory. In *Group Theory* (Singapore, 1987), pp. 29–53. de Gruyter, Berlin.

Robinson, D. J. S. (1989b). *Reflections on the Constructive Theory of Polycyclic Groups* (Proceedings of Groups—Pusan, Korea, 1988), pp. 163–9, Lecture Notes in Mathematics, 1398, Springer, Berlin.

Robinson, D. J. S. (1990a). Deciding if an automorphism of an infinite soluble group is inner. *Glasgow Math. J.*, **32**(3), 265–72.

Robinson, D. J. S. (1990b). Algorithmic problems for automorphisms and endomorphisms of infinite soluble groups. In *Algebraic Structures and Number Theory* (Hong Kong, 1988), pp. 243–54, World Scientific Publishing, Teaneck, NJ.

Robinson, D. J. S. (1996). *A Course in the Theory of Groups*, 2nd edn. Graduate Texts in Mathematics, 80. Springer-Verlag, New York and Berlin.

Robinson, D. J. S. (2002). Derivations and the permutability of subgroups in polycyclic-by-finite groups. *Proc. Amer. Math. Soc.*, **130**(12), 3461–4.

Robinson, D. J. S. and Stonehewer, S. E. (1993). Triple factorizations by abelian groups. *Arch. Math. (Basel)*, **60**(3), 223–32.

Robinson, D. J. S. and Stonehewer, S. E. (2000). Minimal normal subgroups of dinilpotent groups. *J. Algebra*, **234**(2), 480–91.

Robinson, D. J. S. and Strebel, R. (1982). Some finitely presented soluble groups which are not nilpotent-by-abelian by finite. *J. London Math. Soc.*, (2) **26**(3), 435–40.

Robinson, D. J. S. and Wilson, J. S. (1984). Soluble groups with many polycyclic quotients. *Proc. London Math. Soc.*, (3) **48**(2), 193–229.

Romanovskiĭ, N. S. (1973). Certain algorithmic problems for solvable groups. (Russian). *Algebra i Logika*, **13**, 26–34, 121.

Romanovskiĭ, N. S. (1980). The embedding problem for abelian-by-nilpotent groups. (Russian). *Sibirsk. Mat. Zh.*, **21**(2), 170–4, 239.

Romanovskiĭ, N. S. (1982). The word problem for centrally metabelian groups. (Russian). *Sibirsk. Mat. Zh.*, **23**(4), 201–5, 222.

Roseblade, J. E. (1962). On certain classes of locally soluble groups. *Proc. Cambridge Philos. Soc.*, **58**, 185–95.

Roseblade, J. E. (1965). On groups in which every subgroup is subnormal. *J. Algebra*, **2**, 402–12.

Roseblade, J. E. (1970). The derived series of a join of subnormal subgroups. *Math. Z.*, **117**, 57–69.

Roseblade, J. E. (1971). The integral group rings of hypercentral groups. *Bull. London Math. Soc.*, **3**, 351–5.

Roseblade, J. E. (1973*a*). Group rings of polycyclic groups. *J. Pure Appl. Algebra*, **3**, 307–28.

Roseblade, J. E. (1973*b*). Polycyclic group rings and the Nullstellensatz. In *Proceedings of the Conference on Group Theory* (University Wisconsin-Parkside, Kenosha, Wisconsin, 1972), pp. 156–67, Lecture Notes in Mathematics Vol. 319. Springer, Berlin.

Roseblade, J. E. (1973*c*). *The Frattini Subgroup in Infinite Soluble Groups.* Three lectures on polycyclic groups, Paper No. 1, 14 pp. Queen Mary College Mathematics Notes, Queen Mary College, London.

Roseblade, J. E. (1976). Applications of the Artin–Rees lemma to group rings. In *Symposia Mathematics*, Vol. XVII (Convegno sui Gruppi Infiniti, INDAM, Rome, 1973), pp. 471–8, Academic Press, London.

Roseblade, J. E. (1978). Prime ideals in group rings of polycyclic groups. *Proc. London Math. Soc.*, (3) **36**(3) 1978, 385–447.

Roseblade, J. E. (1979). Corrigenda: "Prime ideals in group rings of polycyclic groups" *Proc. London Math. Soc.*, (3) **36**(3), 1978, 385–447; **38**(2), 216–18.

Roseblade, J. E. (1986*a*). *Five lectures on Group Rings* (Proceedings of Groups—St. Andrews, 1985), pp. 93–109, London Mathematics Society Lecture Note Series, 121. Cambridge University Press, Cambridge.

Roseblade, J. E. (1986*b*). Group rings. In *Report on the Fifteenth National Conference on Mathematics.* (Persian) (Shiraz, 1984), pp. 217–28, Tehran.

Roseblade, J. E. and Smith, P. F. (1976). A note on hypercentral group rings. *J. London Math. Soc.*, (2) **13**(1), 183–90.

Roseblade, J. E. and Smith, P. F. (1979). A note on the Artin-Rees property of certain polycyclic group algebras. *Bull. London Math. Soc.*, **11**(2), 184–5.

Roseblade, J. E. and Wilson, J. S. (1971). A remark about monolithic groups. *J. London Math. Soc.*, (2) **3**, 361–2.

Sabbagh, G. and Wilson, J. S. (1991). Polycyclic groups, finite images, and elementary equivalence. *Arch. Math. (Basel)*, **57**(3), 221–7.

Schmidt, O. J. (1945). Infinite soluble groups. (Russian). *Math. Sb.*, **17**, 145–62.

Segal, D. (1972). A note on module automorphism groups over Noetherian rings. *Arch. Math. (Basel)*, 23, 594–7.

Segal, D. (1973). Groups of automorphisms of infinite soluble groups. *Proc. London Math. Soc.*, (3) **26**, 630–52.

Segal, D. (1974). A residual property of finitely generated abelian-by-nilpotent groups. *J. Algebra*, **32**, 389–99.

Segal, D. (1975*a*). Groups whose finite quotients are supersoluble. *J. Algebra*, **35**, 56–71.

Segal, D. (1975*b*). On abelian-by-polycyclic groups. *J. London Math. Soc.*, (2) **11**(4), 445–52.

Segal, D. (1976). Unipotent groups of module automorphisms over polycyclic group rings. *Bull. London Math. Soc.*, **8**(2), 174–8.

Segal, D. (1977*a*). Irreducible representations of finitely generated nilpotent groups. *Math. Proc. Cambridge Philos. Soc.*, **81**(2), 201–8.

Segal, D. (1977*b*). On the residual simplicity of certain modules. *Proc. London Math. Soc.*, (3) **34**(2), 327–53.

Segal, D. (1978). Two theorems of polycyclic groups. *Math. Z.*, **164**(2), 185–7.

Segal, D. (1983). *Polycyclic Groups.* Cambridge Tracts in Mathematics, 82. Cambridge University Press, Cambridge.

Segal, D. (1986). Subgroups of finite index in soluble groups. I, II. In *Proceedings of Groups—St. Andrews 1985*, pp. 307–14, 315–19 London Mathematics Society Lecture Note Series, 121. Cambridge University Press, Cambridge, 1986.

Segal, D. (1987). The general polycyclic group, *Bull. London Math. Soc.*, **19**(1), 49–56.

Segal, D. (1990*a*). Decidable properties of polycyclic groups. *Proc. London Math. Soc.*, (3) **61**(3), 497–528.

Segal, D. (1990*b*). On the outer automorphism group of a polycyclic group. Proceedings of the Second International Group Theory Conference (Bressanone, 1989). *Rend. Circ. Mat. Palermo*, (2) (Suppl.) **23**, 265–78.

Segal, D. (1996). A footnote on residually finite groups. *Israel J. Math.*, **94**, 1–5.

Segal, D. (2001*a*). On modules of finite upper rank. *Trans. Amer. Math. Soc.*, **353**(1), 391–410.

Segal, D. (2001*b*). On the group rings of abelian minimax groups. *J. Algebra*, **237**(1), 64–94.

Segal, D. and Shalev, A. (1993). Groups with fractionally exponential subgroup growth. *J. Pure Appl. Algebra*, **88**(1–3), 205–23.

Seidenberg, A. (1974). Constructions in algebra. *Trans. Amer. Math. Soc.*, **197**, 273–313.

Seidenberg, A. (1978). Constructions in a polynomial ring over the ring of integers. *Amer. J. Math.*, **100**(4), 685–703.

Seksenbaev, K. (1965). On the theory of polycyclic groups. (Russian). *Algebra i Logika*, **4**(3), 79–83.

Seksenbaev, K. (1967). The finitary approximability of the finite extension of a nilpotent group with respect to conjugacy. (Russian). *Algebra i Logika*, **6**(6), 29–31.

Seksenbaev, K. (1979). The lower central series of a wreath product of finitely generated abelian groups. (Russian). In *Theoretical and Applied Problems of Mathematics and Mechanics*, pp. 202–5, 245. "Nauka" Kazakh. SSR, Alma-Ata.

Semko, N. N. and Kuzennyĭ, N. F. (1984). The structure of infinite nilpotent periodic meta-Hamiltonian groups. (Russian). In *The Structure of Groups and Subgroup Characterizations*, pp. 101–11, 143, Akad. Nauk Ukrain. SSR, Inst. Mat., Kiev.

Semko, N. N. and Kuzennyĭ, N. F. (1989). The structure of metacyclic meta-Hamiltonian groups. (Russian). In *Current Analysis and its Applications* (Russian), pp. 173–83, 226. "Naukova Dumka", Kiev.

Serre, J.-P. (1980). *Trees.* (Translated from the French by J. Stillwell). Springer-Verlag, Berlin and New York.

Shirvani, M. and Wehrfritz, B. A. F. (1986). *Skew Linear Groups.* London Mathematical Society Lecture Note Series, 118. Cambridge University Press, Cambridge.

Shumyatsky, P. V. (1991). A four-group of automorphisms with a small number of fixed points. *Algebra and Logic*, **30**(6), 481–9.

Shumyatsky, P. V. (1992). On periodic soluble groups and the fixed point groups of operators. *Comm. Algebra*, **20**(10), 2815–20.

Shumyatsky, P. V. (1994). On periodic solvable groups having automorphisms with nilpotent fixed point groups. *Israel J. Math.*, **87**(1–3), 111–16

Silcock, H. L. (1973). Metanilpotent groups satisfying the minimal condition for normal subgroups. *Math. Z.*, **135**, 165–73.

Silcock, H. L. (1975). On the construction of soluble groups satisfying the minimal condition for normal subgroups. *Bull. Austral. Math. Soc.*, **12**, 231–57.

Silcock, H. L. (1976). Representations of metabelian groups satisfying the minimal condition for normal subgroups. *Bull. Austral. Math. Soc.*, **14**(2), 267–78.

Šmel'kin, A. L. (1962). Nilpotent products and nilpotent torsion-free groups. (Russian). *Sibirsk. Mat. Ž.*, **3**, 625–40.

Šmel'kin, A. L. (1963). On the isomorphism of nilpotent decompositions of torsion-free nilpotent groups. (Russian). *Sibirsk. Mat. Ž.*, **4**, 1412–25.

Šmel'kin, A. L. (1964). Free polynilpotent groups. (Russian). *Izv. Akad. Nauk SSSR Ser. Mat.*, **28**, 91–122.

Šmel'kin, A. L. (1965). On soluble products of groups. (Russian). *Sibirsk. Mat. Ž.*, **6**, 212–20.

Šmel'kin, A. L. (1967). Complete nilpotent groups. (Russian). *Algebra i Logika*, **6**(2), 111–14.

Šmel'kin, A. L. (1968). Polycyclic groups. (Russian). *Sibirsk. Mat. Ž.*, **9**, 234–5.

Smirnov, D. M. (1951). On the theory of locally nilpotent groups. (Russian). *Dokl. Akad. Nauk SSSR (N.S.)*, **76**, 643–6.

Smirnov, D. M. (1953). On groups of automorphisms of soluble groups. (Russian). *Mat. Sbornik N.S.*, **32**(74), 365–84.

Smith, G. C. (1985). Compressibility in nilpotent groups. *Bull. London Math. Soc.*, **17**(5), 453–7.

Smith, H. (1983). Hypercentral groups with all subgroups subnormal. *Bull. London Math. Soc.*, **15**(3), 229–34.

Smith, H. (1984). Some remarks on locally nilpotent groups of finite rank. *Proc. Amer. Math. Soc.*, **92**(3), 339–41.

Smith, H. (1986). Hypercentral groups with all subgroups subnormal. II. *Bull. London Math. Soc.*, **18**(4), 343–8.

Smith, H. (1987). Group theoretic properties inherited by lower central factors. *Glasgow Math. J.*, **29**(1), 89–91.

Smith, H. (1989). On torsion-free hypercentral groups with all subgroups subnormal. *Glasgow Math. J.*, **31**(2), 193–4.

Smith, H. (1994). On the normal cores of certain subgroups of nilpotent groups. *Glasgow Math. J.*, **36**(1), 113–15.

Smith, H. (1995*a*). Groups with finitely many conjugacy classes of subgroups of large derived length. *Boll. Un. Mat. Ital. A*, (7) **9**(1), 167–75.

Smith, H. (1995*b*). On infinite polynilpotent groups. *Proc. Roy. Irish Acad. Sect. A*, **95**(1), 39–46.

Smith, H. (1999). Residually finite groups with all subgroups subnormal. *Bull. London Math. Soc.*, **31**(6), 679–80.

Smith, H. (2000). Nilpotent-by-(finite exponent) groups with all subgroups subnormal. *J. Group Theory*, **3**(1), 47–56.

Smith, H. (2001*a*). Torsion-free groups with all subgroups subnormal. *Arch. Math. (Basel)*, **76**(1), 1–6.

Smith, H. (2001*b*). Hypercentral groups with all subgroups subnormal. III. *Bull. London Math. Soc.*, **33**(5), 591–8.

Smith, H. (2001*c*). Residually nilpotent groups with all subgroups subnormal. *J. Algebra*, **244**(2), 845–50.

Smith, H. (2001*d*). On non-nilpotent groups with all subgroups subnormal. *Ricerche Mat.*, **50**(2), 217–21.

Smith, H. (2002*a*). Bounded Engel groups with all subgroups subnormal. *Comm. Algebra*, **30**(2), 907–9.

Smith, H. (2002*b*). Torsion-free groups with all non-nilpotent subgroups subnormal. *Quaderni di Matematica*, **8**, 297–308.

Smith, H. (2002*c*). Groups with all non-nilpotent subgroups subnormal. *Quaderni di Matematica*, **8**, 309–26.

Smith, H. and Wiegold, J. (1997). Groups which are isomorphic to their non-abelian subgroups. *Rend. Sem. Mat. University Padova*, **97**, 7–16.

Smith, H. and Wiegold, J. (1998). Groups isomorphic to their non-nilpotent subgroups. *Glasgow Math. J.*, **40**(2), 257–62.

Smith, H. and Wiegold, J. (1999). Soluble groups isomorphic to their non-nilpotent subgroups. *J. Austral. Math. Soc. Ser. A*, **67**(3), 399–411.

Stammbach, U. (1970). On the weak homological dimension of the group algebra of solvable groups. *J. London Math. Soc.*, (2) **2**, 567–70.

Stammbach, U. (1973). *Homology in Group Theory*. Lecture Notes in Mathematics, Vol. 359. Springer-Verlag, Berlin and New York.

Stewart, A. G. R. (1966). On the class of certain nilpotent groups. *Proc. Roy. Soc. Ser. A*, **292**, 374–9.

Stonehewer, S. E. (1970). The join of finitely many subnormal subgroups. *Bull. London Math. Soc.*, **2**, 77–82.

Stonehewer, S. E. (1974). Nilpotent residuals of subnormal subgroups. *Math. Z.*, **139**, 45–54.

Stonehewer, S. E. (1987). Subnormal subgroups of factorised groups. In *Group Theory* (Bressanone, 1986), pp. 158–75, Lecture Notes in Mathematics, 1281. Springer, Berlin.

Strebel, R. (1978). On finitely related soluble groups. In *Topology and Algebra* (Proceedings of Colloqium, Eidgenöss. Tech. Hochsch., Zurich, 1977), pp. 243–50, Monograph. Enseign. Math., 26, University Genève, Geneva.

Strebel, R. (1981a). On Finitely Related Abelian-by-nilpotent Groups. (Preprint).

Strebel, R. (1981b). On one-relator soluble groups. *Comment. Math. Helv.*, **56**(1), 123–31.

Strebel, R. (1983). Subgroups of finitely presented centre-by-metabelian groups. *J. London Math. Soc.*, (2) **28**(3), 481–91.

Strebel, R. (1984). Finitely presented soluble groups. In *Group Theory*, pp. 257–314. Academic Press, London.

Subbotin, I. Ya. and Kuzennyĭ, N. F. (1987). Solvable groups with a quasicentralizer condition for non-abelian normal subgroups. (Russian). *Izv. Vyssh. Uchebn. Zaved. Mat.*, **10**, 68–70, 75.

Suprunenko, D. A. (1963). *Soluble and Nilpotent Linear Groups.* American Mathematical Society, Providence, RI.

Swan, R. G. (1967). Representations of polycyclic groups. *Proc. Amer. Math. Soc.*, **18**, 573–4.

Tits, J. (1972). Free subgroups in linear groups. *J. Algebra*, **20**, 250–70.

Tushev, A. (1986). On a class of locally supersolvable groups. (Russian). In *The Structure of Groups and the Properties of their Subgroups*, pp. 119–27, vi. Akad. Nauk Ukrain. SSR, Inst. Mat., Kiev.

Tushev, A. (1990). The Min-$\infty - n$ condition and related representations of solvable groups. *Ukrainian Math. J.*, **42**(10), 1233–8.

Tushev, A. (1991a). Noetherian modules over abelian groups of finite free rank. *Ukrainian Math. J.*, **43**(7–8), 975–81.

Tushev, A. (1991b). Irreducible representations of locally polycyclic groups over an absolute-valued field. *Ukrainian Math. J.*, **42**(10), 1233–8.

Tushev, A. (1993a). Minimally infinite modules over locally polycyclic groups of finite rank. (Russian). In *Infinite Groups and Related Algebraic Structures*, pp. 312–25. Akad. Nauk Ukrainy, Inst. Mat., Kiev.

Tushev, A. (1993b). Solvable groups all of whose proper quotient groups have finite rank. *Ukrainian Math. J.*, **45**(9), 1430–7.

Tushev, A. (1994). On solvable groups whose finite homomorphic images have bounded rank. *Math. Notes*, **56**(5–6), 1190–92.

Tushev, A. (1999a). *On Modules over Group Rings of Soluble Groups of Finite Rank* (Proceedings of Groups—St. Andrews, 1997) Bath II, pp. 718–27, London Mathematics Society Lecture Note Series, 261. Cambridge University Press, Cambridge.

Tushev, A. (1999b). Induced modules over group algebras of metabelian groups of finite rank. *Comm. Algebra*, **27**(12), 5921–38.

Tushev, A. (2001). On the Fitting subgroup of soluble groups just of infinite rank. *Dopov. Nats. Akad. Nauk Ukr. Mat. Prirodozn. Tekh. Nauki*, **10**, 45–7.

Tushev, A. (2002). Minimally infinite modules over metabelian groups of finite rank. *Sb. Math.*, **193**(5–6), 761–78.

Wehrfritz, B. A. F. (1969). A residual property of free metabelian groups. *Arch. Math. (Basel)*, **20**, 248–50.

Wehrfritz, B. A. F. (1970). Groups of automorphisms of soluble groups. *Proc. London Math. Soc.*, (3) **20**, 101–22.

Wehrfritz, B. A. F. (1971a). Supersoluble and locally supersoluble linear groups. *J. Algebra*, **17**, 41–58.

Wehrfritz, B. A. F. (1971b). Remarks on centrality and cyclicity in linear groups. *J. Algebra*, **18**, 229–36.

Wehrfritz, B. A. F. (1972). A note on residual properties of nilpotent groups. *J. London Math. Soc.*, (2) **5**, 1–7.

Wehrfritz, B. A. F. (1973a). *Infinite Linear Groups. An Account of the Group-theoretic Properties of Infinite Groups of Matrices.* Ergebnisse der Mathematik und ihrer Grenzgebiete, Band 76. Springer-Verlag, New York and Heidelberg.

Wehrfritz, B. A. F. (1973b). Two examples of soluble groups that are not conjugacy separable. *J. London Math. Soc.*, (2) **7**, 312–16.

Wehrfritz, B. A. F. (1973c). *The Holomorph of a Polycyclic Group.* Three lectures on polycyclic groups, Paper No. 3, 14 pp. Queen Mary College Mathematics Notes, Queen Mary College, London.

Wehrfritz, B. A. F. (1974). On the holomorphs of soluble groups of finite rank. *J. Pure Appl. Algebra*, **4**, 55–69.

Wehrfritz, B. A. F. (1975). Representations of holomorphs of group extensions with abelian kernels. *Math. Proc. Cambridge Philos. Soc.*, **78**(3), 357–68.

Wehrfritz, B. A. F. (1976a). Another example of a soluble group that is not conjugacy separable. *J. London Math. Soc.*, (2) **14**(2), 381–2.

Wehrfritz, B. A. F. (1976b). On locally supersoluble groups. *J. Algebra* **43**(2), 665–9.

Wehrfritz, B. A. F. (1978). On the Lie-Kolchin-Mal'cev theorem. *J. Austral. Math. Soc. Ser. A*, **26**(3), 270–76.

Wehrfritz, B. A. F. (1980a). Finitely generated modules over polycyclic groups. *Quart. J. Math. Oxford Ser.*, (2) **31**(121), 109–27.

Wehrfritz, B. A. F. (1980b). On finitely generated soluble linear groups. *Math. Z.*, **170**(2), 155–67.

Wehrfritz, B. A. F. (1983). Endomorphisms of polycyclic groups. *Math. Z.*, **184**(1), 97–9.

Wehrfritz, B. A. F. (1984a). On division rings generated by polycyclic groups. *Israel J. Math.*, **47**(2–3), 154–64.

Wehrfritz, B. A. F. (1984*b*). Isolators in soluble groups of finite rank. *Rocky Mountain J. Math.*, **14**(2), 415–21.

Wehrfritz, B. A. F. (1984*c*). Faithful representations of finitely generated abelian-by- polycyclic groups over division rings. *Quart. J. Math. Oxford Ser.*, (2) **35**(139), 361–72.

Wehrfritz, B. A. F. (1991). On rings of quotients of group algebras of soluble groups of finite rank. *J. Pure Appl. Algebra*, **74**(1), 95–107.

Wehrfritz, B. A. F. (1994). Two remarks on polycyclic groups. *Bull. London Math. Soc.*, **26**(6), 543–8.

Wells, C. F. (1971). Automorphisms of group extensions. *Trans. Amer. Math. Soc.*, **155**, 189–94.

Wiegold, J. (1959). Nilpotent products of groups with amalgamations. *Publ. Math. Debrecen*, **6**, 131–68.

Wiegold, J. (1963). Adjunction of elements to nilpotent groups. *J. London Math. Soc.*, **38**, 17–26.

Wiegold, J. (1964). Groups with many nilpotent subgroups. *Arch. Math.*, **15**, 241–50.

Wiegold, J. (1965). Soluble embeddings of group amalgams. *Publ. Math. Debrecen*, **12**, 227–30.

Wielandt, H. (1937). Eine Kennzeichnung der direkten Produkte von *p*-Gruppen. *Math. Z.*, **41**, 281–2.

Wilson, J. S. (1982*a*). Abelian subgroups of polycyclic groups. *J. Reine Angew. Math.*, **331**, 162–80.

Wilson, J. S. (1982*b*). Large nilpotent subgroups of polycyclic groups. *Arch. Math. (Basel)*, **39**(1), 1–4.

Wilson, J. S. (1983). Polycyclic groups and topology. *Rend. Sem. Mat. Fis. Milano*, **51**, 17–28.

Wilson, J. S. (1985). On products of soluble groups of finite rank. *Comment. Math. Helv.*, **60**(3), 337–53.

Wilson, J. S. (1986). Some recent contributions to the theory of infinite soluble groups. (Italian). *Rend. Sem. Mat. Fis. Milano*, **53**, 187–94.

Wilson, J. S. (1987). On elliptically embedded subgroups of soluble groups. *Canad. J. Math.*, **39**(4), 956–68.

Wilson, J. S. (1988*a*). Soluble groups which are products of minimax groups. *Arch. Math. (Basel)*, **50**(3), 193–8.

Wilson, J. S. (1988*b*). Soluble products of minimax groups and nearly surjective derivations. *J. Pure Appl. Algebra*, **53**(3), 297–318.

Wilson, J. S. (1989). Soluble groups which are products of groups of finite rank. *J. London Math. Soc.* (2) **40**(3), 405–19.

Wilson, J. S. (1993). On finitely presented soluble groups with small abelian-by-finite images. *J. London Math. Soc.*, (2) **48**(2), 229–48.

Wilson, J. S. (1996*a*). Soluble groups of deficiency 1. *Bull. London Math. Soc.*, **28**(5), 476–80.

Wilson, J. S. (1996*b*). Finitely presented soluble groups, Hilbert–Serre dimension, and integer polynomials. In *Algebra* (Krasnoyarsk, 1993), pp. 287–95. de Gruyter, Berlin.

Wilson, J. S. (1998). Finitely presented soluble groups. In *Geometry and Cohomology in Group Theory* (Durham, 1994), 296–316, London Mathematics Soc. Lecture Note Series, 252. Cambridge University Press, Cambridge.

Zaĭcev, D. I. (1967*a*). The existence of infinite solvable and infinite nilpotent subgroups in infinite groups. (Ukrainian). *Dopov Akad. Nauk Ukran. RSR Ser. A*, 772–5.

Zaĭcev, D. I. (1967*b*). Stably nilpotent groups. (Russian). *Mat. Zametki*, **2**, 337–46.

Zaĭcev, D. I. (1968*a*). Groups which satisfy a weak minimality condition. (Russian). *Dokl. Akad. Nauk SSSR*, **178**, 780–82.

Zaĭcev, D. I. (1968*b*). Solvable groups of finite rank. (Russian). *Dokl. Akad. Nauk SSSR*, **181**, 13–14.

Zaĭcev, D. I. (1969*a*). The groups which satisfy a weak minimality condition. (Russian). *Mat. Sb.* (*N.S.*), **78**(120), 323–31.

Zaĭcev, D. I. (1969*b*). Stably solvable groups. (Russian). *Izv. Akad. Nauk SSSR Ser. Mat.*, **33**, 765–80.

Zaĭcev, D. I. (1971*a*). Solvable groups of finite rank. (Russian). In *Groups with Restricted Subgroups*, pp. 115–30. "Naukova Dumka", Kiev.

Zaĭcev, D. I. (1971*b*). On the theory of minimax groups. (Russian). *Ukrain. Mat. Ž.*, **23**, 652–60.

Zaĭcev, D. I. (1971*c*). Groups that satisfy the weak minimum condition for nonabelian subgroups. (Russian). *Ukrain. Mat. Ž.*, **23**, 661–5.

Zaĭcev, D. I. (1974). The exponent of minimality of a group. (Russian). In *Investigation of Groups with Prescribed Properties for Subgroups*, pp. 199–226, 265–6. Inst. Mat., Akad. Nauk Ukrain. SSR, Kiev.

Zaĭcev, D. I. (1976*a*). The existence of direct complements in groups with operators. (Russian). In *Studies in Group Theory*, pp. 26–44, 168. Izdanie Inst. Mat. Akad. Nauk Ukrain. SSR, Kiev.

Zaĭcev, D. I. (1976*b*). A class of torsion-free solvable groups. (Russian). In *Studies in Group Theory*, pp. 162–6, 172. Izdanie Inst. Mat. Akad. Nauk Ukrain. SSR, Kiev.

Zaĭcev, D. I. (1976*c*). The existence of direct complements in groups with operators. (Russian). In *Studies in Group Theory*, pp. 26–44, 168. Izdanie Inst. Mat. Akad. Nauk Ukrain. SSR, Kiev.

Zaĭcev, D. I. (1977). Solvable groups of finite rank. (Russian). *Algebra i Logika*, **16**(3), 300–12, 377.

Zaĭcev, D. I. (1978a). Solvable groups with complemented normal subgroups. (Russian). In *Group-theoretical Studies*, pp. 77–86, 158. Izdat. "Naukova Dumka", Kiev.

Zaĭcev, D. I. (1978b). Locally solvable groups of finite rank. (Russian). *Dokl. Akad. Nauk SSSR*, **240**(2), 257–60.

Zaĭcev, D. I. (1979). Hypercyclic extensions of abelian groups. (Russian). In *Groups Defined by Properties of a System of Subgroups*, pp. 16–37, 152. Akad. Nauk Ukrain. SSR, Inst. Mat., Kiev, 1979.

Zaĭcev, D. I. (1980a). Products of abelian groups. (Russian). *Algebra i Logika* **19**(2), 150–72, 250.

Zaĭcev, D. I. (1980b). Extensions of abelian groups. (Russian). In *Constructive Description of Groups with Properties Prescribed on their Subgroups*, pp. 16–40, 143, Akad. Nauk Ukrain. SSR, Inst. Mat., Kiev.

Zaĭcev, D. I. (1981a). Factorization of polycyclic groups. (Russian). *Mat. Zametki*, **29**(4), 481–90, 632.

Zaĭcev, D. I. (1981b). *On the Splittability of Extensions of Abelian Groups*. (Russian). pp. 14–25, 107. Akad. Nauk Ukrain. SSR, Inst. Mat., Kiev.

Zaĭcev, D. I. (1981c). Nilpotent approximations of metabelian groups. (Russian). *Algebra i Logika*, **20**(6), 638–53, 728.

Zaĭcev, D. I. (1982). Weakly uniform products of polycyclic groups. (Russian). In *Subgroup Characterization of Groups*, pp. 13–26, 111. Akad. Nauk Ukrain. SSR, Inst. Mat., Kiev.

Zaĭcev, D. I. (1983a). Products of minimax groups. (Russian). In *Groups and Systems of their Subgroups*, pp. 15–31. Akad. Nauk Ukrain. SSR, Inst. Mat., Kiev.

Zaĭcev, D. I. (1983b). The Itô theorem and products of groups. (Russian). *Mat. Zametki*, **33**(6), 807–18

Zaĭcev, D. I. (1984). Solvable factorizable groups. (Russian). In *The Structure of Groups and Subgroup Characterizations*, pp. 15–33, 140. Akad. Nauk Ukrain. SSR, Inst. Mat., Kiev.

Zaĭcev, D. I. (1986a). Itô's theorem and products of groups. (Russian). *Ukrain. Mat. Zh.*, **38**(4), 427–31, 541.

Zaĭcev, D. I. (1986b). Splittable extensions of abelian groups. (Russian). In *The Structure of Groups and the Properties of their Subgroups*, pp. 22–31, i. Akad. Nauk Ukrain. SSR, Inst. Mat., Kiev.

Zaĭcev, D. I. (1986c). A class of periodic groups with Abelian commutator group. (Russian). Akad. Nauk Ukrain. SSR Inst. Mat. Preprint No. 24, 31.

Zaĭcev, D. I. (1986d). On a class of locally supersolvable groups. (Russian). In *The Structure of Groups and the Properties of their Subgroups*, pp. 119–27, vi. Akad. Nauk Ukrain. SSR, Inst. Mat., Kiev.

Zaĭcev, D. I. (1988). Hyperfinite extensions of abelian groups. (Russian). In *Investigations of Groups with Restrictions for Subgroups.* (Russian) pp. 17–26, i. Akad. Nauk Ukrain. SSR, Inst. Mat., Kiev.

Zaĭcev, D. I. and Kurdachenko, L. A. (1982). On soluble groups satisfying a weak minimal condition for normal subgroups. In *Proceedings of VIII All Soviet Symposium on Group Theory: Thesis Announcements*, p. 37.

Zaĭcev, D. I., Kurdachenko, L., and Tushev, A. (1985). Modules over nilpotent groups of finite rank. (Russian). *Algebra i Logika*, **24**(6), 631–66, 748.

Zalesskii, A. I. (1970). The group algebras of solvable groups. (Russian). *Vesc. Akad. Navuk BSSR Ser. Fiz.-Mat. Navuk*, **2**, 13–21.

Zalesskii, A. I. (1971*a*). The irreducible representations of finitely generated nilpotent groups without torsion. (Russian). *Mat. Zametki*, **9**, 199–210.

Zalesskii, A. I. (1971*b*). A nilpotent *p*-group possesses an outer automorphism. (Russian). *Dokl. Akad. Nauk SSSR*, **196**, 751–4.

Zalesskii, A. I. (1972). An example of a nilpotent group without torsion that has no outer automorphisms. (Russian). *Mat. Zametki*, **11**, 21–6.

Zappa, G. (1941*a*). Sui gruppi di Hirsch supersolubili. I. (Italian). *Rend. Sem. Mat. University Padova*, **12**, 1–11.

Zappa, G. (1941*b*). Sui gruppi di Hirsch supersolubili. II. (Italian). *Rend. Sem. Mat. University Padova*, **12**, 62–80.

Zappa, G. (1948). Sui sottogruppi finiti dei gruppi di Hirsch. (Italian). *Giorn. Mat. Battaglini*, (4) **2**(78), 55–70.

Zassenhaus, H. J. (1938). Beweis eines Satzes über diskrete Gruppen. *Abh. Math. Sem. Univ. Hamburg*, **12**, 289–312.

INDEX OF AUTHORS

INDEX